Modern Condensed Matter Physics

Modern Condensed Matter Physics

Edited by
Raymond Stevens

WILLFORD PRESS
www.willfordpress.com

Published by Willford Press,
118-35 Queens Blvd., Suite 400,
Forest Hills, NY 11375, USA

ISBN: 978-1-64728-359-9

Cataloging-in-Publication Data

Modern condensed matter physics / edited by Raymond Stevens.
    p. cm.
Includes bibliographical references and index.
ISBN 978-1-64728-359-9
1. Condensed matter. 2. Solid state physics. 3. Physics. I. Stevens, Raymond.
QC173.454 .M63 2022
530.41--dc23

For information on all Willford Press publications
visit our website at www.willfordpress.com

WILLFORD PRESS

Contents

Permissions

List of Contributors

Index

Preface

Every book is initially just a concept; it takes months of research and hard work to give it the final shape in which the readers receive it. In its early stages, this book also went through rigorous reviewing. The notable contributions made by experts from across the globe were first molded into patterned chapters and then arranged in a sensibly sequential manner to bring out the best results.

Condensed matter physics is a branch of physics that studies macroscopic and microscopic physical properties of a matter. It is majorly concerned with the condensed phase of a matter that appears when the number of constituents in a system is large and the interactions among constituents are strong. Solids and liquids are two major examples of condensed phases which arise from the electromagnetic forces between atoms. Laws of quantum mechanics, electromagnetism and statistical mechanics are used in this field to understand solids and liquids phases of condensed matter. This book studies, analyses and upholds the pillars of condensed matter physics and its utmost significance in modern times. It strives to provide a fair idea about this discipline and to help develop a better understanding of the latest advances within this field. This book is a vital tool for all researching and studying this field.

It has been my immense pleasure to be a part of this project and to contribute my years of learning in such a meaningful form. I would like to take this opportunity to thank all the people who have been associated with the completion of this book at any step.

Editor

Superconducting Properties of 3D Low-Density Translation-Invariant Bipolaron Gas

V. D. Lakhno ⓘ

Keldysh Institute of Applied Mathematics, RAS, Miusskaya Sq. 4, Moscow 125047, Russia

Correspondence should be addressed to V. D. Lakhno; lak@impb.ru

Academic Editor: Jörg Fink

Consideration is given to thermodynamical properties of a three-dimensional Bose-condensate of translation-invariant bipolarons (TI-bipolarons). The critical temperature of transition, energy, heat capacity, and the transition heat of ideal TI-bipolaron gas are calculated. The results obtained are used to explain experiments on high-temperature superconductors.

1. Introduction

The theory of superconductivity is one of the finest and oldest subject matters of condensed matter physics which involves both macroscopic and microscopic theories as well as derivation of macroscopic equations of the theory from microscopic description [1]. In this sense the theory was thought to be basically completed and its further development was to have been concerned with further details and consideration of various special cases.

The situation changed when the high-temperature superconductivity (HTSC) was discovered [2]. Surprisingly, it was found that, in oxide ceramics, the correlation length is some orders of magnitude less than that in traditional metal superconductors while the ratio of the energy gap to the temperature of superconducting transition is much greater [3]. The current status of research can be found in books and reviews [4–18].

Today the main problem in this field is to develop a microscopic theory capable of explaining experimental facts which cannot be accounted for by the standard BCS theory. One might expect that development of such a theory would not affect the macroscopic theory based on phenomenological approach.

With all the variety of modern versions of HTSC microscopic descriptions: phonon, plasmon, spin, exciton, and other mechanisms, the central point of constructing the microscopic theory is the effect of electron pairing (Cooper effect). In what follows such a bosonization of electrons provides the basis for the description of their superconducting condensate.

The phenomenon of pairing, in a broad sense, is considered as arising of bielectron states, while, in a narrow sense, if the description is based on phonon mechanism, it is treated as formation of bipolaron states [19]. For a long time this view was hindered by a large correlation length or the size of Cooper pairs in BCS theory. For the same reason, over a long period, the superconductivity was not viewed as a boson condensate (see footnote at p. 1177 in [20]). A significant reason of this lack of understanding was a standard idea that bipolarons are very compact particles.

The most dramatic illustration is the use of the small-radius bipolaron (SRB) theory to describe HTSC [10, 21, 22]. It implies that a stable bound bipolaron state is formed at one node of the lattice and subsequently such small-radius bipolarons are considered as a gas of charged bosons (as a variant individual SRP are formed and then are considered within BCS of creation of the bosonic states). Despite the elegance of such a picture, its actual realization for HTSC comes up against inextricable difficulties caused by impossibility to meet antagonistic requirements. On the one hand, the constant of electron-phonon interaction (EPI) should be large for bipolaron states of small radius to form. On the other hand, it should be small for the bipolaron mass (on which the superconducting temperature depends [23–28]) to be small too. Obviously, the HTSC theory based on SRP concept which uses any other (nonphonon) interaction mechanism mentioned above will run into the same problems.

Alternatively, in describing HTSC one can believe that the role of a fundamental charged boson particle can be played by large-radius bipolarons (LRB) [30–34]. Historically just this assumption was made by Ogg [30] and Schafroth [35] long before the development of the SRP theory. When viewing Cooper pairs as a peculiar kind of large-radius bipolaron states, one might expect that the LRP theory should be used to solve the HTSC problem.

As pointed out above, the main obstacle to consistent use of the LRP theory for explaining high-temperature superconductivity was an idea that electron pairs are localized in a small region, the constant of electron-phonon coupling should be large, and, as a consequence, the effective mass of electron pairs should be large.

In the light of the latest advances in the theory of LRP and LRB, namely, in view of development of an all-new concept of delocalized polaron and bipolaron states, translation-invariant polarons (TI-polarons) and bipolarons (TI-bipolarons) [36–42], it seems appropriate to consider their role in the HTSC theory in a new angle.

We recall the main results of the theory of TI-polarons and bipolarons obtained in [36–42]. Notice that consideration of just electron-phonon interaction is not essential for the theory and can be generalized to any type of interaction.

In what follows we will deal only with the main points of the theory important for the HTSC theory. The main result of papers [36–42] is construction of delocalized polaron and bipolaron states in the limit of strong electron-phonon interaction. The theory of TI-bipolarons is based on the theory of TI-polarons [36, 37] and retains the validity of basic statements proved for TI-polarons. The chief of them is the theorem of analytic properties of the ground state of a TI-polaron (accordingly TI-bipolaron) depending on the constant of electron-phonon interaction α. The main implication of this statement is the absence of a critical value of the EPI constant α_c, below which the bipolaron state becomes impossible since it decays into independent polaron states. In other words, if there exists a value of α_c, at which the TI-state becomes energetically disadvantageous with respect to its decay into individual polarons, then nothing occurs at this point but for $\alpha < \alpha_c$ and the state becomes metastable. Hence, over the whole range of α variation we can consider TI-polarons as charged bosons capable of forming a superconducting condensate.

Another important property of TI-bipolarons is the possibility of changing the correlation length over the whole range of $[0, \infty]$ depending on the Hamiltonian parameters [39]. Hence, it can be both much larger (as is the case in metals) and much less than the characteristic size between the electrons in an electron gas (as happens with ceramics).

A detailed description of the theory of TI-polarons and bipolarons and description of their various properties is given in review [42].

An outstandingly important property of TI-polarons and bipolarons is the availability of an energy gap between their ground and excited states (Section 3).

The above-indicated characteristics can be used to develop a microscopic HTSC theory on the basis of TI-bipolarons.

The paper is arranged as follows. In Section 2 we take Pekar-Frochlich Hamiltonian for a bipolaron as an initial Hamiltonian. The results of three canonical transformations, such as Heisenberg transformation, Lee-Low-Pines transformation, and that of Bogolyubov-Tyablikov are briefly outlined. Equations determining the TI-bipolaron spectrum are derived.

In Section 3 we analyze solutions of the equations for the TI-bipolaron spectrum. It is shown that the spectrum has a gap separating the ground state of a TI-bipolaron from its excited states which form a quasicontinuous spectrum. The concept of an ideal gas of TI-bipolarons is substantiated.

With the use of the spectrum obtained, in Section 4, we consider thermodynamic characteristics of an ideal gas of TI-bipolarons. For various values of the parameters, namely, phonon frequencies, we calculate the values of critical temperatures of Bose condensation, latent heat of transition into the condensed state, heat capacity, and heat capacity jumps at the point of transition.

In Section 5 we discuss the nature of current states in Bose-condensate of TI-bipolarons. It is shown that the transition from a currentless state to a current one is sharp.

In Section 6 the results obtained are compared with the experiment.

In Section 7 we consider the problems of expanding the theory which would enable one to make a more detailed comparison with experimental data on HTSC materials.

In Section 8 we sum up the results obtained.

2. Pekar-Froehlich Hamiltonian: Canonical Transformations

Following [38–42], in describing bipolarons, we will proceed from Pekar-Froehlich Hamiltonian:

$$H = -\frac{\hbar^2}{2m^*}\Delta_{r_1} - \frac{\hbar^2}{2m^*}\Delta_{r_2} + \sum_k \hbar\omega_k^0 a_k^+ a_k$$
$$+ U\left(\left|\vec{r}_1 - \vec{r}_2\right|\right)$$
$$+ \sum_k \left(V_k e^{i\vec{k}\vec{r}_1} a_k + V_k e^{i\vec{k}\vec{r}_2} a_k + H.c.\right), \quad (1)$$

$$U\left(\left|\vec{r}_1 - \vec{r}_2\right|\right) = \frac{e^2}{\epsilon_\infty \left|\vec{r}_1 - \vec{r}_2\right|},$$

where $\vec{r}_1, \vec{r}_2$ are coordinates of the first and second electrons, respectively; a_k^+, a_k are operators of the birth and annihilation of the field quanta with energy $\hbar\omega_k^0 = \hbar\omega_0$; m^* is the electron effective mass; the quantity U describes Coulomb repulsion between the electrons; V_k is the function of the wave vector k:

$$V_k = \frac{e}{k}\sqrt{\frac{2\pi\hbar\omega_0}{\tilde{\epsilon}V}} = \frac{\hbar\omega_0}{ku^{1/2}}\left(\frac{4\pi\alpha}{V}\right)^{1/2},$$

$$u = \left(\frac{2m^* \omega_0}{\hbar}\right)^{1/2}, \quad \alpha = \frac{1}{2}\frac{e^2 u}{\hbar \omega_0 \tilde{\epsilon}}, \quad \tilde{\epsilon}^{-1} = \epsilon_\infty^{-1} - \epsilon_0^{-1},$$

$$(2)$$

where e is the electron charge; ϵ_∞ and ϵ_0 are high-frequency and static dielectric permittivities; α is the constant of electron-phonon interaction; V is the systems volume.

In the system of the center of mass Hamiltonian (1) takes the form:

$$H = -\frac{\hbar^2}{2M_e}\Delta_R - \frac{\hbar^2}{2\mu_e}\Delta_r + \sum_k \hbar \omega_k^0 a_k^+ a_k + U\left(\left|\vec{r}\right|\right)$$

$$+ \sum_k 2V_k \cos\frac{\vec{k}\vec{r}}{2}\left(a_k e^{i\vec{k}\vec{R}} + H.c.\right),$$

$$(3)$$

$$\vec{R} = \frac{(\vec{r}_1 + \vec{r}_2)}{2}, \quad \vec{r} = \vec{r}_1 - \vec{r}_2, \quad M_e = 2m^*, \quad \mu_e = \frac{m^*}{2}.$$

In what follows in this section we will believe $\hbar = 1$, $\omega_k^0 = 1$, $M_e = 1$ (accordingly $\mu_e = 1/4$).

The coordinates of the center of mass $\vec{R}$ can be excluded from Hamiltonian (3) using Heisenberg's canonical transformation [43]:

$$S_1 = \exp\left\{-i\sum_k \vec{k} a_k^+ a_k\right\}\vec{R},$$

$$(4)$$

$$S_1^{-1} a_k S_1 = a_k e^{-i\vec{k}\vec{R}},$$

$$S_1^{-1} a_k^+ S_1 = a_k^+ e^{i\vec{k}\vec{R}}.$$

Accordingly, the transformed Hamiltonian will be written as

$$\tilde{H} = S_1^{-1} H S_1$$

$$= -2\Delta_r + U\left(\left|\vec{r}\right|\right) + \sum_k a_k^+ a_k$$

$$(5)$$

$$+ \sum_k 2V_k \cos\frac{\vec{k}\vec{r}}{2}\left(a_k + a_k^+\right) + \frac{1}{2}\left(\sum_k \vec{k} a_k^+ a_k\right)^2.$$

From (5) it follows that the exact solution of the bipolaron function is determined by the wave function $\psi(r)$, which contains only relative coordinates r and, therefore, is translation-invariant.

Averaging of $\tilde{H}$ over $\psi(r)$ yields the Hamiltonian $\overline{H}$:

$$\overline{H} = \frac{1}{2}\left(\sum_k \vec{k} a_k^+ a_k\right)^2 + \sum_k a_k^+ a_k + \sum_k \overline{V}_k\left(a_k + a_k^+\right) + \overline{T} + \overline{U},$$

$$(6)$$

$$\overline{V}_k = 2V_k\left\langle \Psi \left|\cos\frac{\vec{k}\vec{r}}{2}\right| \Psi \right\rangle, \quad \overline{U} = \langle \Psi |U(r)| \Psi \rangle, \quad \overline{T} = -2\left\langle \Psi \left|\Delta_r\right| \Psi \right\rangle.$$

Equation (6) suggests that the bipolaron Hamiltonian differs from the polaron one in that in the latter the quantity V_k is replaced by $\overline{V}_k$ and the constants $\overline{T}$, $\overline{U}$ are added.

With the use of Lee-Low-Pines canonical transformation [44]:

$$S_2 = \exp\left\{\sum_k f(k)\left(a_k^+ - a_k\right)\right\},$$

$$(7)$$

where f_k are variational parameters having the sense of the distance by which the field oscillators are displaced from their equilibrium positions:

$$S_2^{-1} a_k S_2 = a_k + f_k,$$

$$S_2^{-1} a_k^+ S_2 = a_k^+ + f_k,$$

$$(8)$$

for Hamiltonian $\tilde{\tilde{H}}$:

$$\tilde{\tilde{H}} = S_2^{-1}\overline{H}S_2,$$

$$(9)$$

we get

$$\tilde{\tilde{H}} = H_0 + H_1,$$

$$(10)$$

$$H_0 = 2\sum_k \overline{V}_k f_k + \sum_k f_k^2 + \frac{1}{2}\left(\sum_k \vec{k} f_k^2\right)^2 + \mathcal{H}_0 + \overline{T} + \overline{U},$$

$$(11)$$

$$\mathcal{H}_0 = \sum_k \omega_k a_k^+ a_k + \frac{1}{2}\sum_{k,k'} \vec{k}\vec{k}' f_k f_{k'}\left(a_k a_{k'} + a_k^+ a_{k'}^+ + a_k^+ a_{k'} + a_{k'}^+ a_k\right), \quad \omega_k = 1 + \frac{k^2}{2} + \vec{k}\sum_{k'}\vec{k}' f_{k'}^2.$$

Hamiltonian H_1 contains linear, threefold, and fourfold terms in the birth and annihilation operators. Its explicit form is given in [36–38].

Then, as is shown in [36, 37], the use of Bogolyubov-Tyablikov canonical transformation [45] for passing on from operators a_k^+, a_k to new operators α_k^+, α_k:

$$a_k = \sum_{k'} M_{1kk'} \alpha_{k'} + \sum_{k'} M_{2kk'}^* \alpha_{k'}^+$$

$$a_k^+ = \sum_{k'} M_{1kk'}^* \alpha_{k'}^+ + \sum_{k'} M_{2kk'} \alpha_{k'} \qquad (12)$$

(in which $\mathcal{H}_0$ is a diagonal operator), makes mathematical expectation of H_1 equal to zero.

In the new operators α_k^+, α_k Hamiltonian (11) takes on the form $\widetilde{\widetilde{H}}$:

$$\widetilde{\widetilde{H}} = E_{bp} + \sum_k \nu_k \alpha_k^+ \alpha_k, \qquad (13)$$

$$E_{bp} = \Delta E_r + 2 \sum_k \overline{V}_k f_k + \sum_k f_k^2 + \overline{T} + \overline{U}, \qquad (14)$$

where ΔE_r is the so-called recoil energy. The general expression for $\Delta E_r = \Delta E_r\{f_k\}$ was obtained in [37]. Actually, calculation of the ground state energy E_{bp} was performed in [41] by minimization of (14) in f_k and in ψ.

Notice that in the theory of a polaron with broken symmetry a diagonalized electron-phonon Hamiltonian has the form of (13)-(14) [46]. This Hamiltonian can be treated as a Hamiltonian of a polaron and a system of its associated renormalized real phonons or as a Hamiltonian whose quasiparticle excitations spectrum is determined by (13)-(14) [47]. In the latter case excited states of a polaron are Fermi quasiparticles.

In the case of a bipolaron the situation is qualitatively different since a bipolaron is a boson quasiparticle whose spectrum is determined by (13)-(14). Obviously, a gas of such quasiparticles can experience Bose-Einstein condensation (BEC). Treatment of (13)-(14) as a bipolaron and its associated renormalized phonons does not prevent their BEC since maintenance of the number of particles required in this case takes place automatically due to commutation of the total number of renormalized phonons with Hamiltonian (13)-(14).

Renormalized frequencies ν_k involved in (13)-(14), according to [36, 37], are determined by the equation for s:

$$1 = \frac{2}{3} \sum_k \frac{k^2 f_k^2 \omega_k}{s - \omega_k^2}, \qquad (15)$$

solutions of which yield the spectrum of $s = \{\nu_k^2\}$ values.

3. Energy Spectrum of a TI-Bipolaron

Hamiltonian (13)-(14) is conveniently presented in the form:

$$\widetilde{\widetilde{H}} = \sum_{n=0,1,2,\dots} E_n \alpha_n^+ \alpha_n, \qquad (16)$$

$$E_n = \begin{cases} E_{bp}, & n = 0; \\ \nu_n = E_{bp} + \omega_0 + \dfrac{k_n^2}{2}, & n \neq 0, \end{cases} \qquad (17)$$

where in the case of a three-dimensional ionic crystal is

$$k_{n_i} = \pm \frac{2\pi (n_i - 1)}{N_{a_i}},$$

$$\qquad (18)$$

$$n_i = 1, 2, \dots, \frac{N_{a_i}}{2} + 1, \ i = x, y, z,$$

where N_{a_i} is the number of atoms along the ith crystallographic axis.

Let us prove the validity of the expression for the spectrum (16), (17). Since operators α_n^+, α_n obey Bose commutation relations:

$$[\alpha_n, \alpha_{n'}^+] = \alpha_n \alpha_{n'}^+ - \alpha_{n'}^+ \alpha_n = \delta_{n,n'}, \qquad (19)$$

they can be considered to be operators of birth and annihilation of TI-bipolarons. The energy spectrum of TI-bipolarons, according to (15), is determined by the equation

$$F(s) = 1, \qquad (20)$$

where

$$F(s) = \frac{2}{3} \sum_n \frac{k_n^2 f_{k_n}^2 \omega_{k_n}^2}{s - \omega_{k_n}^2}. \qquad (21)$$

It is convenient to solve (20) graphically (Figure 1).

Figure 1 suggests that the frequencies ν_{k_n} (index i is omitted) lie between the frequencies ω_{k_n} and $\omega_{k_{n+1}}$. Hence, the spectrum ν_{k_n} as well as the spectrum ω_{k_n} are quasicontinuous: $\nu_{k_n} - \omega_{k_n} = O(N^{-1})$, which just proves the validity of (16), (17).

It follows that the spectrum of a TI-bipolaron has a gap between the ground state E_{bp} and the quasicontinuous spectrum, equal to ω_0.

Below we will consider the case of low concentration of TI-bipolarons in a crystal. Then they can adequately be considered as an ideal Bose gas, whose properties are determined by Hamiltonian (16).

4. Statistical Thermodynamics of Low-Density TI-Bipolaron Gas

Let us consider an ideal Bose gas of TI-bipolarons which represents a system of N particles occurring in some volume V. Let us write N_0 for the number of particles in the lower one-particle state and N' for the number of particles in higher states. Then

$$N = \sum_{n=0,1,2,\dots} \overline{m}_n = \sum_n \frac{1}{e^{(E_n - \mu)/T} - 1}, \qquad (22)$$

$$N = N_0 + N',$$

$$N_0 = \frac{1}{e^{(E_{bp} - \mu)/T} - 1}, \qquad (23)$$

$$N' = \sum_{n \neq 0} \frac{1}{e^{(E_n - \mu)/T} - 1}.$$

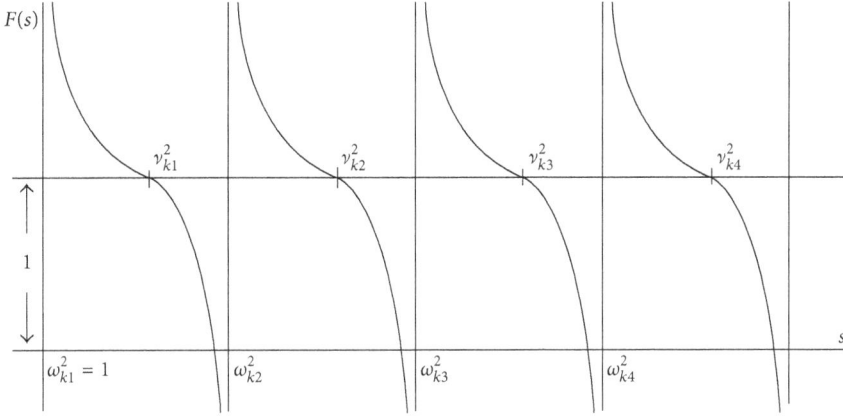

FIGURE 1: Graphical solution of (20).

In expression N' (23), we will perform integration over quasicontinuous spectrum (instead of summation) (16), (17) and assume $\mu = E_{bp}$. As a result, from (22), (23) we get an equation for determining the critical temperature of Bose condensation T_c:

$$C_{bp} = f_{\widetilde{\omega}}\left(\widetilde{T}_c\right),\qquad(24)$$

$$f_{\widetilde{\omega}}\left(\widetilde{T}_c\right) = \widetilde{T}_c^{3/2} F_{3/2}\left(\widetilde{\omega}/\widetilde{T}_c\right),$$

$$F_{3/2}\left(\alpha\right) = \frac{2}{\sqrt{\pi}} \int_0^{\infty} \frac{x^{1/2} dx}{e^{x+\alpha} - 1},$$

$$C_{bp} = \left(\frac{n^{2/3} 2\pi\hbar^2}{M_e \omega^*}\right)^{3/2},\qquad(25)$$

$$\widetilde{\omega} = \frac{\omega_0}{\omega^*},$$

$$\widetilde{T}_c = \frac{T_c}{\omega^*},$$

where $n = N/V$. Figure 2 shows a graphical solution of (24) for the values of parameters $M_e = 2m^* = 2m_0$, where m_0 is the mass of a free electron in vacuum, $\omega^* = 5\,\text{meV}$ (≈ 58 K), $n = 10^{21}\,\text{cm}^{-3}$, and the values $\widetilde{\omega}_1 = 0,2$; $\widetilde{\omega}_2 = 1$; $\widetilde{\omega}_3 = 2$; $\widetilde{\omega}_4 = 10$; $\widetilde{\omega}_5 = 15$; $\widetilde{\omega}_6 = 20$.

It is seen from Figure 2 that the critical temperature grows with increasing phonon frequency ω_0. The relations of critical temperatures T_{ci}/ω_{0i} corresponding to the chosen parameter values are given in Table 1. Table 1 suggests that the critical temperature of a TI-bipolaron gas is always higher than that of ideal Bose gas (IBG). It is also evident from Figure 2 that an increase in the concentration of TI-bipolarons n will lead to an increase in the critical temperature, while a gain in the electron mass m^* to its decrease. For $\widetilde{\omega} = 0$ the results go over into the limit of IBG. In particular, (24), for $\widetilde{\omega} = 0$, yields the expression for the critical temperature of IBG:

$$T_c = \frac{3,31\hbar^2 n^{2/3}}{M_e}.\qquad(26)$$

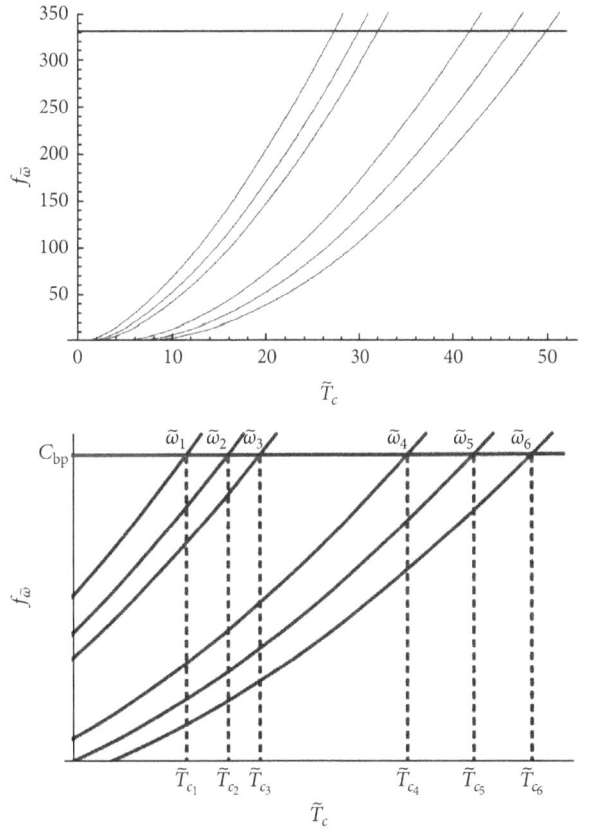

FIGURE 2: Solutions of (24) with $C_{bp} = 331,3$ and $\widetilde{\omega}_i = \{0,2; 1; 2; 10; 15; 20\}$, which correspond to $\widetilde{T}_{c_i}$: $\widetilde{T}_{c_1} = 27,3$; $\widetilde{T}_{c_2} = 30$; $\widetilde{T}_{c_3} = 32$; $\widetilde{T}_{c_4} = 42$; $\widetilde{T}_{c_5} = 46,2$; $\widetilde{T}_{c_6} = 50$.

It should be stressed, however, that (26) involves $M_e = 2m^*$, rather than the bipolaron mass. This resolves the problem of the low temperature of condensation which arises both in the SRP theory and in the LRP theory in which expression (26) involves the bipolaron mass [31–34]. Another important result is that the critical temperature T_c for the parameter values considerably exceeds the gap energy ω_0.

TABLE 1: Calculated characteristics of Bose gas of TI-bipolarons with concentration $n = 10^{21}$ cm^{-3}.

i	0	1	2	3	4	5	6
$\tilde{\omega}_i$	0	0,2	1	2	10	15	20
T_{ci}/ω_{oi}	∞	136,6	30	16	4,2	3	2,5
q_i/T_{ci}	1,3	1,44	1,64	1,8	2,5	2,8	3
$-\Delta(\partial C_{v,i}/\partial\tilde{T})$	0,11	0,12	0,12	0,13	0,14	0,15	0,15
$C_{v,i}(T_c - 0)$	1,9	2,16	2,46	2,7	3,74	4,2	1,6
$(C_s - C_n)/C_n$	0	0,16	0,36	0,52	1,23	1,53	1,8
$n_{bp_i} \cdot$ cm^3	$16 \cdot 10^{19}$	$9,4 \cdot 10^{18}$	$4,2 \cdot 10^{18}$	$2,0 \cdot 10^{18}$	$1,2 \cdot 10^{17}$	$5,2 \cdot 10^{14}$	$2,3 \cdot 10^{13}$

$\tilde{\omega}_i = \omega_i/\omega^*$, $\omega^* = 5$ meV, ω_i is the energy of an optical phonon; T_{c_i} is the critical temperature of the transition, q_i is the latent heat of the transition from condensate to supracondensate state; $-\Delta(\partial C_{v,i}/\partial\tilde{T}) = \partial C_{v,i}/\partial\tilde{T}|_{\tilde{T}=\tilde{T}_{c_i}+0} - \partial C_{v,i}/\partial\tilde{T}|_{\tilde{T}=\tilde{T}_{c_i}-0}$ is the jump in heat capacity during SC transition, $\tilde{T} = T/\omega^*$; $C_{v,i}(T_c - 0)$ is the heat capacity in the SC phase at the critical point; $C_s = C_v(T_c - 0)$, $C_n = C_v(T_c + 0)$. The calculations are carried out for the concentration of TI-bipolarons $n = 10^{21}$ cm^{-3} and the effective mass of a band electron $m^* = m_0$. The table also lists the values of concentrations of TI-bipolarons n_{bpi} for HTSC $YBa_2Cu_3O_7$, based on the experimental value of the transition temperature $T_c = 93$K (Section 6).

From (22), (23), it follows that

$$\frac{N'(\tilde{\omega})}{N} = \frac{\tilde{T}^{3/2}}{C_{bp}}F_{3/2}\left(\frac{\tilde{\omega}}{\tilde{T}}\right),$$

$$\frac{N_0(\tilde{\omega})}{N} = 1 - \frac{N'(\tilde{\omega})}{N} \quad (27)$$

Figure 3 shows temperature dependencies of the number of supracondensate particles N' and the number of particles N_0 occurring in the condensate for the above-indicated parameter values $\tilde{\omega}_i$.

Figure 3 suggests that, as could be expected, the number of particles in the condensate grows as the gap ω_i increases.

The energy of a TI-bipolaron gas E is determined by the expression

$$E = \sum_{n=0,1,2,...} \overline{m}_n E_n = E_{bp}N_0 + \sum_{n\neq0}\overline{m}_n E_n. \quad (28)$$

With the use of (16), (17), and (28) the specific energy (i.e., the energy per one TI-bipolaron) $\tilde{E}(\tilde{T}) = E/N\omega^*$, $\tilde{E}_{bp} = E_{bp}/\omega^*$ will be

$$\tilde{E}(\tilde{T})$$
$$= \tilde{E}_{bp} \quad (29)$$
$$+ \frac{\tilde{T}^{5/2}}{C_{bp}}F_{3/2}\left(\frac{\tilde{\omega}-\tilde{\mu}}{\tilde{T}}\right)\left[\frac{\tilde{\omega}}{\tilde{T}} + \frac{F_{5/2}\left((\tilde{\omega}-\tilde{\mu})/\tilde{T}\right)}{F_{3/2}\left((\tilde{\omega}-\tilde{\mu})/\tilde{T}\right)}\right],$$

$$F_{5/2}(\alpha) = \frac{2}{\sqrt{\pi}}\int_0^\infty \frac{x^{3/2}dx}{e^{x+\alpha}-1}, \quad (30)$$

where $\tilde{\mu}$ is determined from the equation

$$\tilde{T}^{3/2}F_{3/2}\left(\frac{\tilde{\omega}-\tilde{\mu}}{\tilde{T}}\right) = C_{bp}, \quad (31)$$

$$\tilde{\mu} = \begin{cases} 0, & \tilde{T} \leq \tilde{T}_c; \\ \tilde{\mu}(\tilde{T}), & \tilde{T} \geq \tilde{T}_c. \end{cases} \quad (32)$$

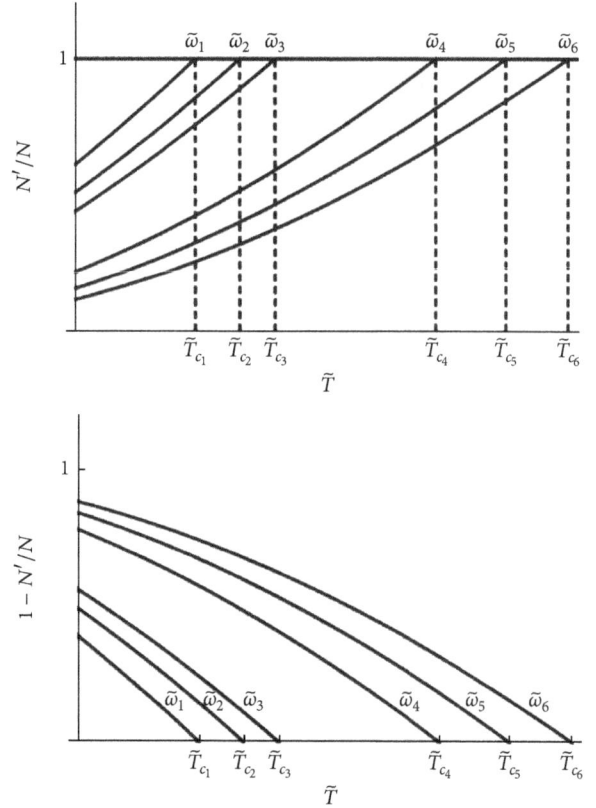

FIGURE 3: Temperature dependencies of the relative number of supracondensate particles N'/N and the particles occurring in the condensate $N_0/N = 1 - N'/N$ for the parameter values $\tilde{\omega}_i$, given in Figure 2.

Relation of $\tilde{\mu}$ with the chemical potential of the system μ is given by the formula $\tilde{\mu} = (\mu - E_{bp})/\omega^*$. From (29)-(31) expressions for the free energy $\Delta F = -(2/3)\Delta E$, $\Delta F = F - E_{bp}N$, $\Delta E = E - E_{bp}N$ and entropy $S = -\partial F/\partial T$ also follow.

Figure 4 illustrates temperature dependencies $\Delta E = \tilde{E} - \tilde{E}_{bp}$ for the above-indicated parameter values ω_i. Break points on the curves $\Delta E_i(\tilde{T})$ correspond to the values of critical temperatures T_{c_i}.

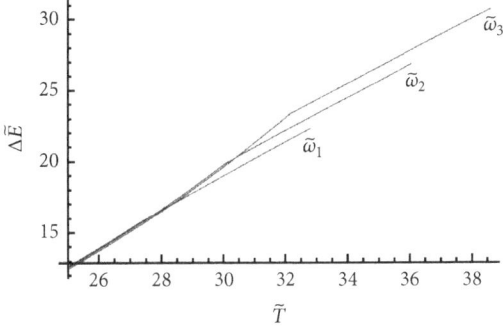

FIGURE 4: Temperature dependencies $\Delta E(\widetilde{T}) = \widetilde{E}(\widetilde{T}) - \widetilde{E}_{bp}$ for the parameter values $\widetilde{\omega}_i$ presented in Figures 2 and 3.

The dependencies obtained enable us to find the heat capacity of a TI-bipolaron gas: $C_v(\widetilde{T}) = d\widetilde{E}/d\widetilde{T}$. With the use of (29) for $\widetilde{T} \leq \widetilde{T}_c$, we express $C_v(\widetilde{T})$ as

$$C_v\left(\widetilde{T}\right) = \frac{\widetilde{T}^{3/2}}{2C_{bp}}\left[\frac{\widetilde{\omega}^2}{\widetilde{T}^2}F_{1/2}\left(\frac{\widetilde{\omega}}{\widetilde{T}}\right) + 6\left(\frac{\widetilde{\omega}}{\widetilde{T}}\right)F_{3/2}\left(\frac{\widetilde{\omega}}{\widetilde{T}}\right)\right.$$
$$\left. + 5F_{5/2}\left(\frac{\widetilde{\omega}}{\widetilde{T}}\right)\right], \tag{33}$$

$$F_{1/2}\left(\alpha\right) = \frac{2}{\sqrt{\pi}}\int_0^\infty \frac{1}{\sqrt{x}}\frac{dx}{e^{x+\alpha}-1}. \tag{34}$$

Expression (33) yields a well-known exponential dependence of the heat capacity at low temperatures $C_v \sim \exp(-\omega_0/T)$, caused by the availability of the energy gap ω_0.

Figure 5 shows temperature dependencies of the heat capacity $C_v(\widetilde{T})$ for the above-indicated parameter values $\widetilde{\omega}_i$. Table 1 lists the values of the heat capacity jumps:

$$\Delta\frac{\partial C_v\left(\widetilde{T}\right)}{\partial\widetilde{T}} = \left.\frac{\partial C_v\left(\widetilde{T}\right)}{\partial\widetilde{T}}\right|_{\widetilde{T}=\widetilde{T}_c+0} - \left.\frac{\partial C_v\left(\widetilde{T}\right)}{\partial\widetilde{T}}\right|_{\widetilde{T}=\widetilde{T}_c-0} \tag{35}$$

at the transition points for the parameter values $\widetilde{\omega}_i$.

The dependencies obtained will enable one to find the latent heat of transition $q = TS$, where S is the entropy of supracondensate particles. At the point of transition this value is $q = 2T_cC_v(T_c - 0)/3$, where $C_v(T)$ is determined by formula (33). For the above-indicated parameter values ω_i, it is given in Table 1.

5. Current States of a TI-Bipolaron Gas

In the foregoing we have considered equilibrium properties of a TI-bipolaron gas. The formation of Bose-condensate per se does not mean that it has superconducting properties. To demonstrate such a possibility let us consider the total momentum of a TI-bipolaron:

$$\overrightarrow{\mathscr{P}} = \widehat{\overrightarrow{P}}_1 + \widehat{\overrightarrow{P}}_2 + \sum\overrightarrow{k}a_k^+a_k, \tag{36}$$

where $\widehat{\overrightarrow{P}}_1$ and $\widehat{\overrightarrow{P}}_2$ are the momenta of the first and second electron, respectively. It is easy to check that $\overrightarrow{\mathscr{P}}$ commutes

with Hamiltonian (1) and therefore is a constant value, that is, c-number.

For this reason, to consider nonequilibrium properties and, particularly current states, we can use generalization of Heisenberg transformation (4) such that

$$S_1\left(\mathscr{P}\right) = \exp\left\{i\left(\overrightarrow{\mathscr{P}} - \sum_k\overrightarrow{k}a_k^+a_k\right)\right\}\overrightarrow{R}. \tag{37}$$

The general expression for the functional of the total energy of a TI-bipolaron for $\overrightarrow{\mathscr{P}} \neq 0$ is given in [36]. TI-bipolarons occurring in condensed state have a common wave function for the whole condensate and do not thermalize in the state when $\overrightarrow{\mathscr{P}} \neq 0$. In the supracondensate part, phonons whose wave vectors contribute to $\overrightarrow{\mathscr{P}}$ (36) are thermalized and their total momentum will be equal to zero.

Hence, all the changes arising in considering the case of $\overrightarrow{\mathscr{P}} \neq 0$ concern only the expression for the ground state energy which becomes dependent on $\overrightarrow{\mathscr{P}}$, while the spectrum of excited states (16), (17) remains unchanged. It follows that, in passing over the critical point, the currentless state suddenly becomes current which is in agreement with the experiment.

6. Comparison with the Experiment

Figure 4 shows typical dependencies of $E(\widetilde{T})$. They suggest that at the point of transition the energy is a continuous function of $\widetilde{T}$. This means that the transition per se occurs without energy expenditure being a phase transition of the second kind in complete agreement with the experiment. At the same time transition of Bose particles from a condensate state to a supracondensate one occurs with consumption of energy which is determined by the value q (Section 4, Table 1), determining the latent heat of transition of a Bose gas which makes it a phase transition of the 1st kind.

By way of example let us consider HTSC $YBa_2Cu_3O_7$ with the temperature of transition $90 \div 93$K, volume of the unit cell $0,1734\cdot10^{-21}$ cm^{-3}, and concentration of holes $n \approx 10^{21}$ cm^{-3}. According to estimates [48], Fermi energy is equal to $\epsilon_F = 0,37$ eV. Concentration of TI-bipolarons in $YBa_2Cu_3O_7$ is found from (24):

$$\frac{n_{bp}}{n}C_{bp} = f_{\widetilde{\omega}}\left(\widetilde{T}_c\right) \tag{38}$$

with $\widetilde{T}_c = 1,6$. Table 1 lists the values of $n_{bp,i}$ for the values of parameters $\widetilde{\omega}_i$ given in Section 4. It follows from Table 1 that $n_{bp,i} \ll n$. Hence, only a small part of charge carriers is in a bipolaron state which justifies the approximation of a low-density TI-bipolaron gas used by us. This fact correlates with recent experiments [49] where it was shown that only a small part of electrons are in SC state. The energy levels of such TI-bipolarons lie near Fermi surface and are described by the wave function:

$$\Psi\left(\overrightarrow{r}\right) = e^{i\overrightarrow{k}_F\overrightarrow{r}}\varphi\left(\overrightarrow{r}\right), \tag{39}$$

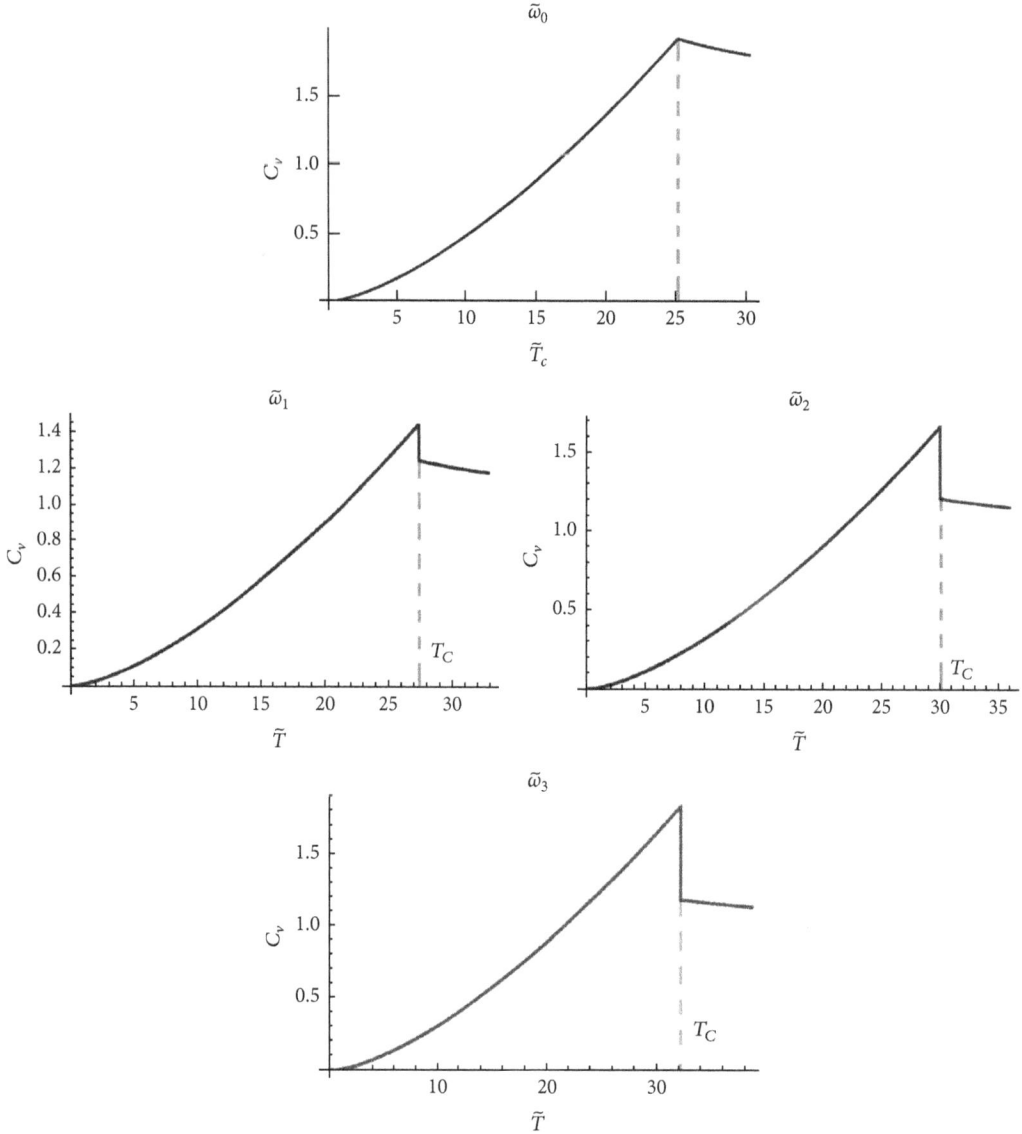

FIGURE 5: Temperature dependencies of the heat capacity for various values of the parameters ω_i: $\omega_0 = 0$; $\tilde{T}_{C_0} = 25, 2$; $C_v(\tilde{T}_{c1}) = 2$; $\omega_1 = 0, 2$; $\tilde{T}_{C_1} = 27, 3$; $C_v(\tilde{T}_{c1} - 0) = 2, 16$; $C_v(\tilde{T}_{c1} + 0) = 1, 9$; $\omega_2 = 1$; $\tilde{T}_{C_2} = 30$; $C_v(\tilde{T}_{c2} - 0) = 2, 46$; $C_v(\tilde{T}_{c2} + 0) = 1, 8$; $\omega_3 = 2$; $\tilde{T}_{C_3} = 32, 1$; $C_v(\tilde{T}_{c3} - 0) = 2, 7$; $C_v(\tilde{T}_{c3} + 0) = 1, 78$.

which leads to replacement of $\overline{T}$, involved in (6), by

$$\overline{T} = -2 \langle \varphi | \Delta_r | \varphi \rangle + 2k_F^2, \qquad (40)$$

that is, reckoning of the energy from Fermi level (the last term in the right-hand side of (40) in dimensional units is equal to $2\epsilon_F$, where $\epsilon_F = \hbar^2 k_F^2 / 2m^*$).

According to our approach, superconductivity arises when coupled states are formed. The condition for the formation of such states has the form

$$E_{bp} < 0, \qquad (41)$$

where E_p is the energy of a TI-polaron [41]. Condition (41) determines the value of a pseudogap:

$$\Delta_1 = |E_{bp}|. \qquad (42)$$

For $\Delta_1 \gg \omega_0$ the value of a pseudogap can greatly exceed both T_c and the energy of the gap (i.e., ω_0). The expression for the spectrums E_{bp} and E_p (16)-(17) suggests that the angular dependence of superconducting gap is completely determined by the symmetry of the isoenergetic surface of the $\omega_0(\vec{k})$. Earlier this conclusion was made by Bennet [50], who proved that the main source of anisotropy of superconducting properties is the angular dependence of the phonon spectrum, though some contribution is also made by the anisotropy of Fermi surface.

It follows from what has been said that formation of a pseudogap is a phase transition preceding the phase transition to the superconducting state. Recent experiments [51] also testify in favor of this statement.

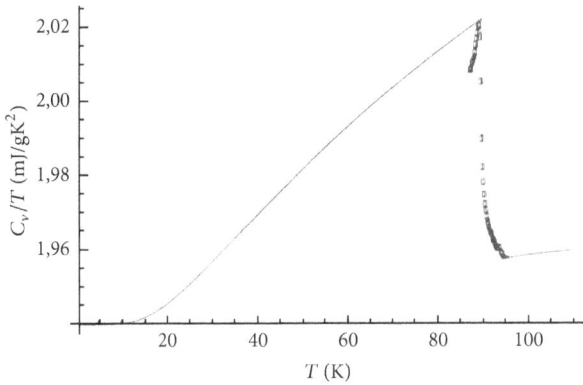

FIGURE 6: Comparison of the theoretical (solid line) and experimental [29] (broken line) dependencies in the region of the heat capacity jump.

In paper [39] correlation length for TI-bipolarons was calculated. According to [39], in HTSC materials its value can vary from several angstroms to several tens of angstroms, which is also in agreement with the experiment.

Of special interest is to determine the characteristic energy of phonons responsible for the formation of TI-bipolarons and superconducting properties of oxide ceramics. To do this let us compare the calculated values of the heat capacity jumps with experimental data.

As is known, in BCS theory, a jump in the heat capacity is equal to

$$\left. \frac{C_s - C_n}{C_n} \right|_{T_c} = 1,43, \qquad (43)$$

where C_s is the heat capacity in the superconducting phase and C_n is the heat capacity in the normal phase and is independent of the parameters of the model Hamiltonian. As it follows from numerical calculations shown in Figure 5 and in Table 1, as distinct from the BCS theory, the value of the jump depends on the phonon frequency. Hence, the approach presented predicts the existence of an isotopy effect for the heat capacity jump.

As it is seen from Figure 6, the heat capacity jump calculated theoretically (Section 4) coincides with the experimental value in $YBa_2Cu_3O_7$ [29], for $\tilde{\omega} = 1,5$, that is, for $\omega = 7,5$ meV. This corresponds to the concentration of TI-bipolarons equal to $n_{bp} = 2,6 \cdot 10^{18}$ cm^{-3}. Hence, in contrast to the widespread notion that in oxide ceramics superconductivity is determined by high-energy phonons (with energy $70 \div 80$ meV [52]), actually, the superconductivity in HTSC materials should be determined by soft phonon modes.

Notice that in calculations of the temperature of transition it was believed that the effective mass M_e in (24) is independent of the direction of the wave vector; that is, isotropic case was dealt with.

In the anisotropic case, choosing principal axes of vector k, as coordinate axes, we will get the quantity $(M_{ex}M_{ey}M_{ez})^{1/3}$ instead of the effective mass M_e. In complex HTSC materials the values of effective masses lying in the plane of layers M_{ex}, M_{ey} are close in value. Assuming in

this case $M_e = M_{ex} = M_{ey} = M_\parallel$, $M_{ez} = M_\perp$, we will get instead of C_{bp}, determined by (24), the value $\widetilde{C}_{bp} = C_{bp}/\gamma$; $\gamma^2 = M_\perp/M_\parallel$ is the parameter of anisotropy. Hence anisotropy of effective masses gives for the concentration n_{bp} the value $\bar{n}_{bp} = \gamma n_{bp}$. Therefore taking account of anisotropy can, in an order of magnitude, enhance the estimate of the concentration of TI-bipolarons. If for $YBa_2Cu_3O_7$ we take the estimate $\gamma^2 = 30$ [52], then for the concentration of TI-bipolarons we will get $\bar{n}_{bp} = 1,4 \cdot 10^{19}$ cm^{-3}, which holds valid the general conclusion: in the case under consideration only a small number of charge carriers are in TI-bipolaron state. The situation can change if the anisotropy parameter is very large. Thus, for example, in layered HTSC Bi-Sr-Ca-Cu-O the anisotropy parameter is $\gamma > 100$; accordingly, the concentration of TI-bipolarons in these compounds can have the same order of magnitude as the total concentration of charge carriers.

Another important conclusion emerging from taking account of the anisotropy of effective masses is that the temperature of the transition T_c depends not on n_{bp} and $M_\parallel$ individually, but on their relation which straightforwardly follows from (24).

7. Essential Generalizations of the Theory

In the foregoing we considered the case of an ideal TI-bipolaron gas. At small concentration of TI-bipolarons their Coulomb interaction will be greatly screened which justifies the use of the model of an ideal gas.

If the concentration of TI-bipolarons is large (e.g., $n = 10^{21}$ cm^{-3}, as was believed in Section 4), then such Bose gas can no longer be considered to be ideal. Taking account of Coulomb interaction between bipolarons becomes necessary. For this purpose we can use Bogolyubov theory [53] for weakly imperfect Bose gas, which implies that the spectrum of elementary excitations with the momentum k will be determined by the expression:

$$E(k) = \sqrt{k^2 u^2(k) + \left(\frac{k^2}{2m_B}\right)^2}, \qquad (44)$$

where $u(k) = \sqrt{n_B V(k)/m_B}$, $V(k)$ is the Fourier component of the potential of pairwise interaction between charged bosons: $V(k) = 4\pi e_B^2/\epsilon_0 k^2$, e_B and m_B are the charge and mass of a boson, and n_B is the concentration of bosons.

Hence it follows that

$$E(k) = \sqrt{(\hbar\omega_p)^2 + \left(\frac{k^2}{2m_B}\right)^2}, \quad \omega_p = \sqrt{\frac{4\pi e^2 n_B}{\epsilon_0 m_B}}. \qquad (45)$$

ω_p is the plasma frequency. According to (45), the spectrum of excitations of quasiparticles of an imperfect gas is characterized by a finite energy gap which in the long-wavelength limit is equal to ω_p. The same result can be arrived at if the Coulomb interaction between the electrons is considered as a result of electron-plasmon interaction.

According to [54], Hamiltonian of electron-plasmon interaction coincides in structure with Froehlich Hamiltonian which involves plasmon frequency instead of phonon frequency. This straightforwardly leads to the energy gap equal to the plasmon frequency ω_p, if in (45) we put $e_B = 2e$, $m_B = 2m^*$, $n_B = n/2$, where n is the electron concentration. Hence, for $\omega_p < \omega_0$, the energy gap will be determined by ω_p, while for $\omega_p > \omega_0$ it will still be determined by ω_0.

Actually, in real HTSC, there are not only phonon and plasmon branches, but also some other elementary excitations which can take part in electron pairing. An example is spin fluctuations. In generalization of the theory to the case of interaction with various brunches of excitations which will contribute to the ground state energy, the value of the gap and the dispersion law of a TI-bipolaron are a topical problem. Obviously, taking account of this interaction will lead to an increase in the coupling energy of both TI-bipolarons and TI-polarons. Therefore, a priori, without any particular calculations one cannot say anything of how the condition of stability of TI-bipolaron states determined by (41) will change.

Another important problem is generalization of the theory to the case of intermediate coupling of electron-phonon interaction. Formally, the expression for the functional of the ground state energy of a TI-bipolaron (14) is valid for any value of the electron-phonon coupling constant. For this reason such a calculation will not change the spectrum of a TI-bipolaron; however it will alter the criteria of fulfillment of the conditions of a TI-bipolaron stability (41).

8. Conclusive Remarks

In this paper we have presented conclusions emerging from consistent translation-invariant consideration of EPI. It implies that pairing of electrons, for any coupling constant, leads to a concept of TI-polarons and TI-bipolarons. Being bosons, TI-bipolarons can experience Bose condensation leading to superconductivity. Let us list the main results following from this approach. First and foremost the theory resolves the problem of the great value of the bipolaron effective mass (Section 4). As a consequence, formal limitations on the value of the critical temperature of the transition are eliminated too. The theory quantitatively explains such thermodynamic properties of HTSC-conductors as availability (Section 4) and value (Section 6) of the jump in the heat capacity lacking in the theory of Bose condensation of an ideal gas. The theory also gives an insight into the occurrence of a great ratio between the width of the pseudogap and T_c (Section 6). It accounts for the small value of the correlation length [39] and explains the availability of a gap (Section 3) and a pseudogap (Section 6) in HTSC materials. The angular dependence of the gap and pseudogap gets a natural explanation (Section 6).

Accordingly, isotopic effect automatically follows from expression (24), where the phonon frequency ω_0 acts as a gap. The conclusion of the dependence of the temperature of the transition T_c on the relation $n_{bp}/M_\parallel$ (Section 6) correlates with Uemura law universal for all HTSC materials which

implies that the temperature of the transition is related to the concentration of charge carriers divided by its effective mass [55, 56]. The TI-bipolaron theory of superconductivity developed in this paper gives an answer to the question of paper [49], namely, where most of electrons disappear in the superconductors analyzed. It lies in the fact that only a small part of electrons are paired. The results obtained suggest that in order to raise the critical temperature T_c one should increase the concentration of bipolarons.

Application of the theory to 1D and 2D systems leads to qualitatively new results since the occurrence of a gap in the TI-bipolaron spectrum automatically removes divergences at small momenta, inherent in the theory of ideal Bose gas. These problems were considered by the author in [57]. The case of 1D-polaron was discussed in [58].

Conflicts of Interest

The author declares that there are no conflicts of interest regarding the publication of this paper.

Acknowledgments

The work was done with the support from the Russian Foundation for Basic Research (RFBR Project no. 16-07-00305).

References

[1] E. M. Lifshitz and L. P. Pitaevskii, *Statistical Physics, Part 2: Theory of the Condensed State*, Butterworth-Heinemann, 1st edition, 1980.

[2] J. G. Bednorz and K. A. Müller, "Possible high T_c superconductivity in the Ba-La-Cu-O system," *Zeitschrift für Physik B*, vol. 64, pp. 189–193, 1986.

[3] V. L. Ginzburg, "Superconductivity: the day before yesterday — yesterday — today — tomorrow," *Physics-Uspekhi*, vol. 43, p. 573, 2000.

[4] S. L. Kakani and S. Kakani, *Superconductivity*, Anshan, China, 2009.

[5] T. Tohyama, "Recent progress in physics of high-temperature superconductors," *Japanese Journal of Applied Physics*, vol. 51, Article ID 010004, 2012.

[6] S. Kruchinin, H. Nagao, and S. Aono, *Modern Aspects of superconductivity. Theory of superconductivity*, World Scientific Publishing Co. Pte. Ltd., River Edge, NJ, USA, 2011.

[7] K. P. Sinha and S. L. Kakani, "Fermion local charged boson model and cuprate superconductors," *Proceedings – National Academy of Sciences, India. Section A, Physical Sciences*, vol. 72, p. 153, 2002.

[8] K. H. Benneman and J. B. Ketterson, *Superconductivity: Conventional and Unconventional Superconductors 1-2*, Springer, New York, NY, UK, 2008.

[9] J. R. Schrieffer, *Theory of Superconductivity*, Westview Press, Oxford, UK, 1999.

[10] A. S. Alexandrov, *Theory of Superconductivity from weak to strong oupling*, IOP, publishing, Bristol, UK, 2003.

[11] N. M. Plakida, "High temperature cuprate superconductors: experiment," in *Theory and Applications*, vol. 166 of *Springer*

Series in Solid-State Sciences, pp. 479–494, Springer, Berlin, Germany, 2010.

[12] I. Askerzade, "Physical Properties of Unconventional Superconductors," in *Unconventional Superconductors*, vol. 153 of *Springer Series in Materials Science*, pp. 1–26, Springer Berlin Heidelberg, Berlin, Germany, 2012.

[13] O. Gunnarsson and O. Rosch, "Interplay between electron–phonon and Coulomb interactions in cuprates," *Journal of Physics*, vol. 20, Article ID 043201, 2008.

[14] T. Moriya and K. Ueda, "Spin fluctuations and high temperature superconductivity," *Advances in Physics*, vol. 49, no. 5, pp. 556–606, 2000.

[15] D. Manske, *Theory of Unconventional Superconductors*, Springer, Heidelberg, Germany, 2004.

[16] R. Szczęśniak and A. P. Durajski, "Anisotropy of the gap parameter in the hole-doped cuprates," *Superconductor Science and Technology*, vol. 27, no. 12, Article ID 125004, 2014.

[17] A. P. Durajski, "Anisotropic evolution of energy gap in Bi2212 superconductor," *Frontiers of Physics*, vol. 11, no. 5, Article ID 117408, 2016.

[18] R. Szczęśniak, A. P. Durajski, and A. M. Duda, "Pseudogap in Eliashberg approach based on electron-phonon and electron-electron-phonon interaction," *Annalen der Physik*, vol. 529, Article ID 1600254, 2017.

[19] V. D. Lakhno, "Phonon interaction of electrons in the translation-invariant strong-coupling theory," *Modern Physics Letters B*, vol. 30, Article ID 1650031, 2016.

[20] J. Bardeen, L. N. Cooper, and J. R. Schrieffer, "Theory of superconductivity," *Physical Review Letters*, vol. 108, pp. 1175–1204, 1957.

[21] A. S. Alexandrov and N. Mott, *Polarons Bipolarons, World Sci*, Pub. CO Inc, Singapore, Singapore, 1996.

[22] A. S. Alexandrov and A. B. Krebs, "Reviews of topical problems," *Soviet Physics—Uspekhi*, vol. 35, no. 5, pp. 345–383, 1992.

[23] P. W. Anderson, "Condensed matter: the continuous revolution," *Physics World*, vol. 8, no. 12, p. 37, 1995.

[24] N. F. Mott, "High-temperature superconductivity debate heats up," *Physics World*, vol. 9, no. 1, p. 16, 1996.

[25] P. W. Anderson, *Physics World*, vol. 9, no. 1, p. 16, 1996.

[26] E. V. de Mello and J. Ranninger, "Dynamical properties of small polarons," *Physical Review B: Condensed Matter and Materials Physics*, vol. 55, no. 22, pp. 14872–14885, 1997.

[27] Y. A. Firsov, V. V. Kabanov, E. K. Kudinov, and A. S. Alexandrov, "Comment on "Dynamical properties of small polarons"," *Physical Review B: Condensed Matter and Materials Physics*, vol. 59, p. 12132, 1999.

[28] E. V. de Mello and J. Ranninger, "Reply to "Comment on 'Dynamical properties of small polarons' "," *Physical Review B: Condensed Matter and Materials Physics*, vol. 59, no. 18, pp. 12135-12136, 1999.

[29] N. Overend, M. A. Howson, and I. D. Lawrie, "3D X-Y scaling of the specific heat of YBa$_2$Cu$_3$O$_{7-\delta}$ single crystals," *Physical Review Letters*, vol. 72, p. 3238, 1994.

[30] R. A. Ogg, "Superconductivity in solid metal-ammonia solutions," *Physical Review*, vol. 70, p. 93, 1946.

[31] V. L. Vinetskii and E. A. Pashitskii, "Superfluidity of charged Bose-gas and bipolaron mechanism of superconductivity," *Ukrainian Journal of Physics*, vol. 20, p. 338, 1975.

[32] E. A. Pashitskii and V. L. Vinetskii, "Plasmon and bipolaron mechanisms of high-temperature superconductivity," *JETP Letters*, vol. 46, p. S104, 1987.

[33] D. Emin, "Formation, motion, and high-temperature superconductivity of large bipolarons," *Physical Review Letters*, vol. 62, no. 13, pp. 1544–1547, 1989.

[34] V. L. Vinetskii, N. I. Kashirina, and E. A. Pashitskii, "Bipolaron states in ion crystals and the problem of high temperature superconductivity," *Ukrainian Journal of Physics*, vol. 37, no. 76, 1992.

[35] M. R. Schafroth, "Superconductivity of a charged boson gas," *Physical Review*, vol. 96, no. 4, p. 1149, 1954.

[36] A. V. Tulub, "Recoil accounting in nonrelativistic quantum field theory," *Vestnik Leningrad University*, vol. 15, no. 22, p. 104, 1960.

[37] A. V. Tulub, "Slow electrons in polar crystals," *Soviet Physics—JETP*, vol. 14, no. 1301, 1962.

[38] V. D. Lakhno, "Energy and critical ionic-bond parameter of a 3D large-radius bipolaron," *Journal of Experimental and Theoretical Physics*, vol. 110, no. 5, pp. 811–815, 2010.

[39] V. D. Lakhno, "Translation-invariant bipolarons and the problem of high-temperature superconductivity," *Solid State Communications*, vol. 152, p. 621, 2012.

[40] N. I. Kashirina, V. D. Lakhno, and A. V. Tulub, "The virial theorem and the ground state problem in polaron theory," *Journal of Experimental and Theoretical Physics*, vol. 114, no. 5, pp. 867–869, 2012.

[41] V. D. Lakhno, "Translation invariant theory of polaron (bipolaron) and the problem of quantizing near the classical solution," *Journal of Experimental and Theoretical Physics*, vol. 116, no. 6, pp. 892–896, 2013.

[42] V. D. Lakhno, "Pekar's ansatz and the strong coupling problem in polaron theory," *Physics-Uspekhi*, vol. 58, p. 295, 2015.

[43] W. Heisenberg, "Die Selbstenergie des Elektrons," *Zeitschrift für Physik*, vol. 65, no. 1-2, pp. 4–13, 1930.

[44] T. D. Lee, F. E. Low, and D. Pines, "The motion of slow electrons in a polar crystal," *Physical Review*, vol. 90, no. 2, pp. 297–302, 1953.

[45] S. V. Tyablikov, *Methods of Quantum Theory of Magnetizm*, Moscow, Russia, 1975.

[46] S. J. Miyake, *in Polarons and Applications*, V. D. Lakhno, Ed., Wiley, Chichester, 1994.

[47] I. B. Levinson and É. I. Rashba, "Threshold phenomena and bound states in the polaron problem," *Soviet Physics—Uspekhi*, vol. 16, no. 6, pp. 892–912, 1974.

[48] L. P. Gor'kov and N. B. Kopnin, "High-Tc superconductors from the experimental point of view," *Soviet Physics—Uspekhi*, vol. 31, no. 9, pp. 850–860, 1988.

[49] I. Božović, X. He, J. Wu, and A. T. Bollinger, "Dependence of the critical temperature in overdoped copper oxides on superfluid density," *Nature*, vol. 536, no. 7616, pp. 309–311, 2016.

[50] A. J. Bennett, "Theory of the anisotropic energy gap in superconducting lead," *Physical Review*, vol. 140, p. A1902, 1965.

[51] R.-H. He, M. Hashimoto, H. Karapetyan et al., "From a single-band metal to a high-temperature superconductor via two thermal phase transitions," *Science*, vol. 331, p. 1579, 2011.

[52] A. Marouchkine, *Room-Temperature Superconductivity*, vol. 7, Cambridge Int. Sci. Publ., Cambridge, UK, 2004.

[53] N. N. Bogolyubov, "On the theory of superfluidity," *Journal of Physics USSR*, vol. 11, p. 77, 1947.

[54] V. D. Lakhno, *In Polarons and Applications*, V. D. Lakhno, Ed., Wiley, Chichester, 1994.

[55] Y. J. Uemura, G. M. Luke, B. J. Sternlieb et al., "Universal correlations between T_c and n_s/m^* (carrier density over effective mass)

in High-T_c cuprate superconductors," *Physical Review Letters*, vol. 62, p. 2317, 1989.

[56] Y. J. Uemura, L. P. Le, G. M. Luke et al., "Basic similarities among cuprate, bismuthate, organic, Chevrel-phase, and heavy-fermion superconductors shown by penetration-depth measurements," *Physical Review Letters*, vol. 66, p. 2665, 1991.

[57] V. D. Lakhno, "A translation invariant bipolaron in the Holstein model and superconductivity," *SpringerPlus*, vol. 5, p. 1277, 2016.

[58] V. D. Lakhno, "Large-radius Holstein polaron and the problem of spontaneous symmetry breaking," *Progress of Theoretical and Experimental Physics*, vol. 2014, no. 7, Article ID 073101, 2014.

Light Leakage of Multidomain Vertical Alignment LCDs using a Colorimetric Model in the Dark State

Chuen-Lin Tien ⓘ,[1,2] **Rong-Ji Lin,**[2] **and Shang-Min Yeh** ⓘ[3]

[1]*Department of Electrical Engineering, Feng Chia University, Taichung 40724, Taiwan*
[2]*Ph.D. Program of Electrical and Communications Engineering, Feng Chia University, Taichung 40724, Taiwan*
[3]*Department of Optometry, Central Taiwan University of Science and Technology, Taichung 40601, Taiwan*

Correspondence should be addressed to Shang-Min Yeh; optom.yap@gmail.com

Academic Editor: Jia-De Lin

Light leakage from liquid crystal displays in the dark state is relatively larger and leads to a degraded contrast ratio and color shift. This work describes a novel colorimetric model based on the Muller matrix that includes depolarization of light propagating through liquid crystal molecules, polarizers, and color filters. In this proposed model, the chromaticity can be estimated in the bump and no-bump regions of an LCD. We indicate that the difference between simulation and measurement of chromaticity is about 0.01. Light leakage in the bump region is three times that in no-bump region in the dark state.

1. Introduction

Many wide viewing-angle LCDs have been investigated and produced. In-Plane Switching (IPS) [1, 2] and Multidomain Vertical Alignment (MVA) [3–5] are widely used for high-end LCD products. The MVA LCD has a superior contrast ratio at normal viewing directions, wide viewing angle, and poor color dispersion. However, the light leakages in the dark state and color shift at an inclined angle have not yet been improved. Therefore, enhancing the contrast ratio and eliminating the light leakage in the dark state are important goals [6, 7]. Many conditions result in the light leakage such as light scattering from liquid crystals and the color filter, misalignment of crossed polarizers, and retardation of phase from liquid crystals [8, 9]. Polarization light scattering from the liquid crystals due to thermally excited orientation-based fluctuations of liquid crystal directors and light depolarization from the color filter due to pigment scattering have to be discussed [10, 11]. Investigating the colorimetric and photometric characteristics of LCDs, which include their major optical components, such as crossed polarizers, liquid crystals, color filters, and backlight modules, is interesting [12–14].

In this colorimetric model based on the Muller matrix, the transmission spectrum of each component is imported

and the light scattering from liquid crystals is ignored. The chromaticity coordinates can be estimated by measuring the degree of depolarization of the polarizer and the color filter. The error of chromaticity can be analyzed by introducing three backlight units and color resists. Finally, this work presents simulation and measurement results for a LCD module in the dark state, which demonstrate that a minimum difference exists in chromaticity, and light leakage in the bump region is three times larger than in no-bump region.

2. Analyzing Light Leakage in the Dark State

Figure 1 shows some LCD parts—the polarizer, analyzer, liquid crystal, and color filter. Two major parts in pixel design exist for an MVA: one is the bump region and the other is the no-bump region. In the bump region, due to the shape of the bump, the orientation of liquid crystal molecules is affected by the boundary condition. When linear polarized light passes through the bump region, phase retardation results in nonlinear polarization light. Finally, light leakage can exist in the bump region. Furthermore, transmittance in the bump region also influences chromaticity of an LCD. We should be concerned with calculating individually the optical performance of a pixel in the bump and no-bump region. Figure 2 presents vertical images of the subpixel

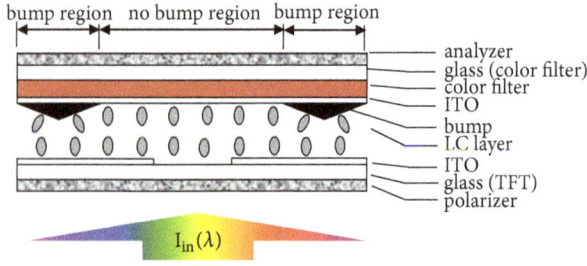

FIGURE 1: Scheme of MVA liquid crystal module.

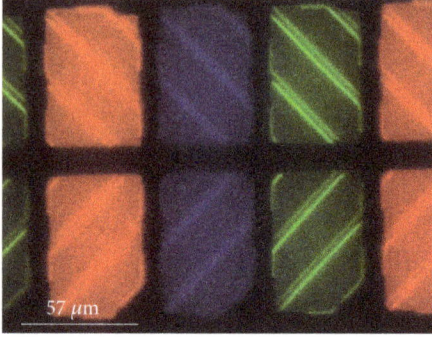

FIGURE 2: The vertical view images of subpixel color filter in dark state.

color filter at a normal direction in the dark state. The blue, green, and red lights are generated by the color filter with the existence of different color pigments. The oblique lines in each grid (subpixel) indicate the bump region, which is brighter than the no-bump region in the dark state due to light leakage. Light leakage can reduce colorimetric and photometric performance of liquid crystal displays in the dark state.

3. Colorimetric Calculation by Matrix Method

Depolarization occurs with the reduction in degree of polarization when light scatters through optical elements such as the polarizer or color filter. In (1), the light emanating out of the LCD through the polarizer, analyzer, liquid crystals, and color filter can be described using a matrix of optics.

$$P\left(\theta_1\right) D\left(d\right) P\left(\theta_2\right) \begin{pmatrix} I \\ Q \\ U \\ V \end{pmatrix} = \begin{pmatrix} I' \\ Q' \\ U' \\ V' \end{pmatrix}, \quad (1)$$

where $P(\theta)$ is the Muller matrix of polarizer; θ is the direction of the transmit axis and the subscript numbers of θ are the polarizer and analyzer, respectively; $D(d)$ is the Muller matrix of the depolarizer (including the polarizer and color filter); d is the degree of depolarization. Notably, I, Q, U, and V are known as stoke parameters of a quasimonochromatic plane wave that represents total intensity, horizontally or vertically polarized state, linear polarization along directions at angle

$\varphi = \pm 45°$ to the x-axis, and the right or left circularly polarized state, respectively. Superscripts of stoke parameters describe the conditions of light propagating through the LCD module. Values of Q, U, and V vanish in the unpolarized state.

We assume the polarizer and analyzer comprise an ideal polarizer. The Muller matrix of the polarizer can be presented as follows:

$$p\left(\theta\right) = \frac{1}{2} \begin{pmatrix} 1 & \cos 2\theta & \sin 2\theta & 0 \\ \cos 2\theta & \dfrac{1 + \cos 4\theta}{2} & \dfrac{\sin 4\theta}{2} & 0 \\ \sin 2\theta & \dfrac{\sin 4\theta}{2} & \dfrac{1 - \cos 4\theta}{2} & 0 \\ 0 & 0 & 0 & 0 \end{pmatrix}, \quad (2)$$

where the Muller matrix in different directions of the transmit axes can be obtained by changing the value of θ.

Equation (3) derives the contrast ratio determined by the intensity of outgoing light with parallel polarizers ($\theta_1 = \theta_2 = 0$) divided by that with crossed polarizers ($\theta_1 = 0, \theta_2 = \pi/2$) in the dark state.

$$\mathrm{CR}_{\mathrm{PDP}} = \frac{I'_{//}}{I'_{\perp}} = \frac{(2 - d)/4}{d/4} = \frac{2 - d}{d}, \quad (3)$$

where $I'_{//}$ and $I'_{\perp}$ are the total intensity of outgoing light with the polarization of polarizers parallel or orthogonally intersected with each other, respectively.

The colorimetric and photometric performance of LCDs depends on wavelength. That is, all mechanisms of light leakage should be determined in wavelength-related terms. Thus, considering the transmittance of color filters and polarizers, the degree of depolarization is function of wavelength. The $d(\lambda)$ can be estimated by (3) after measuring the contrast ratio. The intensity of outgoing light from the bump and no-bump regions is as follows:

$$I'_{\mathrm{BUMP}}\left(\lambda\right) = \frac{d\left(\lambda\right)}{4} \times T_{\mathrm{CF}} \times T_{\mathrm{BUMP}} \times T_{\mathrm{LC\_BUMP}}$$
$$\times T_{\mathrm{PF}} \times T_{\mathrm{TFT}} \times I_{\mathrm{in}}\left(\lambda\right)$$
$$\quad (4)$$
$$I'_{\mathrm{No-BUMP}}\left(\lambda\right) = \frac{d\left(\lambda\right)}{4} \times T_{\mathrm{CF}} \times T_{\mathrm{LC\_No-BUMP}} \times T_{\mathrm{PF}}$$
$$\times T_{\mathrm{TFT}} \times I_{\mathrm{in}}\left(\lambda\right),$$

where T is transmittance of optical components, CF is the color filter, BUMP is the region with a bump, LC_BUMP is the region of the liquid crystal (LC) with a bump, PF is the polarizer or analyzer, TFT is thin-film transistors, and $I_{\mathrm{in}}(\lambda)$ is light source spectrum.

To determine the chromaticity of the dark state, X_{bump}, Y_{bump}, and Z_{bump} tristimulus values of the Commission

International de l'Éclairage (CIE) XYZ color space through the bump region can be written in the following form [14]:

$$X_{\text{bump}} = k \int \frac{d(\lambda)}{4} \times T_{\text{CF}}(\lambda) \times T_{\text{BUMP}}(\lambda)$$
$$\times T_{\text{LC\_BUMP}}(\lambda) \times T_{\text{PF}}(\lambda) \times T_{\text{TFT}}(\lambda) \times \bar{x}(\lambda)$$
$$\times I_{\text{in}}(\lambda) \, d\lambda$$

$$Y_{\text{bump}} = k \int \frac{d(\lambda)}{4} \times T_{\text{CF}}(\lambda) \times T_{\text{BUMP}}(\lambda)$$
$$\times T_{\text{LC\_BUMP}}(\lambda) \times T_{\text{PF}}(\lambda) \times T_{\text{TFT}}(\lambda) \times \bar{y}(\lambda) \qquad (5)$$
$$\times I_{\text{in}}(\lambda) \, d\lambda$$

$$Z_{\text{bump}} = k \int \frac{d(\lambda)}{4} \times T_{\text{CF}}(\lambda) \times T_{\text{BUMP}}(\lambda)$$
$$\times T_{\text{LC\_BUMP}}(\lambda) \times T_{\text{PF}}(\lambda) \times T_{\text{TFT}}(\lambda) \times \bar{z}(\lambda)$$
$$\times I_{\text{in}}(\lambda) \, d\lambda,$$

where $\bar{x}(\lambda)$, $\bar{y}(\lambda)$, and $\bar{z}(\lambda)$ are the tristimulus values of the monochromatic stimuli and k is a constant.

Furthermore, the tristimulus values of light through the no-bump region $X_{\text{no\_bump}}$, $Y_{\text{no\_bump}}$, and $Z_{\text{no\_bump}}$ can be written as follows:

$$X_{\text{No\_bump}} = k \int \frac{d(\lambda)}{4} \times T_{\text{CF}}(\lambda) \times T_{\text{LC\_NoBUMP}}(\lambda)$$
$$\times T_{\text{PF}}(\lambda) \times T_{\text{TFT}}(\lambda) \times \bar{x}(\lambda) \times I_{\text{in}}(\lambda) \, d\lambda$$

$$Y_{\text{No\_bump}} = k \int \frac{d(\lambda)}{4} \times T_{\text{CF}}(\lambda) \times T_{\text{LC\_NoBUMP}}(\lambda) \qquad (6)$$
$$\times T_{\text{PF}}(\lambda) \times T_{\text{TFT}}(\lambda) \times \bar{y}(\lambda) \times I_{\text{in}}(\lambda) \, d\lambda$$

$$Z_{\text{No\_bump}} = k \int \frac{d(\lambda)}{4} \times T_{\text{CF}}(\lambda) \times T_{\text{LC\_NoBUMP}}(\lambda)$$
$$\times T_{\text{PF}}(\lambda) \times T_{\text{TFT}}(\lambda) \times \bar{z}(\lambda) \times I_{\text{in}}(\lambda) \, d\lambda$$

When analyzing the colorimetric and photometric performance of LCDs, the X, Y, and Z values are difficult to interpret as the bump and no-bump regions coexist. The chromaticity coordinates x, y, and z have been developed and established by the Commission International de l'Éclairage (CIE). Equation (7) shows the relationship between chromaticity and tristimulus; as $x + y + z = 1$, describing chromaticity of the stimulus using chromaticity coordinates (x, y) is reasonable. To calculate optical performance of a pixel, the bump and no-bump regions should be investigated individually. This work sets ratio A, which is the bump area divided by the aperture area of a pixel, and the chromaticity (x, y) of module can be derived in the dark state using (8).

$$x = \frac{X}{X + Y + Z},$$
$$y = \frac{Y}{X + Y + Z}, \qquad (7)$$
$$z = \frac{Z}{X + Y + Z}$$

$$x = \frac{X_{\text{bump}} \times A + X_{\text{No\_bump}} \times (1 - A)}{\left(X_{\text{bump}} + Y_{\text{bump}} + Z_{\text{bump}}\right) \times A + \left(X_{\text{No\_bump}} + Y_{\text{No\_bump}} + Z_{\text{No\_bump}}\right) \times (1 - A)}$$
$$\qquad (8)$$
$$y = \frac{Y_{\text{bump}} \times A + Y_{\text{No\_bump}} \times (1 - A)}{\left(X_{\text{bump}} + Y_{\text{bump}} + Z_{\text{bump}}\right) \times A + \left(X_{\text{No\_bump}} + Y_{\text{No\_bump}} + Z_{\text{No\_bump}}\right) \times (1 - A)}$$

4. Measurement

Figure 3 shows the measurement system, in which a pair of polarizers are superimposed onto both sides of the prepared sample; the polarization of polarizers is parallel or orthogonally intersected. The light source is a backlight unit with cold cathode fluorescent lamp (CCFL) (AUO, 59.23M01.002), which can provide luminance of >7500 cd/m². The light from the backlight unit passes through the polarizer (Nitto, NPF-SEG1224DU), color resists (JSR Co., R874, G894, B877), analyzer, and spectrometer (Minolta, CS-1000), sequentially. The color filter is composed of three different color resists (R874, G894, B877 of R, G, and B, respectively). The distance between the first polarizer and backlight unit is 2cm such that the incident angle is <10°, and the distance between

the two polarizers is 2cm. The colorimetric model can be implemented when the degree of depolarization $d(\lambda)$ is measured. This measurement must identify the contrast ratio that is equal to the luminance of the light passing through the parallel polarizers divided by that crossing the polarizers.

5. Simulation Results of Light Leakage in the Dark State

The degree of depolarization of RGB color resists is measured without addressing the transmittance of the black matrix. The contrast ratio of the color filter can be measured using various RGB color resists. If the liquid crystal and TFT layers are absent, the spectrum of outgoing light depends

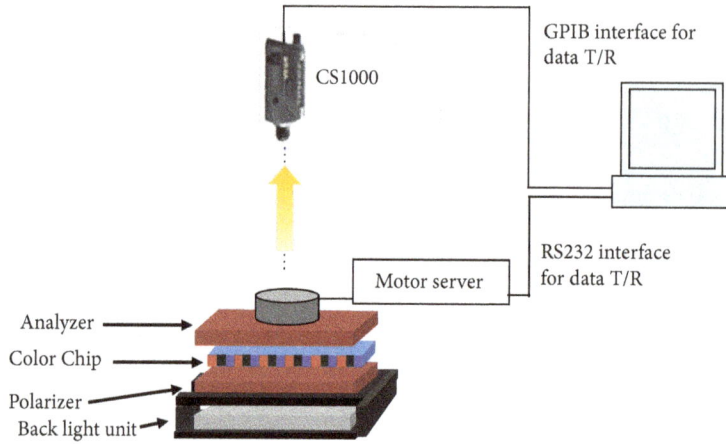

FIGURE 3: Auto-experimental system of measurement for contrast ratio of color resists.

	RED			GREEN			BLUE			White		
BUMP	x	y	Y	x	y	Y	x	y	Y	x	y	Y
	0.5711	0.3782	0.0093	0.2487	0.5269	0.0258	0.1454	0.0563	0.0040	0.253	0.271	0.013

	RED			GREEN			BLUE			White		
No_BUMP	x	y	Y	x	y	Y	x	y	Y	x	y	Y
	0.5015	0.4216	0.0054	0.1798	0.3679	0.0061	0.1552	0.0431	0.0009	0.251	0.245	0.004

FIGURE 4: Simulated result of chromaticity and photometry (Y) in bump and no-bump region.

on the transmission spectrum of the color filter, degree of depolarization, and spectrum of the backlight unit in (4). In the colorimetric model, the values of $Y_{parallel}$ and Y_{cross} are 14.05 and 0.002765, respectively, under the white point condition. This model indicates that the value of contrast ratio is 5110 for a color filter. The contrast ratio of modeling is close to the actual measurement result of 5500. It is a rapid way of evaluating optical performance of color filters when developing new color resists.

The colorimetric and photometric performance (x, y, Y) can be found (Figure 4) by considering the transmittance of the TFT, bump, and LC using (5) and (6). Without addressing the ratio of the bump area over a pixel area in this model, light leakage in the bump region is three times larger than no-bump region. By improving the contrast ratio of LCDs, light leakage caused by LC retardation in the bump region can be reduced.

Figure 5 shows the output spectra of measurement (solid line) and simulations (dotted line) of an LCD module in the dark state. The spectra peaks at about 440 nm, 550 nm, and 610 nm, represent the R, G, and B colors, respectively. The 485 nm and 585 nm peaks generated overlap transmission spectra between the B and G color filters. The 710 nm peak is from the backlight emitter, which has high transmittance in the R region. Figure 5 demonstrates that measurement and simulation results are similar. Figure 6 shows the measurement and simulation results for the colorimetric model in the dark state and bright state with three panel positions. The solid and dotted lines represent the measurement and simulation results, respectively. The points in the upper-right (down-left) side of figure are the bright state (dark

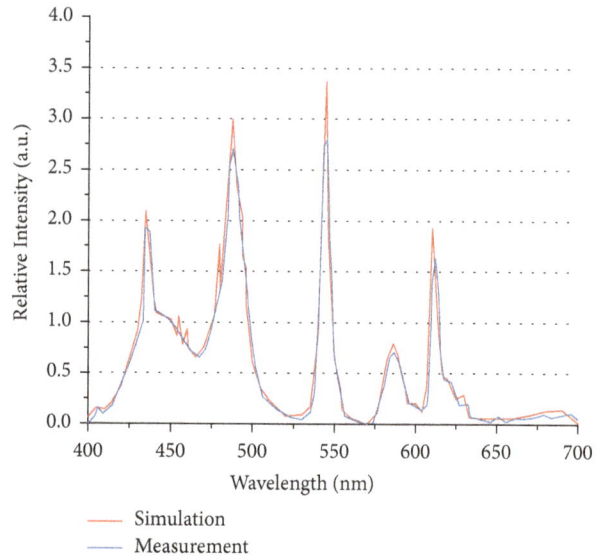

FIGURE 5: Output spectra of measurement and simulation of LCD module in the dark state.

state). In the bright state, the error of (x, y) chromaticity is about 0.003; in the dark state, chromaticity error is about 0.01. The large error in the dark state has two causes. One is the degree of depolarization in this colorimetric model. When the polarization of polarizers is crossed, light intensity is quite low, which results in fluctuations in the degree of depolarization in terms of wavelength. The other cause is error caused by manufacturing variation, such as polarizer

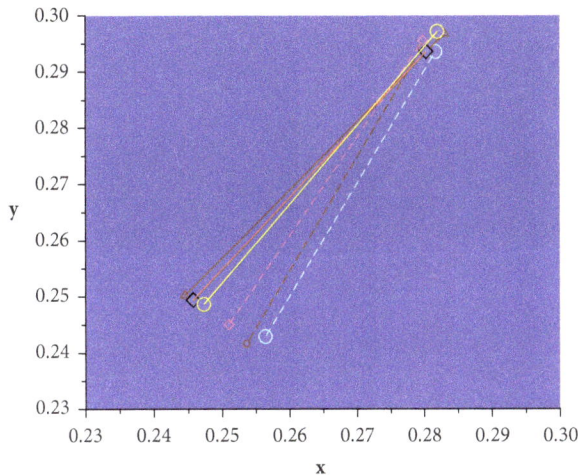

FIGURE 6: The chromaticity coordinates of measured and simulated results in the dark state and bright state.

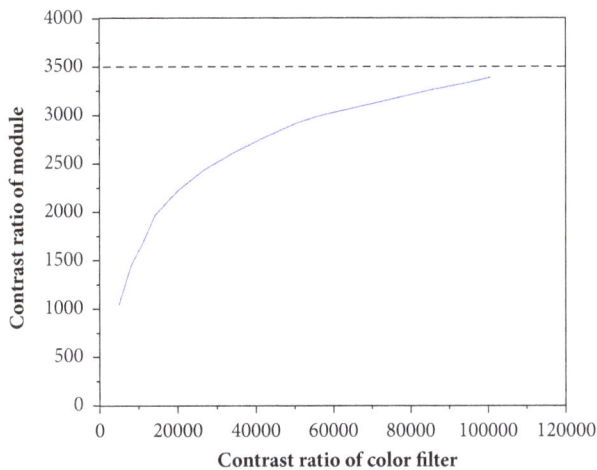

FIGURE 7: Simulation result for the contrast ratio performance of color filter and module.

6. Conclusions

This work investigated the colorimetric and photometric properties of LCDs using a novel colorimetric model. This work demonstrates that the chromaticity coordinates of an LCD can be simulated accurately, and the proposed model can be utilized to optimize color LCDs with different optical components (excluding liquid crystals). Most importantly, this work thoroughly analyzes the colorimetric properties of LCDs and, in doing so, demonstrates that properly optimized LCDs use different backlights and color resists. Finally, the depolarization and retardation effects can be reduced to improve contrast ratio performance in an MVA-type LCD module.

Conflicts of Interest

The authors declare that they have no conflicts of interest.

Acknowledgments

This research was supported by the Ministry of Science and Technology (MOST) of Taiwan under Grants MOST 106-2221-E-035-072-MY2.

References

[1] S. Aratani, H. Klausmann, M. Oh-e et al., "Complete suppression of color shift in in-plane switching mode liquid crystal displays with a multidomain structure obtained by unidirectional rubbing," in *Proceedings of the 1997 Pacific Rim Conference on Lasers and Electro-Optics, CLEO/Pacific Rim*, p. 93, July 1997.

[2] C.-H. Chen, F.-C. Lin, and H.-P. D. Shieh, "A Field Sequential Color LCD Based on Color Fields Arrangement for Color Breakup and Flicker Reduction," *Journal of Display Technology*, vol. 5, no. 1, pp. 34–39, 2009.

[3] H. Chen, R. Zhu, M.-C. Li, S.-L. Lee, and S.-T. Wu, "Pixel-by-pixel local dimming for high-dynamic-range liquid crystal displays," *Optics Express*, vol. 25, no. 3, pp. 1973–1984, 2017.

[4] H. Chen, G. Tan, M.-C. Li, S.-L. Lee, and S.-T. Wu, "Depolarization effect in liquid crystal displays," *Optics Express*, vol. 25, no. 10, pp. 11315–11328, 2017.

[5] S. H. Hong, Y. H. Jeong, H. Y. Kim, H. M. Cho, W. G. Lee, and S. H. Lee, "Electro-optic characteristics of 4-domain vertical alignment nematic liquid crystal display with interdigital electrode," *Journal of Applied Physics*, vol. 87, no. 12, pp. 8259–8263, 2000.

[6] T.-H. Yoon, G.-D. Lee, and J. C. Kim, "Nontwist quarter-wave liquid-crystal cell for a high-contrast reflective display," *Optics Expresss*, vol. 25, no. 20, pp. 1547–1549, 2000.

[7] Q. Hong, T. X. Wu, X. Zhu, R. Lu, and S.-T. Wu, "Designs of wide-view and broadband circular polarizers," *Optics Express*, vol. 13, no. 20, pp. 8318–8331, 2005.

[8] K. Okumoto, T. Tsuchiya, K. Yonemura et al., "A novel simulation method in In-Plane switching mode panel with considering light scattering behavior," in *Proceedings of the 2015 SID International Symposium*, pp. 1555–1558, usa.

[9] R. Lu, X. Zhu, S. Wu, Q. Hong, and T. Wu, "Ultrawide-View

misalignment or space in the cell gap variation. The low light intensity can be improved using a relatively brighter continuous light source. When degree of depolarization noise is minimized, estimating chromaticity of an LCD in the dark state is possible by measuring the degree of depolarization of color resists.

The contrast ratio is defined by Y_{bright}/Y_{dark}. Not only is the depolarization of the color filter an influential factor, but the light leakage in the bump area has an important role when calculating the contrast ratio. This model can evaluate the excellence of the contrast ratio of the LCD module when only improvements to the contrast ratio of the color filter are considered. Figure 7 shows the contrast ratio of the color filter, which increases with the contrast ratio of the LCD module when the extreme ratio of the LCD model is assumed to be about 3500. Actually, for a dispersion-type color filter, improving the contrast ratio of the color filter by 200% is difficult. Finally, the primary cause of contrast ratio deterioration of LCDs is light leakage from the bump region.

Liquid Crystal Displays," *Journal of Display Technology*, vol. 1, no. 1, pp. 3–14, 2005.

[10] M. Yoneya, Y. Utsumi, and Y. Umeda, "Depolarized light scattering from liquid crystals as a factor for black level light leakage in liquid-crystal displays," *Journal of Applied Physics*, vol. 98, no. 1, Article ID 016106, 2005.

[11] H. Chen, F. Peng, Z. Luo et al., "High performance liquid crystal displays with a low dielectric constant material," *Optical Materials Express*, vol. 4, no. 11, pp. 2262–2273, 2014.

[12] J. L. Pezzaniti, S. C. McClain, R. A. Chipman, and S.-Y. Lu, "Depolarization in liquid-crystal televisions," *Optics Expresss*, vol. 18, no. 23, pp. 2071–2073, 1993.

[13] M. Bass, *Handbook of Optics*, vol. 3, 1995.

[14] G. Wyszecki and W. S. Stiles, *Color Science. Concepts and Methods, Quantitative Data and Formulae*, 2000.

3

Absorption Spectrum and Density of States of Square, Rectangular, and Triangular Frenkel Exciton Systems with Gaussian Diagonal Disorder

Ibrahim Avgin[1] and David Huber[2]

[1]*Department of Electrical and Electronic Engineering, Ege University, Bornova, 3500 Izmir, Turkey*
[2]*Physics Department, University of Wisconsin-Madison, Madison, WI 53706, USA*

Correspondence should be addressed to David Huber; dhuber@wisc.edu

Academic Editor: Oleg Derzhko

Using the coherent potential approximation, we investigate the effects of disorder on the optical absorption and the density of states of Frenkel exciton systems on square, rectangular, and triangular lattices with nearest-neighbor interactions and a Gaussian distribution of transition energies. The analysis is based on an elliptic integral approach that gives results over the entire spectrum. The results for the square lattice are in good agreement with the finite-array calculations of Schreiber and Toyozawa. Our findings suggest that the coherent potential approximation can be useful in interpreting the optical properties of two-dimensional systems with dominant nearest-neighbor interactions and Gaussian diagonal disorder provided the optically active states are Frenkel excitons.

1. Introduction

In a series of recent papers [1–3], we have applied the coherent potential approximation (CPA) to the calculation of the optical absorption and density of states of the Frenkel exciton model in one- and three-dimensional arrays with nearest-neighbor interactions and Gaussian disorder associated with the single-site transition energies. In the case of the one-dimensional systems we have shown that the results for the density of states are in excellent agreement with numerical calculations carried out on arrays of 10^7–10^8 sites [1]. The accuracy of the CPA for the optical absorption in one dimension [2] was tested in a comparison with data obtained from finite-array calculations by Schreiber [4]. Good agreement was obtained with the ensemble average of data from arrays of 199 sites. The CPA was applied to cubic lattices in [3]. In the case of the simple cubic lattice, good agreement was obtained with the corresponding finite-array calculations of Schreiber and Toyozawa [5, 6] for the optical absorption and the density

of states. Along with the simple cubic data, corresponding CPA results were reported for the body-centered and face-centered cubic lattices.

A preliminary CPA analysis of the optical absorption and the density of states for the square lattice was reported in [7]. The approach followed was based on a large-energy expansion of Green's function, and the calculations were limited to energies below the absorption edge of the ideal system. Like the simple cubic analysis, the results were in good agreement with findings reported in [5, 6]. In this paper, we investigate the square lattice using an approach for the calculation of Green's function that is based on the evaluation of a complete elliptic integral of the first kind. Unlike the previous approach, we obtain results that are applicable over the entire absorption band. We also extend the theory to rectangular and triangular lattices for which there are no finite-array results to compare with.

In the CPA, the starting point in all three dimensions is lattice Green's function, $G_0(E)$, which in turn depends on

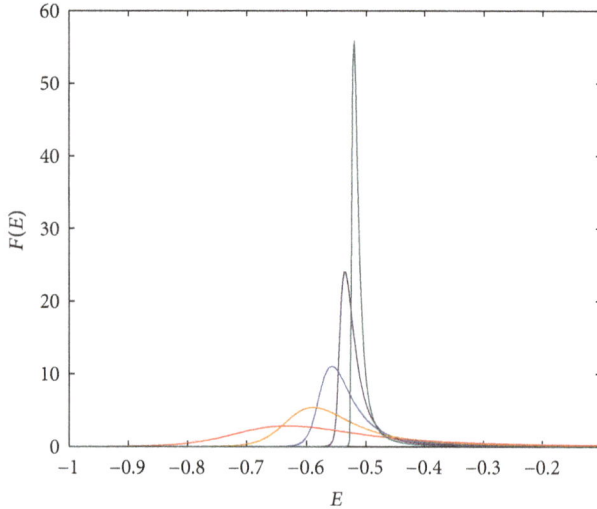

FIGURE 1: Normalized optical absorption for the square lattice. From left to right along the energy axis, the curves correspond to $\sigma = 0.263$, 0.186, 0.131, 0.093, and 0.066. In this and all other figures, energy is in units of the band width.

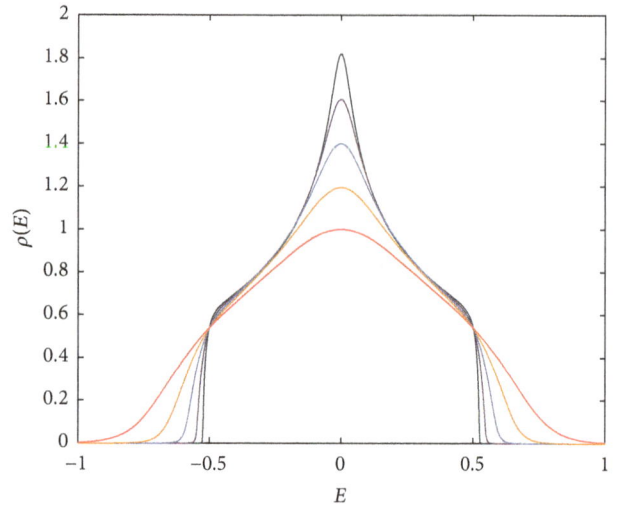

FIGURE 2: Normalized density of states for the square lattice. From left to right along the energy axis, the curves correspond to $\sigma = 0.263$, 0.186, 0.131, 0.093, and 0.066.

the structure of the lattice. The calculation of the optical absorption and the density of states follow once one knows CPA Green's function which is obtained from lattice Green's function by the replacement $E \rightarrow E - V_C(E)$, where $V_C(E)$ is the coherent potential. Equations relating to the calculation of $V_C(E)$ and the connection of CPA Green's function to the absorption and the density of states are given in [1–3] and will not be reproduced here.

2. Square Lattice

In analyzing the square, rectangular, and triangular systems, we take the unit of energy to be the width of the ideal (no disorder) exciton band and assume the absorption edge is at the bottom of the band. In the case of the square lattice, this leads to the exciton energy

$$E_{\mathbf{k}} = -\left(\frac{1}{4}\right)\left(\cos k_x + \cos k_y\right) \qquad (1)$$

in the absence of disorder. The corresponding expression for ideal Green's function for complex energy takes the form [8]

$$G_0(E) = \left(\frac{2}{\pi E}\right) \mathbf{K}\left(2E^{-1}\right), \qquad (2)$$

where $\mathbf{K}(m)$ denotes the complete elliptic integral with modulus m.

The Gaussian averaging associated with the diagonal disorder is discussed in detail in [1–3]. It depends on the variance, σ^2. Following [4–6], we carried out calculations for the square lattices with $\sigma = 0.263$, 0.186, 0.131, 0.093, and 0.066, where σ is in units of the ideal band width. Our results for the normalized optical absorption, $F(E)$, and the normalized density of states, $\rho(E)$, for the square lattice are shown in Figures 1 and 2, respectively. They are seen

to be in good agreement with the finite-array calculations of Schreiber and Toyozawa as displayed in Figure 2 in [5] (absorption) and Figure 5(b) in [6] (density of states). This result is also consistent with [7], where a similar level of agreement was established for $E < -0.5$.

3. Rectangular Lattice

In the absence of disorder, the exciton energy for a rectangular lattice with unit band width takes the form

$$E_{\mathbf{k}} = -\frac{\left(\lambda \cos k_x + \cos k_y\right)}{2\left(\lambda + 1\right)}, \qquad (3)$$

where λ is the interaction between neighboring sites on the x-axis. The corresponding expression for Green's function for complex energy is given in the Appendix to [8]

$$G_0(E) = 2\left(\lambda + 1\right)\left(\pi \lambda^{1/2}\right)^{-1} k_1(E)\,\mathbf{K}\left(k_1(E)\right), \qquad (4)$$

where

$$k_1(E) = \left\{ \frac{4\lambda}{\left[4\left(\lambda + 1\right)^2 E^2 - \left(\lambda - 1\right)^2\right]} \right\}^{1/2}. \qquad (5)$$

Since the absorption is weakly affected by changes in λ, we focus on results for the density of states where we take $\sigma = 0.131$ and $\lambda = 0.5$, 1.0, and 1.5 (Figure 3). It is apparent that the density of states near the center of the band is strongly affected by changes in λ. The lower curve shows the results for $\lambda = 0.5$, the middle curve for $\lambda = 1.5$, and the upper curve for $\lambda = 1.0$. The dip at the center of the band, which grows stronger as with increasing anisotropy, reflects the fact that, in the limits $\lambda \gg 1$ and $\lambda \rightarrow 0$, the system approaches an array of decoupled chains with the consequence that the optical properties and the density of states become characteristic of one-dimensional arrays [1, 2].

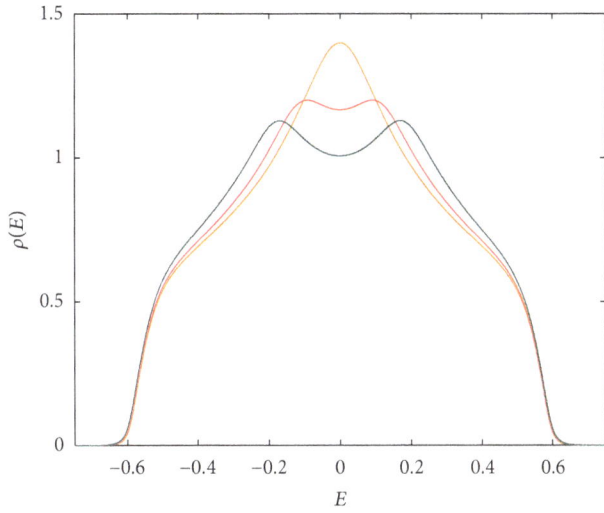

FIGURE 3: Normalized density of states for the rectangular lattice with $\sigma = 0.131$. From bottom to top at $E = 0$, the curves are for $\lambda = 0.5, 1.5$, and 1.0.

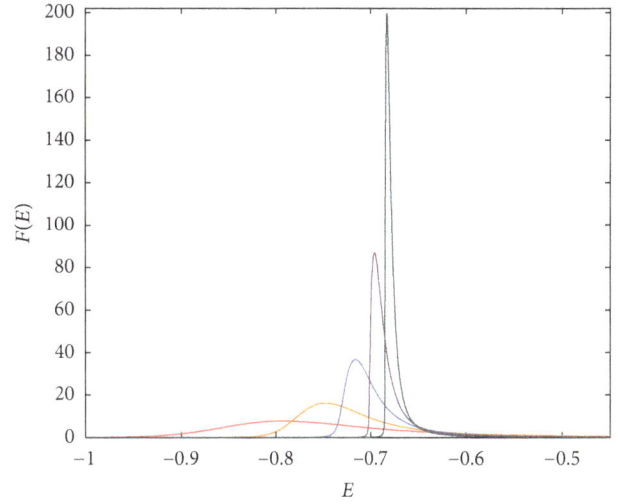

FIGURE 4: Normalized optical absorption for the triangular lattice. From left to right along the energy axis, the curves correspond to $\sigma = 0.263, 0.186, 0.131, 0.093$, and 0.066.

4. Triangular Lattice

In the absence of disorder, the exciton energy for the triangular lattice with unit band width has the form [9]

$$E_{\mathbf{k}} = -\left(\frac{2}{9}\right)\left(\cos 2k_x + 2\cos k_x \cos k_y\right) \qquad (6)$$

with the optical absorption edge at $-2/3$ and the upper edge at $1/3$. Analogous to the face-centered cubic lattice and unlike the square and rectangular lattices, the exciton band does not have a reflection point for the density of states. In units of the band width, the corresponding Green function for complex energy with negative real part has the form

$$G_0(E) = \left(\frac{9}{4\pi}\right) g(E)\,\mathbf{K}\left(k_2(E)\right), \qquad (7)$$

where

$$g(E) = -\frac{8}{\left[(-9E+3)^{1/2}-1\right]^{3/2}\left[(-9E+3)^{1/2}+3\right]^{1/2}}, \qquad (8)$$

$$k_2(E) = \frac{4(-9E+3)^{1/4}}{\left[(-9E+3)^{1/2}-1\right]^{3/2}\left[(-9E+3)^{1/2}+3\right]^{1/2}}.$$

The results for the optical absorption and the density of states are shown in Figures 4 and 5, respectively. For small disorder, the absorption peaks are near the ideal (no disorder) band edge at $-2/3$. With increasing disorder, the absorption peak broadens and shifts to lower energy as occurs with the other lattices. In the case of the density of states, the ideal lattice has band edges at $E = -2/3$ and $1/3$ as well as a singularity at $E = 2/9$. The behavior of the density of states below the absorption edge is shown in greater detail in Figure 6. It is similar to the corresponding results for the square lattice shown in [6, 7].

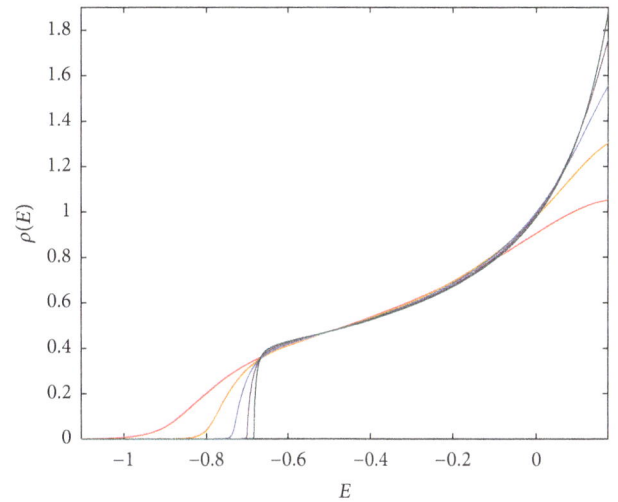

FIGURE 5: Normalized density of states for the triangular lattice. From left to right along the energy axis, the curves correspond to $\sigma = 0.263, 0.186, 0.131, 0.093$, and 0.066.

5. Anisotropic Triangular Lattice

For the triangular lattice discussed above all nearest-neighbor interactions are the same. In [10], Horiguchi investigated an anisotropic triangular lattice in which interactions along two of the three connecting lines took on the value γ while the interaction on the third line was equal to 1. When the optical edge is at the bottom of the band, the exciton energy is given by

$$E_{\mathbf{k}} = -\left[\cos(2ak_x) + 2\gamma\cos(ak_x)\cos(bk_y)\right], \qquad (9)$$

where $a = 1/2$ and $b = \sqrt{3}/2$ when the length of the side of the triangle is 1 [10]. The lower and upper band edges and the band width (BW) are given in Table 1.

TABLE 1: The locations of the band edges and the band width for the ideal anisotropic triangular lattice. The nearest-neighbor interaction on two of the three lines of sites is equal to γ while it is equal to 1 on the third line [10].

Range of γ	Lower band edge	Upper band edge	Band width (BW)
$0 < \gamma < 2$	$-(1 + 2\gamma)$	$1 + \gamma^2/2$	$2 + 2\gamma + \gamma^2/2$
$\gamma = 2$	-5	3	8
$\gamma > 2$	$-(1 + 2\gamma)$	$2\gamma - 1$	4γ

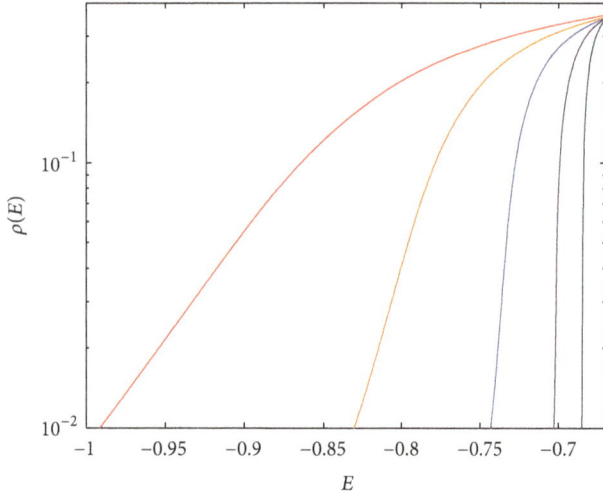

FIGURE 6: Semilog plot of the density of states below the absorption edge for the triangular lattice. From left to right along the energy axis, the curves correspond to $\sigma = 0.263$, 0.186, 0.131, 0.093, and 0.066.

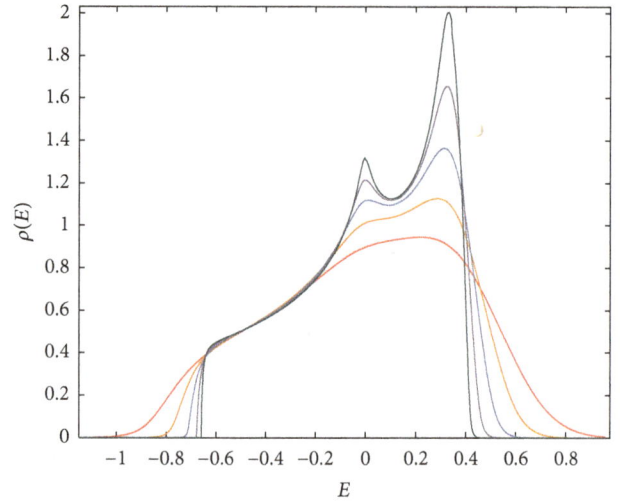

FIGURE 7: Normalized density of states for the anisotropic triangular lattice with $\gamma = 1/2$. From left to right along the energy axis, the curves correspond to $\sigma = 0.263$, 0.186, 0.131, 0.093, and 0.066.

The density of states is strongly affected by the ratio of the interaction strength. Apart from the band edges, the singularities in the ideal Green function for the anisotropic triangular lattice are associated with γ-dependent critical points on the energy axis and corresponding peaks in the density of states [10]. In units of the band width, BW, the peaks are at $(2\gamma - 1)/\text{BW}$ and $1/\text{BW}$ for $0 < \gamma < 1$. For $\gamma = 1$, there is a single peak at $1/\text{BW} = 2/9$. For $1 < \gamma < 2$, the peaks are also at $(2\gamma - 1)/\text{BW}$ and $1/\text{BW}$, and when $\gamma \geq 2$ there is a single peak at $1/\text{BW}$. In Figure 7, we show the effects of disorder on the density of states for $\gamma = 1/2$. In this calculation, we made use of the expression for the Green function given in [10]. In the absence of disorder, the band edges are at $-4/7$ and $3/7$, and the peaks are at 0 and $2/7$.

6. Conclusions

The results for the square lattice extend the earlier work [7] to the entire spectrum. As we mentioned previously, our findings are consistent with numerical studies of the square lattice [5, 6]. The results for the rectangular arrays are expected to be reasonably accurate since the coherent potential approximation works well in the limiting cases $\lambda \gg 1$ and $\lambda \to 0$, where the rectangular array reduces to decoupled chains. In the case of the triangular lattice, finite-array calculations are needed to test the accuracy of the

coherent potential approximation. If this is established, one has confidence that the approximation works reasonably well for the anisotropic triangular lattice since, in the limit $\gamma = 0$, the system decouples into independent chains, whereas when $\gamma \gg 1$, each site is strongly coupled to four nearest neighbors with the consequence that the behavior approaches that of the square array.

In the field of optical spectroscopy, a two-parameter assessment of the characteristics of the absorption line is provided by the peak position (E_{Peak}) and the full width at half maximum (FWHM). In Figure 8 we show the behavior of the two parameters with the increasing σ for both the square and triangular lattices. It is evident that the peak positions for the two lattices approach a common value for large σ whereas the increase in the linewidth occurs more rapidly for the square lattice. The latter behavior may reflect the fact that the effects of the disorder in the triangular lattice may be weaker due to an averaging of the contributions from a larger number of nearest neighbors.

Our results along with earlier studies [6, 7] suggest that the coherent potential approximation can be useful in interpreting the optical properties of two-dimensional systems with dominant nearest-neighbor interactions and Gaussian diagonal disorder provided the optically active states are Frenkel excitons. The findings reported in this paper are applicable to optically active monolayers. In addition,

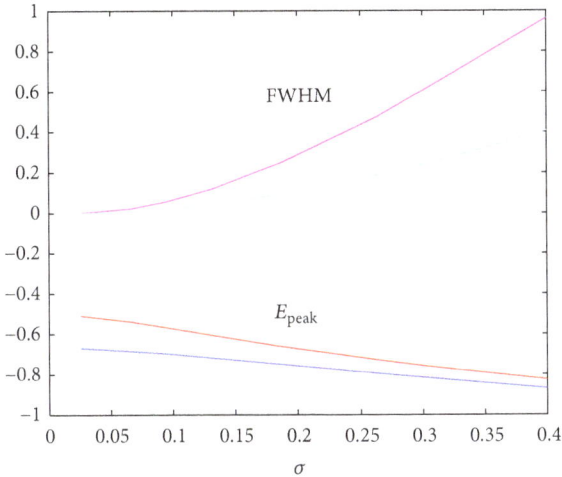

FIGURE 8: Peak position (E_{Peak}) and full width at half maximum (FWHM) versus σ for the square and triangular lattices. Energy and σ are in units of the band width. From top to bottom, square FWHM, triangle FWHM, square E_{Peak}, and triangle E_{Peak}.

there are quasi-two-dimensional magnetic exciton systems where the nearest-neighbor approximation is appropriate.

Conflicts of Interest

The authors declare that there are no conflicts of interest regarding the publication of this paper.

References

[1] I. Avgin, A. Boukahil, and D. L. Huber, "Accuracy of the coherent potential approximation for a one-dimensional Frenkel exciton system with a Gaussian distribution of fluctuations in the optical transition frequency," *Physica E: Low-Dimensional Systems and Nanostructures*, vol. 42, no. 9, pp. 2331–2334, 2010.

[2] A. Boukahil, I. Avgin, and D. L. Huber, "Assessment of the coherent potential approximation for the absorption spectra of a one-dimensional Frenkel exciton system with a Gaussian distribution of fluctuations in the optical transition frequency," *Physica E: Low-Dimensional Systems and Nanostructures*, vol. 65, pp. 141–143, 2015.

[3] I. Avgin, A. Boukahil, and D. L. Huber, "Coherent potential approximation for the absorption spectra and the densities of states of cubic Frenkel exciton systems with Gaussian diagonal disorder," *Physica B: Condensed Matter*, vol. 477, pp. 83–86, 2015.

[4] M. Schreiber, "Exciton absorption tails in one-dimensional systems," *Physical Review B*, vol. 34, no. 4, pp. 2914–2916, 1986.

[5] M. Schreiber and Y. Toyozawa, "Lineshape of the Exciton under Lattice Vibrations. III. The Urbach Rule," *Journal of the Physical Society of Japan*, vol. 51, no. 5, pp. 1544–1550, 1982.

[6] M. Schreiber and Y. Toyozawa, "Numerical Experiments on the Absorption Lineshape of the Exciton Under Lattice Vibrations. IV. The Indirect Edge," *Journal of the Physical Society of Japan*, vol. 52, no. 1, pp. 318–326, 1983.

[7] A. Boukahil and D. L. Huber, "Coherent potential approximation for the absorption spectra of a two-dimensional Frenkel exciton system with Gaussian diagonal disorder," *Modern Physics Letters B*, vol. 28, no. 32, Article ID 1450251, 2014.

[8] T. Morita and T. Horiguchi, "Calculation of the lattice green's function for the bcc, fcc, and rectangular lattices," *Journal of Mathematical Physics*, vol. 12, pp. 986–992, 1971.

[9] T. Horiguchi, "Lattice Green's functions for the triangular and honeycomb lattices," *Journal of Mathematical Physics*, vol. 13, pp. 1411–1419, 1972.

[10] T. Horiguchi, "Lattice Green's function for anisotropic triangular lattice," *Physica A. Statistical and Theoretical Physics*, vol. 178, no. 2, pp. 351–363, 1991.

4

Lattice Dynamics and Transport Properties of Multiferroic DyMn$_2$O$_5$

Javed Ahmad,[1] **Syed Hamad Bukhari,**[1] **M. Tufiq Jamil,**[1] **Mehr Khalid Rehmani,**[1]
Hammad Ahmad,[2] **and Tahir Sultan**[3]

[1]*Department of Physics, Bahauddin Zakariya University, Multan 60800, Pakistan*
[2]*Nanoscience and Technology Department, National Center for Physics, Quaid-i-Azam University Campus, Islamabad 45320, Pakistan*
[3]*Department of Civil Engineering, Bahauddin Zakariya University, Multan 60800, Pakistan*

Correspondence should be addressed to Javed Ahmad; dr.j.ahmad@gmail.com

Academic Editor: Mohindar S. Seehra

We have investigated the optical and electrical properties of polycrystalline DyMn$_2$O$_5$ synthesized by sol-gel method. Analysis of the reflectivity spectrum has led to the observation of 18 infrared (IR) active phonon modes out of 36 predicted ones. We discuss the results in terms of different phonon bands originated as a result of atomic vibrations. Moreover, the optical energy band gap of $E_{g(OC)} \sim 1.78$ eV has been estimated from optical conductivity ($\sigma_1(\omega)$) spectrum. The energy band gap and optical transitions were also determined from UV-visible absorption spectrum and band gap of $E_{g(UV)} \sim 1.57$ eV was estimated. Moreover, DC electrical resistivity shows the p-type polaronic conduction above room temperature.

1. Introduction

Manganites RMn_2O_5, typical type-II multiferroics, usually show large coupling between spin (magnetism), charge (ferroelectricity), and lattice (structure). RMn_2O_5 compounds show cascade of phase transitions with characteristic temperature at Néel transition $T_N = 40$–45 K and ferroelectric transition at $T_C = 28$–39 K and ordering of rare-earth moments occurs below 10 K [1]. Until recently, it is well known that RMn_2O_5 crystallize in the orthorhombic structure with $Pbam$ space group at room temperature [2]. However, very recently it is proposed that RMn_2O_5 crystallize in two possible monoclinic space groups Pm and $P2$ depending on the existence of polarization in ab plane and along c axis, respectively [3].

Among the family of RMn_2O_5, DyMn$_2$O$_5$ exhibits remarkable magnetodielectric behavior as compared to other RMn_2O_5 compounds [4]. An unconventional behavior of phonon anomalies in RMn_2O_5 (R = Bi, Eu, Dy) has been observed in the paramagnetic phase [5]. They have suggested that these anomalies at new characteristics temperature $T^* \sim 1.5T_N$ are related to the spin-phonon coupling and signaling a transition between states. Moreover, spin-phonon coupling has been observed slightly above T_N indicating the drastic change in phonon frequency due to magnetic field and temperature [6, 7]. Remarkably, recently we have observed an unconventional magnetodielectric effect in DyMn$_2$O$_5$ above T_N, where the infrared and Raman phonons have shown spin-phonon coupling and confirm the strong correlation between spin, charge, and lattice degree of freedom [8]. Such a strong interplay between many degrees of freedom in the paramagnetic phase is a special characteristic of RMn_2O_5 family [5–9]. The knowledge of lattice vibrations and their correlation to different conduction mechanisms is of crucial importance for engineering the materials for various technological applications. However, currently there is no report on the correlation between IR active phonon and the conduction mechanism for DyMn$_2$O$_5$.

In this work, we have measured the IR reflectivity spectrum of DyMn$_2$O$_5$ and assign the IR active phonon modes with theoretically calculated modes. In addition, we have performed UV-visible spectroscopy and temperature dependent electrical response to observe the microscopic conduction mechanism. The main objective of this paper is to investigate the lattice vibration in DyMn$_2$O$_5$ and the possible correlation to electronic transport mechanism through energy band gap and activation energy.

2. Experiment

The sample was synthesized by using the sol-gel technique as described previously [9]. Stoichiometric amounts of Gd and Mn nitrates were dissolved into distilled water and stirred for 2 h along with heating at 80°C until viscous gel was formed. The obtained gel was dried at 150°C for 2 h and finally the obtained black powder was sintered at 1000°C for 6 h. Single phase character of prepared sample has been checked by using X-ray diffraction (XRD) pattern. The measurements were performed with the help of Bruker D8 Advance diffractometer with Cu-Kα source having wavelength 1.5405 Å. Microstructure was observed by using the scanning electron microscopy (SEM) and image was taken at accelerating voltage of 10 KV and magnification of 80 K. The powder was uniaxially pressed into pellets of 13 mm diameter under the pressure of 30 KN by using the Paul-Otto Weber hydraulic press to obtain high density ceramics for the reflectivity measurement. One of the surfaces of the disc shaped sample was made smooth prior to spectroscopic measurements. Room temperature infrared reflectivity has been measured using Fourier transform infrared spectrometer (Vertex 80v) at near-normal incidence mode. The frequency measurements in the mid-range (550–7500 cm^{-1}) and far (30–680 cm^{-1}) infrared regions were performed using KBR-DLaTGS and Mylar 6 μm DLaTGS beam splitter-detector combinations, respectively. Mid and far infrared spectra were obtained under vacuum purges, by averaging 100 scans, with spectral resolution of 2 cm^{-1}. A gold mirror was initially used as reference spectrum. The reflectivity spectra in the mid and far infrared regions matched well in the superposition region (550–680 cm^{-1}). The absorption spectra in the 200–1000 nm wavelength region have been obtained using Perkin Elmer Lambda 950 UV/vis/NIR spectrophotometer. The electrical measurements were made by using Keithley source meter 2400, while applying the voltage range of 0–20 V in the temperature range 300–475 K.

3. Results and Discussion

3.1. Structural Analysis. X-ray diffraction (XRD) was performed to check the single phase character of the prepared sample at room temperature and the obtained XRD pattern of DyMn$_2$O$_5$ is shown in Figure 1. JANA2006 program was used for the Rietveld refinement of the XRD pattern assuming orthorhombic structure with *Pbam* space group (Figure 2). All the XRD peaks were well fitted confirming the single phase formation and no trace of any impurity peak was found. The obtained unit cell parameters are *a* = 7.251 Å, *b* = 8.445 Å, and *c* = 5.647 Å, which are in good agreement with theoretically calculated values (7.270, 8.518, and 5.600 Å, resp.) as well as experimentally observed values (7.285, 8.487, and 5.668 Å, resp.) [10].

In addition, we have also calculated the strain present in the material and crystallite size by using the Williamson-Hall (W-H) model considering the uniform isotropic crystal for

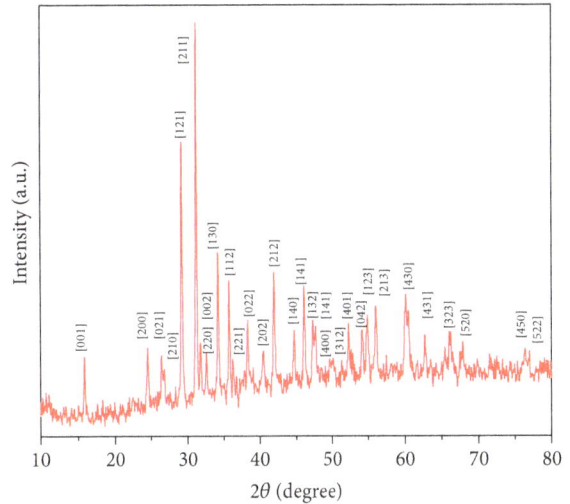

FIGURE 1: Room temperature XRD pattern of DyMn$_2$O$_5$.

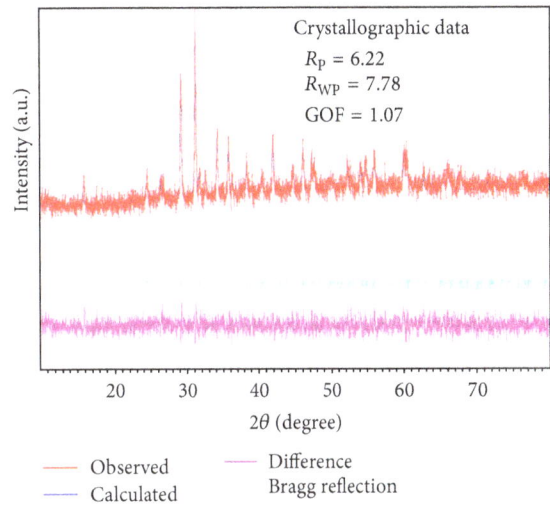

FIGURE 2: Rietveld refinement of XRD pattern where solid (red line) is the observed data and solid curve (blue line) is the best fit, vertical marks indicate the Bragg peaks, and the bottom curve shows the difference between observed and calculated intensities.

the total peak broadening using the relation [11]

$$\beta_{hkl} \cos \theta_{hkl} = \frac{K\lambda}{D_v} = 4\varepsilon \sin \theta_{hkl}, \quad (1)$$

where D_v is the crystallite size, K is the shape factor (K = 0.9), λ is the wavelength of X-rays, and $4\varepsilon \sin \theta_{hkl}$ is the strain induced line broadening. It is easy to calculate the strain and crystallite size by extracting the slope and the intercept of the linear fit as shown in Figure 3. The obtained values of strain and crystallite size are 4.27×10^{-3} and 50 nm, respectively.

Figure 4 shows the SEM micrograph of DyMn$_2$O$_5$. Large grains with less grain boundaries can be clearly seen from micrograph. However, grains show no perfect alignment, which is a typical characteristic of polycrystalline sample. The individual grains shown are of the round shape with an

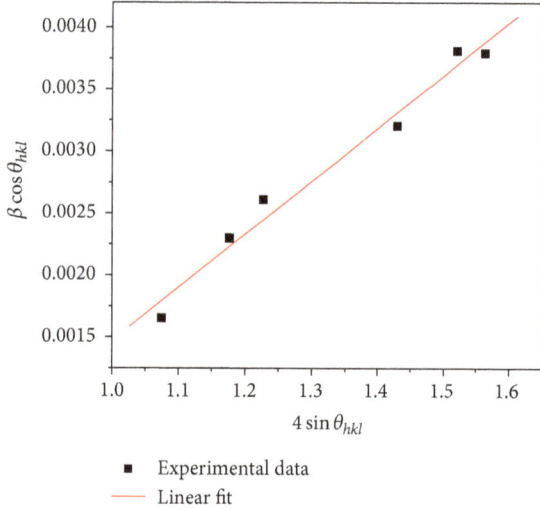

FIGURE 3: W-H plot of polycrystalline $DyMn_2O_5$.

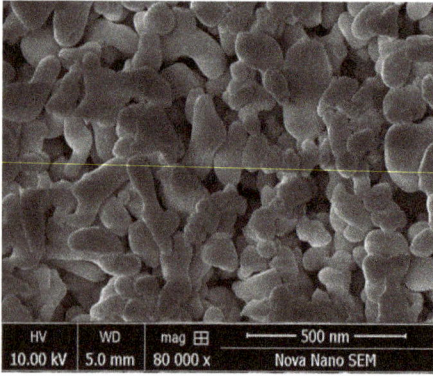

FIGURE 4: SEM image of $DyMn_2O_5$ synthesized by sol-gel method.

average grain size ~40 nm, which is consistent with crystallite size of ~38 nm obtained from XRD analysis by using the well-known Debye-Scherrer formula.

3.2. Infrared Reflectivity. Figure 5 shows the near-normal reflectivity of $DyMn_2O_5$ in frequency range 30–1000 cm^{-1} at room temperature. Considering the quasi-normal incidence angle, the optical reflectivity spectrum (R) is related to the dielectric function by Fresnel's formula

$$R(\omega) = \left| \frac{\sqrt{\epsilon(\omega)} - 1}{\sqrt{\epsilon(\omega)} + 1} \right|^2. \qquad (2)$$

To quantify the infrared phonon contribution to the dielectric function $\epsilon(\omega)$ is defined as

$$\epsilon(\omega) = \epsilon_\infty + \sum_j \frac{\omega_{TO,j}^2 S_j}{\omega_{TO,j}^2 - \omega^2 - i\omega\gamma_j}, \qquad (3)$$

where ϵ_∞ is the high frequency dielectric constant indicating the contribution to the electronic polarization. $\omega_{TO(j)}$, S_j, and γ_j are the optical phonon frequency, oscillator strength,

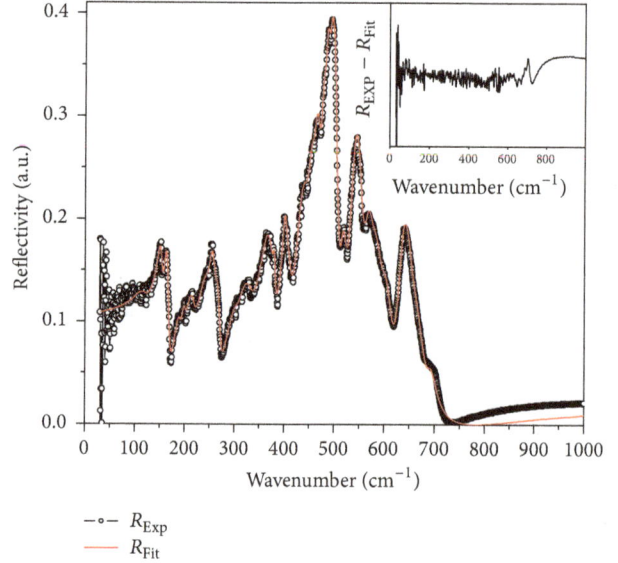

FIGURE 5: Infrared reflectivity spectra of $DyMn_2O_5$ measured at room temperature. The black circles and red line show the experimental data and fitting curve by Lorentz oscillator model, respectively. The inset shows the deviation between measured (R_{EXP}) and fitted (R_{FIT}) reflectivity.

and damping factor of jth phonon, respectively. Both (2) and (3) together can give the measured reflectivity spectrum. Following above relations, the well-known Lyddane-Sachs-Teller (LST) relation becomes as $\epsilon_o/\epsilon_\infty = \omega_{LO}^2/\omega_{TO}^2$, used to calculate the longitudinal optical phonon modes (ω_{LO}).

According to group theoretical analysis, there are 36 infrared active vibrational phonon modes in the paraelectric phase of RMn_2O_5 at the Γ point ($8B_{1u} + 14B_{2u} + 14B_{3u}$) [7, 12]. These modes are classified into different phonon bands which correspond to relative motion of rare-earth, manganese, and oxygen ions. The observed phonon frequencies obtained from the fit are assigned via comparison with those found in lattice dynamics calculations (Table 1) and are in agreement with reported data [7, 12]. Among the 36 theoretically predicted infrared active modes, we are able to observe only 18 modes. The origin of this discrepancy will be discussed later.

As far as the microscopic motion of the active phonon modes is concerned, the phonon modes at 151, 167, 194, 217, and 251 cm^{-1} involve the relative motion of Dy ions with respect to manganese ions of MnO_5 and MnO_6 polyhedra and assigned to the calculated modes between 95 and 245 cm^{-1}, respectively, as can be seen in Figures 6(a)–6(d). Moreover, high frequency modes between 260 and 684 cm^{-1} are attributed to Mn-O bending and stretching motion of the distorted octahedra and square pyramidal building block units. These modes may correspond to the calculated modes between 283 and 728 cm^{-1} and mainly arise from the Mn-O stretching motion within the equatorial plan of the MnO_6 octahedra and Mn-O stretching motion in the axial direction of MnO_5 square pyramids as shown in Figures 6(e)–6(i) [12]. Similar stretching and bending vibration of oxygen atoms at

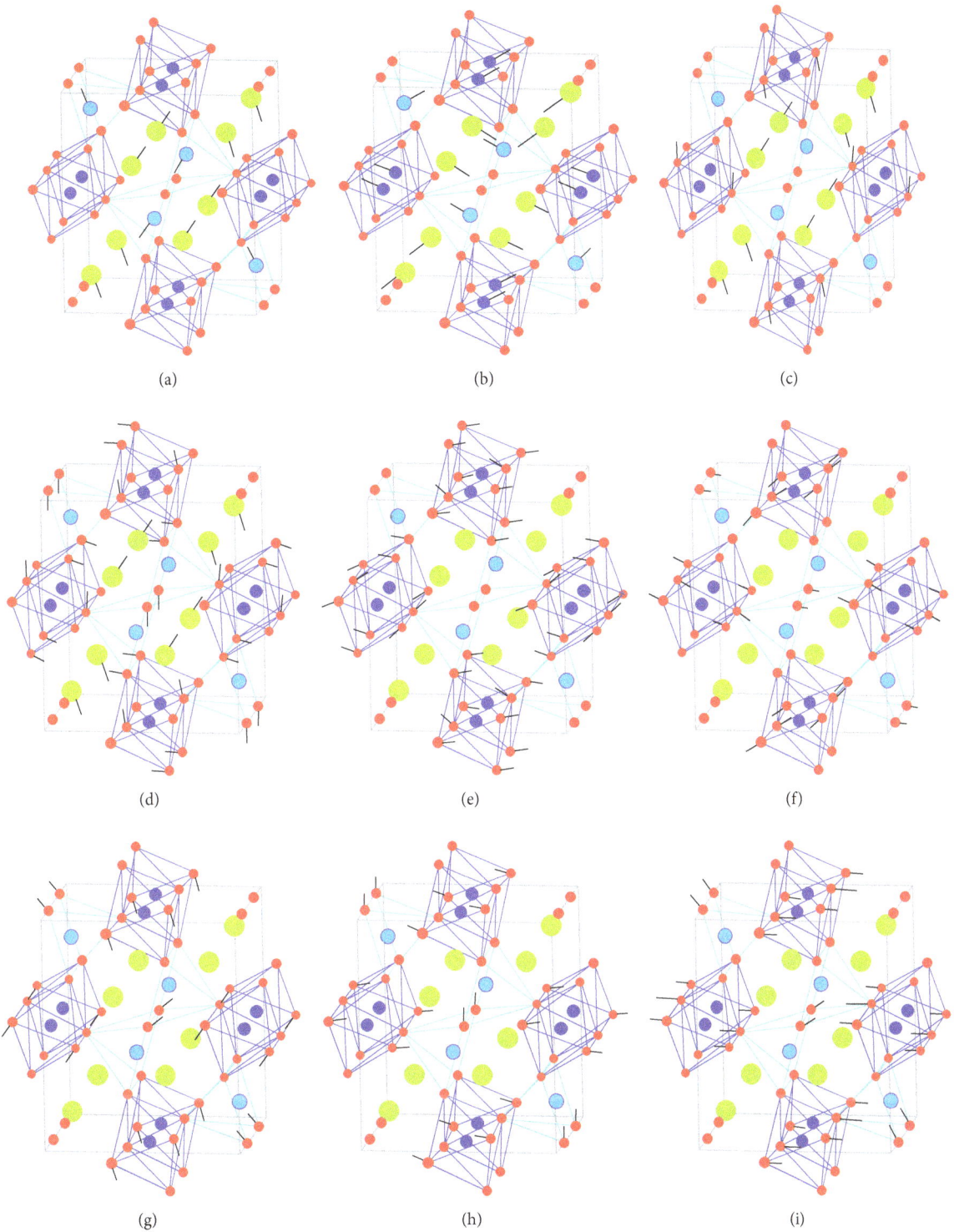

FIGURE 6: Panels (a), (b), (c), (d), (e), (f), (g), (h), and (i) are the representative mode displacement obtained from lattice dynamics calculation [7] with calculated frequencies at 170, 176, 208, 231, 283, 567, 576, 626, and 728 cm^{-1}, respectively.

>300 cm^{-1} has also been observed through Raman spectra in RMn$_2$O$_5$ (R = Bi, Tb, Eu, Dy, Ho) [5, 13].

As mentioned previously, out of 36 theoretically predicted modes, we have only observed 18 modes. The origin of this discrepancy is due to the polycrystalline nature of our sample which has mixed ab-plane and c axis response. For example, let us consider the infrared modes of 14B$_{2u}$ and 14B$_{3u}$ that occur in close pair such that at least 10 of them have wavenumber difference less than 10 cm^{-1}, which is due to the typical damping value. This clearly reflects that they

TABLE 1: Phonon-fit parameters of the infrared reflectivity spectrum of $DyMn_2O_5$ at room temperature and calculated modes [6] for comparison. The wavenumbers $\omega_{TO(LO)}$ are given in cm^{-1}. 18 observed modes: $\varepsilon_\infty = 2.78$, $\varepsilon_\circ = 5$.

$\omega_{TO(LO)}$	S_j	γ_j	B_{3u}	B_{2u}	B_{1u}	Assignment
151(204)	0.20	9.29	95	104	117	Relative motion of Mn polyhedra and Dy ions
167(226)	0.28	8.51	170	176	—	
194(262)	0.11	15.4	189	184	—	
217(294)	0.02	4.46	208	—	—	Relative motion of Dy ions and oxygens
251(341)	0.05	4.73	—	231	245	
260(352)	0.22	18.3	—	283	—	Mn-O bending motions in MnO_6 octahedra
291(395)	0.01	10.8	310	—	—	
327(443)	0.05	9.7	336	339	325	Mn-O twisting motions in Mn polyhedra
371(503)	0.27	27.3	382	387	368	
402(544)	0.10	12.72	403	—	—	
433(586)	0.12	27.46	—	441	456	
462(626)	0.10	18.08	475	464	473	Mn-O bending motions within equatorial MnO_2 planes in MnO_6 octahedra
484(655)	0.20	18.25	486	475	—	
511(692)	0.01	7.15	—	—	509	
534(723)	0.16	17.31	567	576	—	Mn-O stretching motions in MnO_6 octahedra and MnO_5 square pyramids
565(765)	0.23	36.58	585	589	—	
628(850)	0.08	23.22	617	626	655	
684(926)	0.02	32.57	762	728	—	

can not be resolved separately by using unpolarized light in polycrystal and that they have been considered together in the fitting procedure.

For quantitative analysis of phonon contribution to the infrared spectrum of $DyMn_2O_5$, we have calculated real part of optical conductivity $\sigma_1(\omega) = \omega\varepsilon_2/4\pi$ spectrum by using Kramers-Krönig transformation, as shown in Figure 7. $\sigma_1(0)$ is zero suggesting no contribution from free carriers at low frequency range and suggesting the charge carriers are rather localized. Moreover, the optical energy band gap has been calculated from the spectrum of optical conductivity [see inset of Figure 7]. An optical band gap can be observed as a gradual onset of the spectral intensity showing a sharp rise, as determined by the crossing point of the energy axis [14, 15]. The band gap distinctly seen at $E_{g(OC)} \sim 1.78\,eV$ arises due to lowest energy oscillator, close to the band gap ($\sim$1.70 eV) observed for $TbMn_2O_5$ through ellipsometry technique [16].

3.3. UV-Visible Spectroscopy.
We have measured UV-visible absorption spectrum of $DyMn_2O_5$ in the wavelength range 200–1000 nm as shown in Figure 8. The diffuse reflectance spectrum has been transformed into Kubelka-Munk function or absorption $(F(R))$ by using the Kubelka-Munk theory [17, 18]. Such strong absorption with cutoff wavelength at 1000 nm indicates that $DyMn_2O_5$ can efficiently absorb light over a broad range of wavelengths including the complete visible range (390–780 nm). Moreover, we have observed two strong absorption peaks at about 485 nm and 525 nm which reflects that there is a probability of e_g electrons from Mn^{3+}

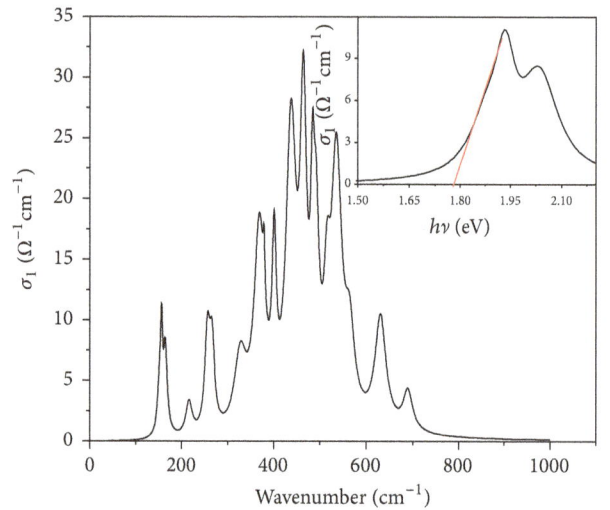

FIGURE 7: Real part of optical conductivity of $DyMn_2O_5$ at room temperature. Inset shows the estimated energy band gap, where the linear portion of the lowest energy extrapolated to the energy axis to obtain the energy band gap ($\sim$1.78 eV).

$(t_{2g}^3 e_g^1)$ ions tunnel to Mn^{4+} $(t_{2g}^3 e_g^0)$ ions, giving rise to the local distortion near these ions [19]. These strong absorption peaks may be attributed to the $3d$-electronic transitions of Mn^{3+}/Mn^{4+}.

In order to verify the energy band gap estimated from optical conductivity, we have calculated the energy band gap

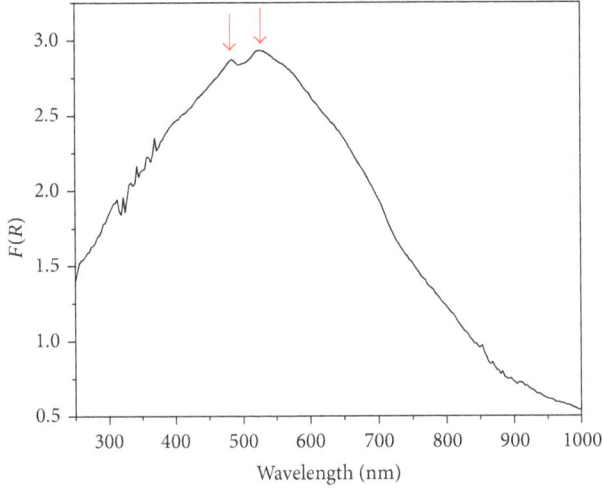

FIGURE 8: Kubelka-Munk function versus wavelength for $DyMn_2O_5$.

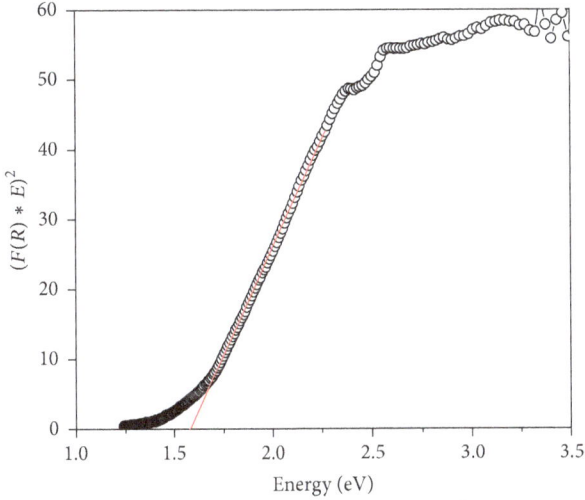

FIGURE 10: I-V characteristics of $DyMn_2O_5$ at different temperatures. Inset shows variation of current with temperature.

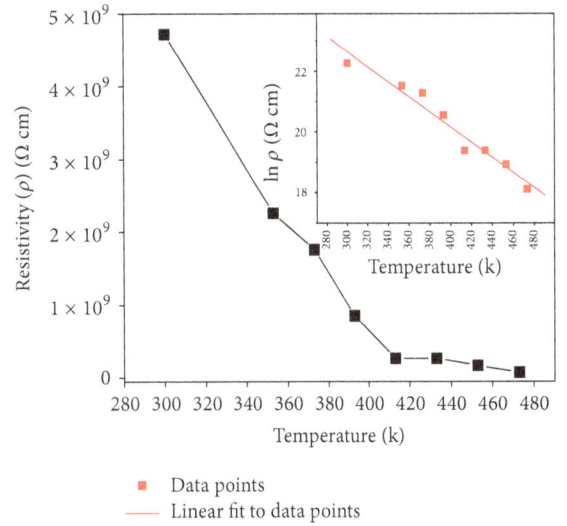

FIGURE 9: Determination of band gap from the inflection point of the $(F(R) * E)^2$ versus E.

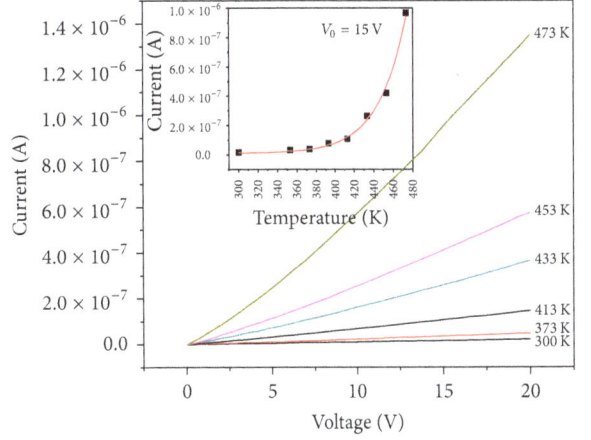

FIGURE 11: Variation of resistivity with temperature of $DyMn_2O_5$. Inset shows logarithmic variation of resistivity with temperature.

by using the relation [20]

$$(F(R)h\upsilon)^n = A\left(h\upsilon - E_g\right), \qquad (4)$$

where $h\upsilon$ is the energy of the incident photon, $F(R)$ is the K-M function, A is a characteristic parameter, and E_g is the band gap. Exponent n specifies the type of transition and it may be 1/2 or 2 for the direct forbidden or indirect transition, respectively. Here, we have estimated the indirect band gap from the plot of $(F(R)h\upsilon)^2$ versus energy ($E = h\upsilon$) by extrapolating the linear portion to the energy axis (i.e., $F(R) = 0$) to determine the energy band gap of about $E_{g(UV)} \sim 1.57$ eV as shown in Figure 9. We have observed a comparable energy band gap estimated through optical conductivity ($E_{g(OC)} \sim 1.78$ eV) and UV-visible ($E_{g(UV)} \sim 1.57$ eV). Despite the slight difference in energy band gap ($\Delta = 0.21$ eV), $DyMn_2O_5$ exhibits a semiconducting behavior. The obtained value of band gap is close to the band gap

observed for similar compounds such as $TbMn_2O_5 \sim 1.70$ eV, $YMn_2O_5 \sim 1.21$ eV [16, 21].

3.4. Electrical Properties. In order to study the conduction mechanism, we have performed temperature dependent I-V characteristics of $DyMn_2O_5$ which demonstrate a smooth increase in current with increase in applied voltage (0–20 V) as shown in Figure 10. Moreover, at a certain voltage (i.e., $V_0 = 15$ V) current increases with increase in temperature supporting a semiconducting nature of the material [see inset of Figure 10]. Temperature dependent DC electrical resistivity of $DyMn_2O_5$ system has been measured in the temperature range (300–473 K), as shown in Figure 11. The logarithmic variation of resistivity with temperature is also shown [see inset of Figure 11]. It follows the Arrhenius rule with positive slope as shown in Figure 12. The plot in Figure 12

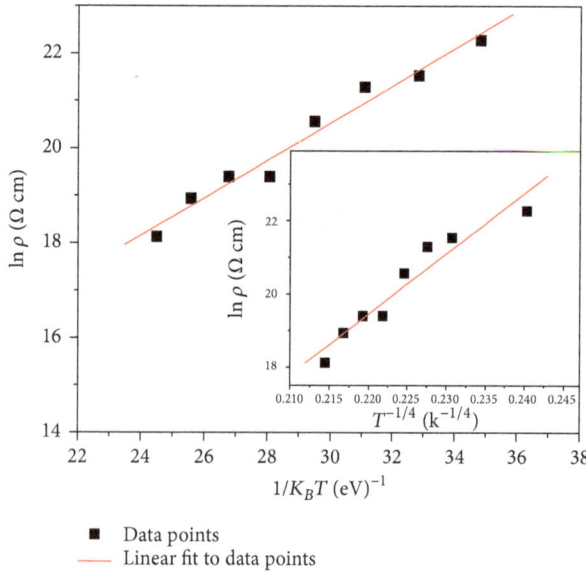

FIGURE 12: Arrhenius plot of $DyMn_2O_5$. Inset shows logarithmic variation of resistivity versus $T^{-1/4}$.

is extracted from the given equation

$$\rho = \rho_0 \exp\left(\frac{-\Delta E}{K_B T}\right), \qquad (5)$$

where ΔE is activation energy, T is absolute temperature, and K_B is Boltzmann constant.

The activation energy ~0.759 eV for $DyMn_2O_5$ is calculated from slope of $\ln \rho$ versus $1/K_B T$. This may be explained in terms of dominant character of grain boundaries to find out the resistive properties of this system. Generally, activation energies for the oxide ionic conductors are >0.9 eV. It is usually <0.2 eV for n-type polaronic conduction of electrons and >0.2 eV for the polaronic conduction of holes [22]. In our case, observed value of activation energy suggests a p-type polaronic conduction of holes above room temperature in $DyMn_2O_5$. Moreover, the temperature dependent DC resistivity (see Figure 11) shows decrease in resistivity with the increase in temperature reflecting the role of thermally activated charge carriers according to the hopping conduction mechanism [23–25]. We have tested the validity of this polaronic hopping conduction mechanism in the same temperature range (300–473 K) by Mott's variable-range-hopping (VRH) conduction rule [26–28]

$$\rho = \rho_0 \exp\left(\frac{T_0}{T}\right)^{1/4}. \qquad (6)$$

It is observed that variation of $\ln \rho$ with $T^{-1/4}$ is a straight line as depicted in inset of Figure 12. The linear dependence in our case confirms the presence of hopping type conduction in given temperature range [29–31]. Moreover, the obtained value of ΔE is almost equal to half of the optical energy band gap (1.78 eV), confirming the correlation between thermally activated charge carriers with localized free-carrier response. The obtained value of activation energy (0.759 eV)

of $DyMn_2O_5$ in our investigated compound is comparable with Mn based compound such as $SmMn_2O_5$ for which activation energy is 0.59 eV [32].

4. Conclusion

We have investigated the infrared reflectivity, UV-visible absorption spectra, and electrical conduction mechanism of polycrystalline $DyMn_2O_5$. The experimentally observed phonon modes have been assigned to the different lattice dynamics and crystal symmetry. In addition, we have calculated optical conductivity, and no contribution of free carriers has been observed at low frequency. On comparison, an agreement found between energy band gap estimated through UV-visible measurement and optical conductivity and is associated with the semiconducting behavior of $DyMn_2O_5$. Indeed, the semiconducting mechanism has also been confirmed from existence of p-type polaronic conduction above room temperature. The results conclude $DyMn_2O_5$ as a typical semiconductor.

Conflicts of Interest

The authors declare that there are no conflicts of interest regarding the publication of this paper.

References

[1] M. Tachibana, K. Akiyama, H. Kawaji, and T. Atake, "Lattice effects in multiferroic R Mn_2O_5 (R = Sm – Dy, Y)," *Physical Review B—Condensed Matter and Materials Physics*, vol. 72, no. 22, Article ID 224425, 2005.

[2] P. G. Radaelli and L. C. Chapon, "A neutron diffraction study of RMn2O5 multiferroics," *Journal of Physics Condensed Matter*, vol. 20, no. 43, Article ID 434213, 2008.

[3] V. Balédent, S. Chattopadhyay, P. Fertey et al., "Evidence for room temperature electric polarization in RMn_2O_5 multiferroics," *Physical Review Letters*, vol. 114, no. 11, Article ID 117601, 2015.

[4] N. Hur, S. Park, P. A. Sharma, S. Guha, and S.-W. Cheong, "Colossal magnetodielectric effects in DyMn2O5," *Physical Review Letters*, vol. 93, no. 10, Article ID 107207, 2004.

[5] A. F. García-Flores, E. Granado, H. Martinho et al., "Anomalous phonon shifts in the paramagnetic phase of multiferroic RMn_2O_5 (R= Bi, Eu, Dy): possible manifestations of unconventional magnetic correlations," *Physical Review B - Condensed Matter and Materials Physics*, vol. 73, no. 10, Article ID 104411, 2006.

[6] J. Cao, L. I. Vergara, J. L. Musfeldt et al., "Spin-lattice interactions mediated by magnetic field," *Physical Review Letters*, vol. 100, no. 17, Article ID 177205, 2008.

[7] J. Cao, L. I. Vergara, J. L. Musfeldt et al., "Magnetoelastic coupling in DyMn2O5 via infrared spectroscopy," *Physical Review B*, vol. 78, no. 6, Article ID 064307, 2008.

[8] S. H. Bukhari and J. Ahmad, "Evidence for magnetic correlation in the paramagnetic phase of $DyMn_2O_5$," *Physica B: Condensed Matter*, vol. 503, pp. 179–182, 2016.

[9] S. H. Bukhari and J. Ahmad, "Emergent excitation in the paramagnetic phase of geometrically frustrated GdMn2O5," *Physica B: Condensed Matter*, vol. 492, pp. 39–44, 2016.

[10] G. R. Blake, L. C. Chapon, P. G. Radaelli et al., "Spin structure and magnetic frustration in multiferroic RMn2O5 (R=Tb,Ho,Dy)," *Physical Review B*, vol. 71, no. 21, Article ID 214402, 2005.

[11] K. Venkateswarlu, A. Chandra Bose, and N. Rameshbabu, "X-ray peak broadening studies of nanocrystalline hydroxyapatite by Williamson-Hall analysis," *Physica B: Condensed Matter*, vol. 405, no. 20, pp. 4256–4261, 2010.

[12] R. Valdés Aguilar, A. B. Sushkov, S. Park, S.-W. Cheong, and H. D. Drew, "Infrared phonon signatures of multiferroicity in TbMn2O5," *Physical Review B*, vol. 74, no. 18, Article ID 184404, 2006.

[13] B. Mihailova, M. M. Gospodinov, B. Güttler, F. Yen, A. P. Litvinchuk, and M. N. Iliev, "Temperature-dependent Raman spectra of Ho Mn_2O_5 and $TbMn_2O_5$," *Physical Review B - Condensed Matter and Materials Physics*, vol. 71, no. 17, Article ID 172301, 2005.

[14] K. P. Ong, P. Blaha, and P. Wu, "Origin of the light green color and electronic ground state of La Cr O_3," *Physical Review B—Condensed Matter and Materials Physics*, vol. 77, no. 7, Article ID 073102, 2008.

[15] T. Arima, Y. Tokura, and J. B. Torrance, "Variation of optical gaps in perovskite-type 3d transition-metal oxides," *Physical Review B*, vol. 48, no. 23, pp. 17006–17009, 1993.

[16] A. S. Moskvin and R. V. Pisarev, "Charge-transfer transitions in mixed-valent multiferroic TbMn2O5," *Physical Review B*, vol. 77, no. 6, Article ID 060102, 2008.

[17] P. Kubelka, "New contributions to the optics of intensely light-scattering materials. I," *Journal of the Optical Society of America*, vol. 38, no. 5, pp. 448–457, 1948.

[18] S. Thota, J. H. Shim, and M. S. Seehra, "Size-dependent shifts of the Néel temperature and optical band-gap in NiO nanoparticles," *Journal of Applied Physics*, vol. 114, no. 21, Article ID 214307, 2013.

[19] B. K. Khannanov, V. A. Sanina, E. I. Golovenchits, and M. P. Scheglov, "Room-temperature electric polarization induced by phase separation in multiferroic GdMn2O5," *JETP Letters*, vol. 103, no. 4, pp. 248–253, 2016.

[20] V. Sasca and A. Popa, "Band-gap energy of heteropoly compounds containing kegging polyanion-[PVxMo12¡xO40]¡(3+x) relates to counter-cations and temperature studied by uv-vis diffuse reflectance spectroscopy," *Journal of Applied Physics*, vol. 114, no. 13, pp. 133503–133510, 2013.

[21] H. Yang, S. F. Wang, T. Xian, Z. Q. Wei, and W. J. Feng, "Fabrication and photocatalytic activity of YMn_2O_5 nanoparticles," *Materials Letters*, vol. 65, no. 5, pp. 884–886, 2011.

[22] M. Idrees, M. Nadeem, and M. M. Hassan, "Investigation of conduction and relaxation phenomena in LaFe 0.9Ni0.1O3 by impedance spectroscopy," *Journal of Physics D: Applied Physics*, vol. 43, no. 15, Article ID 155401, 2010.

[23] D. R. Patil and B. K. Chougule, "Effect of resistivity on magnetoelectric effect in (x)NiFe2O4-(1 - x)Ba0.9Sr0.1TiO3 ME composites," *Journal of Alloys and Compounds*, vol. 470, no. 1-2, pp. 531–535, 2009.

[24] J. Ryu, S. Priya, K. Uchino, and H.-E. Kim, "Magnetoelectric effect in composites of magnetostrictive and piezoelectric materials," *Journal of Electroceramics*, vol. 8, no. 2, pp. 107–119, 2002.

[25] C. M. Kanamadi, G. Seeta Rama Raju, H. K. Yang, B. C. Choi, and J. H. Jeong, "Conduction mechanism and magnetic properties of (x)Ni0.8Cu0.2Fe2O4 + (1 - x)Ba0.8Pb0.2Ti0.8Zr0.2O3 multiferroics," *Journal of Alloys and Compounds*, vol. 479, no. 1-2, pp. 807–811, 2009.

[26] J. M. D. Coey, M. Viret, L. Ranno, and K. Ounadjela, "Electron localization in mixed-valence manganites," *Physical Review Letters*, vol. 75, no. 21, pp. 3910–3913, 1995.

[27] J. M. De Teresa, M. R. Ibarra, J. Blasco et al., "Spontaneous behavior and magnetic field and pressure effects on La2/3Ca1/3MnO3 perovskite," *Physical Review B*, vol. 54, no. 2, pp. 1187–1193, 1996.

[28] M. Viret, L. Ranno, and J. M. D. Coey, "Magnetic localization in mixed-valence manganites," *Physical Review B - Condensed Matter and Materials Physics*, vol. 55, no. 13, 1997.

[29] M. F. Wasiq, M. Y. Nadeem, K. Mahmood, M. F. Warsi, and M. A. Khan, "Impact of metal electrode on charge transport behavior of metal-Gd2O3 systems," *Journal of Alloys and Compounds*, vol. 648, pp. 577–580, 2015.

[30] W.-H. Jung, "Evaluation of Mott's parameters for hopping conduction in La0.67Ca0.33MnO3 above Tc," *Journal of Materials Science Letters*, vol. 17, no. 15, pp. 1317–1319, 1998.

[31] M. Ziese and C. Srinitiwarawong, "Polaronic effects on the resistivity of manganite thin films," *Physical Review B - Condensed Matter and Materials Physics*, vol. 58, no. 17, pp. 11519–11525, 1998.

[32] K. Saravana Kumar, N. Aparnadevi, A. Muthukumaran, and C. Venkateswaran, "Phase Stabilization of Fe Substituted $SmMn_2O_5$: $SmFeMnO_5$," *Transactions of the Indian Institute of Metals*, vol. 68, no. 5, pp. 693–698, 2015.

5

Resonance Raman Scattering in TlGaSe$_2$ Crystals

N. N. Syrbu,[1] **A. V. Tiron,**[1] **V. V. Zalamai,**[2] **and N. P. Bejan**[1]

[1]*Department of Telecommunication, Technical University of Moldova, Chişinău, Moldova*
[2]*Laboratory of Materials for Photovoltaics and Photonics, Institute of Applied Physics, Academy of Sciences of Moldova, Chişinău, Moldova*

Correspondence should be addressed to N. N. Syrbu; sirbunn@yahoo.com

Academic Editor: Gary Wysin

The resonance Raman scattering for geometries $Y(YX)Z$ and $Y(ZX)Z$ at temperature 10 K and infrared reflection spectra in $E \parallel a$ and $E \parallel b$ polarizations at 300 K were investigated. The number of A_a (B_a) and A_u (B_u) symmetry vibrational modes observed experimentally and calculated theoretically agree better in this case than when TlGa$_2$Se$_4$ crystals belong to D_{2h} symmetry group. The emission of resonance Raman scattering and excitonic levels luminescence spectra overlap. The lines in resonance Raman spectra were identified as a combination of optical phonons in Brillouin zone center.

1. Introduction

TlGaSe$_2$ crystals are triple thallium chalcogenides with a layered structure [1, 2]. One of these crystals features is the strong anisotropy of physical characteristics due to the specificity of the crystals lattice [1–3]. Optical spectra in the absorption edge region [4–11] and resonance Raman scattering for different geometries and temperatures (77–400 K) [12] were investigated in TlGaSe$_2$ crystals. Reflection spectra for the 50–4000 cm^{-1} region were studied and polar vibrational modes LO and TO and their parameters were determined. Such crystals had an effect of switching of current-voltage and acoustooptic characteristics [13–15]. There are a lot of materials dedicated to the investigations of these materials (see [4–16] and the references therein). But resonance Raman scattering in TlGaSe$_2$ crystals has not been investigated.

2. Experimental Methods

Raman scattering spectra of TlGaSe$_2$ crystals were measured on double high-aperture spectrometers DFS-32 with linear dispersion of 5 Å/mm and relative aperture of 1 : 5 and resonance Raman scattering spectra on spectrometer SDL-1 with dispersion of 7 Å/mm and relative aperture of 1 : 2. The photomultiplier working in the photon counting regime

was used as a detector. Resonance Raman spectra had an accuracy of ±0.5 meV. Reflection spectra in $E \parallel a$ and $E \parallel b$ polarizations in the range 50–400 cm^{-1} were measured on a vacuum spectrometer KSDI-82 using an acoustooptical receiver with an accuracy of ±1 cm^{-1}. Cleft crystals of TlGaSe$_2$ with different thicknesses mounted on a cold finger of a closed-circuit helium cryostat LTS-22 C 330 optical cryogenic system were used in the measurements. The Raman scattering was excited by 6328 Å line of a He-Ne laser. The resonance Raman scattering was excited by lines 4579 Å and 5145 Å of an Ar$^+$ laser.

3. Experimental Results and Discussions

According to the crystallographic data, the TlGaSe$_2$ structure is described by the space group $C2/c$ ($C_{2h}{}^6$). The unit cell contains 8 formula units of TlGaSe$_2$. The main motive of the structure is formed by tetrahedral polyhedrons of Ga$_4$Se$_{10}$, consisting of 4 tetrahedrons of GaSe$_4$. These tetrahedrons have common atoms of selenium on the tops of the octahedron [1–3]. These tetrahedral polyhedrons have common vertices of 4 selenium atoms and take up layered positions perpendicular to the c axis. The layers are rotated to each other at 90°. The edges of polyhedrons lie in the xy plain and are situated along the diagonal of the base square. Thus, the TlGaSe$_2$ compound has a monoclinic pseudotetragonal

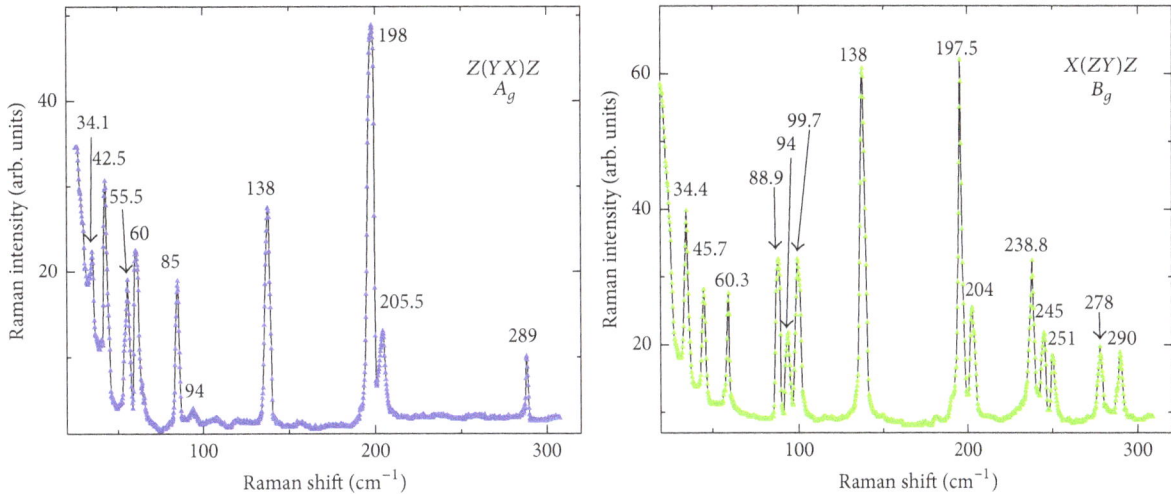

FIGURE 1: Raman scattering of TlGaSe$_2$ crystals.

structure with the following parameters: $a = b = 10.75$ Å, $c = 15.56$ Å, and $\beta = 100.0°$. The distances between Tl-Se, Se-Se, and Tl-Tl are equal to 3.45 Å, 3.92 Å, and 3.42 Å, respectively. This distance corresponds to the sum of ionic radiuses Tl^1-Se (3.38 Å) [1–3].

The next vibrational modes should be observed in Brillouin zone center of the above-mentioned crystals:

$$\Gamma = \left(A_u + 2B_u\right)_{ac} + 23A_g + 25B_g + 22A_u + 23B_u. \quad (1)$$

The phonons of A_g and B_g symmetry should be observed in Raman spectra and A_u and B_u in IR reflection spectra in $E \parallel a$ and $E \parallel b$ polarizations, respectively. The scattering tensors are next:

$$A_g = \begin{vmatrix} a & d & 0 \\ d & b & 0 \\ 0 & 0 & c \end{vmatrix},$$
$$B_g = \begin{vmatrix} 0 & 0 & e \\ 0 & 0 & f \\ e & f & 0 \end{vmatrix}. \quad (2)$$

Figure 1 shows the Raman scattering of TlGaSe$_2$ crystals measured at a temperature of 10 K and in $X(ZY)Z$ and $Z(YX)Z$ geometries. The structure of vibrational modes depends on polarization. It was reported in [12] that 8 and 6 modes of A_g symmetry were recognized at temperatures of 77 K and 300 K, respectively. One can see from the analysis of the above-mentioned spectra that, even at 10 K, the number of experimentally observed modes is smaller than the number of modes theoretically predicted by group-theoretic calculations. 14 modes of B_g symmetry and 10 modes of A_g symmetry were observed. Hence, the amount of the observed vibrational modes as in IR reflection spectra as in Raman spectra is lower than expected according to the theory.

The most intensive modes in reflection spectra for both polarizations are high-frequency modes (see Figure 2). Thus, in reflection spectra of TlGaSe$_2$ crystals, 23 and 22 modes are expected in the region of single-phonons vibrational modes, but only 5 modes and 8 modes have been observed in $E \parallel a$ and $E \parallel b$ polarizations, respectively (Figure 2).

Compounds TlGaS$_2$ and TlGaSe$_2$ belong to thallium based crystals. This group of crystals (TlMX$_2$, where M = Ga, In and X = S, Se, Te) has a family likeness of optical spectra and energy band structures. The analogs are observed in all well-studied compounds (Si, Ge, $A^{III}B^V$ and $A^{II}B^{VI}$). The results of band structure calculations for TlMX$_2$ crystals have a common character and reflect only its main features. This leads to proximity of the above-mentioned compounds lattices. Crystal structures of TlGaS$_2$ and TlGaSe$_2$ compounds are different only in the replacement of S atoms by Se atoms in the crystal lattice. The structure of the layer TlGaS$_2$ in [12] is symmetrized by an insignificant shift of atoms inside the layer to achieve a tetragonal structure with D_{2d}^5 space group. The hypothetic structure with space group D_{2h}^{15} with unit cell comprising two layers was achieved by the authors of [12], introducing the interlayer inversion operation with preserving the elements of layer symmetry. Similarly, we analyze and investigate TlGaSe$_2$ crystals. Using the same assumption, one can obtain a better agreement in the number of theoretically predicted and experimentally measured vibration modes in the case of TlGaSe$_2$ crystals. Based on the analysis of polarization dependences of Raman and IR reflection spectra, the TlGaS$_2$ and TlGaSe$_2$ crystals can be attributed to symmetry group D_{2h} or D_{4h}.

The emission lines (1–20) of resonance Raman scattering in the region of excitonic resonances at excitation of 514.5 nm laser line of TlGaSe$_2$ crystals at a temperature of 10 K and $Z(YY)Z$ geometry were observed (see Figure 3 and Table 1). These lines (1–20) skirt the broad emission lines at 2.17–2.19 eV, 2.30 eV, and 2.39 eV in resonance Raman scattering spectra of TlGaSe$_2$ crystals. These broad lines are caused by emission of ground states A, B, and C excitons.

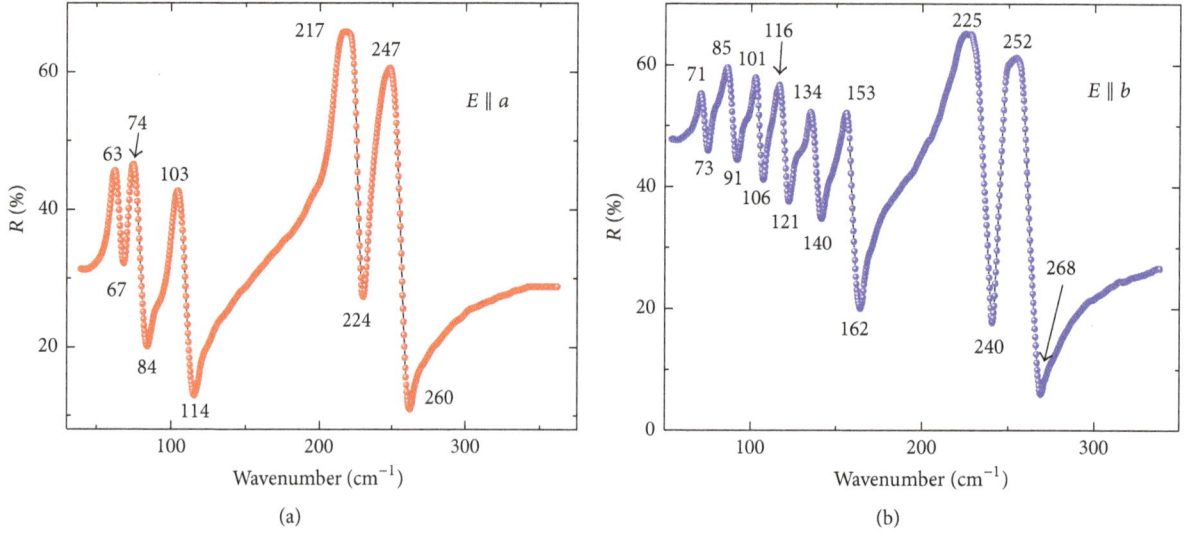

FIGURE 2: Reflection spectra of TlGaSe$_2$ crystals in $E \parallel a$ (a) and $E \parallel b$ (b) polarizations.

TABLE 1: Emission lines of resonance Raman scattering in TlGaSe$_2$ crystals measured at 10 K and excited by 514.5 nm Ar$^+$ laser line in $Z(YY)Z$ geometry and possible phonons combinations (symmetry, modes, and frequencies) responsible for resonance scattering.

Line, n	Wavenumber shift, cm^{-1}	Assignment
1	120	A_u(LO), 121;
2	134	B_u(LO), 67 + B_u(LO), 67;
3	140	A_u(LO), 140;
4	151	B_u(LO), 67 + B_u(LO), 84;
5	163	A_u(LO), 73 + A_u(LO), 91;
6	187	A_u(LO), 106 + B_u(LO), 84; B_u(LO), 114 + A_u(LO), 73;
7	238	B_u(LO), 114 + A_u(LO), 121;
8	313	A_u(LO), 240 + A_u(LO), 73;
9	436	$2A_u$(LO), 106 + B_u(LO), 224;
10	480	A_u(LO), 240 + A_u(LO), 240;
12	590	B_u(LO), 260 + A_u(LO), 91 + A_u(LO), 240;
13	634	B_u(LO), 260 + B_u(LO), 260 + B_u(LO), 114;
14	732	A_u(LO), 240 + B_u(LO), 224 + A_u(LO), 268;
15	960	A_u(LO), 240 + A_u(LO), 240 + A_u(LO), 240 + A_u(LO), 240;
16	999	A_u(LO), 240 + A_u(LO), 240 + B_u(LO), 260 + B_u(LO), 260;
17	1075	A_u(L), 73 + A_u(LO), 240 + A_u(LO), 240 + B_u(LO), 260 + B_u(LO), 260;

Figure 3 shows resonance Raman scattering spectra in TlGaSe$_2$ crystals measured at a temperature of 10 K in $Z(YY)Z$ geometry and excited by 496.5 nm Ar$^+$ laser line. The narrow lines (1–17) that skirted the line at 2.4 eV were observed in these spectra. The observed lines of resonance Raman scattering and possible combination of phonons responsible for these emission lines are presented in Tables 1 and 2. At high frequencies, these data do not include all possible combinations of phonons responsible for lines of resonance Raman scattering.

4. Conclusions

The resonance Raman scattering in $Y(YX)Z$ and $Y(ZX)Z$ geometries excited by He-Ne laser was investigated at a temperature of 10 K. The energies of phonons with A_g and B_g symmetries were determined. It was shown that the amount of modes in Raman scattering and IR reflection spectra measured at 10 K is half the expected according to group theory calculations. The experimental and theoretical results coincide if the crystal is described by symmetry

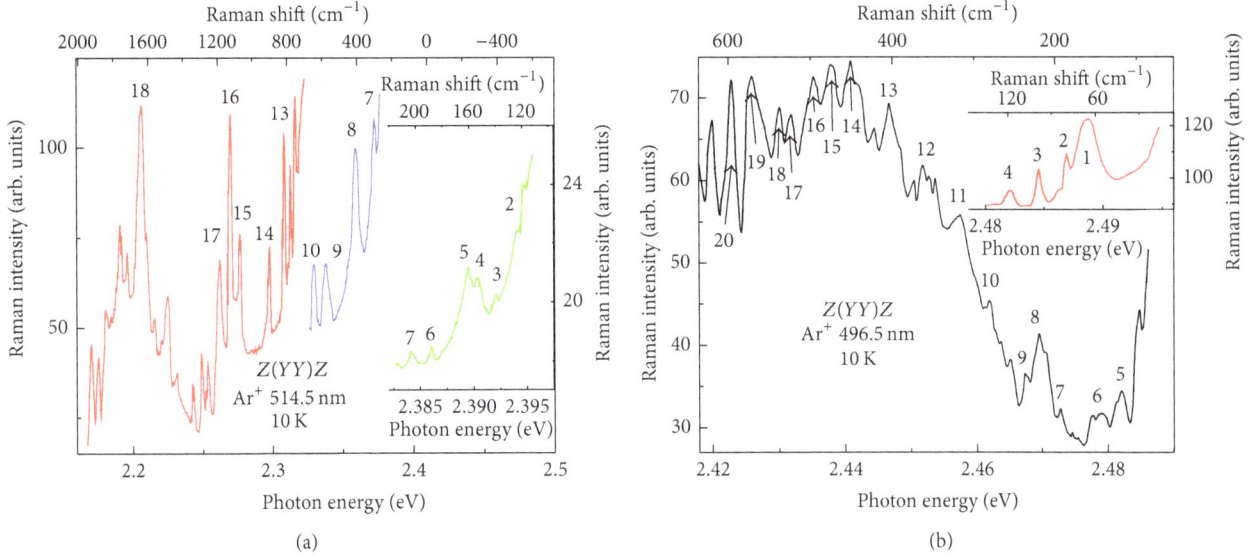

FIGURE 3: Resonance Raman scattering in TlGaSe$_2$ crystals measured at a temperature of 10 K and excited by lines 514.5 nm (a) and 496.5 nm (b) of an Ar$^+$ laser in $Z(YY)Z$ geometry.

TABLE 2: Emission lines of resonance Raman scattering in TlGaSe$_2$ crystals measured at 10 K and excited by 496.5 nm Ar$^+$ laser line in $Z(YY)Z$ geometry and possible phonons combinations (symmetry, modes, and frequencies) responsible for resonance scattering.

Line, n	Wavenumber shift, cm^{-1}	Assignment
1	66	$B_u(LO)$, 67;
2	75	$A_u(LO)$, 73;
3	85	$B_u(LO)$, 84;
4	96	$A_u(LO)$, 91;
5	117	$A_u(LO)$, 121; $A_u(TO)$, 116; $B_u(LO)$, 114;
6	142	$A_u(LO)$, 140; $B_u(LO)$, 67 + $A_u(LO)$, 73;
7	194	$B_u(LO)$, 114 + $B_u(LO)$, 84; $A_u(LO)$, 121 + $A_u(LO)$, 73;
8	217	$A_u(LO)$, 106 + $B_u(LO)$, 106;
9	236	$A_u(LO)$, 121 + $B_u(LO)$, 114; $A_u(LO)$, 162 + $A_u(LO)$, 73;
10	279	$B_u(LO)$, 114 + $A_u(LO)$, 162;
11	322	$A_u(LO)$, 240 + $B_u(LO)$, 84; $A_u(LO)$, 162 + $A_u(LO)$, 162;
12	355	$A_u(LO)$, 268 + $B_u(L)$, 84;
13	401	$A_u(LO)$, 240 + $B_u(LO)$, 260;
14	446	$B_u(LO)$, 224 + $B_u(LO)$, 224;
15	482	$B_u(LO)$, 260 + $B_u(LO)$, 224;
16	493	$A_u(LO)$, 268 + $B_u(LO)$, 224;
17	525	$B_u(LO)$, 260 + $A_u(LO)$, 268;

group D_{2h}. The superposition of excitonic luminescence with resonance Raman scattering emission was observed. The lines of resonance Raman emission were identified and attributed to optical phonons in Brillouin zone center.

Conflicts of Interest

The authors declare that they have no conflicts of interest.

References

[1] D. Muller, E. Poltmann, and H. Hahn, "Zur structur ternarer chalkogenide des thalliums mit aluminium and indium," *Zeitschrift für Naturforschung*, vol. 29, no. 1, pp. 117-118, 1974.

[2] D. Müller and H. Hahn, "Untersuchungen über ternäre Chalkogenide. XXIV. Zur Struktur des TlGaSe$_2$," *Zeitschrift für anorganische und allgemeine Chemie*, vol. 432, p. 258, 1978.

[3] K. J. Range, G. Maheberd, and S. Obenland, "High Pressure Phases of TlAlSe$_2$ and TlGaSe$_2$ with TlSe-type Structure," *Zeitschrift für Naturforschung*, vol. 32a, p. 1354, 1977.

[4] S. N. Mustafaeva, E. M. Kerimova, and N. Z. Gasanov, "Exciton characteristics of intercalated TlGaSe$_2$ single crystal," *Semiconductors*, vol. 32, no. 2, p. 145, 1998.

[5] A. V. Sheleg, O. B. Plusch, and V. A. Aliev, "X-ray investigation of incommensurate phase in β-TlInS$_2$ crystals," *Solid State Physics*, vol. 36, p. 245, 1994.

[6] E. M. Kerimova, S. N. Mustafaeva, R. N. Kerimov, and G. A. Gadjieva, "Photo- and x-ray conductibilty of (TlGaS$_2$)$_{1-x}$(TlInSe$_2$)$_x$ compounds," *Inorganic Materials*, vol. 35, no. 11, p. 1313, 1999.

[7] S. B. Vahrusev, B. B. Zdanov, B. E. Kviatkovscii, N. M. Ocuneva, K. P. Alahverdiev, and P. M. Sardarli, "Incommensurate phase transitions in TlInS$_2$ crystals," *Technical Physics Letters*, vol. 39, p. 245, 1984.

[8] S. G. Guseinov, G. D. Guseinov, N. Z. Gasanov, and S. B. Kyazimov, "Special features of exciton absorption spectra of A^{III}B^{III}X$_2$^{VI}-type layer-semiconductor crystals," *Physica Status Solidi (b)*, vol. 133, no. 1, pp. K25–K30, 1986.

[9] S. G. Abdulaeva, S. S. Abdinbekov, and G. G. Guseinov, "Electroabsorption in monocrystals TlGaSe$_2$ and TlGaS$_2$," *Doklady Academii Nauk AzUSSR*, vol. 36, no. 8, p. 34, 1980.

[10] N. M. Gasanly, B. N. Mavrin, K. E. Sterin, V. I. Tagirov, and Z. D. Khalafov, "Raman study of layer TlGaS$_2$, β-TlInS$_2$, and TlGaSe$_2$ crystals," *Physica Status Solidi (b)*, vol. 86, no. 1, pp. K49–K53, 1978.

[11] N. M. Gasanli, N. N. Melnik, A. S. Ragimov, and V. I. Tagirov, "Raman study of layer TlGaS$_2$, β-TlInS$_2$, and TlGaSe$_2$," *Solid State Physics*, vol. 26, no. 2, p. 558, 1984.

[12] N. N. Syrbu, V. E. Lvin, I. B. Zadnipru, H. Neumann, H. Sobotta, and V. Riede, "Raman and infrared vibration spectra in TlGaS$_2$ crystals," *Semiconductors*, vol. 26, no. 2, p. 232, 1992.

[13] K. R. Allakhverdiev, T. G. Mamedov, R. A. Suleymanov, and N. Z. Gasasnov, "Pressure and temperature effects on electronic spectra of TlGaSe$_2$ type crystals," *Fizika*, vol. 8, p. 44, 2002.

[14] S. A. Husein, G. Attia, S. R. Alharbi, A. A. AlGhamdi, F. S. AlHaxmi, and S. E. AlGarni, "Investigation of the switching phenomena in TlGaSe$_2$ single crystal," *Journal of King Abdulaziz University-Science*, vol. 21, p. 27, 2009.

[15] V. Grivickas, V. Bigbajevas, V. Gavriusinas, and J. Linnros, "Photoacoustic pulse generation in TlGaSe$_2$ layered crystals," *Materials Science (Medžiagotyra)*, vol. 12, p. 4, 2006.

[16] R. Paucar, H. Itsuwa, K. Wakita, Y. Shim, O. Alekperov, and N. Mamedov, "Phase transitions and Raman scattering spectra of TlGaSe$_2$," *Journal of Physics: Conference Series*, vol. 619, no. 4, Article ID 012018, 2015.

6

Gravitation at the Josephson Junction

Victor Atanasov ⓘ

Department of Condensed Matter Physics, Sofia University, 5 Boul. J. Bourchier, 1164 Sofia, Bulgaria

Correspondence should be addressed to Victor Atanasov; vatanaso@gmail.com

Academic Editor: Jörg Fink

A geometric potential from the kinetic term of a constrained to a curved hyperplane of space-time quantum superconducting condensate is derived. An energy conservation relation involving the geometric field at every material point in the superconductor is demonstrated. At a Josephson junction the energy conservation relation implies the possibility of transforming electric energy into geometric field energy, that is, curvature of space-time. Experimental procedures to verify that the Josephson junction can act as a voltage-to-curvature converter are discussed.

1. Introduction

The success of the Laser Interferometer Gravitational Wave Observatory (LIGO) and its sister collaboration, VIRGO [1, 2], in observing the geometric field's ripples in space-time has, in addition to opening a new experimental method of astrophysical observation, proved a practical scheme for measuring the space-time metric; that is, LIGO had seen the death spiral of a pair of black holes, through the dislocation and the associated change in optical (and physical) path travelled by light beams in the arms of an interferometer as the gravitational perturbation travels between the reflection points. In theory, the geometry of space-time, that is, the components of the Riemann curvature tensor, can be reconstructed by taking measurements of the deviation of two adjacent light paths (geodesics) [3, 4].

Bearing in mind this spectacular success in experimental astrophysics, we pose the following questions: *(i) Are there other experimentally relevant strategies for detecting space-time geometry? (ii) Is there other extremely sensitive measurement technique besides two-light beam interference that can potentially sense the geometric field?* The present paper aims to answer both questions and is organised as follows. The first section discusses the appearance of a geometric potential term from the kinetic energy term in the Schrödinger equation. In order to improve on the detection possibilities we assume this constrained to a hyperplane quantum dynamics to concern a superconducting condensate. The

second section focuses on the hydrodynamic interpretation of the governing quantum equation and reveals that the geometric field enters it on an equal footing with the "quantum potential" (in Bohm's views), thus making its way as a real force moving the condensate superfluid. The third section contains the proof that the emergent geometric field enters an energy conservation relation valid at each point of the superconducting condensate. Based on this conservation relation applied to a Josephson junction, an experimentally verifiable voltage-to-curvature conversion effect is proposed in the fourth section. The fifth section discusses possible experimental methodologies to test the reality of the effect.

2. The Geometric Field in the Schrödinger Equation

The general theory of relativity conveyed an understanding of phenomena such as the distortion of time-space by a gravitational or acceleration field [5]. However, the manner in which the curvature of space-time, that is, the Riemannian space, affects the electronic properties of condensed matters systems on a microscopic scale is largely unknown and due to its experimental accessibility is of great interest [6, 7].

Riemannian geometric effects in a quantum system, which can either be free or constrained, stem from the dependance of the kinetic term on the metric of the embedding space or the metric of the submanifold onto which the quantum system is constrained by a confining potential (rigid chemical bond; electrostatic attraction).

The problem of constraining particle motion to a curved submanifold embedded in a Euclidean space $\mathbb{R}^n$ can be resolved in one of two alternative ways. (i) In the *intrinsic* quantization approach, the motion is constrained to the curved submanifold in the first place. A Hamiltonian is constructed from generalised coordinates and momenta intrinsic to the submanifold and the system is quantized canonically. As a result, the embedding space is inaccessible and the quantum system depends only on the geometry intrinsic to the submanifold/hyperplane [8–11]. (ii) In the *confining* potential approach, a free in the embedding space quantum particle is subjected by a normal force to the submanifold force that in effect confines the dynamics in it. The effective Hamiltonian depends on the intrinsic geometry and on the way this submanifold is immersed in the embedding space.

On one hand, the intrinsic quantization procedure is plagued with ordering ambiguities that allow for multiple consistent quantization procedures different by a term proportional to the curvature of the submanifold [10–14]. On the other hand, the confining potential procedure leads to a unique effective Hamiltonian that depends on the constraint. In real microscopic quantum systems, constrained motion is a result of a strong confining force (electrostatic, rigid chemical bonds, etc.). Therefore, confining potential formalism seems a physically more realistic approach to constraints [15–25].

The nonrelativistic quantum mechanics in a three-dimensional hyperplane of the four-dimensional curved space-time can be treated in a well established manner [14]. In this case the embedding space-time is non-Euclidean but equipped with a metric. This four-dimensional metric is related to the matter distribution by the Einstein equation. Suppose the four-dimensional space-time M is topologically the product $M \cong \mathbb{R} \times \Sigma$, where Σ represents a space-like three-dimensional hyperplane. We can then foliate M by a one-parameter family of embeddings given by the map $\tau_t : \Sigma \to M$ such that $\Sigma_t = \tau_t(\Sigma) \subset M$, which means that Σ_t is the image of the map τ in M for a fixed "time" t. We assume that the leaves Σ_t are space-like with respect to the metric in M. As a result, there exists in M a time-like field normal to the leaves Σ_t, and therefore there is a notion of future and past. This time evolution vector field, $t^a = (\partial/\partial t)^a$, satisfies $t^a \nabla_a t = 1$, so that local coordinates t, x^1, x^2, x^3 (satisfying $t^a \nabla_a x^b = 0$, for $b = 1, 2, 3$) can be introduced. In effect, the space-time is splittable into $3 + 1$ dimensions and the induced Riemannian metric g_{ij} onto the three-dimensional Σ_t can be used to write the Laplace-Beltrami operator Δ_{LB}, which is the kinetic energy term in the Schrödinger equation for the subjected to the geometric field quantum particle or quantum condensate:

$$\Delta_{\mathrm{LB}}\Psi = \frac{1}{\sqrt{|g|}}\partial_j\left(\sqrt{|g|}g^{jk}\partial_k\Psi\right)$$
$$= g^{jk}\partial_j\partial_k\Psi - g^{jk}\Gamma^l_{jk}\partial_l\Psi. \qquad (1)$$

The emergence of the geometric field from the kinetic term can be made clearer in the vicinity of the origin where the following Taylor expansion of the induced metric in normal coordinates applies: $g_{ij} = \delta_{ij} - (1/3)R_{ikjl}x^k x^l + O(|x|^3)$ (in fact, Riemann used the expansion of a metric in normal coordinates to originally define the curvature tensor; in normal coordinates there is no necessity to distinguish between co- and contravariant indices; see [26, 27]), and expanding the square root of the determinant of the metric yields

$$\sqrt{|g|} = 1 - \frac{1}{6}R_{jk}x^j x^k + O\left(|x|^3\right). \qquad (2)$$

Using a standard renormalisation of the wave-function $\Psi = \psi/|g|^{1/4}$ and keeping the lowest order terms (the only relevant for the quantum dynamics) in the Taylor expansion we get for the kinetic term in the Schrödinger equation

$$-\frac{\hbar^2}{2m}\Delta_{\mathrm{LB}}\frac{\psi}{|g|^{1/4}}$$
$$= \frac{1}{|g|^{1/4}}\left(-\frac{\hbar^2}{2m}\Delta\psi + \frac{\hbar^2}{4m}\frac{g^{lk}\partial_l\partial_k\sqrt{|g|}}{\sqrt{|g|}}\psi\right)$$
$$+ O\left(|x|\right)$$
$$= \frac{1}{|g|^{1/4}}\left(-\frac{\hbar^2}{2m}\Delta\psi - \frac{\hbar^2}{24m}R\psi\right) + O\left(|x|\right). \qquad (3)$$

Here Δ is the Laplacian on flat space. Adding an additional potential $U(x^1, x^2, x^3)$ that may act in the system we convey the complete symbolic equation with which we will further work:

$$-\frac{\hbar^2}{2m}\Delta\psi + \left(V_{\mathrm{Geom}} + U\right)\psi = i\hbar\partial_t\psi. \qquad (4)$$

Here

$$V_{\mathrm{Geom}} = -\frac{\hbar^2}{2m}\alpha R, \qquad (5)$$

where R is the three-dimensional Ricci scalar curvature and $\alpha = 1/12$ is a numeric coefficient. The emergence of a geometric potential from the kinetic term is obvious. Such a term is a standard coupling term between curvature and a quantum field in quantum field theory in curved space-time.

Note that the particular form of the geometric potential may vary and in the case of a constraining potential approach it takes the expressions: (i) $V_{\mathrm{Geom}} = -(\hbar^2/8m)\kappa^2$, where κ is the principle curvature of a space curve embedded in $\mathbb{R}^3$ [17, 18, 20, 21]; (ii) $V_{\mathrm{Geom}} = -(\hbar^2/8m)(\kappa_1 - \kappa_2)^2$, where κ_i, for $i = 1, 2$, are the principle curvatures of a surface embedded in $\mathbb{R}^3$ [15–18]; (iii) $V_{\mathrm{Geom}} = -(\hbar^2/8m)[\kappa_3(\kappa_3 - 2(\kappa_1 + \kappa_2)) + (\kappa_1 - \kappa_2)^2]$, where κ_i, for $i = 1, 2, 3$, are the principle curvatures of a three-dimensional manifold embedded in $\mathbb{R}^4$ [23–25].

When electric field (defined with the potential V) and magnetic field, defined through the vector potential $\vec{A}$, are present the Schrödinger equation takes the following form:

$$\frac{1}{2m}\left(\frac{\hbar}{i}\nabla - q\vec{A}\right) \cdot \left(\frac{\hbar}{i}\nabla - q\vec{A}\right)\psi + qV\psi$$
$$+ \left(V_{\text{Geom}} + U\right)\psi = i\hbar\partial_t\psi. \tag{6}$$

3. Hydrodynamic Interpretation of the Condensate Wave-Function

Suppose that we deal with a Cooper pair condensate inside a superconductor. The Schrödinger equation for the Cooper pair will be (6) with $q = 2e$, that is, twice the charge of the electron. This equation will describe the state of the entire condensate. Therefore, we may write $\psi = \sqrt{\rho(\vec{r})}e^{i\theta(\vec{r})}$, where $\rho(\vec{r})$ is the charge density of the condensate and $\theta(\vec{r})$ is its phase. Upon substitution of this form of the wave-function into (6) we can separate the real and imaginary part of the equation to arrive at slightly modified standard result:

$$\frac{\partial\rho}{\partial t} = -\nabla\cdot\vec{J}, \quad \vec{J} = \vec{v}\rho = \frac{1}{m}\left(\hbar\nabla\theta - q\vec{A}\right)\rho, \tag{7}$$

$$\hbar\frac{\partial\theta}{\partial t} = -qV - \frac{1}{2m}\left(\hbar\nabla\theta - q\vec{A}\right)^2$$
$$+ \frac{\hbar^2}{2m}\left(\frac{\Delta\sqrt{\rho}}{\sqrt{\rho}} + \alpha R\right). \tag{8}$$

Here $\vec{J}$ is the current density, which in the case of a superconducting condensate stands also for the probability current. The generalised momentum is contained in the expression for $\vec{J}$: $\vec{p} = \hbar\nabla\theta - q\vec{A}$, and therefore the current density is just the velocity of the superconducting current times the charge density.

Taking the gradient of the whole equation (8) and expressing $\nabla\theta$ from (7) (akin to [28]) we obtain the modified version of the hydrodynamic interpretation of the quantum condensate dynamics:

$$\frac{d\vec{v}}{dt} = \frac{\partial\vec{v}}{\partial t} + \vec{v}\cdot\nabla\vec{v}$$
$$= \frac{1}{m}\vec{F}_L + \frac{1}{m}\nabla\left[\frac{\hbar^2}{2m}\left(\frac{\Delta\sqrt{\rho}}{\sqrt{\rho}} + \alpha R\right)\right], \tag{9}$$

$$\nabla\times\vec{v} = -\frac{q}{m}\vec{B}, \tag{10}$$

where $\vec{F}_L = q\vec{E} + q\vec{v}\times\vec{B}$ is the Lorentz force acting on the charged Cooper pairs. These two equations are the equations of motion of the superconducting Cooper pair fluid in the presence of an induced curvature (from the embedding space-time), which in this paper is referred to as geometric field. Note that the geometric field enters the gradient of the mystical quantum mechanical potential, recognised by Bohm as a unique interaction with the ψ-field itself [29, 30]. As a result, the geometric field and the ψ-field (in view of Bohmian quantum mechanics) have similar and competing action.

Next we recall the London equations for the quantum superconducting current density [31]

$$\vec{J} = \widehat{\Pi}\vec{A}, \tag{11}$$

where $\widehat{\Pi} = -\rho q/m$ (see eq. (21.20) in [28]). For brevity we will call the introduced quantity $\widehat{\Pi}$, which can be either of scalar or tensorial character, the polarisation operator. A correct microscopic theory of superconductivity can produce an expression for it in terms of the energy gap and critical temperature [32, 33].

Note that an important issue needs to be addressed, namely, to what extent the London equations hold in curved space-time. The above London equation is implicitly contained in the Schrödinger equation within the form of the canonical momentum. However, in the curved space case, the canonical momentum (7) coincides with the flat space case (see [28]); therefore we will not seek any generalisation of (11). An additional reinforcement of this choice comes from the original London brothers' derivation; namely, the supercurrent is being accelerated under the influence of external electromagnetic fields as if made up of free charged particles. Therefore in curved space-time $J^\nu = \partial_\mu\sqrt{-g}F^{\mu\nu}$, where $F^{\mu\nu}$ is the electromagnetic tensor [34]. Reducing the above to the space part and using (2) within the zero-th order in the vicinity of the origin (the same approximation as the one used in the derivation of (6)) we obtain $J^\nu = \partial_\mu(1 + O(x^2))F^{\mu\nu} \approx \partial_\mu F^{\mu\nu}$, which coincides with the flat space case; therefore the second London equation (produced by taking a curl from this one) should also coincide with the flat space one, that is, (11). London equations are grounded in the electrodynamics of the superconductor and more specifically the phenomenological description of its ideal diamagnetism. We do not have any indication that this material property is rendered invalid in curved space-time.

In addition, the London theory can be viewed as a limit (the London limit) of the phenomenological Ginzburg-Landau theory, which in the case of curved space-time is extended with an extra term encoding the interaction with the geometric field, besides the standard extension of the covariant derivatives, to include the Christoffel symbols [35, 36]. The supercurrent operator emerging from this approach coincides with (7) (see eq. (24) from [35, 36]); therefore the above conclusion on the validity of the London theory in curved space-time is preserved. London theory remains valid also in the case of the gravitoelectromagnetic approximation to the Einstein field equations [37, 38].

Next, we divide both sides of (11) by the current density ρ and then differentiate with respect to time

$$\frac{d\vec{v}}{dt} = \frac{d}{dt}\frac{\vec{J}}{\rho} = \frac{d}{dt}\frac{\widehat{\Pi}\vec{A}}{\rho}, \tag{12}$$

only to equate the r.h.s. of (9) with the r.h.s. of (12):

$$\vec{F}_L + \nabla\left[\frac{\hbar^2}{2m}\left(\frac{\Delta\sqrt{\rho}}{\sqrt{\rho}} + \alpha R\right)\right] = \frac{d}{dt}m\frac{\widehat{\Pi}\vec{A}}{\rho}. \tag{13}$$

4. The Geometric Effect

In the case when the superconducting state is robust, we may assume that

$$\frac{d\widehat{\Pi}}{dt} \approx 0,$$

$$\frac{d\rho}{dt} \approx 0, \tag{14}$$

$$\Delta \sqrt{\rho} \approx 0;$$

the current density in the superconductor is approximately constant as well as the polarisation operator (no internal changes in the microscopic mechanism). The ideal diamagnetism of the superconducting state reduces the Lorentz force to its electrostatic part, which is nonvanishing only in the case when a Josephson junction is present (two separated conducting domains at different electrostatic potentials). Finally, the above simplifications yield

$$\nabla \left[\frac{\hbar^2}{2m} \alpha R \right] = -q\vec{E} + m\frac{\widehat{\Pi}}{\rho} \frac{d\vec{A}}{dt}, \tag{15}$$

which is simply an expression for the conservation of energy.

Let us take a line integral of the above along an open path from point A to point B:

$$\frac{\hbar^2}{2m} \alpha \int_A^B \nabla R \cdot d\vec{l} = -\int_A^B q\vec{E} \cdot d\vec{l} + m\frac{\widehat{\Pi}}{\rho} \int_A^B \frac{d\vec{A}}{dt} \\ \cdot d\vec{l}. \tag{16}$$

Next we introduce the geometric field energy

$$W_g\left(\vec{r}\right) = \alpha \frac{\hbar^2}{2m} R\left(\vec{r}\right). \tag{17}$$

Recall that $E_{ind} = -\partial\vec{A}/\partial t$, which means that $q\int_A^B \vec{E}_{ind} \cdot d\vec{l} = \mathscr{E}_{ind}(B) - \mathscr{E}_{ind}(A)$ is the electromotive potential difference between the two points and

$$\frac{d\vec{A}}{dt} = \frac{\partial\vec{A}}{\partial t} + \vec{v} \cdot \nabla\vec{A}$$

$$\frac{dA_i}{dt} = \frac{\partial A_i}{\partial t} + \frac{\partial r_j}{\partial t} \frac{\partial A_i}{\partial r_j}. \tag{18}$$

Finally (16) can be rewritten using (7) as

$$W_g(B) - W_g(A) = q\left[U_{stat}(B) - U_{stat}(A)\right]$$

$$- \frac{m}{q}\frac{\widehat{\Pi}}{\rho}\left[\mathscr{E}_{ind}(B) - \mathscr{E}_{ind}(A)\right] \tag{19}$$

$$+ \frac{m}{q}\frac{\widehat{\Pi}}{\rho}q\int_A^B \left(\vec{v} \cdot \nabla\vec{A}\right) \cdot d\vec{l}.$$

Now, suppose the Cooper pair charge velocity is constant at the two adjacent points; then the last integral quantity

measures the difference in the interaction energy δW_{int} between the Cooper pairs and the vector potential at the two points:

$$q\int_A^B \left(\vec{v} \cdot \nabla\vec{A}\right) \cdot d\vec{l} = q\vec{v} \cdot \vec{A}(B) - q\vec{v} \cdot \vec{A}(A) \tag{20}$$

$$= \delta W_{int}.$$

Introducing the electrostatic energy $W_{stat} = qU_{stat}(\vec{r})$ we can put (16) in its final form

$$W_g(B) - W_{stat}(B) + \frac{m}{q}\frac{\widehat{\Pi}}{\rho}\left[\mathscr{E}_{ind}(B) - W_{int}(B)\right]$$

$$= W_g(A) - W_{stat}(A) \tag{21}$$

$$+ \frac{m}{q}\frac{\widehat{\Pi}}{\rho}\left[\mathscr{E}_{ind}(A) - W_{int}(A)\right].$$

As a result of introducing the scalar polarisation operator from (11) and [39], the following conserved quantity at each material point of the superconductor emerges:

$$W_g\left(\vec{r}\right) - \mathscr{E}\left(\vec{r}\right) + W_{int}\left(\vec{r}\right) = \text{const}. \tag{22}$$

Here $\mathscr{E}(\vec{r}) = W_{stat}(\vec{r}) - \mathscr{E}_{ind}(\vec{r})$ is the electrical energy of the Cooper pairs.

5. Direct and Reversed Effect

In the previous section we have seen that the geometric field is equivalent to an electric field in the superconductor (15) the statement of which is analogous to the law of conservation of energy (25). Therefore, provided the superconducting element is homogeneous, we can expect $W_{int}(\vec{r}) = \text{const}$ and as a result of the perfect conductor aspect of the superconducting state, we can also expect a redistribution of the shift by the geometric field charges inside the superconductor in order to maintain the superconductor at a constant potential.

A completely different behaviour can be expected at the Josephson junction. We will discuss two cases of the junction, one between superconducting sides made from the same superconductor (symmetric) and one between two different superconductors (asymmetric).

Clearly, in the symmetric junction, the interaction energy $W_{int}(\vec{r})$ will be the same on both sides. However the electrostatic potential on the two sides can be different and the voltage drop U can be equated to the geometric potential, that is, the curvature scalar R itself. In effect, the difference in the geometric field between the two sides δR can produce a voltage drop at the junction (direct effect) or the voltage drop across the junction can produce curvature difference (reversed effect):

$$\alpha\frac{\hbar^2}{2m}\delta R = qU. \tag{23}$$

In this case, we may regard the Josephson junction as a curvature-to-voltage converter with the following ratio (m is the free electron mass):

$$1 \, [\text{V}] \approx 6.3 \times 10^{20} \, \left[\text{m}^{-2}\right]. \tag{24}$$

Note, the only difference the asymmetric junction can introduce is the difference in the interaction energy δW_{int} between the supercurrent and the electromagnetic field at the two sides of the junction. Suppose the supercurrent flows in the junction at vanishing potential difference $U \rightarrow 0$; then the interaction energy gradient can produce rippling in the geometric field according to

$$\alpha \frac{\hbar^2}{2m} \delta R \approx -\delta W_{\text{int}}. \tag{25}$$

The interaction energy (20) is a function of the supercurrent drift velocity and one may view the asymmetric effect as produced by a sharp change in momentum, which converts into rippling of the geometric field. We may expect a backreaction on the entire junction as well. The kinetic energy of the bulk material can certainly be included in (25). The backreaction will increase with the increase in the difference between the interaction energy of the supercurrent with the electromagnetic field on both sides.

6. Proposed Experimental Verification

An argument that the Josephson junction can act as a reversible curvature-to-voltage converter was presented in the previous section. In effect, the argument is prone to experimental testing and now we will discuss how and to what extent. Note that the conversion factor (25) points to the impossibility to observe the travelling ripples in space-time, that is, gravitational waves, with a Josephson junction. Along the span of the junction (few angstrom [Å]) the expected difference in the induced scalar curvature is very small $\delta R > 10^{-21} \, [\text{m}^{-2}]$; therefore according to (25) we may not hope for potential difference greater than $10^{-40} \, [\text{V}]$ which is unmeasurable. We are in a position to answer the questions from the introduction. We are unable to sense the geometric field produced by a gravitational wave at the Josephson junction.

However, since we expect that the effect is reversible, we may attempt to create a geometric field at the junction by an electric discharge between its sides. The greater the potential difference that can be created between the sides, the greater the geometric field that could be created. We are unaware of the dynamics of the created geometric field and do not have any governing equations at the present stage of discussion on its propagation. Nevertheless, we can propose two detection methods which in theory can confirm the proposed effect.

The first approach stems from the reversibility argument; see Figure 1(a). Suppose we have two junctions in close proximity. We have no firm reason to choose a particular set-up but choose to discuss the idea of the experiment with the two junctions placed along a line in such a way that the plane of the junctions (the insulating layer) is normal to the

imaginary line connecting them. Both junctions should be magnetically and electrically shielded from each other and the surrounding environment. One of them will serve as an emitter and the other as detector. A high-voltage discharge should be conducted in the emitter and an induced voltage drop should be recorded at the detector. Provided such an electric potential difference is recorded via proper coincidence scheme (an additional detector junction involved), we may confirm that the geometric effect at the Josephson junction is a physical reality.

The second approach involves the ability of the geometric field to impart motion to objects with inertia; see Figure 1(b). In the geometric field a measure of the curvature of space-time is an acceleration field as well or better induces force (in the lab frame) on a free object of inertia via the Newton's second law $\vec{F} = m\vec{a}$, where $\vec{a}$ is the imparted acceleration. We can give a rough estimate of the imparted acceleration in order to come up with an experimental procedure to verify the effect. There are two possibilities to arrive at an estimate. The first approach involves the use of Gauss's law for gravitation $\nabla \cdot \vec{a} = -4\pi G \rho_{\text{matter}}$ and the (t, t) component of Einstein's field equations for a perfect fluid (where $T_{tt} = -\rho_{\text{matter}} c^2$): $G_{tt} = 8\pi G \rho_{\text{matter}}/c^2$. Here c is the velocity of light in vacuum, G is Newton's gravitation constant, and ρ_{matter} is the matter density. Next we make use of an exact result valid for a 3 + 1 decomposition of space-time: $G_{tt} = R/2$, where R is the scalar Ricci curvature of the three-dimensional hypersurface [40]. Combining these relations we end up with $\nabla \cdot \vec{a} = Rc^2/4$, which is equivalent to

$$a = \frac{Rc^2}{4} \delta x, \tag{26}$$

in the case when the imparted acceleration is in one dimension along the span of δx. Interestingly, a similar result is obtained if one takes up the geodesic deviation equation $D^2 x^\mu/d\tau^2 = R^\mu_{ij\nu} T^i T^j x^\nu$, where T^i is the 4-velocity of an object travelling along the geodesic and x^μ is the deviation vector. Note, the Riemann tensor enters the relation directly. The geodesic deviation measures the acceleration with which two neighbouring geodesics deviate from each other in the curved geometry. Suppose T^i is a unit vector in the time direction and $x^\nu = x_0^\nu + a^\mu t^2/2 + O(t^3)$, where x_0^ν is a constant; then the geodesic deviation equation in one dimension is approximately $a \approx Rc^2 \delta x$ upon a substitution of the Riemann tensor component with the Ricci scalar curvature. Most importantly, the two expressions agree in the order of magnitude and the expected acceleration in a geometric field pulse with a magnitude of $R \sim 10^{20} \, [\text{m}^{-2}]$ and a width of the size of the Josephson junction $10^{-10} \, [\text{m}]$ that can be imparted to an object along its path is enormous $a \sim 10^{25} g$. The dislocation such a pulse can cause is proportional to the time of its duration squared. However, we have no estimate of this quantity, but given the large value for the acceleration even a femtosecond pulse can lead to substantial dislocation of the order of meters. We also have no estimate of the spread with distance of this geometric field pulse and believe the assumed

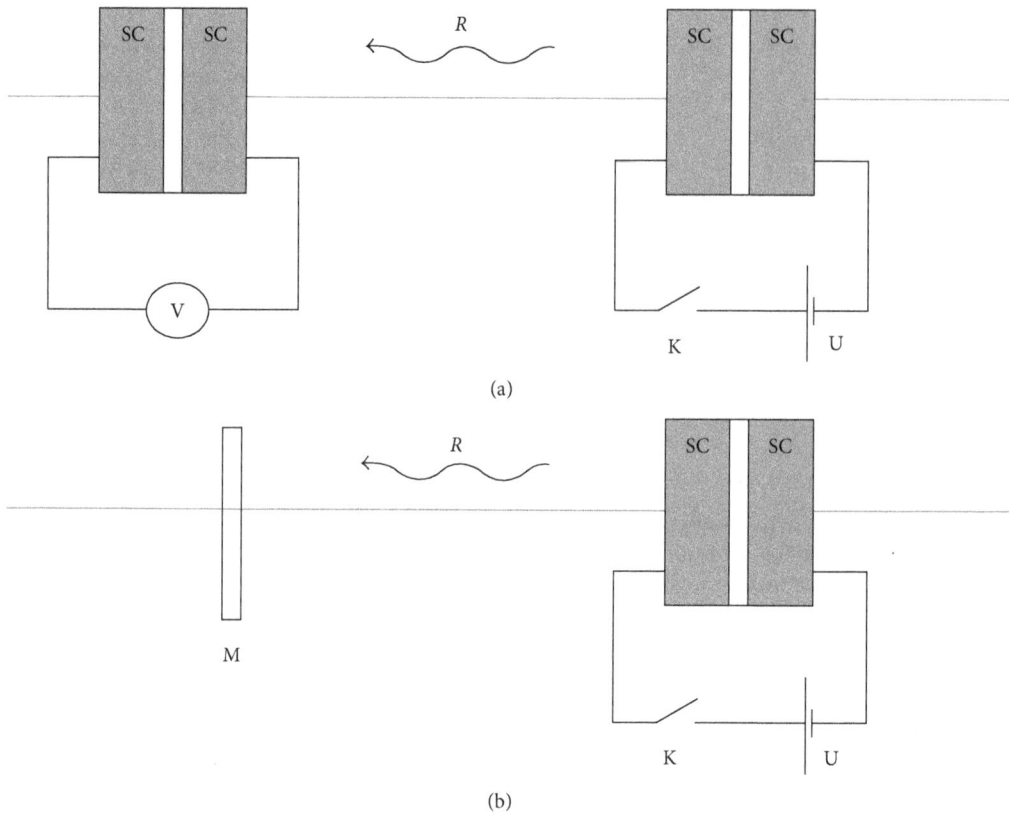

FIGURE 1: Detection schemes. (a) Detection based on the reversibility of the effect. Here a discharge with e.m.f. U is conducted in one of the junctions as the circuit is closed via key K. The emitted geometric field R is detected with the second Josephson junction with the voltage induced between its sides as the geometric pulse impinge on it. (b) Detection based on the dislocation induced by the geometric pulse R on a mirror M part of an interferometer measurement circuit.

value for the scalar curvature and imparted to a material body acceleration to be largely exaggerated.

Now suppose we conduct a high-voltage discharge in a Josephson junction and try to measure the dislocation of a mirror, mechanically shielded from the junction. Such a dislocation will be induced by the emitted geometric field (at the junction). The position of the mirror with respect to the junction is unknown; therefore few geometric set-ups should be tried. Next, in order to increase the sensitivity of the experiment, we suggest the inclusion of the detection mirror as a primary or secondary mirror in a Michelson interferometer. Provided a dislocation in the mirror is recorded, we may confirm the generation of a geometric field in a Josephson junction.

7. Conclusion

In conclusion we would like to point out the origin of the geometric potential from the kinetic term of a constrained to a curved three-dimensional hyperplane of space-time quantum mechanical condensate (superconductor). This potential makes its way into the hydrodynamic interpretation of the Schrödinger equation and enters it on an equal footing with Bohm's "quantum potential." When external electromagnetic field is included in the dynamics and suitable simplifications

are applied, one is able to derive an obvious energy conservation relation at every material point in the superconductor. This conservation relation includes a geometric field part associated with the curvature of the hyperplane. It turns out that at a tunnelling junction (Josephson junction) the energy conservation relation implies the possibility to transform electric energy into geometric energy, that is, create curvature in the hyperplane and vice versa. In effect, it turns out that the Josephson junction can act as a voltage-to-curvature converter. Experimental procedures are discussed in hope the present study invites experimental effort to verify the effect.

Conflicts of Interest

The author declares that there are no conflicts of interest regarding the publication of this paper.

References

[1] B. P. Abbott and LIGO Scientific Collaboration and Virgo Collaboration, "Observation of gravitational waves from a binary black hole merger," *Physical Review Letters*, vol. 116, no. 6, p. 061102, 2016.

[2] B. P. Abbott and LIGO Scientific Collaboration and Virgo Collaboration, "GW170817: Observation of Gravitational Waves

from a Binary Neutron Star Inspiral," *Physical Review Letters*, vol. 119, Article ID 161101, 2017.

[3] H. Ohanian, *Gravitation and Spacetime*, p. 271, 1st edition, 1976.

[4] S. Carroll, "Spacetime and Geometry," p. 144, 2004.

[5] A. Einstein, "Die Grundlage der allgemeinen Relativitätstheorie," *Annalen der Physik (Leipzig)*, vol. 49, p. 769, 1916.

[6] H. Shima, H. Yoshioka, and J. Onoe, "Geometry-driven shift in the Tomonaga-Luttinger exponent of deformed cylinders," *Physical Review B: Condensed Matter and Materials Physics*, vol. 79, no. 20, Article ID 201401(R), 2009.

[7] J. Onoe, T. Ito, H. Shima, H. Yoshioka, and S.-I. Kimura, "Observation of Riemannian geometric effects on electronic states," *EPL (Europhysics Letters)*, vol. 98, no. 2, Article ID 27001, 2012.

[8] B. Leaf, "Momentum operators for curvilinear coordinate systems," *American Journal of Physics*, vol. 47, p. 811, 1979.

[9] Q. H. Liu, J. Zhang, D. K. Lian, L. D. Hu, and Z. Li, "Generalized centripetal force law and quantization of motion constrained on 2D surfaces," *Physica E: Low-dimensional Systems and Nanostructures*, vol. 87, pp. 123–128, 2017.

[10] B. De Witt, "Point transformations in quantum mechanics," *Physical Review*, vol. 85, p. 635, 1952.

[11] P. J. Camp and J. L. Safko, "Quantization conditions in curved spacetime and uncertainty-driven inflation," *International Journal of Theoretical Physics*, vol. 39, no. 6, pp. 1643–1668, 2000.

[12] N. D. Birrell and P. C. W. Davies, *Quantum Fields in Curved Space*, Cambridge University Press, Cambridge, UK, 1982.

[13] S. Hollands and R. M. Wald, "Quantum fields in curved spacetime," *Physics Reports*, vol. 574, pp. 1–35, 2015.

[14] R. M. Wald, *Quantum field theory in Curved Spacetime and Black Hole Thermodynamics*, University of Chicago Press, Chicago, Ill, USA, 1994.

[15] H. Jensen and H. Koppe, "Quantum mechanics with constraints," *Annals of Physics*, vol. 63, no. 2, pp. 586–591, 1971.

[16] L. C. da Silva, C. C. Bastos, and F. G. Ribeiro, "Quantum mechanics of a constrained particle and the problem of prescribed geometry-induced potential," *Annals of Physics*, vol. 379, pp. 13–33, 2017.

[17] R. C. da Costa, "Quantum mechanics of a constrained particle," *Physical Review. A. General Physics. Third Series*, vol. 23, no. 4, pp. 1982–1987, 1981.

[18] F. T. Brandt and J. A. Sánchez-Monroy, "Quantum dynamics of spinless particles on a brane coupled to a bulk gauge field," *Classical and Quantum Gravity*, vol. 34, no. 7, p. 075010, 2017.

[19] N. Ogawa, K. Fujii, and A. Kobushukin, "Quantum mechanics in Riemannian manifold," *Progress of Theoretical and Experimental Physics*, vol. 83, no. 5, pp. 894–905, 1990.

[20] J. Goldstone and R. L. Jaffe, "Bound states in twisting tubes," *Physical Review B: Condensed Matter and Materials Physics*, vol. 45, no. 24, pp. 14100–14107, 1992.

[21] G.-H. Liang, Y.-L. Wang, L. Du, H. Jiang, G.-Z. Kang, and H.-S. Zong, "Coherent electron transport in a helical nanotube," *Physica E: Low-dimensional Systems and Nanostructures*, vol. 83, pp. 246–255, 2016.

[22] P. Duclos, P. Exner, and D. Krejcirik, "Bound states in curved quantum layers," *Communications in Mathematical Physics*, vol. 223, no. 1, pp. 13–28, 2001.

[23] P. C. Schuster and R. L. Jaffe, "Quantum mechanics on manifolds embedded in Euclidean space," *Annals of Physics*, vol. 307, no. 1, pp. 132–143, 2003.

[24] R. Pincak and J. Smotlacha, "The chiral massive fermions in the graphitic wormhole," *Quantum Matter*, vol. 5, p. 107, 2016.

[25] G. G. Naumis, "Electronic and optical properties of strained graphene and other strained 2D materials: a review," *Reports on Progress in Physics*, vol. 80, p. 096501, 2017.

[26] U. Muller, C. Schubert, and A. E. M. van de Ven, "A closed formula for the Riemann normal coordinate expansion," *General Relativity and Gravitation*, vol. 31, no. 11, pp. 1759–1768, 1999.

[27] G. Herglotz, "Über die Bestimmung eines Linienelementes in Normalkoordinaten aus dem Riemannschen Krümmungstensor," *Mathematische Annalen*, vol. 93, p. 46, 1925.

[28] R. P. Feynman, R. B. Leighton, and M. Sands, *The Feynman Lectures on Physics, Volume III*, Addison-Wesley, Reading, Mass, USA, 1963.

[29] D. Bohm, "A Suggested Interpretation of the Quantum Theory in Terms of "Hidden" Variables. I," *Physical Review A: Atomic, Molecular and Optical Physics*, vol. 85, p. 166, 1952.

[30] D. Bohm, "A Suggested Interpretation of the Quantum Theory in Terms of "Hidden" Variables. II," *Physical Review A: Atomic, Molecular and Optical Physics*, vol. 85, p. 180, 1952.

[31] F. London, "On the problem of the molecular theory of superconductivity," *Physical Review A: Atomic, Molecular and Optical Physics*, vol. 74, no. 5, pp. 562–573, 1948.

[32] J. Bardeen, L. N. Cooper, and J. R. Schrieffer, "Theory of superconductivity," *Physical Review Letters*, vol. 108, pp. 1175–1204, 1957.

[33] R. G. Sharma, "A Review of Theories of Superconductivity," in *Superconductivity*, vol. 214 of *Springer Series in Materials Science*, pp. 109–133, Springer International Publishing, Cham, Switzerland, 2015.

[34] L. D. Landau and E. M. Lifshitz, *The Classical Theory of Fields*, Pergamon, Oxford, UK, 1975.

[35] V. Atanasov, "The geometric field (gravity) as an electrochemical potential in a Ginzburg-Landau theory of superconductivity," *Physica B: Condensed Matter*, vol. 517, pp. 53–58, 2017.

[36] V. Atanasov, "Gravity at a quantum condensate," *Journal of the Physical Society of Japan*, vol. 86, p. 074004, 2017.

[37] D. K. Ross, "The London equations for superconductors in a gravitational field," *Journal of Physics A: Mathematical and General*, vol. 16, no. 6, pp. 1331–1335, 1983.

[38] M. Liu, "Rotating superconductors and the frame-independent London equation," *Physical Review Letters*, vol. 81, p. 3223, 1998.

[39] E. M. Lifshitz and L. P. Pitaevskii, "Statistical Physics, Part 2: Theory of the Condensed State," in *Butterworth-Heinemann*, vol. 9, Part 2, Theory of the Condensed State, 1st edition, 1980.

[40] D. Giulini, Canonical Gravity, and 16th Saalburg Summer School, *Fundamentals and New Methods in Theoretical Physics*, Wolfersdorf, Germany, 2010.

7

Tunable Optical Bistability in One-Dimensional Photonic Crystal with a Nonlinear Defect Coupled by Graphene Sheets

Zhiwei Zheng,[1,2] Leyong Jiang,[2] Jun Guo,[1] Xiaoyu Dai,[1] and Yuanjiang Xiang[1]

[1]*SZU-NUS Collaborative Innovation Center for Optoelectronic Science and Technology,*
 Key Laboratory of Optoelectronic Devices and Systems of Ministry of Education and Guangdong Province,
 College of Optoelectronic Engineering, Shenzhen University, Shenzhen 518060, China
[2]*College of Physics and Information Science, Hunan Normal University, Changsha 410081, China*

Correspondence should be addressed to Jun Guo; guojun@szu.edu.cn and Xiaoyu Dai; xiaoyudai@126.com

Academic Editor: Yan Luo

The optical bistability in one-dimensional photonic crystal (1DPC) with a nonlinear defect is investigated. It is demonstrated that, by introducing graphene layers into the nonlinear defect, the optical bistability in 1DPC can be changed significantly. The hysteresis threshold increases with the number of graphene monolayers and can be lowered or enhanced by tuning the Fermi energy of graphene. On the other hand, the hysteresis width and the nonlinear lateral shift can also be controlled by varying the Femi energy and the number of graphene monolayers. These results may be useful for controlling the optical bistability and nonlinear lateral shift in 1DPCs.

1. Introduction

Optical bistability is a kind of optical phenomenon where one input state can induce two steady transmission states [1]. The input and output intensity in the system can then form a hysteresis loop. One of the simplest examples of bistable systems is a Fabry-Perot cavity filled with a medium which presents saturable absorption or nonlinear dispersion. In recent years, nonlinear photonic crystal formed by introducing Kerr nonlinear material into periodical structure has been proposed to achieve optical bistability [2–5]. Due to the dynamic shifting of the band edge and the strong intensity localized inside the defect mode, the threshold for the onset of optical bistability can be lowered. However, it is hard to control the threshold value in fixed configuration. Hence, the exploration of new optical material with tunable optical properties is important for dynamically tunable optical switches.

Graphene, a single layer of carbon atoms in a hexagonal lattice, has given birth to a new branch of modern optics and new possibilities for manipulating light waves, due to its unique optical and electronic properties [6–9]. Although graphene is atomically thin, it can strongly interact with light over a wide frequency spectrum and has been demonstrated for various photonic applications from photodetectors, ultrafast mode lockers to modulators [10–13]. The linear optical properties in graphene lead to broadband and tunable optical features from IR to visible spectrum [14–16]. The broadband optical property allows graphene to be used as an intrinsically smart optical material for the building block of light controlling system. Recently, the optical bistability of reflection at the interface between graphene and Kerr-type nonlinear substrates was investigated theoretically, and the influence of graphene sheets on the hysteretic response of the nonlinear interface was discussed [17]. It was found that the bistable behavior of the reflected light can be electrically controlled by suitably varying the applied voltage on the graphene. Moreover, the optical bistability in nonlinear photonic crystals exhibits rich nonlinear dynamic behaviors. Hence, the nonlinear photonic crystal coupled with graphene sheets will provide a new scheme to control the hysteresis response of the transmitted light intensity. Moreover, the phase of the transmitted (or reflected) light also exhibits bistable behaviors, thus leading to the hysteresis response of the lateral shift of the transmitted light. We believe that the controllable graphene optical bistable devices could find potential applications in optical all-optical

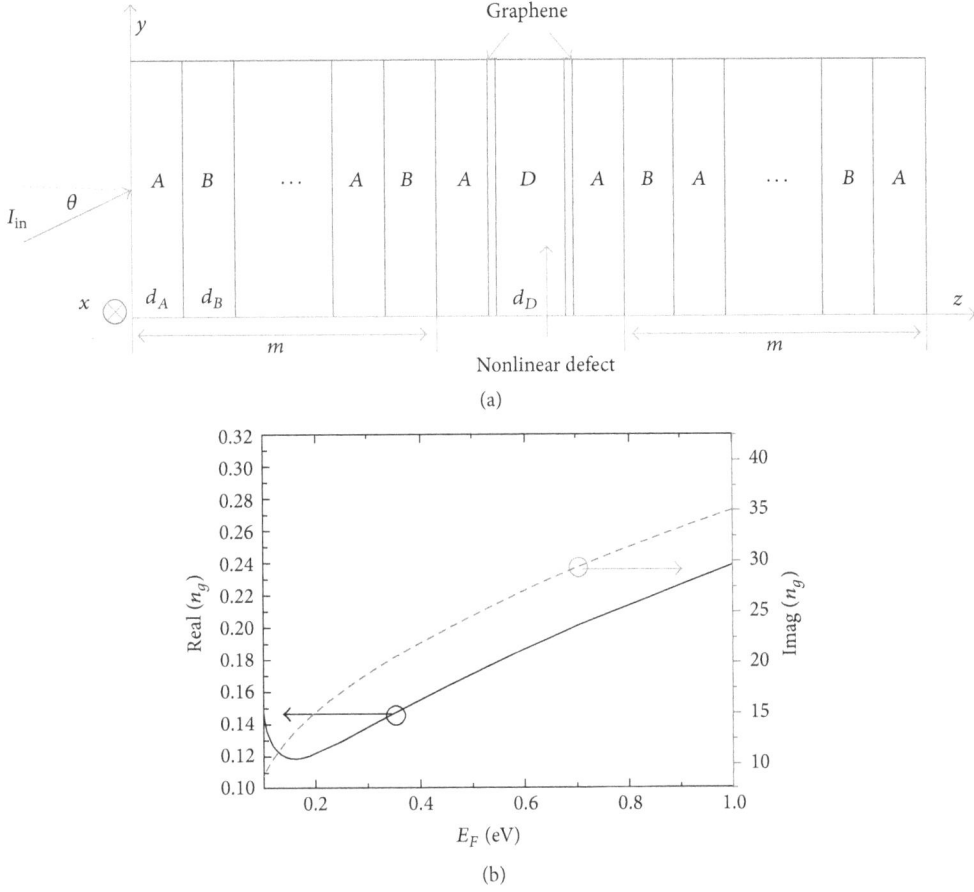

FIGURE 1: (a) The schematic diagram of a 1DPC with a nonlinear defect and graphene layers. m represents the number of periods. (b) Complex effective refractive index of graphene n_g as function of E_F. The solid line and dashed line represent real and imaginary components respectively.

switching [18, 19], optical memory [20], and chemical science [21–23].

2. The Proposed Structure and Simulation Method

This paper is proposed to utilize the tunable features of graphene and to explore the tunable nonlinear transmission features of optical bistability, such as the manipulation of hysteresis threshold, hysteresis width, and nonlinear lateral shift. One-dimensional photonic crystal (1DPC) containing a graphene coupled nonlinear defect is taken as an example.

The structure is shown in Figure 1(a), consisting of two alternate linear layers A and B as 1DPC and a Kerr-type nonlinear layer as defect with effective refraction index $n(I) = n_D + n_2 I$, where n_D is the linear refractive index of the nonlinear defect material, I is the intensity of optical field, and n_2 is the nonlinear refractive. In the following discussion, a normalized unit has been used, which is expressed in units of n_2^{-1}, so that the results will be valid for all Kerr materials with the same n_D and different n_2 [24]. The alternate layers of A, B have high and low linear refractive index n_A, n_B and their thicknesses d_A and d_B satisfy $n_A d_A = n_A d_B = \lambda_{PC}/4$ (the refractive indexes of SiO_2 and TiO_2 are 1.47 and 2.1, resp.). Such a system has a band gap with $2\pi c/\lambda_{PC}$ as the center

frequency for the case of normal incidence. The graphene layers are incorporated into both sides of the nonlinear defect layer as shown in Figure 1(a). Both of their graphene thicknesses are set to be $0.34 \times N$ nm, and N indicates the graphene is multilayered with N monolayer(s). The thickness of a monolayer graphene is chosen to be 0.34 nm [25].

Graphene can be characterized by a complex surface conductivity σ, which is a function of angular frequency $\omega = 2\pi c/\lambda$, Fermi energy E_F, carrier scattering rate Γ, and absolute temperature T of the environment. σ is obtained by intraband and interband $\sigma = \sigma_{intra} + \sigma_{inter}$ terms, which can be expressed according to the Kubo formula [26]:

$$\sigma_{intra} = i \frac{e^2 k_B T}{\pi \hbar^2 (\omega + i\Gamma)} \left[\frac{E_F}{k_B T} + 2 \ln \left(e^{-E_F/k_B T} + 1 \right) \right],$$

$$\sigma_{inter} = i \frac{e^2}{4\pi \hbar} \ln \left[\frac{2E_F - (\omega + i\Gamma)\hbar}{2E_F + (\omega + i\Gamma)\hbar} \right].$$ (1)

In the above formulas, e is the elementary charge, $\hbar = h/2\pi$ is the reduced Planck constant, and k_B is the Boltzmann constant. The Fermi energy of graphene can be manipulated via different approaches, including voltage biasing, exposure to magnetic fields, and chemical doping, which then provide various avenues to control the electronic band property

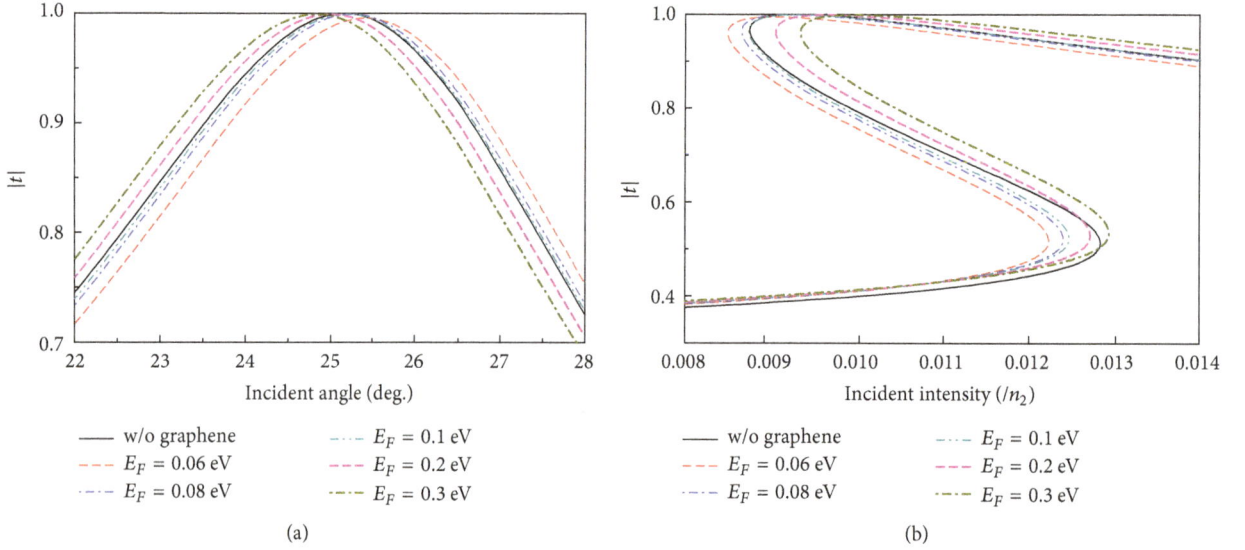

FIGURE 2: (a) $|t|$ versus the angle of incidence for structure without graphene layers (solid curve), with graphene layers of which $E_F = 0.06$ eV (dashed line), $E_F = 0.08$ eV (dash-dotted line), $E_F = 0.1$ eV (dash-dot-dotted line), $E_F = 0.2$ eV (bold dashed line), and $E_F = 0.3$ eV (bold dash-dotted line). (b) $|t|$ versus the incident intensity corresponding to the five situations mentioned in (a).

of graphene [9, 15]. In this research, the graphene carrier scattering rate is assumed to be $\Gamma = 2.4$ THz, the temperature $T = 300$ K, and the incident wavelength $\lambda = 10.6\,\mu m$ (the wavelength of CO_2 lasers). In the simulations, graphene is assumed to be a homogenous medium with small thickness, and then the effective refractive index can be derived by $n_g = \sqrt{1 + i\sigma/(\omega\varepsilon_0 d)}$ [15], where $d = 0.34$ nm is the thickness of monolayer graphene. Note that n_g is independent of N, the layer number of a multilayered graphene. Figure 1(b) shows the complex effective refractive index of graphene n_g as function of E_F at the incident wavelength $\lambda = 10.6\,\mu m$. It can be seen that graphene has complex n_g, indicating that the graphene behaves like a very thin metal layer. The solid line implies the real part of n_g, while the dashed line is the imaginary part of n_g. As shown, both of them increase with E_F, and the imaginary part is larger than the real one.

In this paper, we suppose a TE-polarized wave with wavelength λ incident from vacuum upon a finite 1DPC at angle of 27 degrees. In the following discussion, we consider the symmetric multilayer stack consisting of two alternate linear layers A and B as our 1DPC structure. The middle layer is a Kerr-type nonlinear layer. The nonlinear defect layer is sandwiched between two monolayer graphene. The parameters are set as follows: $n_A = 2.1, n_B = 1.47, n_D = 1.594$, $d_D = 3.5\,\mu m, \theta = 27°, \lambda = 10.6\,\mu m$, and $m = 3$, where d_D and m are the thickness of nonlinear defect layer and the period number of 1DPC, respectively [27]. By applying the transfer matrix method, the characteristic matrix for the nonlinear layer and the composite medium can be calculated. Then the transmission coefficient t can be given by [28]

$$t(k_y)$$
$$= \frac{2p_f(k_y)}{[M_{11} + M_{12}p_f(k_y)]p_f(k_y) + [M_{21} + M_{22}p_f(k_y)]}, \quad (2)$$

where $p_f(k_y) = (k^2 - k_y^2)^{1/2}/k$ and $M_{ij}(k_y)$ are the elements of 2×2 matrix $M(k_y)$.

The phase shift of the transmitted beam with respect to the incident beam is defined as [29]

$$\phi(k_y) = \tan^{-1}\left[\frac{\operatorname{Im} t(k_y)}{\operatorname{Re} t(k_y)}\right], \quad (3)$$

where k_y is the y component of the incident wave vector. Then the lateral shift Δ of the transmitted beam through the multilayered structure is

$$\Delta = \left. \frac{-d\phi(k_y)}{dk_y}\right|_{\theta=\theta_0}. \quad (4)$$

3. Results and Discussion

First, the angular dependence of transmission coefficient with and without graphene is considered. n_g can be manipulated by setting different E_F values, and $N = 5$ is chosen for the calculation. Figure 2(a) shows the angular dependence of transmission coefficient for structures with and without graphene. As shown, although the graphene layers are very thin, their effects on the nonlinear optical response of the entire structure turn out to be significant. The numerical results for the relation between the normalized incident intensity and transmission coefficient for the same cases in Figure 2(a) are also demonstrated in Figure 2(b), in which a typical S-shaped curve indicates that such a system operates in an optical bistable regime. Figure 2(b) shows that when $E_F = 0.06$ eV, the bistability threshold decreases compared to the case without graphene layers. But when $E_F = 0.3$ eV, the bistability threshold increases. To understand the mechanism that accounts for this variation, the structure

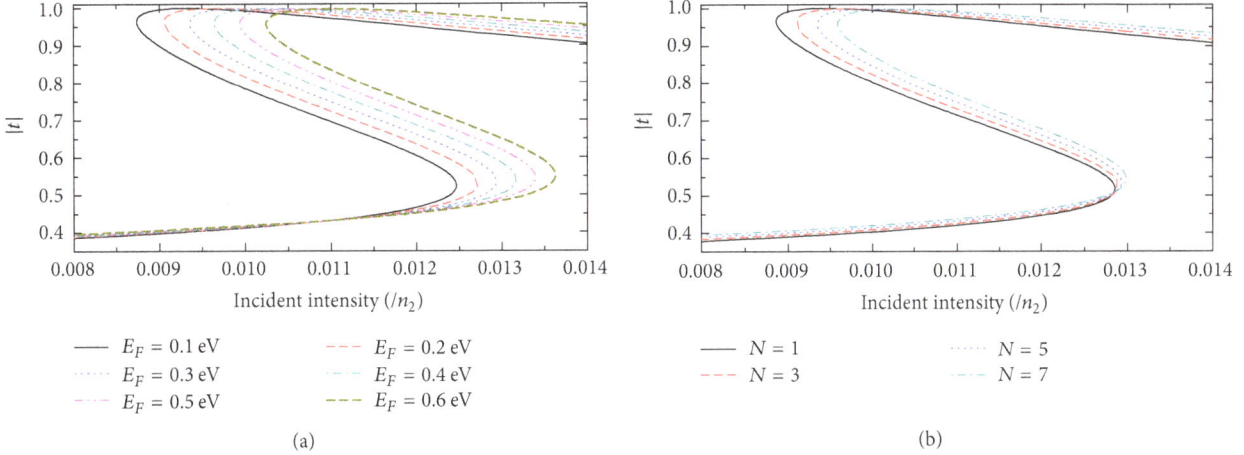

FIGURE 3: $|t|$ versus the incident intensity for graphene with (a) different E_F and $N = 5$ and (b) with different N and $E_F = 0.3$ eV. The incident angle $\theta = 27°$.

in Figure 1(a) can be considered as a resonator filled with nonlinear materials. As shown in Figure 1(b), the effective refractive index of graphene includes the imaginary part and the real part, which indicates that graphene behaves like a thin metal layer. The addition of graphene layers introduces extra positive phase shift and energy loss because of the real part and the imaginary part of n_g. The real part of n_g introduces a positive phase shift, and then the "resonator" length increases. As a result, the intensity dependent nonlinear index change required for switching the system could correspondingly decrease; for example, the intensity required should decrease. The imaginary part of n_g introduces additional energy loss, implying that the input intensity must be higher in order to reach the same "resonator" state. Correspondingly, the intensity required for switching the system should increase. Given that n_g shows both the real part and the imaginary part, each part will impact the nonlinear transmission in an opposite and competitive way, leading to an S shape curve as in Figure 2. As shown in Figure 1(b), both the real part and the imaginary part of n_g increase with E_F, whereas the imaginary part is larger and increases faster than the real part. Therefore, when E_F is smaller, the effect of the real part of n_g is more significant than that of the imaginary part; besides, the intensity required for switching the system decreases compared to the case without graphene. And as E_F becomes larger, the effect of the imaginary part of n_g increases faster than that of the real part. So, with larger E_F, the effect of imaginary part is more significant and the intensity required for switching the system increases. The numerical results shown in Figure 2(b) can prove the above discussion.

Secondly, the effects of graphene layers with different E_F or N on the transmission coefficient are discussed to verify the tunability of graphene layers. Figure 3(a) shows the transmission coefficient dependence on the normalized incident intensity for graphene layers with different E_F. Figure 3(b) is for graphene layers with different N. In Figure 3(a), $N = 5$ while the value of E_F varies. As shown, the hysteresis threshold increases with E_F. It can also be interpreted by the simple resonator analogy mentioned above. When E_F

increases, the effect of the imaginary part of n_g (energy loss) becomes more significant, so that both the switch-up and switch-down thresholds will increase. As the switch-up threshold increases faster than the switch-down threshold, the hysteresis width will increase with E_F. In Figure 3(b), the study sets $E_F = 0.3$ eV and varies N and shows the transmission coefficient versus normalized incident intensity. The results are similar to those in Figure 3(a). However, in this case n_g does not change with different N values. The hysteresis threshold increases as N becomes larger. Although n_g is fixed to be constant, the total thickness of multilayered graphene increases with N, so both the additional positive phase shifts and the energy loss increase. Due to the larger imaginary part of n_g, energy loss would increase faster and the hysteresis threshold will increase subsequently. However, the switch-up and switch-down thresholds will increase at the same speed so the hysteresis width will be kept the same.

Lastly, the effects of graphene layers with different E_F or N values on the lateral shift of the transmitted beam are analyzed. Figure 4(a) shows the lateral shift versus the normalized incident intensity for graphene with different E_F, while Figure 4(b) is for the case with different N. It is clear that the hysteretic effect of lateral shift on the incident intensity occurs. As the incident intensity increases, the lateral shift can be switched to a very large value if the incident intensity is larger than the switch-on threshold intensity, which will then enhance the lateral shift. However, as the incident intensity decreases, the lateral shift can be switched to a very small value when the incident intensity is smaller than the switch-down threshold intensity and hence depress the lateral shift. Moreover, the optical properties of graphene sheets exert an important influence on the hysteretic responses of the lateral shifts. Figure 4(a) sets $N = 5$ and varies the value of E_F. As shown, the variation of Δ with different E_F is similar to the transmission coefficient. Both the hysteresis threshold and the hysteresis width increase with E_F. The maximum value of Δ appears near the switch-down threshold and decreases slightly as E_F increases. Figure 4(b) sets $E_F = 0.3$ eV and varies N. As shown, the hysteresis threshold

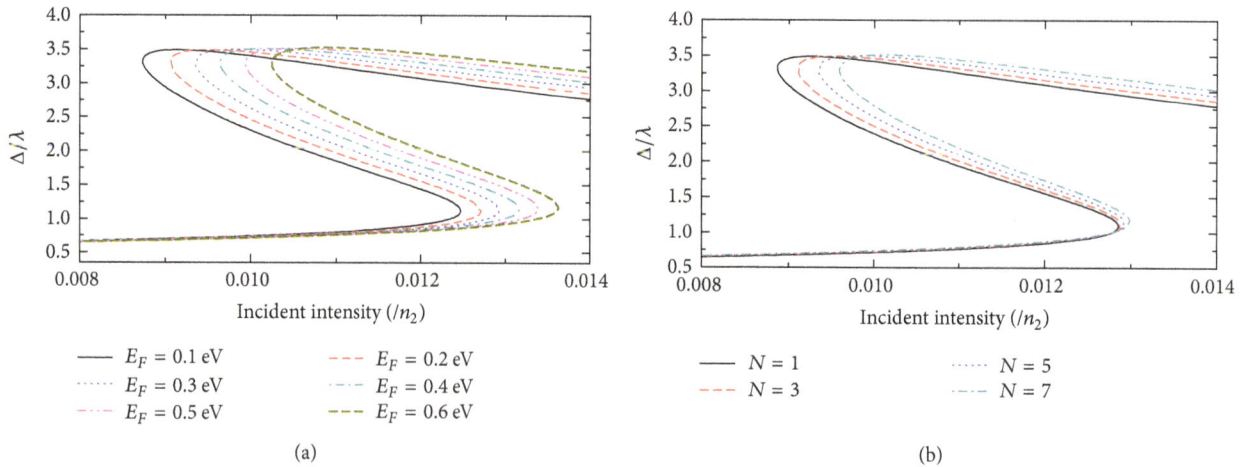

FIGURE 4: The lateral shift versus the normalized incident intensity for graphene with (a) different E_F and $N = 5$ and (b) different N and $E_F = 0.3$ eV. The incident angle $\theta = 27°$.

increases as N becomes larger. Hence, the electrical tunability of optical bistability with graphene could potentially open a new possibility of controlling the lateral shift in a fixed configuration.

4. Conclusions

In summary, this paper mainly explores the effect of multi-layered graphene on a nonlinear 1DPC by attaching graphene layers to both sides of the nonlinear defect. It is found that though the graphene layers are very thin, they can significantly modify the nonlinear transmission response, containing the hysteresis threshold, the hysteresis width, and the nonlinear lateral shift. In addition, the influences of graphene layers with different Femi energy and different number of monolayers are analyzed, and the hysteresis threshold shifts by analog of a resonator filled with nonlinear materials are discussed. The results show that the hysteresis threshold increases with Femi energy and the hysteresis width increases at the same time. Besides, the hysteresis threshold also increases with the number of graphene monolayers. These results may be useful for the control of the optical bistability in 1DPCs.

Conflicts of Interest

The authors declare that there are no conflicts of interest regarding the publication of this paper.

Acknowledgments

This work is partially supported by the National Natural Science Foundation of China (Grant nos. 11647135 and 11704119), Hunan Provincial Natural Science Foundation of China (Grant no. 14JJ6007), Scientific Research Fund of Hunan Provincial Education Department (Grant no. 14B119), and the Project Supported for Excellent Talents in Hunan Normal University (Grant no. ET1502).

References

[1] N. M. Litchinitser, I. R. Gabitov, A. I. Maimistov, and V. M. Shalaev, "Effect of an optical negative index thin film on optical bistability," *Optics Letters*, vol. 32, no. 2, pp. 151–153, 2007.

[2] M. Soljacic and J. D. Joannopoulos, "Enhancement of nonlinear effects using photonic crystals," *Nature Materials*, vol. 3, no. 4, pp. 211–219, 2004.

[3] F. Y. Wang, G. X. Li, H. L. Tam, K. W. Cheah, and S. N. Zhu, "Optical bistability and multistability in one-dimensional periodic metal-dielectric photonic crystal," *Applied Physics Letters*, vol. 92, no. 21, Article ID 211109, 2008.

[4] J. Guo, L. Jiang, Y. Jia, X. Dai, Y. Xiang, and D. Fan, "Low threshold optical bistability in one-dimensional gratings based on graphene plasmonics," *Optics Express*, vol. 25, no. 6, pp. 5972–5981, 2017.

[5] S. Chen, L. Miao, X. Chen et al., "Few-Layer Topological Insulator for All-Optical Signal Processing Using the Nonlinear Kerr Effect," *Advanced Optical Materials*, vol. 3, no. 12, pp. 1769–1778, 2015.

[6] A. K. Geim and K. S. Novoselov, "The rise of graphene," *Nature Materials*, vol. 6, no. 3, pp. 183–191, 2007.

[7] Z. Zheng, C. Zhao, S. Lu et al., "Microwave and optical saturable absorption in graphene," *Optics Express*, vol. 20, no. 21, pp. 23201–23214, 2012.

[8] F. Bonaccorso, Z. Sun, T. Hasan, and A. C. Ferrari, "Graphene photonics and optoelectronics," *Nature Photonics*, vol. 4, no. 9, pp. 611–622, 2010.

[9] A. H. Castro Neto, F. Guinea, N. M. R. Peres, K. S. Novoselov, and A. K. Geim, "The electronic properties of graphene," *Reviews of Modern Physics*, vol. 81, no. 1, pp. 109–162, 2009.

[10] M. Liu, X. Yin, E. Ulin-Avila et al., "A graphene-based broadband optical modulator," *Nature*, vol. 474, no. 7349, pp. 64–67, 2011.

[11] H. Zhang, D. Tang, R. J. Knize, L. Zhao, Q. Bao, and K. P. Loh, "Graphene mode locked, wavelength-tunable, dissipative soliton fiber laser," *Applied Physics Letters*, vol. 96, no. 11, Article ID 111112, 2010.

[12] Y. Chen, G. Jiang, S. Chen et al., "Mechanically exfoliated black phosphorus as a new saturable absorber for both Q-switching

and mode-locking laser operation," *Optics Express*, vol. 23, no. 10, pp. 12823–12833, 2015.

[13] G. Gong, H. Zhang, and M. Yao, "Speckle noise reduction algorithm with total variation regularization in optical coherence tomography," *Optics Express*, vol. 23, no. 19, pp. 24699–24712, 2015.

[14] E. Simsek, "Improving tuning range and sensitivity of localized SPR sensors with graphene," *IEEE Photonics Technology Letters*, vol. 25, no. 9, pp. 867–870, 2013.

[15] W. Zhu, I. D. Rukhlenko, L.-M. Si, and M. Premaratne, "Graphene-enabled tunability of optical fishnet metamaterial," *Applied Physics Letters*, vol. 102, no. 12, Article ID 121911, 2013.

[16] S. H. Mousavi, I. Kholmanov, K. B. Alici et al., "Inductive tuning of fano-resonant metasurfaces using plasmonic response of graphene in the mid-infrared," *Nano Letters*, vol. 13, no. 3, pp. 1111–1117, 2013.

[17] Y. Xiang, X. Dai, J. Guo, S. Wen, and D. Tang, "Tunable optical bistability at the graphene-covered nonlinear interface," *Applied Physics Letters*, vol. 104, no. 5, Article ID 051108, 2014.

[18] D. A. Mazurenko, R. Kerst, J. I. Dijkhuis et al., "Ultrafast optical switching in three-dimensional photonic crystals," *Physical Review Letters*, vol. 91, no. 21, p. 213903/4, 2003.

[19] L. Jiang, J. Guo, L. Wu, X. Dai, and Y. Xiang, "Manipulating the optical bistability at terahertz frequency in the Fabry-Perot cavity with graphene," *Optics Express*, vol. 23, no. 24, pp. 31181–31191, 2015.

[20] H. Nihei and A. Okamoto, "Switching time of optical memory devices composed of photonic crystals with an impurity three-level atom," *Japanese Journal of Applied Physics, Part 1: Regular Papers and Short Notes and Review Papers*, vol. 40, no. 12, pp. 6835–6840, 2001.

[21] J. Liu, L. Yu, Z. Zhao et al., "Potassium-modified molybdenum-containing SBA-15 catalysts for highly efficient production of acetaldehyde and ethylene by the selective oxidation of ethane," *Journal of Catalysis*, vol. 285, no. 1, pp. 134–144, 2012.

[22] Y. Luo, V. K. Guda, E. B. Hassan, P. H. Steele, B. Mitchell, and F. Yu, "Hydrodeoxygenation of oxidized distilled bio-oil for the production of gasoline fuel type," *Energy Conversion and Management*, vol. 112, pp. 319–327, 2016.

[23] Q. Xu, H. Xu, J. Chen, Y. Lv, C. Dong, and T. S. Sreeprasad, "Graphene and graphene oxide: Advanced membranes for gas separation and water purification," *Inorganic Chemistry Frontiers*, vol. 2, no. 5, pp. 417–424, 2015.

[24] P. Hou, Y. Chen, X. Chen, J. Shi, and Q. Wang, "Giant bistable shifts for one-dimensional nonlinear photonic crystals," *Physical Review A - Atomic, Molecular, and Optical Physics*, vol. 75, no. 4, Article ID 045802, 2007.

[25] K. S. Novoselov, A. K. Geim, S. V. Morozov et al., "Electric field in atomically thin carbon films," *Science*, vol. 306, no. 5696, pp. 666–669, 2004.

[26] B. Vasic, M. M. Jakovljević, G. Isić, and R. Gajić, "Tunable metamaterials based on split ring resonators and doped graphene," *Applied Physics Letters*, vol. 103, no. 1, p. 011102, 2013.

[27] W. L. Zhang and S. F. Yu, "Bistable switching using an optical Tamm cavity with a Kerr medium," *Optics Communications*, vol. 283, no. 12, pp. 2622–2626, 2010.

[28] J. A. Porto, L. Martín-Moreno, and F. J. García-Vidal, "Optical bistability in subwavelength slit apertures containing nonlinear media," *Physical Review B - Condensed Matter and Materials Physics*, vol. 70, no. 8, pp. 1–81402, 2004.

[29] L. Jiang, Q. Wang, Y. Xiang, X. Dai, and S. Wen, "Electrically tunable Goos-Hänchen shift of light beam reflected from a graphene-on-dielectric surface," *IEEE Photonics Journal*, vol. 5, no. 3, 2013.

8

Magnetic Phase Separation and Magnetic Moment Alignment in Ordered Alloys $FE_{65}Al_{35-x}M_x$ (M_x = Ga, B; x = 0; 5 at.%)

E. V. Voronina [ORCID],[1] A. K. Arzhnikov,[2] A. I. Chumakov,[3] N. I. Chistyakova,[4] A. G. Ivanova,[1] A. V. Pyataev,[1] and A. V. Korolev[5]

[1]Institute of Physics, Kazan Federal University, Kazan 420008, Russia
[2]Physical-Technical Institute, Ural Branch of Russian Academy of Sciences, Izhevsk 426000, Russia
[3]European Synchrotron Radiation Facility, 38043 Grenoble Cedex 9, France
[4]Faculty of Physics, M. V. Lomonosov Moscow State University, Moscow 119991, Russia
[5]Institute of Metal Physics, Ural Branch of Russian Academy of Sciences, Yekaterinburg 620108, Russia

Correspondence should be addressed to E. V. Voronina; evoronina2005@yandex.ru

Academic Editor: Mohindar S. Seehra

The structure and the magnetic state of ordered $Fe_{65}Al_{35-x}M_x$ (M_x = Ga, B; x = 0; 5 at.%) alloys are investigated using X-ray diffraction, Mössbauer spectroscopy, and magnetic measurements. The behavior of the magnetic characteristics and Mössbauer spectra of the binary alloy $Fe_{65}Al_{35}$ and the ternary alloy with gallium addition $Fe_{65}Al_{30}Ga_5$ is explained in terms of the phase separation into two magnetic phases: a ferromagnetic one and a spin density wave. It is shown that the addition of boron to the initial binary alloy $Fe_{65}Al_{35}$ results in the ferromagnetic behavior of the ternary alloy.

1. Introduction

One of the most intriguing topics of modern solid-state physics is noncollinear disordered [1] and ordered [2–4] structures of magnetic moments. In addition to the need to understand the cause of their existence [2], they are closely related to superconductivity [2–4] and are important for spintronic applications [5]. There are different concepts of the nature of magnetic nanostructures. It is assumed that they may arise as a result of competing exchange interactions, the Fermi surface nesting, and low-lying thermal excitations. Moreover, the very identification of the structures remains a complicated problem. In neutron-diffraction studies of $B2$-ordered Fe-Al alloys, a correlation of the magnetic moments with a coherence length of about 5 nm was revealed [6]. The observed ordering of the magnetic moments was explained in terms of the spin density wave (SDW). It was found that, at a concentration of x = 35 at.% in $Fe_{100-x}Al_x$ alloys, the spin density wave has the largest coherence length. In a study [7] of the magnetotransport properties of quasi-ordered $Fe_{100-x}Al_x$ alloys, x = 30–35 at.%, the anomalies in the behavior of

the transverse magnetoresistance and the Hall constant were explained based on the model of an inhomogeneous magnetic structure. In this context, the quasi-ordered Fe-Al alloys with an Al content from 25 to 40 at.% are of interest as model objects for studying the origin and stabilization of magnetic inhomogeneities in structurally homogeneous magnets, in particular incommensurate long-period spin structures. It is assumed that small additions of a third element, for example, Ga or B, to the $Fe_{65}Al_{35}$ alloy will allow one to follow the changes in the magnetic state of the initial alloy and, thus, to clarify the peculiarities of its magnetic microstructure.

The aim of this work is to study the magnetic state of ternary quasi-ordered $Fe_{65}Al_{35-x}M_x$ (M_x = Ga, B; x = 5 at.%) alloys based on an analysis of the structure, Mössbauer and magnetometric data, and their comparison with the results of analogous studies of the $Fe_{65}Al_{35}$ alloy.

2. Experimental

The binary and ternary quasi-ordered alloys $Fe_{65}Al_{35}$ and $Fe_{65}Al_{35-x}M_x$ (M_x = Ga, B; x = 5 at.%) were obtained by heat treatment of disordered nanocrystalline alloys synthesized by mechanochemistry. The alloys from the original powders (99.98% Fe and 99.99% Al) and a Ga or B (99.98%) additive in the appropriate weight ratio were synthesized in a FRITSCH P-7 planetary ball mill with vials and balls of hardened steel in an Ar atmosphere for 16 hours. Next the scheme and the parameters of heat treatment were selected in such a way as to ensure the structural and chemical homogeneity of the alloys and the required state of structure (a superstructure of the $B2$ or $D0_3$ type). The chemical composition of the alloys was determined by the methods of secondary ion mass spectrometry (SIMS, MC7201) and atomic emission spectrometry (SPECTROFLAME-MODULA D with inductively coupled plasma inductively coupled plasma). The results of these studies showed that the content of additives coincided with that in the initial mixture for mechanosynthesis with an accuracy of 0.5 at.%.

Attestation of the structure was carried out by X-ray diffraction (XRD) at room temperature using a SMARTLAB (Rigaku) diffractometer with Cu K_α-monochromated radiation. The phase composition of the obtained materials was determined by the Rietveld analysis of the diffraction data, using the PDXL 2 software and the ICDD (PDF-2) database. The Mössbauer spectra on the ^{57}Fe nucleus were measured at the Nuclear resonant scattering station ID18 [8] of the European Synchrotron Radiation Facility (ESRF) using a Synchrotron Mössbauer Source [9]. The measurements were carried out at a temperature of 4.2 K, both without an external magnetic field and in an applied external field H_{ext} = 5 T. The direction of the external magnetic field was chosen to be vertical, that is, perpendicular to the direction of the gamma radiation incident on the samples and perpendicular to the direction of the magnetic vector of almost completely (~99%) polarized radiation from the Synchrotron Mössbauer Source [9]. Under these conditions, the measured Mössbauer spectra should contain only the components of hyperfine transitions with a change in the quantum number of the magnetic moment projection, $\Delta m = \pm 1$, which greatly facilitates the interpretation of the results. The Mössbauer spectra were fitted using the software product SpectrRelax [10]. Magnetic measurements were performed at the Magnetometry Center of the IMF UrB RAS on a SQUID magnetometer MPMS-XL5 (Quantum Design) in external magnetic fields up to 5 T at temperatures from 5 to 400 K.

3. Results and Discussion

The X-ray-diffraction analysis data indicate that the alloys investigated are of single phase in structure and are chemically homogeneous. In the diffractograms of all samples of the $Fe_{65}Al_{35}$ and $Fe_{65}Al_{30}Ga_5$ alloys, only reflexes of the $B2$-superstructure are present (Figure 1). No diffuse background indicating the presence of $D0_3$-ordered regions is detected. For the $Fe_{65}Al_{35}B_5$ alloy, superstructural reflections of a $D0_3$-type lattice are revealed. To obtain a single-phase structure of the sample $Fe_{65}Al_{35}B_5$, it was heat-treated at a relatively low temperature of 400°C. Despite the presence of superstructural (110), (200), (311) reflections on the diffractogram of this sample, all the reflections are noticeably broadened. For a

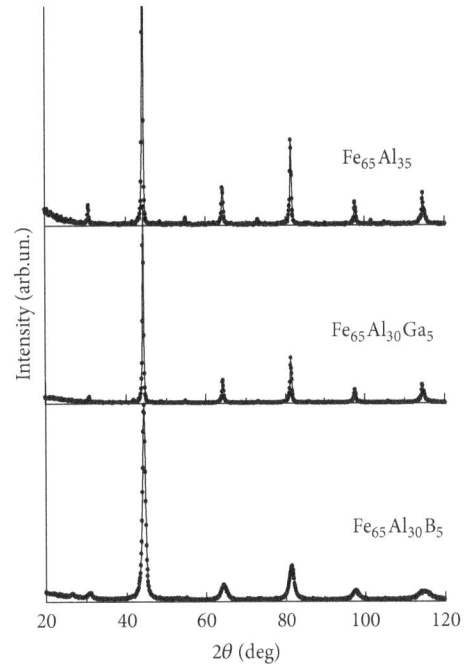

FIGURE 1: X-ray-diffraction patterns of the quasi-ordered $Fe_{65}Al_{35}$ and $Fe_{65}Al_{35-x}M_x$ (M_x = Ga, B; x = 5 at.%) alloys. Cu K_α-radiation, T = 300 K.

sample subjected to prolonged annealing, such broadening is apparently caused by the small size of the regions of X-ray coherent scattering. This, in turn, suggests a higher degree of disordering of this sample in comparison with $Fe_{65}Al_{35}$ and $Fe_{65}Al_{30}Ga_5$. An increase of 0.1% in the bcc lattice parameter of the ternary alloy with a 5 at.% Ga additive is observed as compared to $Fe_{65}Al_{35}$. For the $Fe_{65}Al_{30}B_5$ alloy, the lattice parameter decreases by 0.3% (Table 1).

As seen from Figure 2, for the $Fe_{65}Al_{35}$ and $Fe_{65}Al_{30}Ga_5$ alloys, the magnetization curves do not reach saturation in external magnetic fields up to H_{ext} = 5.0 T. The technical saturation at H_{ext} = 3 T, characteristic for a ferromagnet, is observed only for the $Fe_{65}Al_{30}B_5$ alloy. Hysteresis loops are unbiased and symmetric, and the coercive force values estimated from the hysteresis loop are small (Table 1). The value of the specific saturation magnetization σ_0 was estimated by extrapolating the high-field part of the magnetization curve to H_{ext} = 0 (shown in Figure 2 by dashed lines). The magnetic moment per Fe atom, $\overline{m}_{Fe}$, calculated from σ_0, is also given in Table 1. The results obtained agree with the magnetic measurements at T = 4.2 K in [11], where it was shown that the magnetization does not reach saturation in deformed $Fe_{100-x}Al_x$ alloys with x > 40 at.% in pulsed magnetic fields up to 15.0 T. The values of magnetization in the hysteresis loops in the ZFC and FC cycles of $Fe_{65}Al_{30}Ga_5$ coincide at the relevant temperature (5 K) and field, while the virgin magnetization curve lies well below the hysteresis loop (Figure 2, inset). The susceptibilities of the initial section of the virgin magnetization curve and of the analogous section of the hysteresis loop differ by a factor of two.

TABLE 1: Structural and magnetic parameters of quasi-ordered alloys $Fe_{65}Al_{35}$ and $Fe_{65}Al_{35-x}M_x$ (M_x = Ga, B; x = 5 at.%): the lattice type and parameter; the magnetic ordering temperature T_C; the saturation magnetization σ_0, in brackets—the magnetic moment per Fe atom $\overline{m}_{Fe}$; the coercive force H_C; the average HFF $\langle B_{hf} \rangle$ calculated from $p(B_{hf})$, in brackets—$\langle B_{hf} \rangle$ at H_{ext} = 5 T.

Alloy	Lattice type	Lattice parameter*, nm	Temperature T_C, K	Magnetization σ_0, A·m²/kg ($\overline{m}_{Fe}$, μ_B/Fe)	Coercive force, A/m	$\langle B_{hf} \rangle$, T
$Fe_{65}Al_{35}$	B2	0.2894	340	15 (0.3–0.4)	$5.17 \cdot 10^3$	14.0 (13.1)
$Fe_{65}Al_{30}Ga_5$	B2	0.2897	400	30 (0.7–0.8)	$5.57 \cdot 10^2$	15.5 (13.6)
$Fe_{65}Al_{30}B_5$	D0$_3$	0.5772	620	89 (2.2)	$1.03 \cdot 10^4$	19.8 (16.2)

*The error in determining the lattice parameter by X-ray diffraction: $\pm 1 \cdot 10^{-4}$ nm for $Fe_{65}Al_{35}$ and $Fe_{65}Al_{30}Ga_5$ and $\pm 4 \cdot 10^{-4}$ nm for $Fe_{65}Al_{30}B_5$.

(a)

(b)

(c)

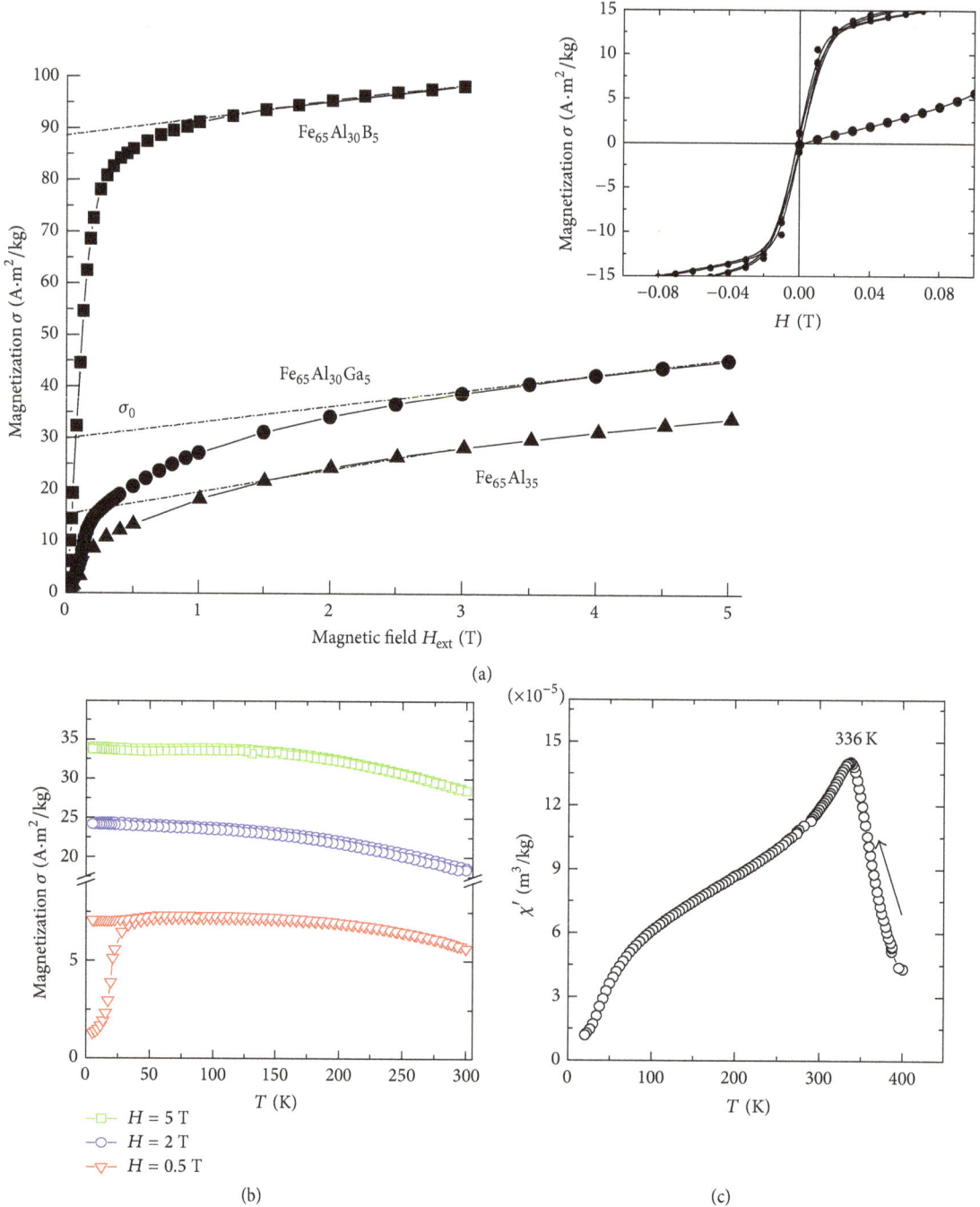

FIGURE 2: (a) Virgin magnetization curves of quasi-ordered alloys $Fe_{65}Al_{35}$ and $Fe_{65}Al_{35-x}M_x$ (M_x = Ga, B; x = 5 at.%). The inset: the virgin magnetization curve and the hysteresis loop for the alloy $Fe_{65}Al_{30}Ga_5$; (b) ZFC and FC cycles for $Fe_{65}Al_{30}Ga_5$ alloy at H_{ext} = 0.5, 2, 5 T; (c) the real part of AC susceptibility of the alloy $Fe_{65}Al_{35}$.

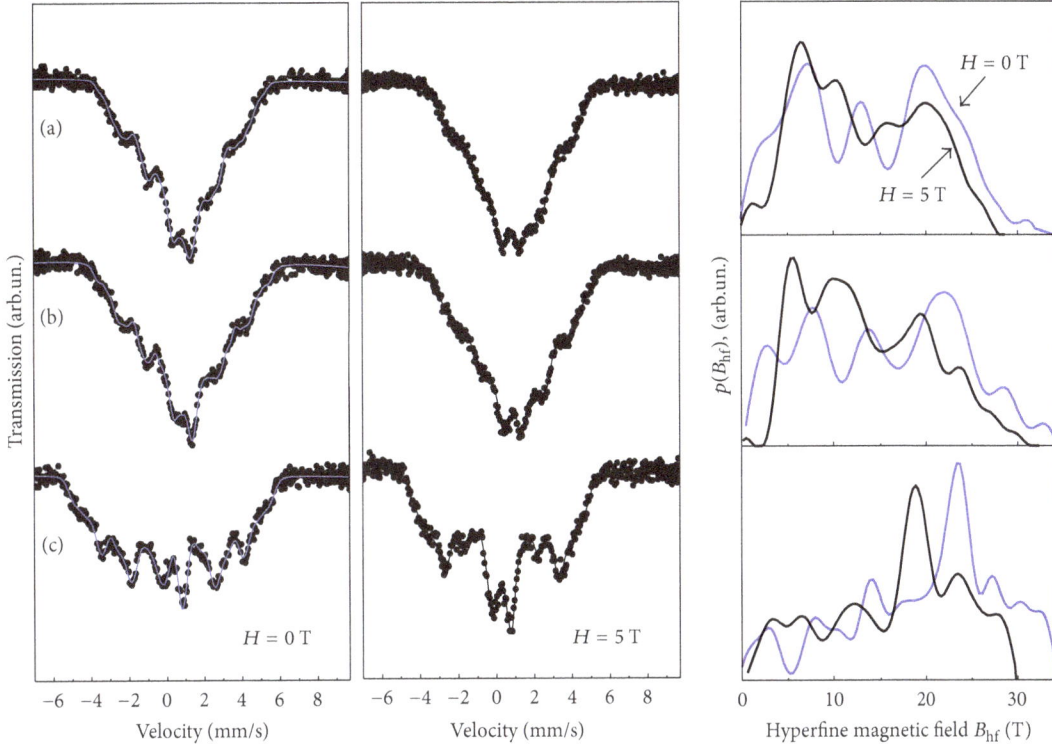

FIGURE 3: Mössbauer spectra and HFF distributions for the alloys: (a) $Fe_{65}Al_{35}$, (b) $Fe_{65}Al_{30}Ga_5$, (c) and $Fe_{65}Al_{30}B_5$. Mössbauer spectra were measured without external field (left column, dotted line) in an applied external magnetic field H_{ext} = 5 T (central column, dotted line). Calculated for HFF distributions $p(B_{hf})$, Mössbauer spectra are represented by solid lines. For HFF distributions $p(B_{hf})$, blue curve in the right column corresponds to Mössbauer spectra measured without external field; black curve in the right column corresponds to the spectra measured in an external magnetic field H_{ext} = 5 T.

The temperature magnetization curves $\sigma(T)$, measured under cooling in a magnetic field and in a zero magnetic field (FC and ZFC), show the presence of hysteresis in small fields (0.005–0.1 T). At higher values of the external field, the hysteresis disappears. A comparison of the $\sigma(H)$ values for the hysteresis loop and the ZFC and FC cycles shows that they are close. Apparently, the observed hysteresis is magnetic rather than thermomagnetic in character. The temperature dependence of the magnetic susceptibility of the investigated alloys has the form characteristic for an antiferromagnet [12] and makes it possible to estimate the temperature of transition to the paramagnetic state: for the $Fe_{65}Al_{30}Ga_5$ alloy, T_C = 400 K, which is higher compared to the binary alloy $Fe_{65}Al_{35}$ ($\approx$340 K). The observed peculiarities of the field and temperature behavior of the magnetic characteristics allow us to state that in the low-temperature range the magnetic state of the alloys investigated cannot be unambiguously assigned to any of the well-known types of magnetic ordering [13].

Figure 3 presents the Mössbauer spectra measured with and without an external field and the corresponding hyperfine magnetic field (HFF) distributions $p(B_{hf})$ (blue and black lines). It is obvious that the addition of gallium to the $Fe_{65}Al_{35}$ alloy leads to an increase in the "integral" average magnetic characteristics: magnetization and the mean HFF. The introduction of a 5 at.% boron additive causes a sharp

increase in the magnetization (by a factor of three) and the mean HFF (by 5.8 T).

According to the magnetic measurements, the magnetic moment for the $Fe_{65}Al_{35}$ alloy is 0.3–0.4 μ_B/Fe atom (Table 1). If this alloy is considered to be a single-phase ferromagnet, then based on the mean HFF value of 14.0 T and using the known coefficient $\langle B_{hf}\rangle/\overline{m}_{Fe}$: 11.0 $\div$ 12.0 T/μ_B [14, 15] for ferromagnetic systems of iron with sp-elements, one can estimate the magnetic moment per Fe atom in the alloy, $\overline{m}_{Fe}$. Thus, the magnetic moment $\overline{m}_{Fe}$ calculated from the Mössbauer measurements is 1.1–1.2 μ_B/Fe for $Fe_{65}Al_{35}$ and 1.3 μ_B/Fe for $Fe_{65}Al_{30}Ga_5$, which is, respectively, 3 and 1.5 times larger than the values obtained from the magnetic measurements.

The resolved hyperfine structure in the Mössbauer spectra of Fe alloys with an Al content above 25 at.% is lacking. In a series of studies, this was considered as the result of an increase in the number of possible local environments of Fe atoms in superstructures of nonstoichiometric composition or a manifestation of the itinerant magnetism. Mössbauer, magnetometric, and neutron diffraction studies [11, 16–19] led to the assumption of the existence of clusters or groups of magnetic moments forming magnetic microinhomogeneities in ordered Fe-Al alloys. However, these revealed magnetic inhomogeneities differ significantly in the size: from one or several unit cells [11] to 2–4 nm [16]. In addition, in a number of papers, the idea of the existence of oppositely oriented magnetic moments in ordered Fe-Al alloys was advanced, which was explained, for example, by antiferromagnetic indirect exchange interaction between the Fe atoms separated by an Al atom [17, 20] or negative exchange interaction due to the RKKY interaction. Theoretical calculations of the electronic structure and local magnetic moments of ordered Fe-Al alloys (29–44 at.% Al) within the density functional theory [21] have shown a possibility for the existence in these alloys of three types of magnetic ordering with a small energy difference between the states: the collinear local antiferromagnetic, ferromagnetic alignment, and the spiral spin density wave. The estimated difference in energy between these states did not exceed 7 mRy/cell. In the ground state, the most energetically preferable are the spin density wave and the local antiferromagnetic state. However, the energy of the ferromagnetic state is so close that, under external influences or with changes in the boundary conditions and lattice parameters, the probability that this state will arise is high. This suggests the possibility of magnetic phase separation in such systems. In addition, the authors of the work noted the dependence of the magnetic moment magnitude and direction on the number of Al atoms in the nearest environment of the Fe atom.

Neutron diffraction studies of the B2-ordered Fe-Al alloys with Al concentrations of 34–43 at.% confirmed the existence of spatial correlations between the magnetic moments with a coherence length of about 5 nm, which were interpreted in terms of spin density wave.

When studying the transverse magnetoresistance and the Hall effect in the ordered alloys $Fe_{100-x}Al_x$, $x = 30$–35 at.%, the authors of [7] revealed anomalies of the magnetotransport properties explained in the context of the spatially inhomogeneous magnetic structure model.

The spectra were fitted in a discrete representation with the number of sextets equal to the number of centers detected in the HFF distribution $p(B_{hf})$, with each center being assigned to the nucleus of the ^{57}Fe atom in a certain local environment. It was also assumed that the parameters of the local atomic environment (the HFF of ^{57}Fe and the configuration probabilities) are close to those of the corresponding local configurations for ordered Fe-Al alloys [22]. However, no satisfactory description of the Mössbauer spectra was obtained within the model of local atomic environments.

It was expected that low-temperature Mössbauer measurements in an external magnetic field would provide additional information for the selection of an adequate model of Mössbauer spectra. It is known that, in the spectra of a polycrystalline ferromagnetic material with $m_{Fe} \uparrow\downarrow B_{hf}$ in a magnetic field whose direction is parallel to that of the magnetic vector of incident radiation, which was realized in this study, two changes are observed: the ratio of the elementary sextet line intensities changes from $3:2:1$ to $3:0:1$, and $p(B_{hf})$ is shifted into the region of smaller values by the amount of the applied magnetic field. For the samples under investigation, a similar behavior (of the ferromagnetic type) occurs only for the $Fe_{65}Al_{30}B_5$ alloy, in the spectrum of which an apparent decrease in the intensity of the second

and fifth lines of the sextet is observed, and the mean HFF decreases by 3.6 T (Figure 4). For the $Fe_{65}Al_{35}$ and $Fe_{65}Al_{30}Ga_5$ alloys, an application of an external magnetic field of 5 T leads to a decrease in the mean value of HFF by 0.9 T and 1.9 T, respectively. The changes observed in $p(B_{hf})$ point to inhomogeneous magnetic structure of the $Fe_{65}Al_{35}$ and $Fe_{65}Al_{30}Ga_5$ alloys. Components are detected, the effective HFF of which decreases when an external field is applied. These components correspond to the Fe atoms, the orientation of the magnetic moments of which coincides with the direction of the applied magnetic field. The values of the effective HFF on the nuclei of such atoms lie in the range from 12.0–15.0 to 33.0 T. A comparison of the HFF distributions (Figure 3) shows that the spectra contain also a component whose behavior is nontypical for the ferromagnetic alignment of magnetic moments. This component is characterized by the HFF values in the range from 0 to 15.0–17.5 T. For the $Fe_{65}Al_{30}B_5$ alloy, a comparison of the Mössbauer spectra and $p(B_{hf})$, measured in an external magnetic field and without it, shows that the magnetic state of this alloy is of the ferromagnetic type.

Based on the above-mentioned experimental and theoretical literature data and the results of measurements carried out in this work, the following model of the magnetic microstructure of the $Fe_{65}Al_{35}$ and $Fe_{65}Al_{30}Ga_5$ alloys is proposed. The spectrum is presented as follows:

(1) There is a contribution from resonant atoms in the ferromagnetic phase, for which the values of the local HFF depend on the local environment and can be represented by the local HFF distribution $p(B_{hf})$.

(2) There is a contribution from resonant atoms for which the values of the local HFF (and, correspondingly, the magnetic moments) vary from site to site of the crystal lattice proportionally to the spin density wave.

There are a lot of examples of estimating the Mössbauer experiment data in the SDW model [23, 24]. The spin density wave is a stationary periodic field of electron spin density in the crystal lattice. It is assumed that the HFF on the nucleus of a resonant atom in the mid position in the crystal lattice is proportional to the SDW amplitude in this position. We assume the SDW to be collinear and antiferromagnetic. In addition, the SDW is assumed to have a symmetry similar to that found earlier, for example, in chromium [25]. The SDW can be described by a series of odd harmonics, and the SDW amplitude (B_{hf}) can be expressed in the form [26]

$$B_{hf}(qx) = \sum_{n=1}^{N} b_{2n-1} \sin[(2n-1)qx], \qquad (1)$$

where the symbol b_{2n-1} denotes the amplitude of the successive harmonics, the variable q is the wave number of the SDW, and the symbol x designates the relative position of the resonance nucleus along the stationary SDW propagation direction. The index N numbers the highest necessary harmonic. By virtue of the SDW periodicity, the argument qx satisfies the following condition: $0 \leq qx \leq 2\pi$. The average amplitude of the SDW, $\langle B_{hf} \rangle$, described by expression (1), is zero. The amplitude of the first harmonic b_1 is positive by definition, since the absolute phase shift between the

TABLE 2: The results of the spectra fitting for the $Fe_{65}Al_{35}$ and $Fe_{65}Al_{30}Ga_5$ alloys without external field and in an applied magnetic field H_{ext} = 5 T: the fraction of the ferromagnetic type phase S_F, the SDW fraction S_{SDW}, the average HFF of the ferromagnetic subsystem $\langle B_{hf}(F)\rangle$, and the harmonic amplitudes b_{2n+1}, describing the SDW by formula (1).

Alloy		S_F, %	S_{SDW}, %	$\langle B_{hf}(F)\rangle$, T	b_1	b_3	b_5	b_7	b_9
$Fe_{65}Al_{35}$	$H = 0\,T$	20–30	80–70	23.0	183.8	−24.0	15.7	0.0	0.0
$Fe_{65}Al_{30}Ga_5$		45–55	55–45	23.3	137.8	−10.1	9.7	0.66	0.0
$Fe_{65}Al_{35}$	$H = 5\,T$	47	53	17.8	138.2	18.3	13.9	10.4	−17.4
$Fe_{65}Al_{30}Ga_5$		60	40	17.6	128.5	11.9	18.8	3.9	−2.6

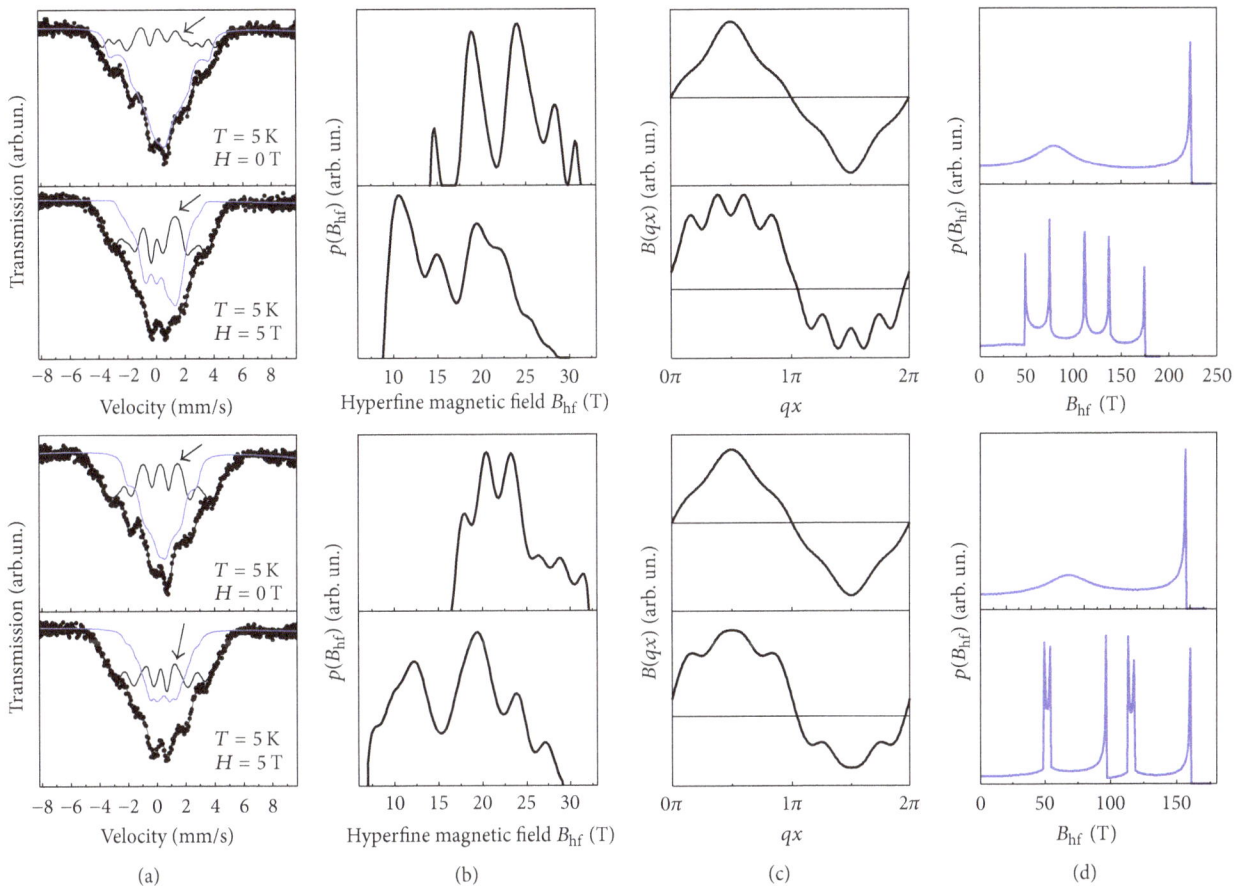

FIGURE 4: (a) Mössbauer spectra measured without external field and in an applied external magnetic field H_{ext} = 5 T for the alloys $Fe_{65}Al_{35}$ and $Fe_{65}Al_{30}Ga_5$. The spectrum components corresponding to the subsystems of the ferromagnetic type (black colour, shown by the arrow) and of the SDW (blue color) are presented, as well as the resulting spectrum envelope; (b) the HFF distributions of the ferromagnetic-type component; (c) the shape of the spin density wave; (d) the HFF distribution resulting from the SDW shape.

SDW and the crystal lattice generally is not observed in the framework of the method used. The amplitudes of the subsequent harmonics in the experimental spectrum processing are variable parameters. The electric quadrupole interaction is considered as a perturbation of the first order of smallness, because it is small compared to the magnetic interaction for all the cases under consideration. The results of MS fitting in the framework of this model are presented in Table 2.

The major contribution to the MS of the $Fe_{65}Al_{35}$ alloy (H_{ext} = 0 T) comes from a long-period magnetic structure of the SDW type, whose fraction is no less than 70%. For the

alloy with gallium addition, this contribution is much smaller but not less than that of the ferromagnetic-type phase. As seen from Figure 4, in these alloys, the SDW has a "triangular" shape, for which the distribution $p(B_{hf})$ is fairly uniform, "flat" [26]. In the spectra measured with H_{ext} = 5 T, the shape of the SDW changes to "rectangular"; higher-order harmonics appear. The distribution $p(B_{hf})$ for the SDW of this form is a Dirac delta function.

In the adopted SDW approximation, we assume a zero contribution to the magnetization from atoms with the SDW-generating magnetic moments. With consideration for the fraction and the average HFF $\langle B_{hf}(F)\rangle$ of the ferromagnetic phase, the magnetic moment $\overline{m}$ is estimated to be $0.42\,\mu_B$/Fe for $Fe_{65}Al_{35}$ and $0.8\,\mu_B$/Fe for $Fe_{65}Al_{30}Ga_5$. It should be noted that in the ferromagnetic phase region one cannot exclude the presence of atoms with magnetic moments, the direction of which is opposite to the total magnetization. The magnitude and direction of these magnetic moments are determined by the characteristics of the local environment of the atom [22]. As the proportion of such atoms is small, the rather rough two-phase approximation makes it possible to reach a satisfactory agreement with the magnetometry data.

When an external magnetic field is applied, the average value of the HFF of the ferromagnetic subsystem is reduced by an amount approximately equal to the applied field. At the same time, due to the decrease of the SDW contribution, the fraction of the ferromagnetic subsystem increases. The observed changes in the contributions to the Mössbauer spectrum make it possible to conclude that when the investigated alloys are exposed to a magnetic field, there occurs a change in the spatial arrangement and the proportion of the two magnetic phases.

Based on the known literature data and the results of magnetometric and Mössbauer investigations in the present work, the magnetic microstructure of the $Fe_{65}Al_{35}$ and $Fe_{65}Al_{30}Ga_5$ alloys in the ground state can be qualitatively represented as a set of regions characterized by two different types of order. In the absence of an external field and at low temperatures, the region with ferromagnetically ordered moments occupies a small part of the material volume. In this region, there may chaotically occur atoms with the magnetic moments, whose direction is opposite to the total magnetization of the region. Most of the crystallite bulk is occupied by regions (or clusters) in which the spin density, the related magnetic moment, and effective HFF on the nuclei of ^{57}Fe atoms vary periodically from site to site. In such regions, covering several cells of the crystal lattice ($10 \div 15$), the arrangement of the collinear magnetic moments is described in the spin density wave representation. It is the presence of these regions that leads to the large paramagnetic contribution to the magnetic susceptibility of the material, observed at low temperatures and small external fields. First, as the external magnetic field increases, the ferromagnetic region gets magnetized. Then the magnetic moments in the SDW region change gradually their orientation to be more favorable with respect to the external magnetic field direction, and SDW region "dissolves" (becomes smaller) in the ferromagnetic region which increases in volume. That is, during the material magnetization, there occurs a change in the spatial arrangement and the proportion of the two magnetic phases in the crystal lattice, which explains the unusual behavior of the virgin magnetization curve relative to the hysteresis loop on magnetization.

In the alloy with the addition of gallium, the fraction of the SDW is noticeably smaller. As possible causes of this experimental evidence, one can mention an increase in the interatomic spacing, composition fluctuations, and/or increasing disordering, which immediately results from a chaotic substitution of Al atoms by Ga atoms in the lattice of the initial alloy. According to the conclusions of the theoretical work [21], even small perturbations of the parameters of the nearest environment in ordered Fe-Al alloys can lead to the realization of the ferromagnetic alignment of the magnetic moments rather than the SDW.

The ferromagnetic behavior of the $Fe_{65}Al_{30}B_5$ alloy, mainly observed in magnetic and Mössbauer experiments, is most likely due to its structural state. The small size of the regions of coherent X-ray scattering, as evidenced by XRD studies, results in a high-degree disorder in the atomic arrangement of the $D0_3$ superstructure, since a large part of the atoms are located directly on or near the crystallite boundaries. In addition, according to our estimate, the size of the coherent-scattering regions for this alloy is of the same order of magnitude as the SDW coherence length in Fe-Al alloys [6]. All this in aggregate may be responsible for the ferromagnetic ordering of the magnetic moments.

4. Conclusions

In this paper, we propose an explanation of the observed behavior of the magnetic characteristics of $Fe_{65}Al_{35-x}M_x$ (M_x = Ga, B; $x = 0$; 5 at.%) alloys within the magnetic phase separation model, which presupposes the existence of at least two magnetic subsystems: ferromagnetic or ferrimagnetic and spin density wave. The accepted model representations make it possible to obtain a satisfactory agreement of the magnetic moment values from Mössbauer and magnetic measurements and to interpret the observed behavior of the alloys in an external magnetic field as a result of changes in the spatial arrangement and the proportions of the magnetic phases. Although the main qualitative changes in the magnetic characteristics of the $Fe_{65}Al_{35}$ and $Fe_{65}Al_{30}Ga_5$ alloys are reproducible in this representation, the chosen model requires further experimental and theoretical study of the features of long-period magnetic structures of the SDW type under the action of temperature and external magnetic field. The $Fe_{65}Al_{30}B_5$ alloy demonstrates the behavior characteristic of a ferromagnet.

Conflicts of Interest

The authors declare that there are no conflicts of interest.

Acknowledgments

The authors are grateful to the ESRF (station ID18) for the opportunity to perform the Mössbauer measurements (Grant HC-1907). This work was funded by the subsidy allocated to Kazan Federal University for the state assignment in the sphere of scientific activities (3.7352.2017/8.9).

References

[1] M. van Schilfgaarde, I. A. Abrikosov, and B. Johansson, "Origin of the Invar effect in iron-nickel alloys," *Nature*, vol. 400, no. 6739, pp. 46–49, 1999.

[2] S. Mühlbauer, B. Binz, F. Jonietz et al., "Skyrmion lattice in a chiral magnet," *Science*, vol. 323, no. 5916, pp. 915–919, 2009.

[3] J. M. Tranquada, B. J. Sternlieb, J. D. Axe, Y. Nakamura, and S. Uchida, "Evidence for stripe correlations of spins and holes in

copper oxide superconductors," *Nature*, vol. 375, no. 6532, pp. 561–563, 1995.

[4] C. de la Cruz, Q. Huang, J. W. Lynn et al., "Magnetic order close to superconductivity in the iron-based layered $LaO_{1-x}F_xFeAs$ systems," *Nature*, vol. 453, no. 7197, pp. 899–902, 2008.

[5] A. Manchon, N. Ryzhanova, A. Vedyayev, and B. Dieny, "Spin-dependent diffraction at ferromagnetic/spin spiral interface," *Journal of Applied Physics*, vol. 103, no. 7, Article ID 07A721, 2008.

[6] D. R. Noakes, A. S. Arrott, M. G. Belk et al., "Incommensurate spin density waves in iron aluminides," *Physical Review Letters*, vol. 91, no. 21, Article ID 217201, pp. 1–4, 2003.

[7] A. E. Elsukova, N. S. Perov, V. N. Prudnikov et al., "Magnetoresistance and the hall effect of the ordered alloys $Fe_{100-x}Al_x$ (25 < x < 35 at. %)," *Physics of the Solid State*, vol. 50, no. 6, pp. 1071–1075, 2008.

[8] R. Rüffer and A. I. Chumakov, "Nuclear resonance beamline at ESRF," *Hyperfine Interactions*, vol. 97, no. 1, pp. 589–604, 1996.

[9] V. Potapkin, A. I. Chumakov, G. V. Smirnov et al., "The ^{57}Fe synchrotron Mössbauer source at the ESRF," *Journal of Synchrotron Radiation*, vol. 19, no. 4, pp. 559–569, 2012.

[10] M. E. Matsnev and V. S. Rusakov, *Mössbauer Spectroscopy in Materials Science-2012: Proceedings of the International Conference MSMS-12*, vol. 1489 of *American Institute of Physics Conference Series*, 2012.

[11] M. J. Besnus, A. Herr, and A. J. P. Meyer, "Magnetization of disordered cold-worked Fe-Al alloys up to 51 at.% Al," *Journal of Physics F: Metal Physics*, vol. 5, no. 11, pp. 2138–2147, 1975.

[12] E. V. Voronina, E. P. Elsukov, S. K. Godovikov, A. V. Korolev, and A. E. Elsukova, "Temperature and field behavior of magnetic properties of ordered alloys $Fe_{100-x}Al_x$ (25 < × < 35 at %)," *The Physics of Metals and Metallography*, vol. 109, no. 5, pp. 417–426, 2010.

[13] C. M. Hurd, "Varieties of magnetic order in solids," *Contemporary Physiscs*, vol. 23, no. 5, pp. 469–493, 1982.

[14] E. P. Elsukov, G. N. Konygin, E. V. Voronina et al., "YAGR-issledovaniya formirovaniya magnitnykh svoystv v neuporyadochennykh sistemakh $Fe_{100-x}M_x$ (M = Al, Si, P)," *Izvestiya Rossiiskoi Akademii Nauk. Seriya Fizicheskaya*, vol. 56, no. 7, pp. 119–123, 1992 (Russian).

[15] E. P. Yelsukov, E. V. Voronina, G. N. Konygin et al., "Structure and magnetic properties of $Fe_{100-x}Sn_x$ (3.2 < x < 62) alloys obtained by mechanical milling," *Journal of Magnetism and Magnetic Materials*, vol. 166, no. 3, pp. 334–348, 1997.

[16] S. Takahashi and Y. Umakoshi, "Superlattice dislocations and magnetic transition in Fe-Al alloys with the B2-type ordered structure," *Journal of Physics: Condensed Matter*, vol. 3, no. 31, pp. 5805–5816, 1991.

[17] J. W. Cable, L. David, and R. Parra, "Neutron study of local environment effects and magnetic clustering in $Fe_{0.7}Al_{0.3}$," *Physical Review B: Condensed Matter and Materials Physics*, vol. 16, no. 3, pp. 1132–1137, 1977.

[18] G. P. Huffman, "Mössbauer study and molecular field theory of the magnetic properties of Fe-Al alloys," *Journal of Applied Physics*, vol. 42, no. 4, pp. 1606-1607, 1971.

[19] I. Vincze and A. T. Aldred, "Mössbauer measurements in iron-base alloys with nontransition elements," *Physical Review B: Condensed Matter and Materials Physics*, vol. 9, no. 9, pp. 3845–3853, 1974.

[20] A. Arrott and H. Sato, "Transitions from ferromagnetism to antiferromagnetism in iron-aluminum alloys. Experimental results," *Physical Review A: Atomic, Molecular and Optical Physics*, vol. 114, no. 6, pp. 1420–1426, 1959.

[21] A. K. Arzhnikov and L. V. Dobysheva, "Magnetic moment of iron atoms in Fe-Al body-centered cubic alloys as a function of the nearest environment," *Physics of the Solid State*, vol. 50, no. 11, pp. 2095–2101, 2008.

[22] E. P. Elsukov, E. V. Voronina, A. V. Korolev, A. E. Elsukova, and S. K. Godovikov, "On the magnetic structure of the ground state of ordered Fe-Al alloys," *The Physics of Metals and Metallography*, vol. 104, no. 1, pp. 35–52, 2007.

[23] A. Olariu, P. Bonville, F. Rullier-Albenque, D. Colson, and A. Forget, "Incommensurate spin density wave versus local magnetic inhomogeneities in $Ba(Fe_{1-x}Ni_x)_2As_2$: a ^{57}Fe Mössbauer spectral study," *New Journal of Physics*, vol. 14, no. 5, Article ID 053044, 2012.

[24] J. Cieślak and S. M. Dubiel, "Influence of charge-density wave parameters on ^{119}Sn Mössbauer spectra: calculations," *Nuclear Instruments and Methods in Physics Research Section B: Beam Interactions with Materials and Atoms*, vol. 101, no. 3, pp. 295–302, 1995.

[25] E. Fawcett, "Spin-density-wave antiferromagnetism in chromium," *Reviews of Modern Physics*, vol. 60, no. 1, pp. 209–283, 1988.

[26] A. Błachowski, K. Ruebenbauer, J. Zukrowski, K. Rogacki, Z. Bukowski, and J. Karpinski, "Shape of spin density wave versus temperature in AFe_2As_2 (A = Ca, Ba, Eu): a Mössbauer study," *Physical Review B: Condensed Matter and Materials Physics*, vol. 83, no. 13, Article ID 134410, 2011.

9

Boson Peak and Superstructural Groups in Na$_2$O-B$_2$O$_3$ Glasses

Armenak A. Osipov [iD] **and Leyla M. Osipova**

Institute of Mineralogy of UB RAS, 456317, Miass, Russia

Correspondence should be addressed to Armenak A. Osipov; armik@mineralogy.ru

Academic Editor: Luis L. Bonilla

Low-frequency Raman spectra of the sodium borate glasses with Na$_2$O content ranging from 0 to 30 mol% were measured and analyzed from the point of view of their structure in the intermediate range order. Our results show that there is a simple linear correlation between correlation length, l_c, and an average size, $<R>$, of the area of ordered arrangement of atoms, obtained via the distribution of the (super)structural units and their representative size. Six types of the areas of ordered arrangement of atoms were determined, being related to the presence of the B$_3$O$_3$Ø$_3$ boroxol rings, B$_3$O$_3$Ø$_4^-$ triborate rings, B$_5$O$_6$Ø$_4^-$ pentaborate groups, B$_8$O$_{10}$Ø$_6^{2-}$ tetraborate groups, B$_4$O$_5$Ø$_4^{2-}$ diborate groups, and B$_3$O$_6^{3-}$ cyclic metaborate anions in the sodium borate glasses. It was shown that the size of the areas of ordered arrangement of atoms can be determined from the simple geometric analysis of the crystallographic data on the superstructural units.

1. Introduction

It is accepted to distinguish two hierarchical structural levels in the structure of borate glasses. These glasses may be considered as an assembly of different boron-oxygen polyhedra at the level of the short range order (SRO) (the first hierarchical level). The polyhedra can be structurally characterized and identified in terms of the bond length and angles. Numerous studies of the local structure of borate glasses and crystals have shown that the boron atoms may be both three-fold and four-fold coordinated by oxygen. It is accepted, in turn, to classify oxygen atoms into two types: bridging oxygen atoms (BO), which form B-O-B bridging bonds belonging to two boron atoms simultaneously, and nonbridging oxygen atoms (NBO or terminal oxygen atoms), which do not form B-O-B bonds and belong only to one boron atom. Thus, five basic structural units are usually distinguished in borates according to the coordination number of the boron atoms and to the number of BOs and NBOs belonging to the given boron atom: BØ$_3$ symmetric triangle (Ø is BO), [BØ$_4$]$^-$ borate tetrahedron, BØ$_2$O$^-$ metaborate triangle, BØO$_2^{2-}$ pyroborate unit, and BO$_3^{3-}$ orthoborate anion. All of these basic structural units can be identified experimentally, e.g., by NMR or vibrational spectroscopies, and to date, the local

structure of borate glasses is well studied both qualitatively and quantitatively. It was established that an addition of alkali oxides to B$_2$O$_3$ leads, first of all, to transformation of BØ$_3$ symmetric triangles into [BØ$_4$]$^-$ tetrahedra and then to the formation of the structural units with NBOs. Thus, the BØ$_3$ symmetric triangles and [BØ$_4$]$^-$ tetrahedra are dominant structural units in structure of all alkali borate glasses with M$_2$O < 25-30 mol% (M = Li, Na, K, Rb and Cs). The concentration of the [BØ$_4$]$^-$ units, N_4, does not depend (or slightly depends) on the type of alkali cation and N_4 as a function of mole fraction of alkali-metal oxide, x, follows $x/(1-x)$ equation in the indicated compositional range [1–3].

In [3–5], however, it was noted that the scale of the short range order does not reflect all the complexity of borate glasses, whose structure is characterized by the presence of the so-called superstructural units (boroxol rings, triborate rings, pentaborate groups, diborate groups, etc.) in the next hierarchical level (intermediate range order (IRO)). The superstructural units consist of well defined arrangements of basic structural units with no internal degrees of freedom in the form of variable bond and/or torsion angles [6–8]. A rigidly defined atomic position in these groups allows interpreting the superstructural units as IRO structures of borate glasses. Investigation of distribution of the superstructural units directly, e.g., with using vibrational spectroscopy or NMR technique, is often problematic due to the ambiguity in identification of these groups [4, 5]. In turn, this leads to the problems of measurement of their concentration in glasses with various compositions. Therefore, experimental data on distribution of superstructural units are typically limited by area of their existence and by general tendencies of variation in their concentrations with change in modifier oxide content [9–11]. This fact does not allow to perform a depth study of the dependence of fraction of superstructural units on a glass composition and type of alkali cation.

Investigation of the low-frequency light scattering (Boson peak) plays important role for study of a glass structure. Today, there is no common opinion on the origin of the Boson peak and this problem is open for discussion. One idea is that the Boson peak is caused from vibrational excitations made up of acoustic phonons (most likely shear or transverse phonons) which scatter strongly from elastic inhomogeneities in the glass structure. The scattering leads to a drastic decrease of the mean free path of vibrations and can increase the vibrational density of states in a certain frequency range. The phonon localization may be viewed in terms of the Ioffe–Regel criteria ($\lambda \sim l$, where λ is phonon wavelength and l is mean free phonon path). These criteria allow to define the correlation length, l_c, through the Boson peak frequency, ω_{BP}, and the sound velocity, v_s [12]:

$$l_c \approx \frac{v_s}{\omega_{BP}} \quad (1)$$

The study of the composition and concentration dependence of the low-frequency light scattering spectra of the borate glasses [13, 14] has shown that both the type of the alkali metal cation and the concentration of modifier oxide strongly affect the spectral properties of the Boson peak (intensity and Boson peak position) (see, e.g., Fig. 18 in ref. [13] or Fig. 7 in ref. [14]) and, hence, the dynamic correlation length l_c. This clearly indicates that IRO structures of various alkali borate glasses are essentially different even if their local structure is qualitatively and quantitatively similar.

Significant progress in understanding of the structure of borate glasses at the level of superstructural units was achieved in the framework of the thermodynamic approach to modeling the properties and structure of the oxide glasses [15]. It was shown that the chemical structure of the lithium and sodium borate glasses and, therefore, the distribution of the superstructural units in these glasses, differ significantly (see, e.g., Fig. 19 in ref. [16] or Fig. 23 and 30 in ref. [8]). At the same time, their local structure is identical at $0 \leq x \leq 25$-30 mol%. Thus, the results of the thermodynamic modeling also testify that the type of alkali cation strongly affects the IRO structures of alkali borate glasses.

The aim of our study is studying of an interrelationship between dynamic correlation length, l_c, and distribution of the superstructural units in sodium borate glasses.

2. Materials and Methods

The glasses with composition $x\text{Na}_2\text{O}$-$(100\text{-}x)\text{B}_2\text{O}_3$ (where x is the molar concentration of Na_2O ranging from 0 to 30 mol%) were prepared by the normal melt quenching method using Na_2CO_3 and H_3BO_3 as sources of sodium oxide and boron trioxide, respectively. The initial reagents were thoroughly mixed in the required ratios and placed in a platinum crucible (Pt). The batch (10 g) in a Pt crucible was placed into electrical furnace, where the temperature was gradually increased up to ~750-1000°C (depending on a glass composition). The fusion was carried out at the given temperatures for 2 hours, whereupon the melting was continued at 1100-1150°C for another one hour. Then, the clear and bubble free liquid was poured into a special steel mould preheated to the temperature close to the glass-transition temperature, T_g, (below 5-10°C than T_g) to obtain the glassy parallelepiped ($7 \times 7 \times 10$ mm, approximately). All glassy samples were annealed at temperature close to T_g for 3 hours. Two samples were prepared for each composition: one of them was used for the measurement of the low-frequency light scattering spectra and the other was used for determination of density (ρ).

The density of each annealed glass sample was measured using Archimedes principle at room temperature using kerosene as a worked liquid. The density was calculated using the formula

$$\rho = \frac{W_A}{W_A - W_B}\rho_B \quad (2)$$

where W_A and W_B are the weights of the sample in air and in kerosene, respectively, and ρ_B is the density of kerosene. The samples were weighted using an electronic digital balance (KERN ALS120-4) of uncertainty ± 0.0001 g.

The low-frequency Raman spectra were measured at a 90° scattering geometry using a double grating DFS-24 monochromator. Solid-state laser (LTI-701) with a wavelength of 532 nm (average power is 500 mW) was used as excitation source of spectra. An uncooled FEU-79 photomultiplier, operating in photon counting regime, was employed to collect the low-frequency Raman spectra. The spectral width of the slit was equal to 2 cm^{-1} in all experiments.

3. Results and Discussion

Nominal glass compositions (in mol%), their designations as well as the velocity of transverse ultrasonic waves, v_t, are presented in Table 1. In addition, the "real" composition of the glasses obtained on the basis of comparison of the density of studied glasses with the data from the literature [17–20] (see Figure 1) is shown in the third column of the Table 1. One can see that the difference between nominal and "real" glass compositions is in the reasonable limits and is less than 0.5 mol%. At the same time, our data are in a good agreement with those published in the literature. The "real" glass composition, therefore, will be used in further discussion.

The examples of the low-frequency Raman spectra for a series of sodium borate glasses are shown in Figure 2. Two main tendencies are observed with variation in modifier oxide content: (i) shift of the Boson peak maximum toward the higher wavenumbers and (ii) decrease in its intensity

Table 1: Samples designation, nominal composition, and "real" composition of xNa_2O-$(100-x)B_2O_3$ (x is given in mol%) glasses as well as the velocity of transverse ultrasonic waves, v_t. The velocities of transverse ultrasonic waves, v_t, were obtained via the method of the piecewise-linear approximation of the corresponding data published in ref. [17].

Sample designation	Nominal composition [mole%]	«Real» composition [mole%]	v_t [ms^{-1}]
NB1	0	0	1900
NB2	2.5	2.5	2080
NB3	5	5	2240
NB4	7.5	7.1	2365
NB5	10	9.5	2478
NB6	12.5	12.5	2595
NB7	15	14.5	2663
NB8	17.5	17.1	2743
NB9	20	20	2842
NB10	22.5	22.2	2928
NB11	25	25	3045
NB12	27.5	27.5	2133
NB13	30	30	3190

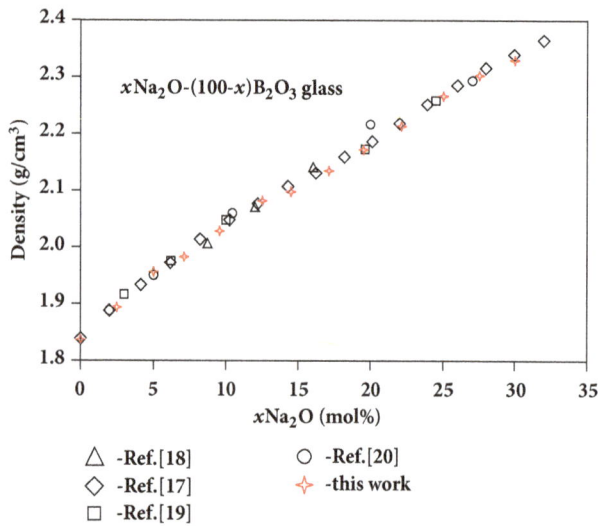

Figure 1: Density of the Na$_2$O-B$_2$O$_3$ glasses as a function of Na$_2$O content.

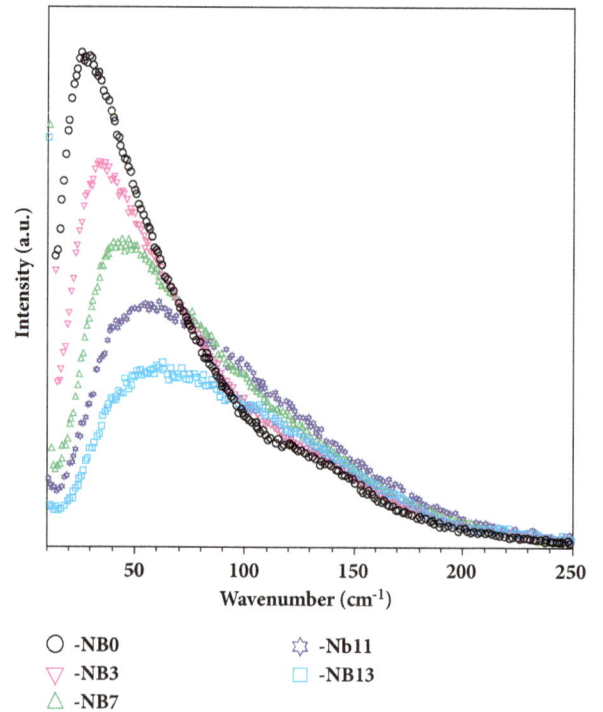

Figure 2: Examples of the low-frequency Raman scattering spectra of Na$_2$O-B$_2$O$_3$ glasses.

with an increase in Na$_2$O content. Similar behavior was found in ref. [21]. Furthermore, a comparison of the Boson peak position showed that our results are in a reasonable accordance (within the experimental errors) with the data published in ref. [21] (see Figure 3). Composition dependence of the dynamic correlation length l_c calculated according to the Eq. (1) is shown in Figure 4. The $l_c(x)$ function was obtained using velocity of transverse ultrasonic waves, v_t, presented in Table 1. As seen from this figure, an increase in sodium oxide content leads to the systematic decrease in the dynamic correlation length l_c from ~ 2.3 nm (g-B$_2$O$_3$) to ~ 1.7-1.8 nm (NB13 glass).

As shown in refs. [13, 21], some characteristic length, which we will denote here as L (L is not the same that l_c), can be determined through the analysis of the low-frequency

Raman scattering spectra. The L is the "short correlation length" according to ref. [13] or it is the intermolecular distance between the centers of interacting groups according to ref. [21]. Both these works show that L systematically changes with the change in the glass composition and give L ~ 7.5-8 Å for g-B$_2$O$_3$. It is reasonable to consider a B$_2$O$_3$ structure in order to understand the meaning of this length.

It is known that the glassy B$_2$O$_3$ consists of BØ$_3$ planar triangles. A major portion of these triangles (up to ~80%

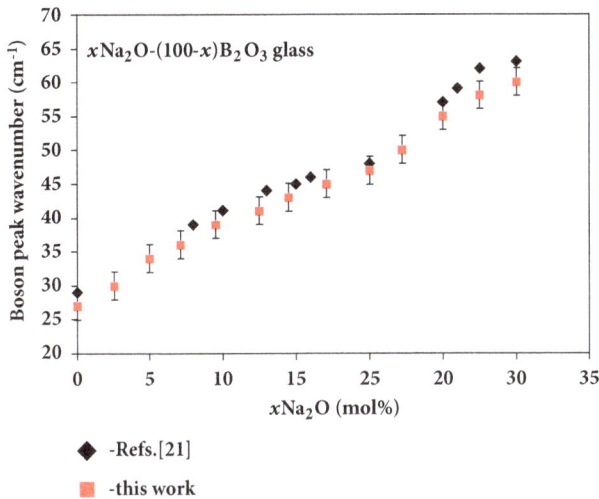

FIGURE 3: Boson peak wavenumber versus glass composition.

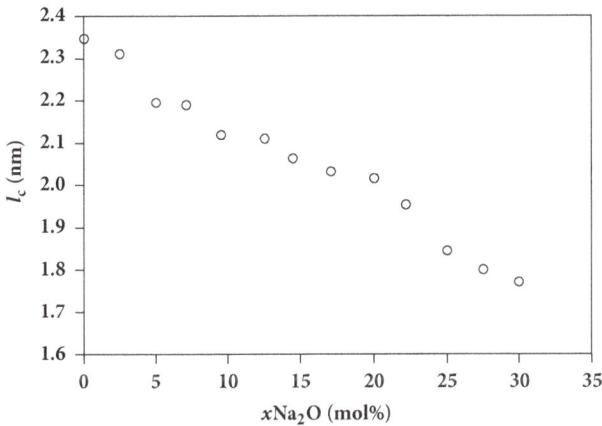

FIGURE 4: Composition dependence of the correlation length l_c.

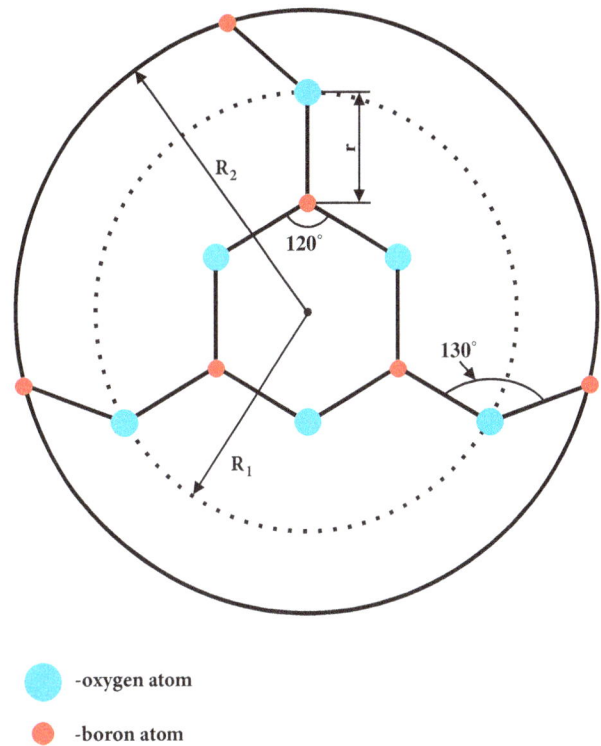

FIGURE 5: Schematic illustration of the $B_3O_3\varnothing_3$ boroxol ring.

according to refs. [7, 22, 23]) is associated into the so-called $B_3O_3\varnothing_3$ boroxol rings, which contain three $B\varnothing_3$ units. The $B_3O_3\varnothing_3$ boroxol ring is a sole type of the superstructural units existing in the g-B_2O_3. Schematic illustration of the $B_3O_3\varnothing_3$ boroxol ring is shown in Figure 5. This figure is useful when considering the correlation between the structure of pure g-B_2O_3 and some characteristic length L. For the fixed length of the B-O bonds ($r = 1.365$ Å [7]) and O-B-O angles (~120°) within the $B\varnothing_3$ units, a diameter of the boroxol ring ($2R_1 = 4r$) will be equal to ~5.5 Å. This value is less than characteristic length, L ~7.5-8 Å [13, 21], for the glassy B_2O_3. Hence it follows that the ordered arrangement of atoms in g-B_2O_3 is not limited by the size of the boroxol ring, but it noticeably in excess of its diameter. It was suggested [21] that the ordered arrangement of atoms in g-B_2O_3 may extend outside the boroxol ring, down to the nearest boron atoms, which are not included into the ring. In the given case, the 7.5 Å corresponds to the $2R_2$ distance (see Figure 5). This distance is obtained if the B-O-B external angle, which is not included into the boroxol ring, is equal to 130°. A good agreement between experimental [13, 21] and theoretical

(obtained via the simple geometric analysis) L values allows us to assume that the B-O-B angle is not changed completely at random. It is evident from Figure 5 that the characteristic length in the glassy B_2O_3 can be interpreted as a distance between the centers of two nearest-neighbor boroxol rings. Such interpretation is in accordance with assumption that the $B_3O_3\varnothing_3$ boroxol rings are not linked to each other completely at random in structure of g-B_2O_3 [13]. On the other hand, the geometric analysis allows interpreting the L length as a distance, characterizing the size of the area of ordered arrangement of atoms in g-B_2O_3. From this point of view, it is obvious that the L distance will change at an increase in concentration of modifier oxide; i.e., L is a function of the glass composition.

As mentioned in "Introduction", addition of the sodium oxide leads to the changes in the SRO and IRO structures in sodium borate glasses. The changes of SRO structures are related to the increase in coordination number of boron atoms from 3 to 4. In turn, the formation of the $[B\varnothing_4]^-$ tetrahedra is reflected in IRO structures: the boroxol rings are transformed into the other superstructural units with $[B\varnothing_4]^-$ tetrahedra. According to the results of the thermodynamic modeling [5, 24], an increase in concentration of Na_2O, first, leads to appearance of the $B_3O_3\varnothing_4^-$ triborate rings and $B_5O_6\varnothing_4^-$ pentaborate groups, simultaneously, and then, the $B_4O_5\varnothing_4^{2-}$ diborate groups and $B_3O_6^{3-}$ cyclic metaborate anions start to appear in the structure of studied glasses. An increase in the fraction of $B_3O_3\varnothing_4^-$ triborate and $B_5O_6\varnothing_4^-$ pentaborate groups is accompanied by a decrease in concentration of the $B_3O_3\varnothing_3$ boroxol rings. In turn, an increase in concentration of $B_4O_5\varnothing_4^{2-}$ diborate and $B_3O_6^{3-}$ metaborate groups is accompanied by a decrease in fraction of pentaborate and triborate groups (see Fig. 5 in ref. [24]).

By the definition given in the "Introduction", each of the above mentioned superstructural units is the area of ordered arrangement of atoms. Thus, for the glass with any given composition, an average size, $<R>$, of the area of ordered arrangement of atoms can be expressed as follows:

$$\langle R \rangle (x) = \sum N_i (x) R_i \qquad (3)$$

Here, N_i and R_i are the concentration and representative size of the area of the ordered arrangement of atoms, respectively, near or within the limits of the superstructural group. Subscript i indicates the type of superstructural units: i = B (boroxol ring), T (triborate ring), P (pentaborate group), D (diborate group), and M (metaborate ring). All necessary data on the concentration of the superstructural units can be readily obtained from refs. [4, 24]. The compositional dependencies of various superstructural units within the range under consideration are shown in Figure 6. One can see that, to calculate the average size $<R>$, four representative sizes (for the $B_3O_3\varnothing_4^-$ triborate ring, $B_5O_6\varnothing_4^-$ pentaborate group, $B_4O_5\varnothing_4^{2-}$ diborate group, and $B_3O_6^{3-}$ cyclic metaborate anion) should be determined in the studied composition range ($0 \le x \le 30$ mol%).

It is evident that the linear size of the ring-type superstructural units, such as triborate and metaborate rings, at the first approximation, can be considered to be equal to the diameter of circle circumscribed around of the boroxol ring, i.e., $R_T = R_M = R_B = 2R_1 \sim 5.5$ Å (see Figure 5). The strict location of atoms in the pentaborate group (Figure 7), as shown in ref. [25], leads to the fact that this group can be unambiguously described via seven independent parameters: four lengths of B-O bonds and three O-B-O angles (see Table 6 in ref. [25]). Based on these data, one can estimate the linear sizes of the pentaborate group. As seen from Figure 7, the pentaborate group has a maximum size along X-axis ($R_X^P = 2R_1^P$), whereas its sizes along Y and Z axes will be equal ($R_Y^P = R_Z^P$) and less than $R_X^P = 2R_1^P$ (for the orientation presented in the figure). Taking into account that the characteristic size of the ring-type superstructural units is the diameter of circle circumscribed around of such groups, it was assumed that, among two specific sizes of the pentaborate group, R_X^P and $R_Y^P = R_Z^P$, the maximum size should be used for calculation of the $<R>(x)$ function according to Eq. (3). Thus, using the data published in ref. [25], we have found that $R_P = R_X^P = 2R_1^P \sim 5.84$ Å.

Diborate group (Figure 8) consists of two $B\varnothing_3$ symmetric triangles and two $[B\varnothing_4]^-$ tetrahedra. The fixed position of atoms in this group is described by nine independent parameters: five lengths of B-O bonds and four O-B-O/B-O-B angles [25, 26]. Figure 8 shows that this group can be described by three linear sizes, R_X^D, R_Y^D and R_Z^D (linear size along Z-axis is not shown), and R_X^D size, obviously, is a maximum size. We have found that the linear size of the diborate group along X-axis, R_X^D, is equal to ~5.7 Å, based on the data presented in Table 7 in ref. [25]. Therefore, the

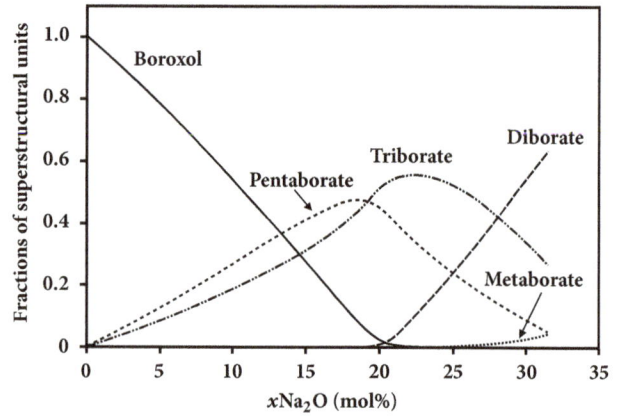

FIGURE 6: Distribution of the superstructural groups in Na_2O-B_2O_3 glasses.

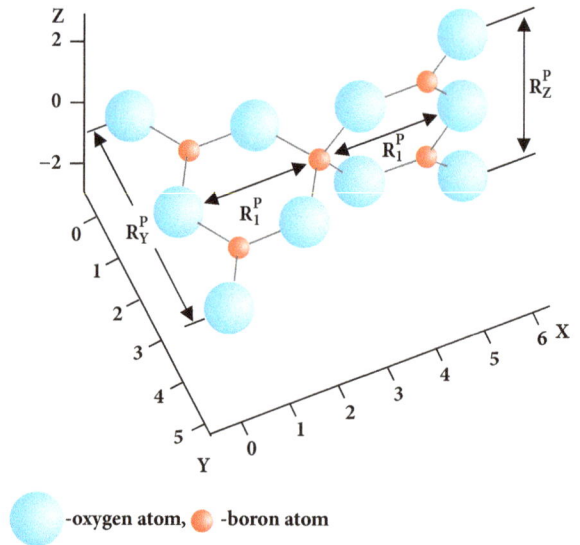

FIGURE 7: Schematic illustration of the $B_5O_6\varnothing_4^-$ pentaborate group.

representative size of the area of ordered arrangement of atoms within the limit of diborate group, R_D, was accepted to be 5.7 Å.

Thus, we have all necessary data (concentrations of superstructural units (Figure 6) and representative sizes of the areas of ordered arrangement of atoms: $R_B = 7.5$ Å, $R_P = 5.84$ Å, $R_D = 5.7$ Å and $R_T = R_M = 5.5$ Å) to calculate the average size $<R>$, according to Eq. (3). The results of calculation are shown in Figure 9 (black dotted line). One can see that the theoretic curve is changed monotonously only in the limited composition ranges ($0 \le x \le 20$ mol%) and it remains practically unchanged with further increasing in Na_2O content. Therefore, a simple linear correlation between $l_c(x)$ and $<R>(x)$ values is observed also in only the limited composition range ($x \le 15$-20 mol%) and strong deviation from the linear behavior is observed at higher concentration of sodium oxide (see Figure 10(a)). This result can be

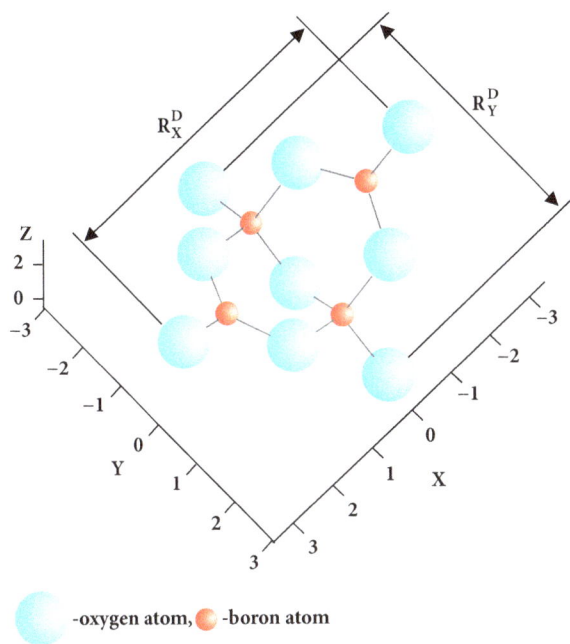

-oxygen atom, -boron atom

FIGURE 8: Schematic illustration of the $B_4O_5\varnothing_4^{2-}$ diborate group.

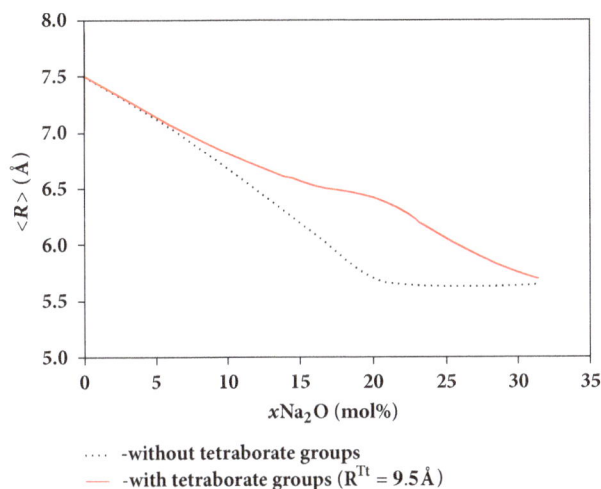

.... -without tetraborate groups

— -with tetraborate groups (R^{Tt} = 9.5Å)

FIGURE 9: Composition dependence of the average size $<R>$ of the area of ordered arrangement of atoms in sodium borate glasses. Black dotted and red solid lines are the results of calculations according to Eq. (3) without and with regard to the tetraborate group, respectively.

explained if we assume that some type of the area of ordered arrangement of atoms was ignored in our calculations.

There are numerous works [9, 11, 14, 27–31] where it was shown that $B_8O_{10}\varnothing_6^{2-}$ tetraborate groups are also present in structure of the sodium borate glasses in addition to those presented in Figure 6. The $B_8O_{10}\varnothing_6^{2-}$ tetraborate group is not a superstructural unit [8, 16], because it consists of $B_5O_6\varnothing_4^{-}$ pentaborate group and $B_3O_3\varnothing_4^{-}$ triborate ring, which are connected to each other by bridging oxygen atom (see Figure 11). For this reason, the tetraborate groups are

absent in the distribution of the superstructural units [24] and, hence, they were ignored initially. Nevertheless, the $B_8O_{10}\varnothing_6^{2-}$ group represents the local area of the ordered arrangement of atoms and, hence, it should be taken into account for calculation of the $<R>(x)$ function according to Eq. (3). Unfortunately, no unambiguous information is known on the concentration of tetraborate groups in structure of the sodium borate glasses. For this reason, at the first approximation, the concentration of the tetraborate group was defined as a product of concentrations of triborate and pentaborate groups. New distribution of the IRO structures obtained under this assumption is shown in Figure 12.

The linear size of the $B_8O_{10}\varnothing_6^{2-}$ tetraborate group may be estimated on the basis of new distribution of IRO structures: R_{Tt} size can be chosen so that the correlation factor, R^2, between l_c, and $<R>$, will be more than any preassigned value ($R^2 \geq 0.98$ in this work). It was found that a good correlation between correlation length, l_c, and average size, $<R>$, is observed at R_{Tt} ~9.5-9.6 Å. It is evident that R_{Tt} value may match to the linear size of the tetraborate group along X-axis for the orientation presented in Figure 11. New $<R>(x)$ dependence is shown in Figure 9 by red solid line.

On the other hand, one can estimate this size based on the data about the structure of crystalline sodium tetraborate [32]. Such estimation provides the maximum linear size of the tetraborate group of 10 Å, approximately. This result is consistent with previously found R_{Tt} of ~9.5-9.6 Å. Thus, the inclusion in consideration of the $B_8O_{10}\varnothing_6^{2-}$ tetraborate group leads to the fact that between $l_c(x)$ and $<R>(x)$ dependencies a simple linear correlation is observed (see Figure 10(b)).

4. Conclusions

The low-frequency light-scattering spectra of a series of sodium borate glasses with sodium oxide content ranging from 0 to 30 mol% were measured. The spectra were analyzed to find a compositional dependence of the correlation length, $l_c(x)$, and its interrelation with the IRO structures in the studied glasses. The correlation length l_c was determined according with Ioffe–Regel criteria and it decreases systematically with the increase in Na2O content. In turn, the average size, $<R>$, of the area of ordered arrangement of atoms was determined via the data of the thermodynamic modeling of the distribution of the superstructural units and the geometric analysis of the superstructural units. Six types of areas of the ordered arrangement of atoms were determined. Each of them was related to the presence of the $B_3O_3\varnothing_3$ boroxol rings, $B_3O_3\varnothing_4^{-}$ triborate rings, $B_5O_6\varnothing_4^{-}$ pentaborate groups, $B_8O_{10}\varnothing_6^{2-}$ tetraborate groups, $B_4O_5\varnothing_4^{2-}$ diborate groups, and $B_3O_6^{3-}$ cyclic metaborate anions. For the ring-type groups, such as boroxol ring, triborate ring, and cyclic metaborate anion, the representative size can be expressed through the diameter of the circle circumscribed around them. If the shape of the large grouping significantly differs from the ring (e.g., pentaborate group, tetraborate group, and diborate group), then the representative size of the area of the ordered arrangement of atoms is

(a)

(b)

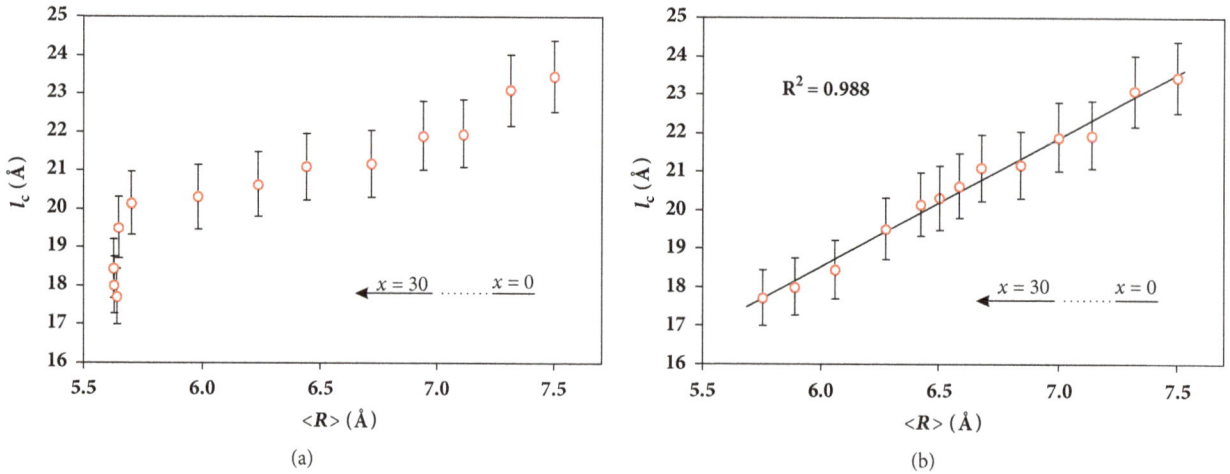

FIGURE 10: Correlations between $l_c(x)$ and $<R>(x)$ functions for the sodium borate glasses without (a) and with regard to (b) the tetraborate group.

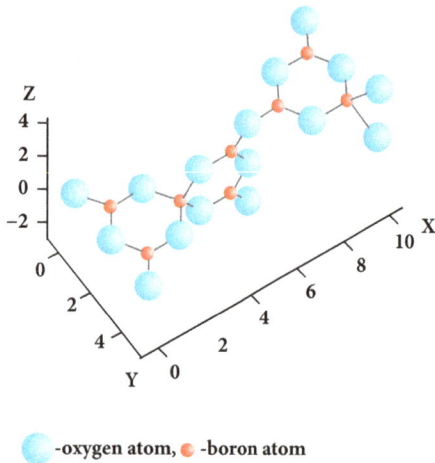

● -oxygen atom, ● -boron atom

FIGURE 11: Schematic illustration of the $B_8O_{10}\varnothing_6^{2-}$ tetraborate group.

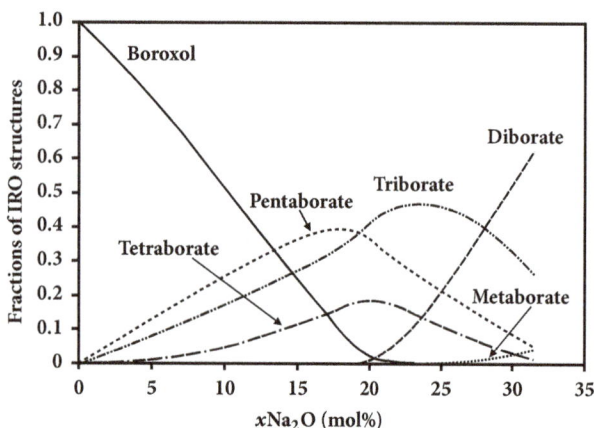

FIGURE 12: Distribution of the IRO structures in Na_2O-B_2O_3 glasses.

determined by the maximum linear size of such groups. In case of the pure g-B_2O_3, the area of the ordered arrangement of atoms (7.5 Å) exceeds the diameter of the boroxol ring that indicates the extension of the ordered arrangement of atoms beyond the ring. The representative size of the area of the ordered arrangement of atoms does not exceed the maximum linear size of the corresponding group in all other cases (5.5, 5.84, 9.4-9.5, 5.7, and 5.5 Å for the triborate ring, pentaborate group, tetraborate group, diborate group, and metaborate ring, resp.). It was found that there is a simple linear correlation between the correlation length, l_c, and the average size, $<R>$, of the area of ordered arrangement of atoms.

Conflicts of Interest

The authors declare that there are no conflicts of interest regarding the publication of this paper.

Acknowledgments

This work was supported by state contract [AAAA-A16-116012510127-9] and state task no. 11.9643.2017/BCh of the Ministry of Education and Science of Russia.

References

[1] V. K. Michaelis, P. M. Aguiar, and S. Kroeker, "Probing alkali coordination environments in alkali borate glasses by multinuclear magnetic resonance," *Journal of Non-Crystalline Solids*, vol. 353, no. 26, pp. 2582–2590, 2007.

[2] W. J. Clarida, J. R. Berryman, M. Affatigato et al., "Dependence of N4 upon alkali modifier in binary borate glasses," *Physics and Chemistry of Glasses*, vol. 44, pp. 215–217, 2003.

[3] S. Kroeker, P. Aguiar, A. Cerquiera et al., "Alkali dependence of tetrahedral boron in alkali borate glasses," *Physics and Chemistry of Glasses: European Journal of Glass Science and Technology Part B*, vol. 47, pp. 393–396, 2006.

[4] N. M. Vedishcheva and A. C. Wright, "Chemical structure of oxide glasses: A concept for establishing structure-property relationships," in *Glass: Selected properties and crystallization*, J. W. P. Schmelzer, Ed., pp. 269–299, De Gruyter, Berlin/Boston, 2014.

[5] A. C. Wright and N. M. Vedishcheva, "Superstructural unit species in vitreous and crystalline alkali, alkaline earth and related borates," *Physics and Chemistry of Glasses: European Journal of Glass Science and Technology Part B*, vol. 54, pp. 147–156, 2013.

[6] A. C. Wright, B. A. Shakhmatkin, and N. M. Vedishcheva, "The chemical structure of oxide glasses: A concept consistent with neutron scattering studies?" *Glass Physics and Chemistry*, vol. 27, no. 2, pp. 97–113, 2001.

[7] A. C. Wright, R. N. Sinclair, D. I. Grimley et al., "Borate glasses, superstructural units and random network theory," *Physics and Chemistry of Glasses*, vol. 22, pp. 268–278, 1996.

[8] A. C. Wright, "Borate structures: crystalline and vitreous," *Physics and Chemistry of Glasses: European Journal of Glass Science and Technology Part B*, vol. 51, pp. 1–39, 2010.

[9] B. N. Meera and J. Ramakrishna, "Raman spectral studies of borate glasses," *Journal of Non-Crystalline Solids*, vol. 159, no. 1-2, pp. 1–21, 1993.

[10] E. I. Kamitsos and M. A. Karakassides, "Structural studies of binary and pseudo binary sodium borate glasses of high sodium content," *Physics and Chemistry of Glasses: European Journal of Glass Science and Technology Part B*, vol. 30, no. 1, pp. 19–26, 1989.

[11] W. L. Konijnendijk and J. M. Stevels, "The structure of borate glasses studied by Raman scattering," *Journal of Non-Crystalline Solids*, vol. 18, no. 3, pp. 307–331, 1975.

[12] J. Schroeder, W. Wu, J. L. Apkarian, M. Lee, L.-G. Hwa, and C. T. Moynihan, "Raman scattering and Boson peaks in glasses: Temperature and pressure effects," *Journal of Non-Crystalline Solids*, vol. 349, no. 1-3, pp. 88–97, 2004.

[13] J. Lorösch, M. Couzi, J. Pelous, R. Vacher, and A. Levasseur, "Brillouin and raman scattering study of borate glasses," *Journal of Non-Crystalline Solids*, vol. 69, no. 1, pp. 1–25, 1984.

[14] B. P. Dwivedi and B. N. Khanna, "Cation dependence of raman scattering in alkali borate glasses," *Journal of Physics and Chemistry of Solids*, vol. 56, no. 1, pp. 39–49, 1995.

[15] B. A. Shakhmatkin and N. M. Vedishcheva, "A thermodynamic approach to the modeling of physical properties of oxide glasses," *Glass Physics and Chemistry*, vol. 24, no. 3, pp. 229–236, 1998.

[16] A. C. Wright, G. Dalba, F. Rocca, and N. M. Vedishcheva, "Borate versus silicate glasses: why are they so different?" *Physics and Chemistry of Glasses: European Journal of Glass Science and Technology Part B*, vol. 51, pp. 233–265, 2010.

[17] M. Kodama and S. Kojima, "Velocity of sound in and elastic properties of alkali metal borate glasses," *Physics and Chemistry of Glasses: European Journal of Glass Science and Technology Part B*, vol. 55, no. 1, pp. 1–12, 2014.

[18] E. Ratai, M. Janssen, and H. Eckert, "Spatial distributions and chemical environments of cations in single- and mixed alkali borate glasses: Evidence from solid state NMR," *Solid State Ionics*, vol. 105, no. 1-4, pp. 25–37, 1998.

[19] V. Mazurin, M. V. Strel'tsina, and T. P. Shvaiko-Shvaikovskaya, *Handbook on the properties of glasses and glass-forming melts*, vol. 4, Nauka, Leningrad, Russia, 1980.

[20] N. S. Abd El-Aal, "Study on ultrasonic velocity and elastic properties of γ-radiated borate glasses," *Bulgarian Journal of Physics*, vol. 28, pp. 275–288, 2001.

[21] A. V. Baranov, T. S. Perova, V. I. Petrov, J. K. Vij, and O. F. Nielsen, "Nature of the boson peak in Raman spectra of sodium borate glass systems: Influence of structural and chemical fluctuations and intermolecular interactions," *Journal of Raman Spectroscopy*, vol. 31, no. 8-9, pp. 819–825, 2000.

[22] G. E. Jr. Jellison, L. W. Panek, P. J. Bray, and G. B. Jr. Rouse, "Determination of structure and bonding in vitreous B_2O_3 by means of ^{10}B and ^{11}B and ^{17}O NMR," *The Journal of Chemical Physics*, vol. 66, pp. 802–812, 1977.

[23] A. C. Hannon, D. I. Grimley, R. A. Hulme, A. C. Wright, and R. N. Sinclair, "Boroxol groups in vitreous boron oxide: new evidence from neutron diffraction and inelastic neutron scattering studies," *Journal of Non-Crystalline Solids*, vol. 177, no. C, pp. 299–316, 1994.

[24] N. M. Vedishcheva, I. G. Polyakova, and A. C. Wright, "Short and intermediate range order in sodium borosilicate glasses: a quantitative thermodynamic approach," *Physics and Chemistry of Glasses: European Journal of Glass Science and Technology Part B*, vol. 55, pp. 225–236, 2014.

[25] A. C. Wright, J. L. Shaw, R. N. Sinclair, N. M. Vedishcheva, B. A. Shakhmatkin, and C. R. Scales, "The use of crystallographic data in interpreting the correlation function for complex glasses," *Journal of Non-Crystalline Solids*, vol. 345-346, pp. 24–33, 2004.

[26] A. C. Wright, R. N. Sinclair, C. E. Stone et al., "A neutron diffraction study of $2M_2O.5B_2O_3$ (M = Li, Na, K, Rb, Cs & Ag) and $2MO.5B_2O_3$ (M = Ca & Ba)," *Physics and Chemistry of Glasses: European Journal of Glass Science and Technology Part B*, vol. 53, pp. 191–204, 2012.

[27] P. J. Bray, S. A. Feller, G. E. Jellison Jr., and Y. H. Yun, "B10 NMR studies of the structure of borate glasses," *Journal of Non-Crystalline Solids*, vol. 38-39, no. 1, pp. 93–98, 1980.

[28] G. E. Jellison Jr. and P. J. Bray, "A structural interpretation of B10 and B11 NMR spectra in sodium borate glasses," *Journal of Non-Crystalline Solids*, vol. 29, no. 2, pp. 187–206, 1978.

[29] J. Krogh-Moe, "Structural interpretation of melting point depression in the sodium borate system," *Physics and Chemistry of Glasses*, vol. 3, pp. 101–110, 1962.

[30] J. Krogh-Moe, "On the structure of boron oxide and alkali borate glasses," *Physics and Chemistry of Glasses*, vol. 1, pp. 26–31, 1960.

[31] J. Krogh-Moe, "Interpretation of the infra-red spectra of boron oxide and alkali borate glasses," *Physics and Chemistry of Glasses: European Journal of Glass Science and Technology Part B*, vol. 6, pp. 46–54, 1965.

[32] A. Hyman, A. Perloff, F. Mauer, and S. Block, "The crystal structure of sodium tetraborate," *Acta Crystallographica*, vol. 22, no. 6, pp. 815–821.

10

Peculiarities of Charge Transfer in SiO_2(Ni)/Si Nanosystems

Egor Yu. Kaniukov [iD],[1] **Dzmitry V. Yakimchuk,**[1] **Victoria D. Bundyukova,**[1]
Alena E. Shumskaya,[1] **Abdulkarim A. Amirov,**[2,3] **and Sergey E. Demyanov**[1]

[1]*Cryogenic Research Division, Scientific-Practical Materials Research Center NAS of Belarus, Minsk 220072, Belarus*
[2]*Center for Functionalized Magnetic Materials (FunMagMa) & Institute of Physics Mathematics and Informational Technologies, Immanuel Kant Baltic Federal University, Kaliningrad 236016, Russia*
[3]*Amirkhanov Institute of Physics Daghestan Scientific Center, Russian Academy of Sciences, Makhachkala 367003, Russia*

Correspondence should be addressed to Egor Yu. Kaniukov; ka.egor@mail.ru

Academic Editor: Yuri Galperin

This work is devoted to study the peculiarities of charge transfer in SiO_2(Ni)/Si nanosystems formed as a result of the electrochemical deposition of nickel into the pores of the ion-track silicon oxide template on silicon. Special attention is given to analysis of the results in the context of the band structure and physical properties of dielectric on semiconductor systems with metallic inclusions in the dielectric matrix. Experimental studies of the current-voltage characteristics of SiO2(Ni)/Si nanostructures demonstrated that value of potential barrier on the Si/metal interface in the pores of the silicon oxide template depended on temperature. On the basis of these results an interpretation of the charge transfer mechanisms in SiO_2(Ni)/Si nanosystems at different temperature ranges was proposed. In the temperature region of ~300–200 K charge carrier motion occurs through the n-Si with an employment of metallic clusters in pores being in a contact with the semiconductor, by means of the overbarrier emission of electrons from higher energy levels of Si conduction band. In the lower temperatures (~200-100 K) a current flow takes place only through the semiconductor due to an increase of resistivity on energy barriers n-Si/metal, which leads to a practically complete exclusion of a participation of the metal in the charge transport process. In the low temperatures (~100–20 K), the variable range hopping conduction between pores on the SiO_2/Si boundary, containing localized states, dominates.

1. Introduction

Nanostructured materials are a special state of condensed matter with properties not usual for materials with mesoscopic or microscopic dimensions. Nanostructures are widely used in biomedicine, chemistry, physics, electronics, and materials science [1–5] and application areas are constantly expanding. It initiates an active search for new approaches to the creation of various types of nanostructures [6–11]. Natural conditions for nanostructures formation are realized using a porous dielectric template on a semiconductor, for the creation of which it is reasonable to use the technology of swift heavy ions tracks [12, 13]. This technology makes it possible to create pores in the silicon oxide layer, the filling of which with appropriate materials, organically adapts resulting heterostructures to the silicon technology

standards [14, 15]. The system complexity, which connected with discreteness of metal particles in contacted with the semiconductor and separated by dielectric interlayers of pores, predetermines the nontriviality of charge transfer processes, the dominant mechanisms of which will be different in a wide range of temperatures and magnetic fields. In such structures, a number of unusual physical effects are noted [16, 17], among which one of the most interesting is the magnetoresistance [17]. Despite intensive studies on nanosized metallic inclusions, a systematic research analyzing those nanoinclusions with respect to electrophysical and galvanometric properties is required. This study is carried out based on the analysis of electrophysical characteristics of systems, containing electrochemically deposited nickel particles in pores of silicon oxide on silicon.

2. Materials and Methods

To create $SiO_2(Ni)/Si$ structures, porous templates of silicon oxide on silicon were used. Si/SiO_2 has the following characteristics: electronic conductivity; doped with phosphorus with donor impurity concentration $N_D = 9 \times 10^{14}$ cm^{-3}, resistance 4.5 $\Omega*cm^{-1}$) obtained using of fast heavy ion tracks technology (irradiation with $^{197}Au^{26+}$ ions with energies of 350 MeV and fluence of 5×10^8 cm^{-2}); through (up to the Si surface) pores with truncated conic shape (base diameters of ~ 300 nm on SiO_2 surface and ~ 200 nm on the boundary with Si) and length corresponding to dielectric layer thickness (~ 350 nm) and average distances between pores of 500 nm. The features of produced $SiO_2(Ni)/Si$ templates are described in detail in our previous work [18].

$SiO_2(Ni)/Si$ nanosystems was made using of boric acid electrolyte (0.5 mol/l H_3BO_3) and nickel sulfate containing solution (0.5 mol/l $NiSO_4$), which was used as cations source. The selected concentration of $NiSO_4$ promoted maximum deposition rate [19, 20]. The electrodeposition potential value was −1 V; it ensured the current efficiency of metal of about 93.3%. The degree of pores filling with metal was controlled by changing process time. The error in potentials measured during deposition did not exceed 1 mV, and the current was not more than 25 nA.

The morphological features of $SiO_2(Ni)/Si$ samples were measured by scanning electron microscopy (SEM, LEO-1455VP, Carl Zeiss (Germany). Elemental analysis was provided by energy dispersive analysis system (EDS, included to LEO-1455VP). The study SiO_2 surface morphology and distribution of outgrown/unreached metal clusters on it was carried out by atomic force microscopy (AFM, NT-206, Microtestmachinery, Belarus) using silicon nitride (Si_3N_4) cantilever with rounding radius of 10 nm. The degree of metallic phase localization in dielectric pores and its outgrowth on SiO_2 surface was determined by X-ray spectral microanalysis (SiLi detector, Röntec).

SEM studies (Figures 1(a), 1(b), and 1(c)) and mapping with X-ray spectral microanalysis (Figures 1(d) and 1(e)) of surface (Figures 1(a), 1(b), and 1(d)) and cross sections (Figures 1(c) and 1(e)) of $Si/SiO_2(Ni)$ samples showed that nickel precipitates uniformly and completely fills the pores, without formation of any outgrown or unreached metal clusters. AFM-scanning of surface showed that change in surface relief does not exceed 40 nm. Carrying out of samples elemental composition analysis by mapping surface and cross sections during X-ray spectral microanalysis, it was established that nickel was localized exclusively in Si/SiO_2 template pores, without nucleation on SiO_2 surface. Our previous studies [21, 22] showed that formed precipitates consist of metal nanoclusters with size of ~ 30-50 nm with stochastic crystallographic orientation.

To study the current-voltage characteristics ($I-V$ curves) of $SiO_2(Ni)/Si$ nanosystems, samples with size of 10×3 mm^2 were prepared. On their surface from the side of $SiO_2(Metal)$, contact area of ~1 mm^2, to which copper electrodes were fed (Figure 2), was applied by ultrasonic soldering with indium. Such contacting method provided mechanically

stable ohmic contact to the sample, which was then verified by thermal cycling over a wide range of temperatures. The $I-V$ curves were investigated in the temperature range of 4-300 K using universal measurement system "Liquid Helium Free High Field Measurement System" (Cryogenic Ltd). The measurement error throughout temperature range did not exceed 5%.

3. Results and Discussion

The value of the specific electrical conductivity of semiconductors, depending on type and concentration of impurities and defects, can change by several orders. In addition, such factors as complexity and heterogeneity of interfaces of contacting materials with substantially different band structure affecting on conductivity are important. It leads to an additional resistance associated with the appearance of an energy barriers and dimensional and edge effects.

To understand a dominant factor having effect on studied system, it is necessary to predict the conditions of current transfer. Schematic representation of current flow paths through the structure was presented in (Figure 2). It is apparent that, at room temperature, presence of silicon substrate, metal-semiconductor interface, and the contact electrical resistivity between the metal clusters in the pores are the predominant factors affecting electrical transfer.

Charge transfer through the pores in dielectric template occurs over concatenation of metal particles pervading entire pore from electrode to boundary with semiconductor. According to [5], the current flow through such clusters grid provides predominantly metallic type of conductivity in spite of the fact that dielectric layers are possible, it is associated with oxides formation on clusters surface. They increase electrical resistivity of material as a whole, but do not affect the conduction mechanism [23, 24].

During current transfer through $SiO_2(Ni)/Si$ nanosystems after the passage of concatenation of metal particles, the charge carriers continue to move through the silicon substrate, where the electronic type of conductivity is predominant. After the passage of silicon, charge transfer is again provided by metal particles located in the pores. The processes of electromigration in silicon were thoroughly studied and described in detail in [25–27].

When charge carriers move from metal to semiconductor and back, depending on thermodynamic work functions ratio, different types of contacts can be realized: neutral, ohmic, nonohmic (Schottky barrier). To determine the contact type, it is necessary to know the peculiarities of $SiO_2(Ni)/Si$ nanosystems band structure of nickel/silicon contact region.

It is known that the work function of nickel $W_{Ni} = 4.9$ eV and n-Si (100) $W_{Si} = 4.3$ eV [28]. Since the relation $W_s < W_{Me}$ is satisfied when the metal is in contact with semiconductor, current of thermionic emission from semiconductor surface is greater than from metal surface. Therefore, in the contact zone of metal particles with silicon, a positive space charge accumulates, and it leads to formation in contact region of an electron-depleted region of depth d_0 (Figure 3(a)) and

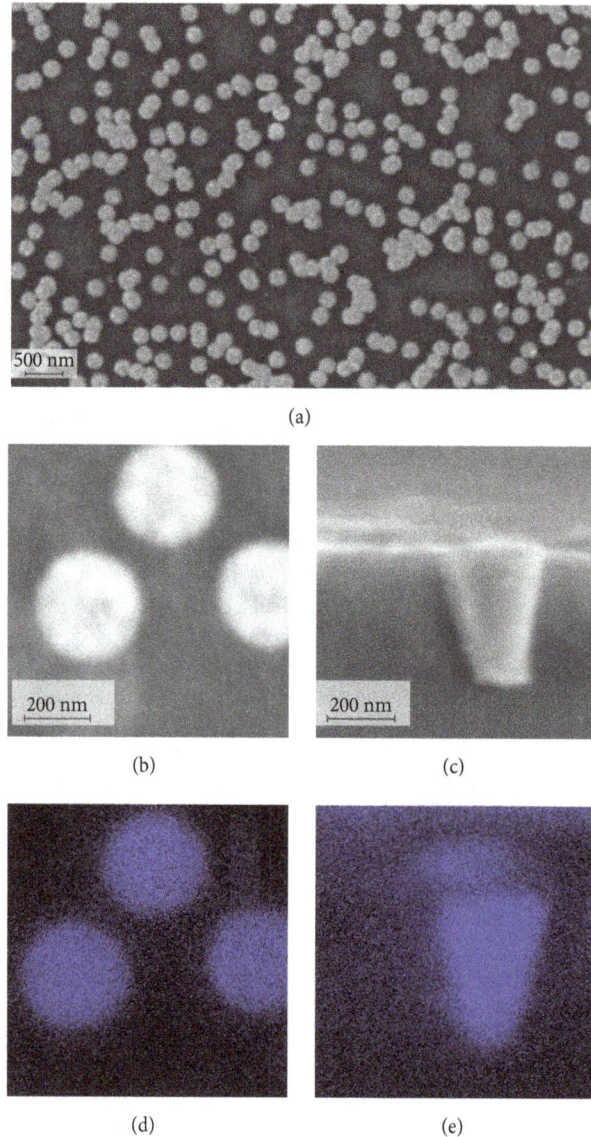

FIGURE 1: SEM images $SiO_2(Ni)/Si$ nanosystems: surface (a, b), cross sections, (c) and EDS mapping images (d, e) of regions (b, c).

bending of energy bands (Figure 3(b)) with the formation of Schottky barrier. The values of barrier height, determined by difference of work function $\varphi_K = W_{Me} - W_s$, are found to be 0.6 eV for nickel.

The application of an external electric field with "+" sign on the metal side (Figure 3(c)) leads to decreasing of potential barrier height and value of electric field in space semiconductor charge region d_+ (Figure 3(d)). The application of E_{ext} in opposite direction (Figure 3(d)) contributes to an increasing of potential barrier and depletion region depth $d-$ (Figure 3(e)).

Thus, during applying voltage in any direction, the current will flow through $SiO_2(Ni)/Si$ nanosystem in the metal-semiconductor-metal scheme. At the same time, on one contact, electrons freely move from metal to semiconductor, and on the other (during transition from semiconductor to metal) they overcome the potential barrier; that is, the contacts

have rectifying properties. Accordingly, in this structure a symmetric $I–V$ curve with two Schottky barriers included towards each other should be observed.

Such property of band structure in the contact region of metallic clusters with n-silicon is observed on an ideal (infinitely large and homogeneous) interface. Accounting of inhomogeneities due to relief and size of contact surface introduces additional corrections to the potential barrier value and $I–V$ curves behavior due to the fact that the actual interface has emission parameters, which substantially differs from corresponding parameters of ideal barrier.

Metallic nanoparticles localized in pores do not have a preferential crystallographic orientation. This fact can affect the height of Schottky barrier in both its increasing and decreasing. The reason for such changes is the difference in work functions for different crystallographic faces, and such difference reaches values of ~ 1 eV (in nickel it fluctuates

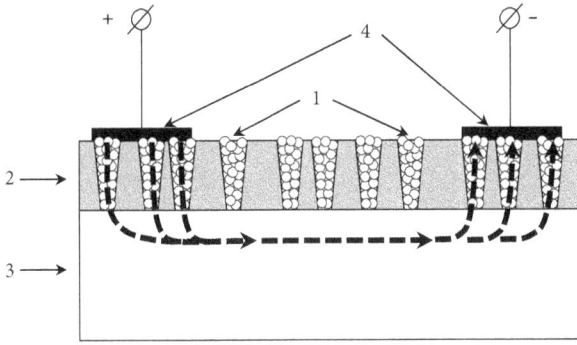

FIGURE 2: Schematic representation of $SiO_2(Ni)/Si$ nanosystems. (1) Metal in the pores, (2) silicon oxide layer, (3) silicon substrate, and (4) electrodes.

within 4.6-5.5 eV [29]). In addition, it is necessary to take into account the small area and inhomogeneity of contact surface. Applying the reverse bias voltage, considerable currents leakage can arise from curvature of space charge region of semiconductor along the contact periphery; it helps to reduce the actual height of potential barrier. Moreover, the number of metal pores filled with current electrodes (Figure 2) is large but the area of electrical contact with semiconductor is a set of parallel connected and electrically noninteracting microcontacts, as a result the potential barrier value has a certain averaged value. For these reasons, it is impossible to correctly estimate the potential barrier without conducting an experimental study of $SiO_2(Ni)/Si$ curves I-V. Typical results of current-voltage characteristics measuring for samples with nickel nanoparticles in pores are shown in Figure 4(a).

These characteristics present two symmetrical Schottky barriers, which are formed in different directions of current flow and charge transfer from a silicon substrate to metal clusters. It is easy to determine the height of potential barrier. The current flowing through the Schottky barrier is determined by the Richardson equation [30]:

$$I_0 = SAT^2 \exp\left(-\frac{\varphi_k}{kT}\right), \qquad (1)$$

where I_0 is the saturation current; S is metal-semiconductor contact area; A is Richardson constant; k is Boltzmann constant.

So equation can be rewritten as follows:

$$\varphi_k = kT \exp\left(-\frac{SAT^2}{I_0}\right). \qquad (2)$$

The contact area S is defined as the effective contact area of one pore (10^4 nm^2) multiplied by the number of pores under the conductive electrode (10^6) and is 10^{-4} cm^2.

The Richardson constant is usually assumed to be 120 A cm^{-2} K^{-2}, but in [31] it is indicated that the experimental values of A may not coincide with the given value. It should be noted that the value of φ_k is not very sensitive to the choice of the Richardson constant, because even at room temperature

the error twice as large in parameter A leads to an error in determining φ_k by only 0.018 eV.

The value of I_0 is determined as follows: the branch of the current-voltage characteristic, plotted in scale $\ln I$ from $V^{1/4}$, for $V \gg kT/e$, is represented by a straight line. By extrapolating the linear part of the dependence on the semilogarithmic scale $\ln I$, the saturation current at $V = 0$ V (Figure 4(b)), which is 1.08×10^{-6} A, is on the ordinate axis. Thus, in accordance with (2), the value of potential barrier was $\varphi_k = 0.56$ eV; it correlates well with the data obtained earlier for the contact of nickel with silicon [32, 33].

The analysis of the I-V curves obtained in the temperature range of 25-300 K (Figure 5(a)) shows that the dependencies are symmetric, qualitatively similar to each other with the preservation of two distinct Schottky barriers in the entire temperature range. As the temperature is lowered, the nonlinearity of the I-V curves begins to appear at higher voltages and has a sharper character, which is associated with increasing of potential barrier and is accompanied by an increase in the resistance at barrier: at $T = 25$ K the resistance of the structure reaches values of the order of 10^8 ohm, becomes of the order of 10^6 ohm at $T = 100$ K, and goes down to $\sim 10^5$ ohm at room temperature. It is explained by the fact that as the temperature increases, electrons in the semiconductor layer near the Schottky barrier are excited to higher energy levels and, accordingly, the tunneling probability of barrier increases (Figures 5(b) and 5(c)).

In the room temperature region, the energy of electrons thermal motion kT is commensurable with band gap, so the electrons go from the valence band to the conduction band with the formation of pair charge carriers (electrons and holes). With temperature decreasing, the number of such electrons decreases, leading to a sharp increase in the resistance. Then there is depletion of the impurity, and the resistance increases due to increasing of silicon band gap width, when temperature decreases.

It should be taken into account that, in the region of metal-semiconductor contact, the relief of valence band and conduction band changes. It leads to competition of tunneling through a potential barrier with charge carriers activated at the percolation level [34, 35]. By the level of percolation it is usually understood the energy, with which the charge carrier moves through the sample from one contact to another contact, bending around the barrier (above-barrier emission). It is clear that at high temperatures the conductivity is predominant in terms of the percolation level, while at low temperatures the tunneling of electrons from one potential well to the other at Fermi level is dominant.

With a resistance at the Schottky barrier less than silicon resistance, charge transfer occurs over the metal, then the charges move in the silicon layer to the next filled track, and the process is repeated in the same way further, leading to a decrease in the system impedance. In addition to such transitions from silicon to metal and back, such mechanisms as the Schottky thermionic emission (at high temperatures) and tunnel emission of charges or hopping conductivity (at low temperatures) can also influence process of charge transfer [17].

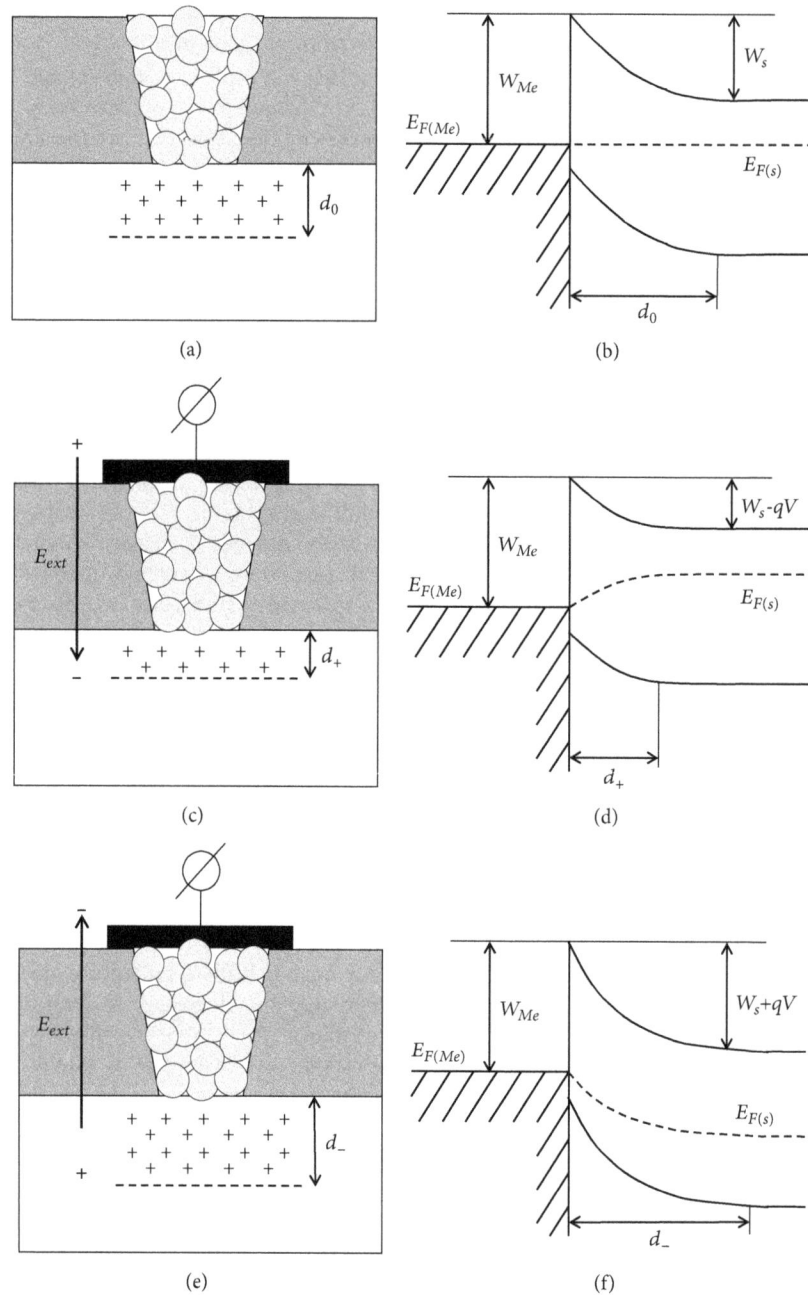

FIGURE 3: Schematic representation of electron-depleted region formation in contact area of metallic particles with n-Si in absence of an external electric field (a), at E_{ext} with "+" sign on metal side (c), with a "–" sign (e); scheme for band structure formation in contact region at $E_{ext} = 0$ (b), $E_{ext} > 0$ (d), $E_{ext} < 0$ (e).

Summarizing the above findings it is possible to propose the following interpretation of the mechanisms of charge transfer in $SiO_2(Ni)/Si$ nanosystems realized in different temperature ranges:

(i) In the region ~300–200 K: charge carrier motion through the n-Si with an employment of metallic clusters in pores being in a contact with the semiconductor, by means of the overbarrier emission of electrons from higher energy levels of Si conduction band

(ii) In the lower temperatures region (~200–100 K): a current flow taking place only through the semiconductor due to an increase in resistivity on energy barriers Si/metal, which leads to a practically complete exclusion of a participation of the metal in the charge transport process

(iii) In the low temperatures region (~100–20 K): a domination of hopping charge transport with a variable hopping length between pores on the SiO_2/Si boundary, containing localized states.

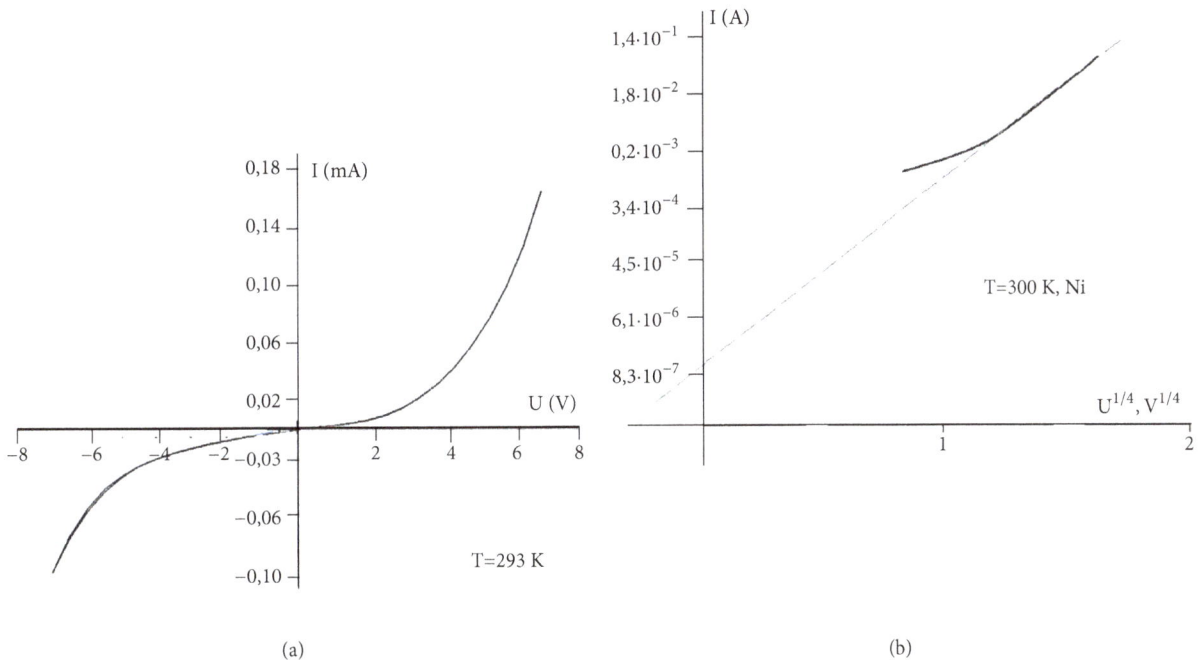

(a)

(b)

FIGURE 4: I–V characteristics curves of SiO_2(Ni)/Si nanosystems at room temperature: I–V curve (a). Dependencies of $\ln I$ on $V^{1/4}$ and the approximating curve for structures (b).

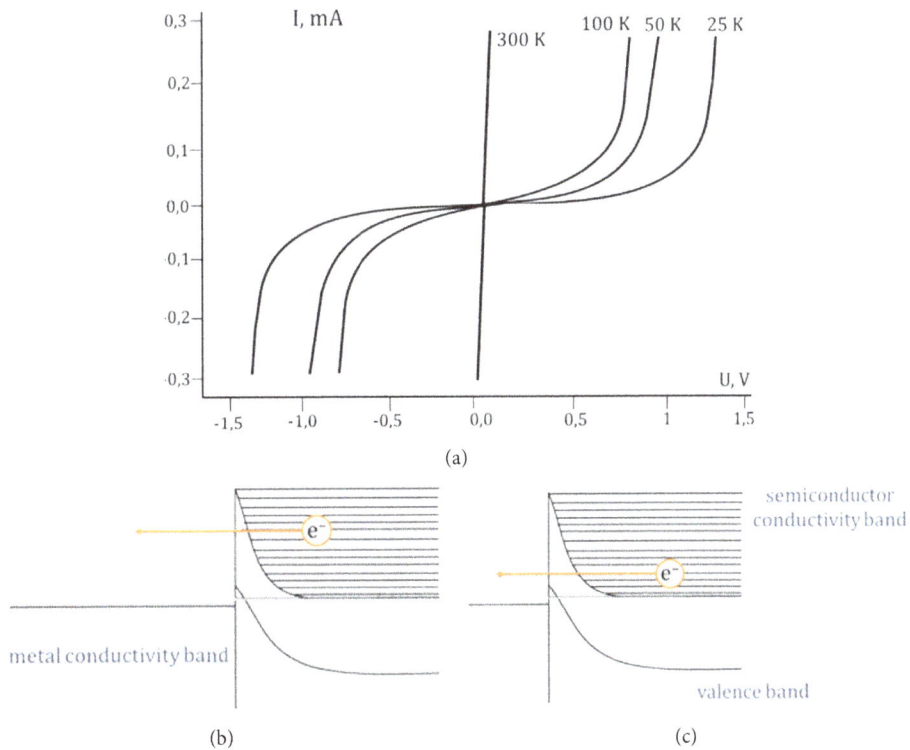

(a)

(b)

(c)

FIGURE 5: (a) I–V curves of SiO_2(Ni)/Si nanosystems for different temperatures. Band structure of n-type semiconductor system, metal with a Schottky barrier: (b) at high temperatures; (c) at low temperatures.

4. Conclusions

Data on the formation of $SiO_2(Ni)/Si$ nanosystems with nickel in the pores of amorphous silicon oxide and the results of its electrical characteristics over a wide temperature range are presented. It is shown that ion-track technology can serve as a connecting link for the adoptive nanosystems to the technological processes of standard silicon technology. It was demonstrated that the selected electrochemical deposition method, due to the high degree of control, allows selectively filling the pores with Ni, at a minimum spread of the protuberances of the metal precipitate over the surface of the template. When considering the mechanisms of charge transfer, it is shown that as the charge carriers move into a pore, a metallic type of conductivity is realized. A Schottky barrier is created at the bottom of the pore thanks to the presence of silicon. Investigations of the current-voltage characteristics show that dependencies are typical for structures with double Schottky barrier. The value of potential barrier at n-Si/Ni interface (0.56 eV) was determined. It was shown that applying of external electric field to metal-semiconductor contact with plus sign on the metal side leads to decreasing of potential barrier height in the region of semiconductor space charge, and applying of inverse field leads to increasing of barrier and charge-depleted region depth. Analyzed current-voltage curve indicates that, over the entire temperature range 25-300 K at high voltages, the nonlinear character of I–V curve takes place, which causes increasing of potential barrier. In this case, the resistance on barrier increases: at T = 25 K structure resistance is of the order of 10^8 ohms, at T = 100 K it is 10^6 ohms, and at room temperatures it decreases to 10^5 ohms.

Disclosure

This material have been presented as poster presentation at the XXII International Scientific Conference of Young Scientists and Specialists (AYSS-2018) 23-27 April 2018.

Conflicts of Interest

The authors declare that they have no conflicts of interest.

Acknowledgments

This work was supported by the Scientific-Technical Program "Technology-SG" (Project no. 3.1.5.1) and Belarusian Foundation for Basic Research (Project no. Ф17М-005).

References

[1] I. Calisir, A. A. Amirov, A. K. Kleppe, and D. A. Hall, "Optimisation of functional properties in lead-free $BiFeO_3$–$BaTiO_3$ ceramics through La^{3+} substitution strategy," *Journal of Materials Chemistry A*, vol. 6, no. 13, pp. 5378–5397, 2018.

[2] D.-H. Kim, E. A. Rozhkova, I. V. Ulasov et al., "Biofunctionalized magnetic-vortex microdiscs for targeted cancer-cell destruction," *Nature Materials*, vol. 9, no. 2, pp. 165–171, 2010.

[3] S. V. Trukhanov, A. V. Trukhanov, V. G. Kostishyn et al., "Magnetic, dielectric and microwave properties of the $BaFe_{12-x}Ga_xO_{19}$ (x ≤ 1.2) solid solutions at room temperature," *Journal of Magnetism and Magnetic Materials*, vol. 442, pp. 300–310, 2017.

[4] D. I. Tishkevich, S. S. Grabchikov, L. S. Tsybulskaya et al., "Electrochemical deposition regimes and critical influence of organic additives on the structure of Bi films," *Journal of Alloys and Compounds*, vol. 735, pp. 1943–1948, 2018.

[5] K. Tamarov, W. Xu, L. Osminkina et al., "Temperature responsive porous silicon nanoparticles for cancer therapy – spatiotemporal triggering through infrared and radiofrequency electromagnetic heating," *Journal of Controlled Release*, vol. 241, pp. 220–228, 2016.

[6] D. Yakimchuk, E. Kaniukov, V. Bundyukova et al., "Silver nanostructures evolution in porous SiO_2/p-Si matrices for wide wavelength surface-enhanced Raman scattering applications," *MRS Communications*, vol. 8, no. 1, pp. 95–99, 2018.

[7] A. L. Kozlovskiy, I. V. Korolkov, G. Kalkabay et al., "Comprehensive study of Ni nanotubes for bioapplications: from synthesis to payloads attaching," *Journal of Nanomaterials*, vol. 2017, Article ID 3060972, 9 pages, 2017.

[8] I. V. Korolkov, O. Güven, A. A. Mashentseva et al., "Radiation induced deposition of copper nanoparticles inside the nanochannels of poly(acrylic acid)-grafted poly(ethylene terephthalate) track-etched membranes," *Radiation Physics and Chemistry*, vol. 130, pp. 480–487, 2017.

[9] M. B. Gongalsky, L. A. Osminkina, A. Pereira et al., "Laser-synthesized oxide-passivated bright Si quantum dots for bioimaging," *Scientific Reports*, vol. 6, no. 1, Article ID 24732, 2016.

[10] M. D. Kutuzau, E. Y. Kaniukov, E. E. Shumskaya et al., "The behavior of Ni nanotubes under the influence of environments with different acidities," *CrystEngComm*, vol. 20, no. 23, pp. 3258–3266, 2018.

[11] E. Kaniukov, A. Shumskaya, D. Yakimchuk et al., "FeNi nanotubes: perspective tool for targeted delivery," *Applied Nanoscience*, pp. 1–10, 2018.

[12] K. Awazu, S. Ishii, K. Shima, S. Roorda, and J. L. Brebner, "Structure of latent tracks created by swift heavy-ion bombardment of amorphous," *Physical Review B: Condensed Matter and Materials Physics*, vol. 62, no. 6, pp. 3689–3698, 2000.

[13] A. Dallanora, T. L. Marcondes, G. G. Bermudez et al., "anoporous SiO_2/Si thin layers produced by ion track etching: dependence on the ion energy and criterion for etchability," *Journal of Applied Physics*, vol. 104, no. 2, Article ID 024307, 8 pages, 2008.

[14] M. Y. Presnyakov, D. A. Sinetskaya, E. Y. Kaniukov, S. E. Demyanov, and E. K. Belonogov, "Growth morphology and structure of PdCu ion-plasma condensate in the pores of SiO_2 and Al_2O_3 amorphous matrices," *Journal of Crystal Growth*, vol. 486, pp. 66–70, 2018.

[15] E. Kaniukov, D. Yakimchuk, G. Arzumanyan et al., "Growth mechanisms of spatially separated copper dendrites in pores of a SiO_2 template," *Philosophical Magazine*, vol. 97, no. 26, pp. 2268–2283, 2017.

[16] S. Sankar, A. E. Berkowitz, and D. J. Smith, "Spin-dependent transport of Co-SiO_2 granular films approaching percolation," *Physical Review B: Condensed Matter and Materials Physics*, vol. 62, no. 21, pp. 14273–14278, 2000.

[17] S. Demyanov, E. Kaniukov, A. Petrov, and V. Sivakov, "Positive magnetoresistive effect in Si/SiO$_2$(Cu/Ni) nanostructures," *Sensors and Actuators A: Physical*, vol. 216, pp. 64–68, 2014.

[18] E. Y. Kaniukov, J. Ustarroz, D. V. Yakimchuk et al., "Tunable nanoporous silicon oxide templates by swift heavy ion tracks technology," *Nanotechnology*, vol. 27, no. 11, Article ID 115305, 2016.

[19] P. Granitzer, K. Rumpf, and H. Krenn, "Micromagnetics of Ni-nanowires filled in nanochannels of porous silicon," *Thin Solid Films*, vol. 515, no. 2, pp. 735–738, 2006.

[20] Y. A. Ivanova, D. K. Ivanou, A. K. Fedotov et al., "Electrochemical deposition of Ni and Cu onto monocrystalline n-Si(100) wafers and into nanopores in Si/SiO2 template," *Journal of Materials Science*, vol. 42, no. 22, pp. 9163–9169, 2007.

[21] S. E. Demyanov, E. Y. Kaniukov, A. V. Petrov et al., "On the morphology of Si/SiO$_2$/Ni nanostructures with swift heavy ion tracks in silicon oxide," *Journal of Surface Investigation. X-Ray, Synchrotron and Neutron Techniques*, vol. 8, no. 4, pp. 805–813, 2014.

[22] S. E. Demyanov, E. Y. Kaniukov, A. V. Petrov, and E. K. Belonogov, "Nanostructures of Si/SiO$_2$/metal systems with tracks of fast heavy ions," *Bulletin of the Russian Academy of Sciences, Physics*, vol. 72, no. 9, pp. 1193–1195, 2008.

[23] B. A. Aronzon, D. Y. Kovalev, A. E. Varfolomeev, A. A. Likal'ter, V. V. Ryl'kov, and M. A. Sedova, "Conductivity, magnetoresistance, and the Hall effect in granular Fe/SiO$_2$ films," *Physics of the Solid State*, vol. 41, no. 6, pp. 857–863, 1999.

[24] S. K. Kulkarni, "Nanolithography," in *Nanotechnology: Principles and Practices*, pp. 241–257, 2015.

[25] S. Yngvesson, "Review of semiconductor physics and devices," in *Microwave Semiconductor Devices*, pp. 1–22, Springer US, Boston, Mass, USA, 1991.

[26] M. Lundstrom, *Fundamentals of Carrier Transport*, Cambridge University Press, Cambridge, UK, 2000.

[27] F. Grund, "Wang, S., Fundamentals of Semiconductor Theory and Device Physics. Englewood Cliffs, Prentice-Hall International 1989. XVI, 864 pp., 33.95. ISBN 0-13-344425-2," *ZAMM—Journal of Applied Mathematics and Mechanics/Zeitschrift für Angewandte Mathematik und Mechanik*, vol. 71, no. 11, p. 464, 1991.

[28] A. Kikuchi, T. Ohshima, and Y. Shiraki, "Schottky barrier height of single-crystal nickel disilicide/silicon interfaces," *Journal of Applied Physics*, vol. 64, no. 9, pp. 4614–4617, 1988.

[29] Y.-J. Chang and J. L. Erskine, "Diffusion layers and the Schottky-barrier height in nickel silicide-silicon interfaces," *Physical Review B: Condensed Matter and Materials Physics*, vol. 28, no. 10, pp. 5766–5773, 1983.

[30] M. Grundmann, *The Physics of Semiconductors*, Springer, Berlin, Germany, 2010.

[31] S. M. Sze and K. K. Ng, *Physics of Semiconductor Devices*, Wiley-Blackwell, 3rd edition, 2006.

[32] C. A. Mead and W. G. Spitzer, "Fermi level position at metal-semiconductor interfaces," *Physical Review A: Atomic, Molecular and Optical Physics*, vol. 134, no. 3A, pp. A713–A716, 1964.

[33] A. Thanailakis and A. Rasul, "Transition-metal contacts to atomically clean silicon," *Journal of Physics C: Solid State Physics*, vol. 9, no. 2, pp. 337–343, 1976.

[34] M. H. Cohen, E. N. Economou, and C. M. Soukoulis, "Electron transport in amorphous semiconductors," *Journal of Non-Crystalline Solids*, vol. 66, no. 1-2, pp. 285–290, 1984.

[35] N. Mott and E. Davis, *Electronic Processes in Non-Crystalline Materials*, Oxford University Press, 2nd edition, 2012.

11

The Dependencies of X-Ray Conductivity and X-Ray Luminescence of ZnSe Crystals on the Excitation Intensity

V. Ya. Degoda,[1] M. Alizadeh ⓘ,[1] N. O. Kovalenko,[2] and N. Yu. Pavlova[3]

[1]*Taras Shevchenko National University of Kyiv, 64 Volodymyrs'ka Street, 01601 Kyiv, Ukraine*
[2]*Institute for Single Crystals NAS of Ukraine, 61001, Nauki Ave, Kharkiv, Ukraine*
[3]*National Pedagogical Dragomanov University, 9 Pyrogova Street, 01601 Kyiv, Ukraine*

Correspondence should be addressed to M. Alizadeh; trefoilsymbol@gmail.com

Academic Editor: Jan A. Jung

This work studies the conductivity and luminescence of ZnSe single crystals under X-ray irradiation. The experimentally derived lux-ampere characteristics of the X-ray conductivity for ZnSe crystals have a sublinear behavior within the temperature range from 8 to 420 K. The theoretical analysis of the conductivity kinetics at X-ray excitation showed that the value of maximum accumulated lightsum at deep traps does not depend on radiation intensity. However, regarding shallow and phosphorescent traps, the strength of accumulated lightsum depends on the intensity of exciting irradiation. Specifically, these shallow traps and phosphorescent traps cause the sublinear behavior of lux-ampere characteristics in the semiconductor material.

1. Introduction

Zinc Selenide (ZnSe) [1–7] is a topic to be investigated. This material was well studied and well researched among the most promising wide-bandgap semiconductors (WBG or WBGS) of the II-VI semiconductor group ($A^{II}B^{VI}$ or A^2B^6). Nowadays, ZnSe is widely used to create short-wave semiconductor electronics and display systems [7, 8]. Primarily, it is due to the recently learned ability to produce high-quality single crystals of a relatively large scale. As a result, another promising direction emerged. These single crystals are used as detectors of indirect ionizing radiation (γ-radiation scintillators) [8–14] and direct conversion of the energy of high-energy radiation into the electric current (semiconductor detectors) [6, 15–17]. Moreover, they are also used in the production of lenses and IR-radiation windows [8, 9, 18, 19]. The application of ZnSe crystals as ionizing radiation detectors has become possible after developing the technology to produce high-quality crystals [20–22] with a low concentration of impurities and high resistivity ~ 10^{10} - 10^{12} Ω·cm. The single crystals have a rather large effective atomic number (Z_{eff} = 32) and a large band gap width (2.7 eV at 300 K), which makes them a promising material to create X-ray detectors, which do not require cooling and can operate at high temperatures (up to 450 K) [17]. These findings make the studies of X-ray conductivity (XRC) and X-ray luminescence (XRL) in ZnSe single crystals topical.

Lux-ampere characteristics of XRC and X-ray luminescence are one of the main features of semiconductor detectors of ionizing radiation. The dependence of conduction current on the irradiation intensity under constant electric field strength (I_X-i_{XRC}) is referred to as the lux-ampere characteristic (LAC), while the dependence of luminescence intensity on the excitation intensity (I_X-J_{XRL}) is referred to as the lux-luminescent characteristic (LLC). Earlier, similar studies were carried out [6, 7] but only at the temperature of liquid nitrogen. There is no comparison between the lux-ampere and lux-luminescent characteristics or the explanation of the obtained nonlinear dependencies.

The work aimed to experimentally investigate the spectra of luminescence and conductivity and their dependencies on the intensity of X-ray excitation for the ZnSe single crystals at various temperatures and explain the sublinear dependencies of the LAC of the X-ray conductivity and LLC of the X-ray luminescence.

2. Experimental Details

In this work, we studied the conductivity and luminescence of ZnSe single crystals under the excitation by X-ray quanta. The specially undoped ZnSe crystals were grown from the pre-cleaned batch, to obtain crystals with a minimum impurity concentration and a maximum resistivity ($\rho \geq 10^{12}$ Ω.cm). The polished samples of 18 x 9 x 2 mm^2 were prepared from different boules. It turned out that the luminescence spectra and the conductivity values did not differ much for these samples; therefore, the main complex of studies was carried out for one sample. We should note that the concentrations of free electrons in such crystals without excitation are 10^2 - 10^4cm^{-3}.

To study the conductivity, the two three-layer metal contacts were sprayed onto one large surface of the sample by the resistive method. The chemical composition of each layer was specially selected to obtain ohmic contacts for the dark conductivity with good adhesion. Copper conductors were soldered to the contacts for the conductivity measurements. The contacts consisted of the strips of rectangular shape 5 mm long and 1 mm wide, with 5 mm distance between them. A stabilized voltage from 0 to 1000 V was applied to one electrode, while another contact was connected to a nanoamperemeter. The nanoamperemeter allowed measuring the value of the conduction current from 1 pA to 10 pA with an accuracy of 10%; from 10 pA to 100 pA with an accuracy of 3%; and from 100 pA to 1 mA with an accuracy of <1%. For all values of the conduction current, the following condition was met: the input impedance of a nanoamperemeter is several orders of magnitude smaller than the electrical resistance of a ZnSe sample. The nanoamperemeter is a part of a specially developed measuring unit that allows selecting the voltage change mode: manual, stepwise, and monotonic, at which the rate of voltage increase can be changed. The study of the conduction current was carried out in a vacuum (<1 Pa).

It was preliminarily established by the thermal EMF method that the samples have the dark n-type conductivity.

The complex experimental studies of X-ray luminescence (XRL) and X-ray conductivity (XRC) were conducted. Their lux-ampere characteristics (dependencies of the conduction current I_X-i_{XRC}; the luminescence intensity of the different luminescence bands I_X-J_{630} and I_X-J_{970} on the intensity of X-ray excitation I_X); the dose dependencies of phosphorescence (P) and current relaxation (CR) at the excitation temperatures of 8, 85, 295, and 420 K were investigated. The sample was placed in the cryostat, which allowed using various temperature modes within 8 – 500 K. The samples were heated using an electric furnace built into the cryostat (800 W) and cooled by liquid nitrogen or helium.

The X-ray excitation was performed by the integral radiation of the X-ray tube BKhV7 (Re, 20 kV, 3 ÷ 25 mA, which corresponds to I_X = 0.0762÷0.635 mW/cm^2) through the beryllium window of the cryostat in the perpendicular direction to the sample surface. All X-ray irradiations were absorbed within the sample. The distance from the anode of the X-ray tube to the sample was 120 mm, which provided for maximum intensity of the X-ray irradiation of 0.635 mW/cm^2. When changing the current of X-ray tube i_{X-tube},

while the voltage is constant, the radiation intensity I_X varies in proportion to the tube current, while the shape of the spectrum remains unchanged.

The conduction current and the luminescence radiation of the sample were recorded simultaneously. The luminescence was registered via two channels: integrally and spectrally. For those passing through one quartz window of the cryostat, the integral glow of the sample (if necessary through an optical filter) was focused by a quartz lens on a photocathode PMT-106. For those passing through another quartz window of the cryostat, the luminescence radiation was directed through a high-speed monochromator MDR-2 with quartz condensers and recorded by the photoelectric multipliers: PMT-106 in the visible region or PMT-83 (in the cooling mode) in the IR region. All spectra were adjusted taking into account the spectral sensitivity of the recording system. And at the translation of the spectra from the wavelength scale (nm) to the quanta energy scale (eV), the spectra were corrected to the spectral radiation density.

The experimental lux-ampere dependencies of luminescence and conductivity (I_X-i_{XRC}, I_X-J_{630} and I_X-J_{970}) were measured in two modes: an increase (5 → 25 mA) and a decrease (25 → 5 mA) of X-ray intensity. Under irradiation for more than 5 minutes for each value of the excitation intensity, the divergence of the LAC curves in both modes is practically absent.

Upon terminating the excitation, both the phosphorescence (by two channels) and the current's relaxation were recorded for 10 minutes. The dose of the sample irradiation (from 0.6 to 4600 mJ/cm^2) was determined by the exposure time (from 5 s to 2 h) and the excitation intensity (tube's current was 5 mA and 25 mA).

The peculiarity of this work measurements was that the registration of the luminescence and conductivity of the sample were carried out simultaneously. This simultaneous registration is essential because it allows obtaining more detailed and reliable information about the processes within the sample.

3. The Results of the Experiments

3.1. The X-Ray Luminescence Spectra of ZnSe Single Crystals. The obtained distinctive spectra of XRL for the samples of ZnSe single crystals in the wavelength range from 400 to 1200 nm at various temperatures and excitation intensity levels are presented in Figures 1, 2, and 3. Under the high temperatures (410 K), due to the temperature quenching of the luminescence, the XRL intensity was so low that it was impossible to register the XRL spectra. At higher sensitivity of the registration system (by two orders of magnitude in the spectral range from 400 to 550 nm) it was impossible to register neither the edge glow nor the emission of the donor-acceptor pairs (DAP) at the temperature of 8 K, when the luminescence is the brightest within the ZnSe [23]. It is also essential to note that the integral luminescence intensity at low temperatures of this ZnSe sample (when there is no temperature quenching of the emission bands) is only several times weaker than the integral intensity of the industrial,

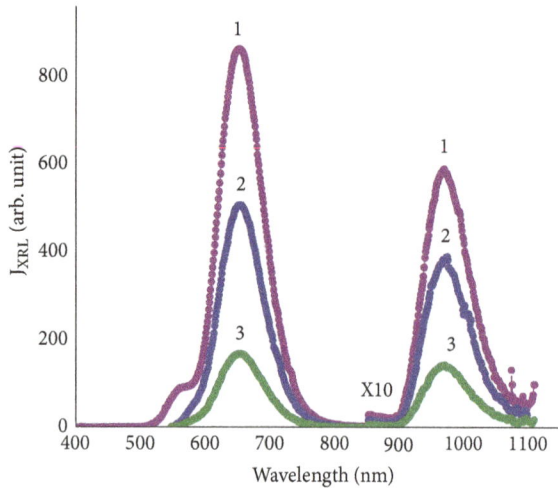

FIGURE 1: The distinctive XRL spectra of the ZnSe single crystal at the temperature of 8 K under different levels of excitation: the current of the X-ray tube, 25mA (1); 15mA (2); and 5mA (3) (the XRL intensity is multiplied by 10 for the 970 nm band).

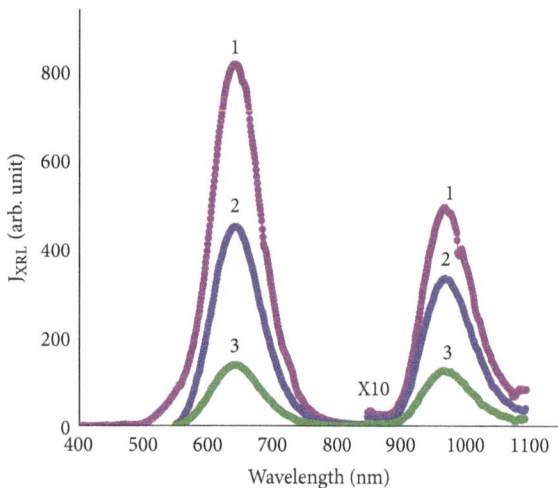

FIGURE 2: The distinctive XRL spectra of the ZnSe single crystal at the temperature of 85 K under different levels of excitation: the current of the X-ray tube, 25mA (1); 15mA (2); and 5mA (3) (the XRL intensity is multiplied by 10 for the 970 nm band).

the brightest X-ray luminescent ZnS-Cu. It indicates a slight nonradiative loss of the excitation energy and, accordingly, high quality of these ZnSe crystals.

The XRL spectra of the investigated samples of ZnSe crystals consist of two main luminescence bands with maximums at 630 nm (1.92 eV) and 970 nm (1.28 eV). The ratio of the intensities of these bands varies for different crystals. According to papers [12, 21–23], the luminescence band with a maximum at 630 nm is due to the crystal complex center, which includes a Zn vacancy, while the luminescence band with the maximum at 970 nm is caused by the complex center with the Se vacancy or the Cu impurity [24, 25].

When the intensity of the X-ray excitation changes almost by an order of magnitude, the spectral positions of the

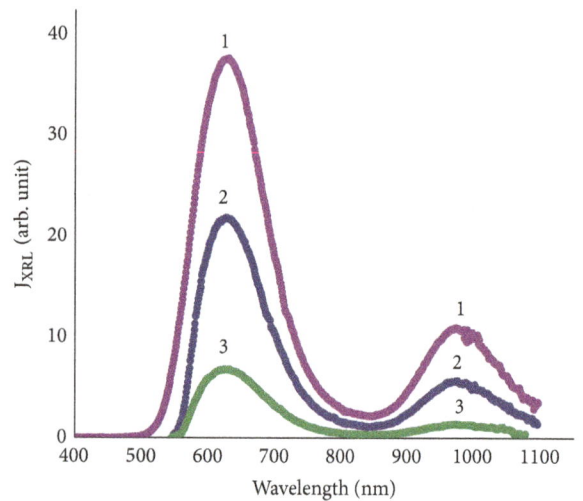

FIGURE 3: The distinctive XRL spectra of the ZnSe single crystal at the temperature of 295 K under different levels of excitation: the current of the X-ray tube, 25mA (1); 15mA (2); and 5mA (3).

maximums and the bands' forms do not change. Therefore, the LLC measurements (IX-J630 and IX-J970) were carried out at the spectral peaks of the bands.

3.2. LAC of the Conductivity and LLC of the Luminescence of the ZnSe Crystals. LAC of the conductivity and LLC of the luminescence show the character of the increase in the conduction current i_{XRC} and the luminescence intensity J_{XRL} under the intensity increase of the X-ray or γ-radiation I_X, while the condition $I_X \sim i_{X\text{-tube}}$ is satisfied. These characteristics are essential for the scintillation and semiconductor detectors of ionizing radiation.

LAC of the conductivity (I_X-i_{XRC}) and LLC of the luminescence (I_X-J_{630} and I_X-J_{970}) for the ZnSe samples were measured at different temperatures. Figure 4 shows the I_X-i_{XRC} dependencies of the nonlinear behavior. Moreover, the XRC has a sublinear dependence. These dependencies were derived at feeble electric field strength (8 V/cm) when the volt-ampere characteristics of the X-ray conductivity are still linear. Thus, such weak electric fields do not influence the luminescence and conductivity [6].

For the X-ray luminescence, the dependencies I_X-J_{630} and I_X-J_{970}, obtained at temperatures of 8, 85 and 295 K, either are linear or have a small superlinearity (Figures 5 and 6). It should be noted that after the X-ray irradiation the phosphorescence and the relaxation of the conduction current are observed in ZnSe samples. Therefore, all the graphs for LAC and LLC are presented after subtracting the phosphorescence and current relaxation.

Similar LLC of photoluminescence is also observed in other crystals [26–29]. The experimental dependencies I_X-i_{XRC}, I_X-J_{630}, and I_X-J_{970} for the ZnSe samples cannot be accounted for by classical theories of the luminescence kinetics [26–28] and the conductivity [29–32]. Moreover, these theories do not take into account the above processes in correlation to each other. Therefore, the processes of

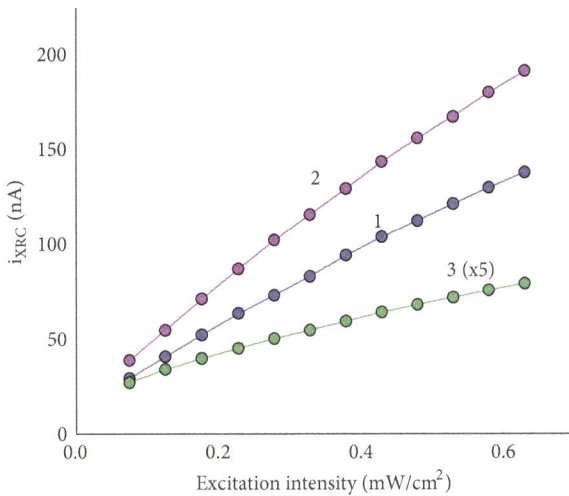

FIGURE 4: Lux-ampere characteristics of XRC of the ZnSe sample at various temperatures: 8 K (1), 85 K (2), and 295 K (3) (the intensity of the field in the sample equals 8 V/cm).

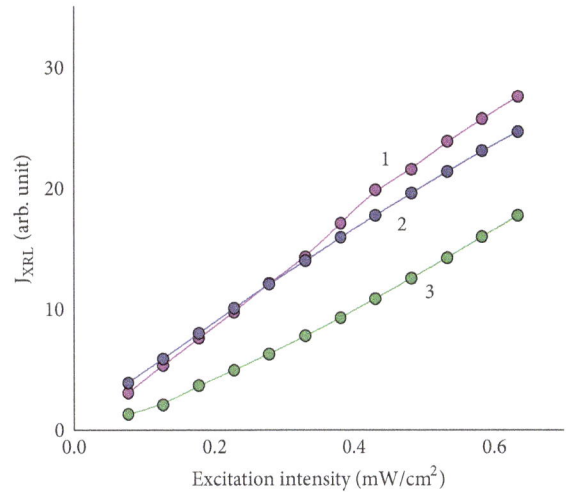

FIGURE 6: Lux-ampere characteristics of XRL of the ZnSe sample at the irradiation wavelength of 970 nm under the temperatures: 8 K (1), 85 K (2), and 295 K (3).

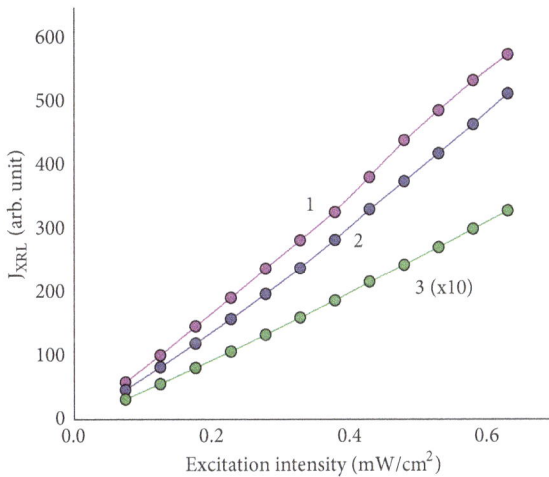

FIGURE 5: Lux-ampere characteristics of XRL of the ZnSe sample at the irradiation wavelength of 630 nm under various temperatures: 8 K (1), 85 K (2), and 295 K (3).

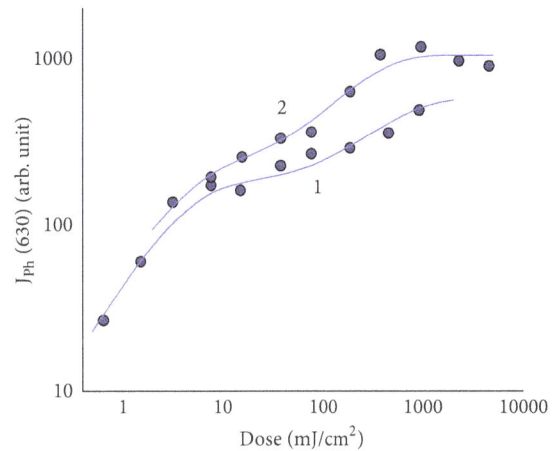

FIGURE 7: Dose dependencies of the phosphorescence intensity J_{Ph} of the ZnSe sample ($t_{ph} = 30$ s) at registration within 630 nm band at the temperature of 85 K under different intensities of the X-ray radiation: $i_{X\text{-tube}} = 5$ mA (1) and 25 mA (2). The solid lines show the theoretical dependency (10).

conductivity and recombination luminescence should be considered in correlation and explain the peculiar features of LAC and LLC, in particular for the high-resistance ZnSe crystals.

3.3. Dose Dependencies of the Phosphorescence Intensity. The accumulated lightsum in the sample during the X-ray excitation is manifested in the form of phosphorescence and current relaxation after excitation, lasting up to tens of minutes. Moreover, the main contribution to the luminescence and current makes one trap, which is called phosphorescent at the excitation temperature. While at further heating of a sample, the lightsum is manifested in the form of TSL and TSC. The accumulated lightsum is the charge carriers localized within the traps and the same number of the recharged recombination centers. It is known [33] that electrons in

ZnSe are the free charge carriers, which determine photoconductivity current.

The most logical is to investigate the dose dependencies of phosphorescence at 85 K. In this case, it has a higher intensity than at 8 and 295 K. Secondly, at this temperature the values of the general concentrations of shallow and deep traps are commensurable (have the same order). It should not be forgotten that, after each dose of irradiation and the registration of phosphorescence, the sample must be heated to 420 K to empty all traps.

Figures 7 and 8 show the dependencies of the phosphorescence intensity J_{Ph} ($t_{ph} = 30$ s after terminating excitation) on the obtained radiation dose. Using two different intensities of the X-ray radiation allows comparing the accumulated lightsums (the concentration of the recharged local centers)

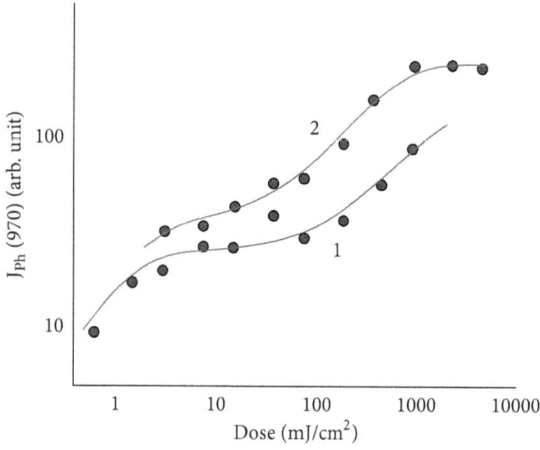

FIGURE 8: Dose dependencies of the phosphorescence intensity J_{Ph} of the ZnSe sample (for 30 s upon terminating excitation) at the registration within 970 nm band at the temperature of 85 K under different intensities of the X-ray radiation: $i_{X\text{-}tube}$ = 5 mA (1) and 25 mA (2). The solid lines show the theoretical dependency (10).

at equal doses of irradiation but obtained for a different period. In Figures 7 and 8 the circles show the experimental data, while solid lines show the theoretical dependency (10).

According to the experimental data, in all cases, the concentration of the recharged centers (even when reaching the saturation level) will be large at the more increased intensity of the X-ray radiation.

4. Analysis of the Experimental Results

As the semiconductor absorbs X-ray radiation, the free electrons and holes are generated, the amount of which is proportional to the energy of the absorbed X-ray quanta. Therefore, the amount of the produced (in the semiconductor) free electrons and holes G per unit of time is proportional to the absorbed energy of the X-ray or γ-radiation [34, 35].

The sublinear behavior of LAC of the X-ray conductivity (I_X-i_{XRC}) in ZnSe crystals indicates that when the intensity of X-ray excitation I_X increases, the concentration of free charge carriers, which determine the conduction current i_{XRC} also increases sublinearly. The concentrations of free electrons and holes are uniquely related to the concentration of generated carriers G via an average lifetime of free electrons (τ^-) and holes (τ^+):

$$N^- = \tau^- G$$
$$\text{and } P^+ = \tau^+ G. \tag{1}$$

Therefore, the sublinear behavior of I_X-i_{XRC} indicates a monotonous decrease of the charge carriers' lifetime by those signs that causes the conduction process. According to [32], these carriers are electrons in the ZnSe crystals. The lifetime of free electrons in the conduction band is determined by concentration of different types of traps v_i, their filling n_i, and

concentration of the recharged recombination centers p_j (i.e., the filled traps for holes):

$$\tau^- = \frac{1}{u^- \left(\sum_i \sigma_i^- \left(v_i - n_i \right) + \sigma_j^- p_j \right)}, \tag{2}$$

where u^-, u^+ are thermal velocities of free electrons and holes, respectively; σ_i^- is the capture cross section of free electrons at the traps of i-type; σ_j^+ is the capture cross section of the free holes at recombination centers; and σ_j^- is the capture cross section of free electrons at the recharged luminescence center (the recombination cross section). Usually, for high-quality crystalline phosphors, the concentrations of luminescence centers are much higher than the traps concentrations ($v_j \gg \Sigma v_i$). At the point of initial radiation, the lifetime of free electrons is determined as follows:

$$\tau^- = \frac{1}{u^- \sum_i \sigma_i^- v_i}. \tag{3}$$

Assuming that the probability of recombination of the free holes with localized electrons can be neglected, in comparison with the probability of their localization at the luminescence centers, the lifetime of free holes can be written as follows:

$$\tau^+ = \frac{1}{u^+ \sigma_j^+ \left(v_j - p_i \right)}$$
$$\text{and } \tau_0^+ = \frac{1}{u^+ \sigma_j^+ v_j}. \tag{4}$$

At $v_j \gg \Sigma v_i$, the lifetime of free holes is much shorter than that of free electrons. During X-ray irradiation, the lightsum is accumulated; i.e., the concentrations of localized electrons at traps n_i and the concentrations of the recharged luminescence centers p_j will increase. However, the law of charge conservation is required to satisfy the balance equation:

$$\sum_i n_i = p_j. \tag{5}$$

The irradiation of the semiconductor causes the increased lifetime of free holes τ^+ in comparison with the reference value τ_0^+. But the execution of the inequality $p_j \ll v_j$ allows ignoring the change of τ^+ during the irradiation. For the free electrons, the denominator in relation (2) can be rewritten as follows:

$$u^- \left(\sum_i \sigma_i^- \left(v_i - n_i \right) + \sigma_j^- p_j \right)$$
$$= u^- \left[\sum_i \sigma_i^- v_i + \left(\sigma_j^- p_j - \sum_i \sigma_i^- n_i \right) \right]. \tag{6}$$

If the capture cross section of the recombination σ_j^- and localization σ_i^- for free electrons would be the same, then, by the balance equation (5), the expression in parentheses (6) should be equal to zero and value of τ^- would not change during the irradiation process. But since $\sigma_j^-/\sigma_i^- > 1$ (due to the extra electric charge of the recharged recombination center), the difference $(\sigma_j^- p_j - \Sigma \sigma_i^- v_i) > 0$. It means that the lifetime of free electrons will decrease as the accumulated lightsum increases, in accordance with the kinetic theory of luminescence and conductivity, for three types of traps (shallow (i-1)-type, phosphorescent i-type, and deep (i+1)-type), as well as one luminescence center [36]. The value of the maximum accumulated lightsum on deep traps for electrons and holes does not depend on the intensity of excitation, if there is no thermal or optical delocalization in the process of excitation. The nonradiative recombination of the electron-hole pairs at nonradiative recombination centers occurs according to the same laws as in the luminescence centers. For a multicenter crystal-phosphorus model [36], when several recombination centers are considered, one of these centers can be considered as nonradiative. It fundamentally does not affect the kinetics of luminescence and conductivity, but it is necessary to establish a recombination mechanism (electron or hole) to be realized at this center. Secondly, in a multicenter crystal-phosphorus model, the nonradiative recombination of free holes in deep traps filled with electrons is taken into account.

The processes of phosphorescence and relaxation of conduction current are observed in ZnSe crystal at temperatures 8 and 85 K. At room temperature, the relaxation of conduction current is observed, and the phosphorescence process is not registered due to the temperature quenching of for both luminescence bands at $T > 100$ K. According to [36], the maximum value for the concentration of electrons $(n_{i\infty})$ localized at traps with i-type:

$$n_{i\infty} = \sqrt{\frac{1}{4}\left[\frac{G}{w_i}\left(1 + \frac{v_i}{v_p}\right)\right]^2 + \frac{Gv_i}{w_i}} - \frac{G}{2w_i}\left(1 + \frac{v_i}{v_p}\right). \quad (7)$$

The carried-out experimental verification (Figures 7 and 8) confirmed the above assumption, since the maximum value of phosphorescence intensities (at registration at both luminescence bands of 630 and 970 nm) is more significant at a higher intensity of excitation. Also, [36] shows the dependence for the kinetics of lightsum accumulation at k-trap depending on the irradiation dose:

$$n_k(t) = n_{k\infty}$$
$$\cdot \frac{1 - \exp\left[-\left(n_{k\infty}/v_k + G/w_k n_{k\infty}\right) w_k t\right]}{1 - w_k n_{k\infty}^2/Gv_k \cdot \exp\left[-\left(n_{k\infty}/v_k + G/w_k n_{k\infty}\right) w_k t\right]}. \quad (8)$$

This ratio for shallow traps of (i-1)-type and phosphorescent trap of i-type at the condition $w_{(i-1)}v_{(i-1)}, w_i v_i > G$ has the following form:

$$n_{(i-1)}(t) = n_{(i-1)\infty}\left[1 - \exp\left(-\frac{t}{\tau_{(i-1)}}\right)\right]$$
$$\text{and } n_i(t) = n_{i\infty}\left[1 - \exp\left(-\frac{t}{\tau_i}\right)\right]. \quad (9)$$

The accumulation of charge carries at shallow traps, and their devastation occurs faster than at deeper traps.

At the phosphorescence, the charges carriers are delocalized from all traps (shallow, phosphorescent, and deep), but

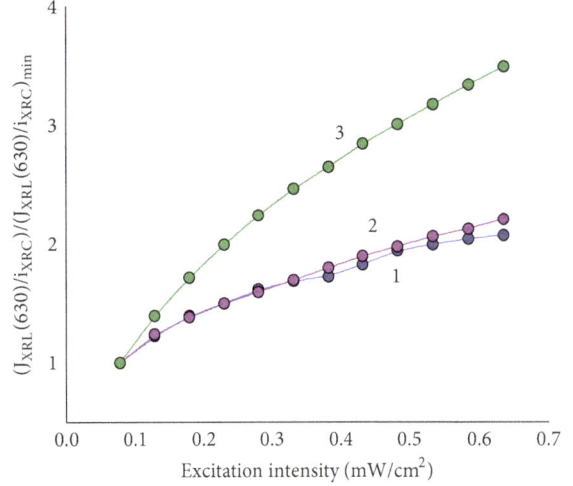

FIGURE 9: Normalized dependencies of the ratio of the intensity of the XRL 630 nm band to the X-ray conductivity current on the excitation intensity at temperatures: 8 K (1), 85 K (2), and 295 K (3).

the registered intensity of the phosphorescence J_{Ph} will be determined, primarily, by the additive sum of the recombinations of delocalized electrons from small and phosphorescent traps:

$$J_{Ph} \sim n_{(i-1)\infty}w_{(i-1)}\left[1 - \exp\left(-\frac{t}{\tau_{(i-1)}}\right)\right]$$
$$+ n_{i\infty}w_i\left[1 - \exp\left(-\frac{t}{\tau_i}\right)\right], \quad (10)$$

where J_{Ph} is the phosphorescence intensity at any point of time after excitation and time t_X in (10) is the duration of X-ray irradiation. The ratio (10) was used for the approximation of experimental dose dependencies of the phosphorescence (Figures 7 and 8).

According to the experimental results (Figures 7 and 8), at a lower intensity of irradiation, the accumulated lightsum in the crystal will be smaller, even at similar doses of radiation. This is true for both types of luminescence centers, which cause broad electronic-vibrational luminescence bands with peaks at 630 and 970 nm.

Another evidence of the reduction of maximum accumulated lightsum at the decreased intensity of excitation during long-term irradiation can be the dependence of the ratio of luminescence intensity to the conduction current value on the intensity of excitation intensity. Since the intensity of luminescence J_{XRL} is proportional to the production of the free charge carriers' concentration and the concentration of the recharged luminescence centers p_j, while the conduction current value i_{XRC} is proportional to the free charge carriers' concentration only, their ratio can be written as follows:

$$\frac{J_{XRL}}{i_{XRC}} \sim p_j(I_X). \quad (11)$$

Figures 9 and 10 show the dependencies of the ratio of luminescence intensity for 630 and 970 nm bands to the X-ray

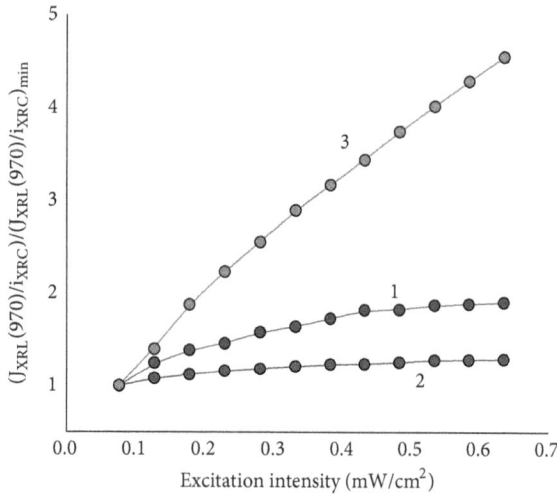

FIGURE 10: Normalized dependencies of the ratio of the intensity of the XRL 970 nm band to the X-ray conductivity current on the excitation intensity at temperatures: 8 K (1), 85 K (2), and 295 K (3).

conduction current on excitation intensity at temperatures: 8, 85, and 295 K for respective luminescence centers. For calculation purposes, the experimental dependencies are shown in Figures 4–6.

In both cases of XRL registration, the increase in the concentration of recharged luminescence centers occurs when the intensity of X-ray excitation increases. It means that when the intensity of the X-ray excitation increases, according to (2), the free electron lifetime in conduction band decreases. As a result, the concentration of free electrons increases unproportionally and sublinearly for the excitation intensity. As a consequence, we derive the sublinear I_X-i_{XRC} behavior in the semiconductor.

The influence of accumulated lightsum within the crystal on the I_X-i_{XRC} behavior indicates that, to obtain a correct experimental dependence, it is necessary to get the maximum accumulated lightsum in the sample for each excitation intensity. Otherwise, different values of current for one excitation intensity can be obtained; i.e., we will observe "hysteresis" at the increase or decrease of the X-ray excitation intensity [24, 25].

The classical theory of photoconductivity [26–32] was developed for a simple crystalline phosphor model (one type of trap and one type of recombination center). In case of an extended period of stationary excitation for the concentration of free charge carriers, it provides for the dependence $N^- \sim \sqrt{I_X}$ and a similar dependence for the concentration of the recharged recombination centers $p_j \sim \sqrt{I_X}$. It ensures a proportional dependence for the luminescence intensity and for the current $- \sqrt{I_X}$. If the experimental dependencies I_X-i_{XRC} are approximated by the provided function, we derive the exponent $\sim 0.8 \div 0.9$, which significantly differ from 0.5.

Therefore, the conduction current dependence on the intensity of the X-ray or γ-radiation can vary in the range from linear $i_{XRC} \sim I_X$ to $i_{XRC} \sim \sqrt{I_X}$. The degree of the sublinearity depends on the traps' concentrations in the sample. The higher the values of the phosphorescence intensities

and relaxation current in the semiconductor are, the more significant the nonlinearity of the lux-ampere characteristic is.

The results of the X-ray and UV excitation of luminescence and conductivity of ZnSe crystals [37] pave the way to study zinc selenide as semiconductor detectors of ionizing radiation for the detection of ionizing radiation.

5. Conclusions

The sublinear behavior of the dependency of the conduction current curves on the intensity of X-ray or γ-radiation (I_X-i_{XRC}) can be explained by the presence of several types of traps for free charge carriers and recombination centers in the semiconductor. According to the theoretical analysis of X-ray conductivity kinetics, the maximum accumulated lightsum at deep traps does not depend on the excitation intensity. Also, for shallow and phosphorescent traps, the accumulated lightsum depends on the excitation intensity: it increases with increasing excitation intensity. According to experimental data analysis, we can assume that the higher the concentration of such defects is, i.e., the higher the intensity of phosphorescence and current relaxation is, the closer the lux-ampere characteristics will be to $i_{XRC} \sim \sqrt{I_X}$. While the concentration of defects (i.e., traps for free charge carriers) decrease, the lux-ampere characteristics will approximate to the linear ones.

Conflicts of Interest

The authors declare that they have no conflicts of interest.

References

[1] A. N. Georgobiani and M. K. Sheinkman, *Physics of AIIBVI Compounds*, Nauka, Moscow, Russia, 1986.

[2] D. D. Nedeoglo and A. V. Simashkevich, *Electric and Luminescence Properties of Zinc Selenide*, Shtiintsa, Kishinev, Moldova, 1984.

[3] V. I. Gavrilenko, A. M. Grekhov, D. V. Korbutyak, and V. G. Litovchenko, *Optical Properties of Semiconductors*, Nauk. Dumka, Kiev, Ukraine, 1987.

[4] N. K. Morozova, V. A. Kuznetsov, and V. D. Ryzhikov, *Zinc Selenide: Receiving and optical properties*, Science, Moscow, Russia, 1992.

[5] G. A. Bordovsky, "X-ray conductivity of High Ohmic Semiconductors," *Sorosovsky Educational Magazine*, vol. 7, no. 3, pp. 84–89, 2001 (Russian).

[6] V. Y. Degoda and A. O. Sofienko, "Specific features of the luminescence and conductivity of zinc selenide on exposure to X-ray and optical excitation," *Semiconductors*, vol. 44, no. 5, pp. 1–7, 2010.

[7] V. Y. Degoda and G. P. Podust, "X-ray conductivity of ZnSe single crystals," *Semiconductors*, vol. 50, no. 5, pp. 579–585, 2016.

[8] V. S. Vavilov, "Physics and applications of wide band gap semiconductors and their practical applications," *Physics-Uspekhi*, no. 3, pp. 287–296, 1994 (Russian).

[9] L. V. Atroshchenko, S. F. Burachas, L. P. Galchinetski, B. V. Grinev, V. D. Ryzhikov, and N. G. Starzhinskii, *Scintillation*

Crystals and Ionization Radiation Detectors on Their Base, Nauk. Dumka, Kiev, Ukraine, 1998.

[10] B. V. Grinyov, V. D. Ryzhikov, and V. P Semynozhenko, *Scintillation Detectors and Radiation Control Systems on Their Basis*, Naukova Dumka, Kiev, Ukraine, 2007.

[11] D. B. Elmurotova and E. M. Ibragimova, "Strengthening electroluminescence of ZnSe single crystals (Te, O) after γ-radiation," *The Physic and The Technology of Semiconductors*, vol. 41, no. 10, 2007.

[12] A. Nasr, A. Aboshosha, and M. Ashour, "Performance Evaluation of Phototransistors and their Behavior under Gamma Radiation Effects," in *Proceedings of the EG0800227 The Second All African IRPA Regional Radiation Protection Congress*, Ismailia, Egypt, 2007.

[13] I. Dafinei, M. Fasoli, F. Ferroni et al., "Low temperature scintillation in ZnSe crystals," *IEEE Transactions on Nuclear Science*, vol. 57, no. 3, pp. 1470–1474, 2010.

[14] N. Starzhinskiy, B. Grinyov, I. Zenya, V. Ryzhikov, L. Gal'chinetskii, and V. Silin, "New trends in the development of AIIBVI-based scintillators," *IEEE Transactions on Nuclear Science*, vol. 55, no. 3, pp. 1542–1546, 2008.

[15] V. D. Ryzhikov et al., "Properties of semiconductor scintillators ZnSe(Te,O) and integrated scintielectronic radiation detectors based thereon," *IEEE Transactions on Nuclear Science*, vol. 48, no. 1, pp. 356–359, 2001.

[16] Y. H. Cho, S. H. Park, W. G. Lee et al., "Comparative study of a CsI and a ZnSe(Te/O) scintillation detector's properties for a gamma-ray measurement," *Journal of Nuclear Science and Technology*, vol. 45, no. 5, pp. 534–537, 2014.

[17] V. Ryzhikov, G. Tamulaitis, N. Starzhinskiy, L. Gal'chinetskii, A. Novickovas, and K. Kazlauskas, "Luminescence dynamics in ZnSeTe scintillators," *Journal of Luminescence*, vol. 101, no. 1-2, pp. 45–53, 2003.

[18] A. O. Sofiienko and V. Y. Degoda, "X-ray induced conductivity of ZnSe sensors at high temperatures," *Radiation Measurements*, vol. 47, no. 1, p. 27, 2012.

[19] A. O. Sofiienko, V. Ya. Degoda, and V. N. Kilin, "Basic Model of the Stationary X-ray Induced Conductivity of Wide-Gap Semiconductors," *Global Journal of Science Frontier Research Physics & Space Science*, vol. 12, no. 3, 2012.

[20] M. S. Brodyn, V. Y. Degoda, A. O. Sofiienko, B. V. Kozhushko, and V. T. Vesna, "Monocrystalline ZnSe as an ionising radiation detector operated over a wide temperature range," *Radiation Measurements*, vol. 65, pp. 36–44, 2014.

[21] E. Krause, H. Hartmann, J. Menninger et al., "Influence of growth non-stoichiometry on optical properties of doped and non-doped ZnSe grown by chemical vapour deposition," *Journal of Crystal Growth*, vol. 138, pp. 75–80, 1994.

[22] V. M. Koshkin, A. Ya. Dulfan, V. D. Ryzhikov, L. P. Gal'chinetskii, and N. G. Starzhinskiyet, "Thermodynamics of isovalent tellurium substitution for selenium in ZnSe semiconductors," *Journal of Functional Materials*, vol. 8, no. 4, pp. 708–713, 2001.

[23] R. N. Bhargava, R. J. Seymour, B. J. Fitzpatrick, and S. P. Herko, "Donor-acceptor pair bands in ZnSe," *Physical Review B*, vol. 20, no. 6, pp. 2407–2419, 1979.

[24] V. Ryzhikov, B. Grinyov, S. Galkin, N. Starzhinskiy, and I. Rybalka, "Growing technology and luminescent characteristics of ZnSe doped crystals," *Journal of Crystal Growth*, vol. 364, pp. 111–117, 2013.

[25] N. K. Morozova, I. A. Karetnikov, V. V. Blinov, and E. M. Gavrishchuk, "A study of luminescence centers related to copper and oxygen in ZnSe," *Semiconductors*, vol. 35, no. 1, pp. 24–32, 2001.

[26] N. K. Morozova, I. A. Karetnikov, V. V. Blinov, and E. M. Gavrishchuk, "Studies of the infrared luminescence of ZnSe doped with copper and oxygen," *Semiconductors*, vol. 35, no. 5, pp. 512–515, 2001.

[27] J. Ji, A. M. Colosimo, W. Anwand et al., "ZnO Luminescence and scintillation studied via photoexcitation, X-ray excitation, and gamma-induced positron spectroscopy," *Scientific Reports*, vol. 6, Article ID 31238, 2016.

[28] A. Janotti and C. G. Van de Walle, "Fundamentals of zinc oxide as a semiconductor," *Reports on Progress in Physics*, vol. 72, no. 12, Article ID 126501, 2009.

[29] C. R. Varney, M. A. Khamehchi, J. Ji, and F. A. Selim, "X-ray luminescence based spectrometer for investigation of scintillation properties," *Review of Scientific Instruments*, vol. 83, no. 10, Article ID 103112, 2012.

[30] F. A. Selim, M. H. Weber, D. Solodovnikov, and K. G. Lynn, "Nature of native defects in ZnO," *Physical Review Letters*, vol. 99, no. 8, Article ID 085502, 2007.

[31] E. I. Adirovich, "Some Problems in the Theory of Luminescence of Crystals," Tech. Rep., State publishing house of technical and theoretical literature, Moscow, Russia, 1956.

[32] M. V. Fock, *Introduction to Kinetics of Luminescence of Phosphor Crystals*, Nauka, Moscow, Russia, 1964.

[33] V. E. Lashkarev, A. V. Lyubchenko, and M. K. Sheinkman, *Nonequilibrium Processes in Photoconductors*, Nauk. Dumka, Kiev, Ukraine, 1981.

[34] V. V. Antonov-Romanovski, *Photoluminescence Kinetics of Phosphor Crystals*, Nauka, Moscow, Russia, 1966.

[35] R. H. Bube, *Photoconductivity of Solids*, John Wiley & Sons, NY, USA, 1960.

[36] V. Y. Degoda, A. F. Gumenyuk, and Y. A. Marazuev, *The Kinetic of Recombination Luminescence and Conductivity of Crystalophosphors*, Taras Shevchenko National University of Kyiv VPT, Kiev, Ukraine, 2016.

[37] V. Y. Degoda, M. Alizadeh, N. O. Kovalenko, and N. Y. Pavlova, "V-I characteristics of X-ray conductivity and UV photoconductivity of ZnSe crystals," *Journal of Applied Physics*, vol. 123, no. 7, Article ID 075702, 2018.

12

Influence of Pressure on the Temperature Dependence of Quantum Oscillation Phenomena in Semiconductors

G. Gulyamov,[1] U. I. Erkaboev,[2] and A. G. Gulyamov[2]

[1]*Namangan Engineering Pedagogical Institute, 160103 Namangan, Uzbekistan*
[2]*Physico-Technical Institute, NGO "Physics-Sun", Academy of Sciences of Uzbekistan, 100084 Tashkent, Uzbekistan*

Correspondence should be addressed to U. I. Erkaboev; erkaboev1983@mail.ru

Academic Editor: Sergio E. Ulloa

The influence of pressure on the oscillations of Shubnikov-de Haas (ShdH) and de Haas-van Alphen (dHvA) in semiconductors is studied. Working formula for the calculation of the influence of hydrostatic pressure on the Landau levels of electrons is obtained. The temperature dependence of quantum oscillations for different pressures is determined. The calculation results are compared with experimental data. It is shown that the effect of pressure on the band gap is manifested to oscillations and ShdH and dHvA effects in semiconductors.

1. Introduction

Currently a variety of experimental methods for the study of influence of pressure on the oscillations ShdH and dHvA are tested in new types of semiconductors. Most quantum oscillation phenomena in semiconductors are due to oscillation density of energy states in a strong magnetic field [1–5].

The works of [6, 7] studied the temperature dependence of the density of states in quantizing magnetic fields. These studies showed that the continuous spectrum of the density measured at the temperature of liquid nitrogen at low temperatures turns into discrete Landau levels. Mathematical modeling of processes using the experimental values of the continuous spectrum of the density of states makes it possible to calculate the discrete Landau levels. The work of [8] examined the effect of temperature on the oscillation dHvA effect using this model. Here, the obtained oscillation dHvA effects take into account the thermal broadening of the Landau levels. However, these studies did not consider the effect of pressure on the effects of oscillations ShdH and dHvA in semiconductors.

The aim of this work is a theoretical study of the influence of hydrostatic pressure on the quantum oscillation phenomena in semiconductors.

2. Method of Calculation of the Density of the Energy States

Consider the dynamics of the free electron gas in a quantizing magnetic field. In the presence of a magnetic field parallel to the z direction, the energies of electrons and holes in the conduction bands and valence bands are as follows:

$$E_c\left(N, k_z\right) = E_g + \left(N + \frac{1}{2}\right)\hbar\omega_c + \frac{\hbar^2 k_z^2}{2m_c} \pm \frac{1}{2}sg\mu_B B \quad (1)$$

$$E_v\left(N, k_z\right) = -\left(N + \frac{1}{2}\right)\hbar\omega_v - \frac{\hbar^2 k_z^2}{2m_v} \pm \frac{1}{2}sg\mu_B B. \quad (2)$$

Here, ω_c and ω_v represent cyclotron frequency of electrons and holes, s represents spin quantum number, and B represents the magnetic field induction.

For parabolic zone [9], $E = \hbar^2 k^2/2m$ and $S = \pi k_\perp^2 = \pi(k^2 - k_z^2)$.

Hence, the cyclotron mass

$$m_c = \frac{\hbar^2}{2\pi}\frac{\partial S}{\partial E} = m. \quad (3)$$

We now find the number of states in the interval between two Landau levels. Using expression (3), let us find the difference between the areas of cross-sections of two equal-energy surfaces, which differ in energy $\Delta E = \hbar \omega_c$:

$$\Delta S = \frac{2\pi m_c}{\hbar^2} \Delta E = \frac{2\pi m_c}{\hbar^2} \hbar \omega_c. \qquad (4)$$

For the determination of the oscillation effects of ShdH and dHvA in the conduction band, primarily we must calculate the oscillations of the density of energy states in quantizing magnetic field. We will consider a box with large but finite sides L_x, L_y, L_z (the main area of the crystal) [9]. As you can see from expressions (1), the second term is called the energy of the electron motion in the plane xy and changing discretely. Hence, the number of states per unit area in a plane $k_x k_y$ will be $L_x L_y / (2\pi)^2$. That is, the number of states between two quantum orbits equals

$$\frac{L_x L_y}{(2\pi)^2} \Delta S = \frac{m \omega_c}{2\pi \hbar} L_x L_y. \qquad (5)$$

From (1) we obtained the following:

$$k_z = \frac{(2 m_n)^{1/2}}{\hbar} \cdot \sqrt{E - \left(E_g + \hbar \omega_c \left(N + \frac{1}{2} \right) \right)}. \qquad (6)$$

Then

$$k_z = \frac{2\pi}{L_z} n_z. \qquad (7)$$

According to expressions (6) and (7) the number of states in the energy range from $(N + 1/2)\hbar \omega_c$ to E is equal to

$$n_z = \frac{(2m)^{1/2}}{\pi \hbar} \cdot \sqrt{E - \left(E_g + \hbar \omega_c \left(N + \frac{1}{2} \right) \right)}. \qquad (8)$$

The dependence of the effective mass from the hydrostatic pressure can be represented by the following expression [10, 11]:

$$m_c^* (P) = m_c^* (0) \cdot \left(1 - \frac{\Delta E_g}{E_g (0)} \right) = m_c^* (0) \cdot \frac{E_g (P)}{E_g (0)}; \qquad (14)$$

or cyclotron frequency depends on the pressure:

$$\omega_c (P) = \frac{eB}{m_c^* (P)}. \qquad (15)$$

Then the total number of quantum states with energies less than E is equal to

$$N (E) = \frac{L_x L_y L_z m^{3/2}}{\pi^2 \hbar^3} \hbar \omega_c$$
$$\cdot \sum_{N=0}^{N_{max}} \left(\frac{E^2 + E_g \left(E - (N + 1/2) \hbar \omega_c \right)}{E_g} \right)^{1/2}. \qquad (9)$$

As a result, we determine the density of the energy states in the presence of a magnetic field to the sample with a parabolic dispersion law:

$$N_S (E, H) = \frac{dN (E)}{dE}$$
$$= \frac{(m)^{3/2}}{(2)^{1/2} \pi^2 \hbar^3} \frac{\hbar \omega_c}{2} \sum_{N=0}^{N_{max}} \frac{1}{\sqrt{E - \left(E_g + \hbar \omega_c (N + 1/2) \right)}}. \qquad (10)$$

As is known, the band gap depends on the magnetic field, temperature, and pressure. The dependence of the semiconductor band gap at hydrostatic pressure changes as follows [10, 11]:

$$E_g (P) = E_g (0) - \beta P. \qquad (11)$$

Here, β represented pressure coefficients, characterizing the change in position of the edges of the valence band and the conduction band with pressure.

The dependence of the Fermi level from pressure can be written in the following form:

$$E_F (P, T) = -\frac{E_g (P)}{2} + \frac{3}{4} kT \ln \left(\frac{m_h^*}{m_e^*} \right). \qquad (12)$$

Then the derivative with respect to the energy from Fermi-Dirac distribution function has the following form:

$$\frac{\partial f_0 \left(E, E_F (P, T), T \right)}{\partial E} = -\frac{1}{kT} \frac{\exp \left(\left(E - \left(-E_g (P) /2 + (3/4) kT \ln \left(m_h^*/m_e^* \right) \right) \right) /kT \right)}{\left[1 + \exp \left(\left(E - \left(-E_g (P) /2 + (3/4) kT \ln \left(m_h^*/m_e^* \right) \right) \right) /kT \right) \right]^2}. \qquad (13)$$

3. Influence of Pressure on the Oscillations Effects of ShdH and dHvA in Semiconductors

It is known that in the case of the density of states Landau quantization is a periodic function of the magnetic field. This leads to oscillations ShdH and dHvA that are periodic in the strong magnetic field. The relaxation time takes the following form: $\tau = \tau_0 E^r$. The exponent r has different values for different scattering mechanisms. For example, in the case of scattering by acoustic vibrations of impurity ions, the exponent is equal to $-1/2$ and $3/2$ [12]. Naturally, the effects of oscillations ShdH and dHvA appear on the change in the density of energy states in semiconductors. Hence, we define

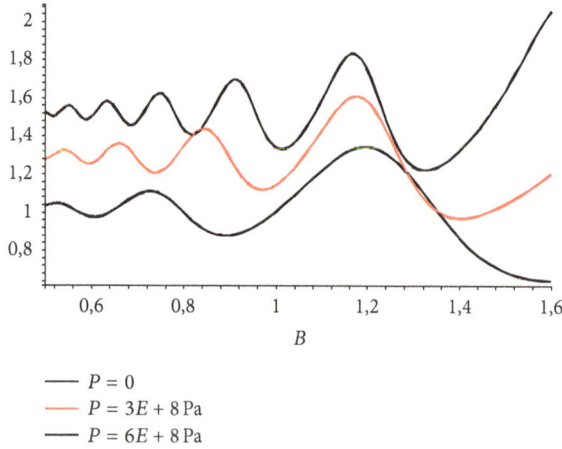

FIGURE 1: Comparison of oscillations effects of ShdH in Si at different pressures.

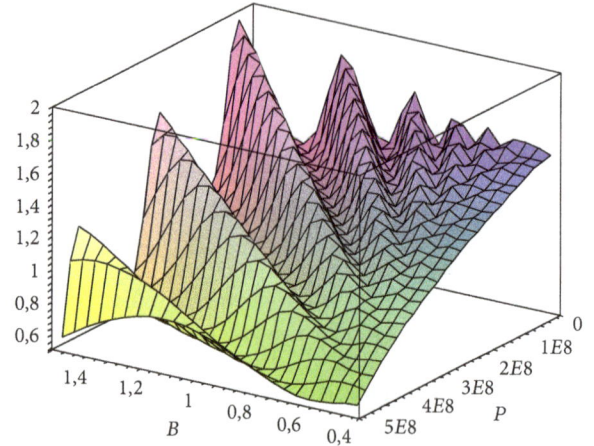

FIGURE 2: The dependence of the longitudinal magnetoresistance on hydrostatic pressure in Si.

the dependence of the oscillation effects of ShdH [9] and dHvA [12] on the full pressure with the help of expressions (10)–(15):

$$\sigma_{zz}(B, T, P) = A \cdot \hbar\omega_c(P)$$

$$\cdot \int_{\hbar\omega_c(P)}^{\infty} \sum_N \frac{\tau_N(E)}{\sqrt{E - \left(E_g(P) + \hbar\omega_c(P)(N + 1/2)\right)}} \quad (16)$$

$$\cdot \left(-\frac{\partial f_0\left(E, E_F(P, T), T\right)}{\partial E}\right) dE,$$

and longitudinal resistance $\rho_{zz}(B, T, P) = 1/\sigma_{zz}(B, T, P)$. Here, $A = -(2m)^{1/2}e^2/\pi^2\hbar^3$.

$$\chi(B, T, P)$$

$$= 2\mu_B^2 \int_0^{\infty} \sum_N \frac{1}{\sqrt{E - \left(E_g(P) + \hbar\omega_c(P)(N + 1/2)\right)}} \quad (17)$$

$$\cdot \left(-\frac{\partial f_0\left(E, E_F(P, T), T\right)}{\partial E}\right) dE.$$

Here, μ_B represents Bohr magneton. χ represents magnetic susceptibility.

Now, we must determine the critical pressure (P_k) in a strong magnetic field. If the pressure is equal to or greater than the critical value ($P \geq P_k$), the Landau levels begin to shift from the edges of the conduction band. For the calculation of critical pressure, consider the simplest case: $N = 0$, $(1/2)\hbar\omega_c - \beta P_k = 0$, or $P_k = (1/2\beta)\hbar\omega_c$. Consider estimation for semiconductor Si: ($\beta = -1.5 \cdot 10^{-11}$ eV/Pa) at $B = 2$ T (20 kGs) [10, 11]. $P_k = (1/2\beta)\hbar\omega_c \approx 2 \cdot 10^8$ Pa. This means that, for Si, if the pressure of $P \geq 2 \cdot 10^8$ Pa, it will change shape oscillations ShdH and dHvA.

Now, we get the graphics effects oscillations ShdH and dHvA by means of formulas (16) and (17). Figure 1 shows the dependence of the oscillations effects of ShdH and dHvA on the hydrostatic pressure in Si at low temperatures. As

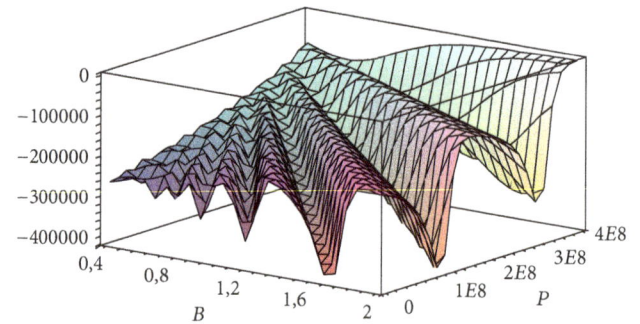

FIGURE 3: Influence of the hydrostatic pressure on the oscillations of magnetosusceptibility in Si.

seen from these figures, with increasing pressure, the shape of the Landau levels strongly changes. In Figures 2 and 3, three-dimensional image oscillations ShdH and dHvA are shown at different pressures in Si. With increasing pressure to $5 \cdot 10^8$ Pa semiconductors Si was observed a decrease the number of Landau levels oscillations ShdH and dHvA at $T = 5$ K. Figure 4 shows the oscillations of the longitudinal magnetoresistance in Si at different temperatures and pressures. With increasing temperature, the pressure of the Landau levels is noticeably reduced. From Figure 4 it is seen that without pressure and at a temperature of 40 K Landau levels are manifested, but $P = 6 \cdot 10^8$ Pa and $T = 40$ K oscillations disappear. Figure 5 shows the influence of pressure on the temperature dependence of the oscillations ShdH in the three-dimensional image.

4. Influence of Nonparabolicity Energy Bands on the Oscillations Longitudinal Magnetoresistance in Narrow-Gap Semiconductors

The work of [4] investigated the oscillations longitudinal magnetoresistance at hydrostatic pressures up to 1 GPa in

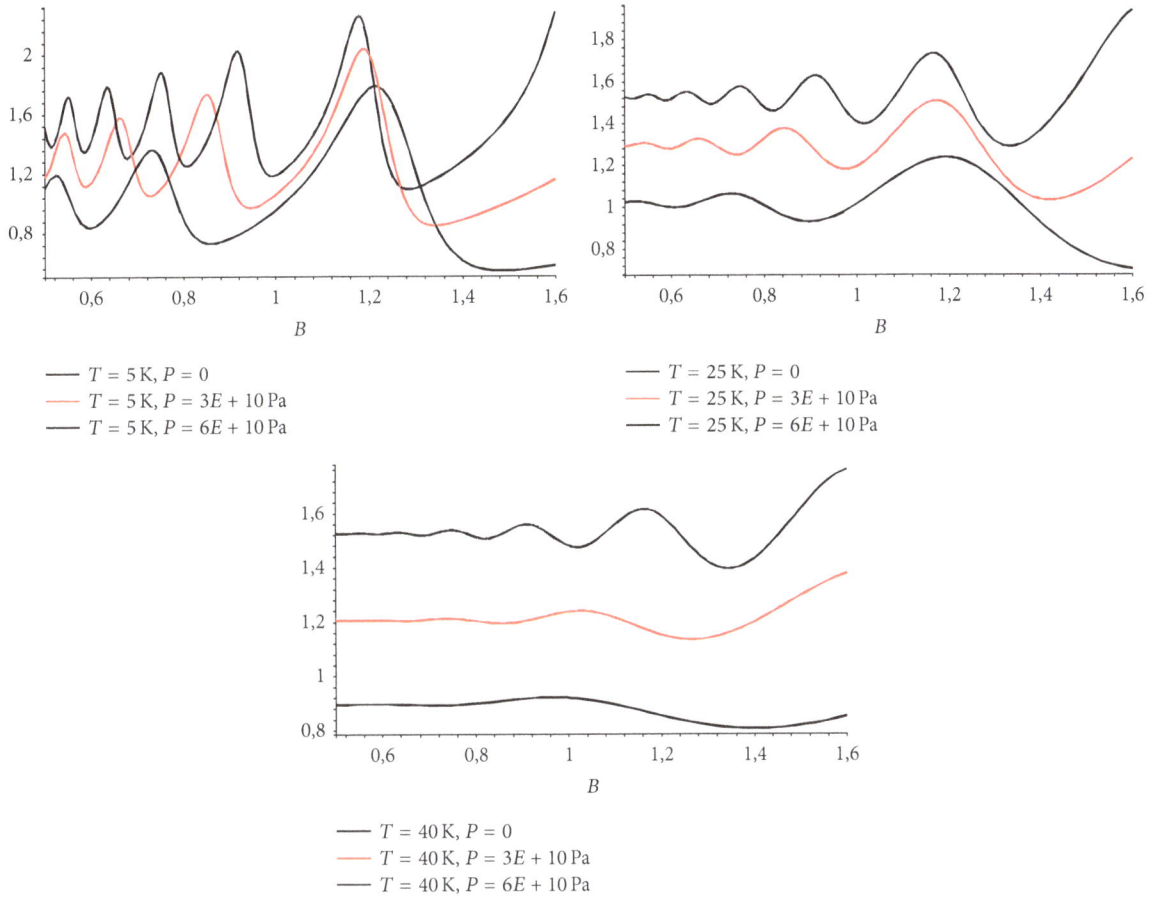

FIGURE 4: Influence of pressure on the temperature dependence of the oscillations ShdH in Si.

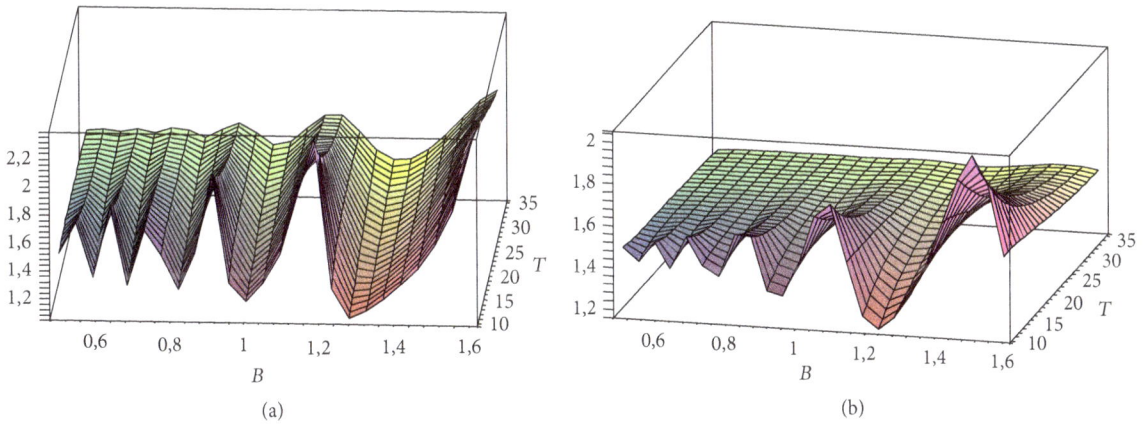

FIGURE 5: Temperature dependence of the oscillations ShdH in Si. (a) At no pressure; (b) at the pressure, $P = 6 \cdot 10^8$ Pa.

$HgSe_{1-x}S_x$. In this work, the quantum oscillations of ShdH were observed at $T = 4.2$ K. Figure 6 shows the oscillations ShdH of pressure $P = 0,38$ GPa in the samples with $x = 0.104$ [4]. In these semiconductors bandgap varies from 0.1 eV to 0.4 eV, for different values of x [4, 13]. An important feature

of the semiconductor with a narrow band gap is a strong nonparabolicity conduction band [13]. The limiting case of Kane's model is realized. The low value of the band gap narrow-gap semiconductors leads to its stronger dependence on pressure, temperature, and external fields. The work of

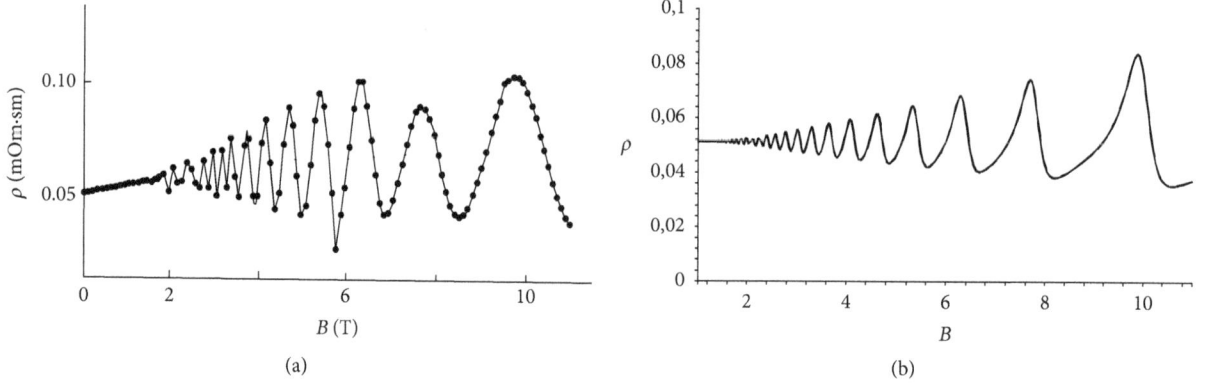

FIGURE 6: Oscillations effects of ShdH in HgSe$_{0.896}$S$_{0.104}$ at $T = 4.2$ K. (a) Experimental data [4]; (b) calculated from formula (19).

[14] obtained a formula for the density of energy states in quantizing magnetic field with Kane dispersion law:

$$N_S^n (E, H)$$

$$= \frac{(m)^{3/2}}{(2)^{1/2} \pi^2 \hbar^3} \frac{\hbar \omega_c}{2} \sum_{N=0}^{N_{max}} \frac{2E/E_g + 1}{\sqrt{E^2/E_g + E - (N + 1/2) \hbar \omega_c}}. \quad (18)$$

Here, $N_S^n(E, H)$ refers to the density of energy states in a quantizing magnetic field with Kane dispersion law. E_g refers to band gap without pressure. E represents energy free electrons in the conduction band. Influence of pressure on the oscillations effect of ShdH with the Kane dispersion law according to formula (18) is determined from expression (16):

$$\sigma_z^n (B, P) = A \cdot \hbar \omega_c (P)$$

$$\cdot \int_{\hbar \omega_c(P)}^{\infty} \sum_N \frac{2E/E_g (P) + 1}{\sqrt{E^2/E_g (P) + E - (N + 1/2) \hbar \omega_c (P)}} \tau_N (E) \quad (19)$$

$$\cdot \left(-\frac{\partial f_0 (E)}{\partial E} \right) dE.$$

Here, $\sigma_z^n(B, P)$ stands for longitudinal conductivity with a nonparabolic dispersion law.

As a result, we obtain graph dependence of the effect of oscillations ShdH on the pressure in HgSe$_{1-x}$S$_x$ with nonquadratic dispersion law. Figure 6(b) shows oscillation longitudinal resistance in HgSe$_{0.896}$S$_{0.104}$ at $P = 0,38$ GPa. These figures show the oscillations effects of ShdH at low constant temperatures. The working equation (19) makes it possible to build charts oscillations ShdH of the sample at different temperatures and pressures.

5. Conclusion

The influence of pressure and temperature on the oscillations effects of ShdH and dHvA is considered in semiconductor. An analytical expression for the longitudinal magnetoresistance in semiconductors with Kane dispersion law for electrons is obtained. The calculation results are compared with experimental data. It is shown that the effect of pressure on the band

gap is manifested to oscillations and ShdH and dHvA effects in semiconductors. The above results are valid when there is not any Lifshitz transition or any other pressure-induced phase transition.

Conflicts of Interest

The authors declare that they have no conflicts of interest.

References

[1] P. D. Grigoriev, "Theory of the Shubnikov-de Haas effect in quasi-two-dimensional metals," *Physical Review B*, vol. 67, no. 14, Article ID 144401, 2003.

[2] I. L. Drichko, A. M. D'yakonov, I. Y. Smirnov, Y. M. Gal'perin, V. V. Preobrazhenskiĭ, and A. I. Toropov, "Role of Si-doped Al$_{0.3}$Ga$_{0.7}$as layers in the high-frequency conductivity of GaAs/Al$_{0.3}$Ga$_{0.7}$ as heterostructures under conditions of the quantum hall effect," *Semiconductors*, vol. 38, no. 6, pp. 702–711, 2004.

[3] I. M. Lifshitz, M. Y. Azbel, and M. I. Kaganov, *Electron Theory of Metals*, part 2, Nauka, Moscow, Russia, 1971 (Russian).

[4] V. V. Shchennikov, A. E. Kar'kin, N. P. Gavaleshko, and V. M. Frasunyak, "Magnetoresistance of HgSeS crystals at hydrostatic pressures of up to 1 GPa," *Physics of the Solid State*, vol. 39, no. 10, pp. 1528–1532, 1997.

[5] É. A. Neĭfel'd, K. M. Demchuk, G. I. Kharus et al., "Shubnikov-de Haas oscillations in HgSefbffkFefbfftand HgSefbffkCofbfftunder hydrostatic pressure," *Semiconductors*, vol. 31, no. 3, pp. 261–264, 1997.

[6] G. Gulyamov, U. I. Erkaboev, and N. Y. Sharibaev, "Effect of temperature on the thermodynamic density of states in a quantizing magnetic field," *Semiconductors*, vol. 48, no. 10, pp. 1287–1292, 2014.

[7] G. Gulyamov, U. I. Erkaboev, and N. Y. Sharibaev, "Simulation of the temperature dependence of the density of states in a strong magnetic field," *Journal of Modern Physics*, vol. 5, no. 8, pp. 680–685, 2014.

[8] G. Gulyamov, U. I. Erkaboev, and N. Y. Sharibaev, "The de Haas-Van Alphen effect at high temperatures and in low magnetic fields in semiconductors," *Modern Physics Letters B*, vol. 30, no. 7, Article ID 1650077, 2016.

[9] I. M. Tsidilkovsky, *Electrons and Holes in Semiconductors*, chapter 5, Nauka, Moscow, Russia, 1972 (Russian).

[10] A. L. Polyakova, *Deformation of Semiconductors and Semiconductor Devices*, chapter 1, Energy, Moscow, Russia, 1979.

[11] G. L. Bir and G. E. Pikus, *Symmetry and Deformation Effects in Semiconductors*, part 6, Nauka, Moscow, Russia, 1972.

[12] A. I. Anselm, *Introduction to the Theory of Semiconductors*, chapter 8, Nauka, Moscow, Russia, 1978.

[13] N. N. Berchenko, V. E. Krevs, and V. G. Sredin, *Semiconductor Solid Solutions and Their Application (Reference Tables)*, chapter 1, Military, Moscow, Russia, 1982.

[14] G. Gulyamov, U. I. Erkaboev, and P. J. Baymatov, "Determination of the density of energy states in a quantizing magnetic field for model Kane," *Advances in Condensed Matter Physics*, vol. 2016, Article ID 5434717, 5 pages, 2016.

13

Droplet Drying Patterns on Solid Substrates: From Hydrophilic to Superhydrophobic Contact to Levitating Drops

Sujata Tarafdar [ID],[1] Yuri Yu. Tarasevich [ID],[2]
Moutushi Dutta Choudhury,[3] Tapati Dutta,[1,4] and Duyang Zang[5]

[1]Condensed Matter Physics Research Center (CMPRC), Physics Department, Jadavpur University, Kolkata 700 032, India
[2]Laboratory of Mathematical Modeling, Astrakhan State University, Astrakhan 414056, Russia
[3]Centre for Advanced Studies in Condensed Matter and Solid State Physics, Department of Physics,
 Savitribai Phule Pune University, Pune 411 007, India
[4]Physics Department, St. Xavier's College, Kolkata 700 016, India
[5]Functional Soft Matter and Materials Group (FS2M), Key Laboratory of Material Physics and Chemistry under Extraordinary
 Conditions, School of Science, Northwestern Polytechnical University, Xi'an, Shaanxi 710129, China

Correspondence should be addressed to Sujata Tarafdar; sujata_tarafdar@hotmail.com

Academic Editor: Charles Rosenblatt

This review is devoted to the simple process of drying a multicomponent droplet of a complex fluid which may contain salt or other inclusions. These processes provide a fascinating subject for study. The explanation of the rich variety of patterns formed is not only an academic challenge but also a problem of practical importance, as applications are growing in medical diagnosis and improvement of coating/printing technology. The fundamental scientific problem is the study of the mechanism of micro- and nanoparticle self-organization in open systems. The specific fundamental problems to be solved, related to this system, are the investigation of the mass transfer processes, the formation and evolution of phase fronts, and the identification of mechanisms of pattern formation. The drops of liquid containing dissolved substances and suspended particles are assumed to be drying on a horizontal solid insoluble smooth substrate. The chemical composition and macroscopic properties of the complex fluid, the concentration and nature of the salt, the surface energy of the substrate, and the interaction between the fluid and substrate which determines the wetting all affect the final morphology of the dried film. The range of our study encompasses the fully wetting case with zero contact angle between the fluid and substrate to the case where the drop is levitated in space, so there is no contact with a substrate and angle of contact can be considered as 180°.

1. Introduction

The study of drying droplets, especially those containing colloidal particles or salts, has become a topic of wide interest in recent years. This is evident from the fact that international conferences on droplets are regularly organized (Droplets Conferences and EMN Droplets) and books exclusively devoted to this subject have been published [1–8]. There are, in addition, several excellent review articles on droplets [9–21]. Different features of this problem, such as the rate of drying, evolution of the drop geometry, and the final pattern formed, depend on a number of parameters, notably the composition of the drying fluid, ambient conditions during drying, and the substrate which supports the drop.

The widespread interest in the drying droplets with inclusions stems from important and innovative applications, mainly in medical science and in technology. When the fluid in the droplet evaporates, the solid material left behind can be distributed in a wide variety of patterns on the substrate. Inclusions may be salts, nanoparticles in the form of nanorods, nanotubes, or any other shape, starches, proteins, and so on. Patterns formed can range from a simple ring at the periphery of the droplet, the so-called coffee ring [22, 23], to multiple rings forming bands, fractal and

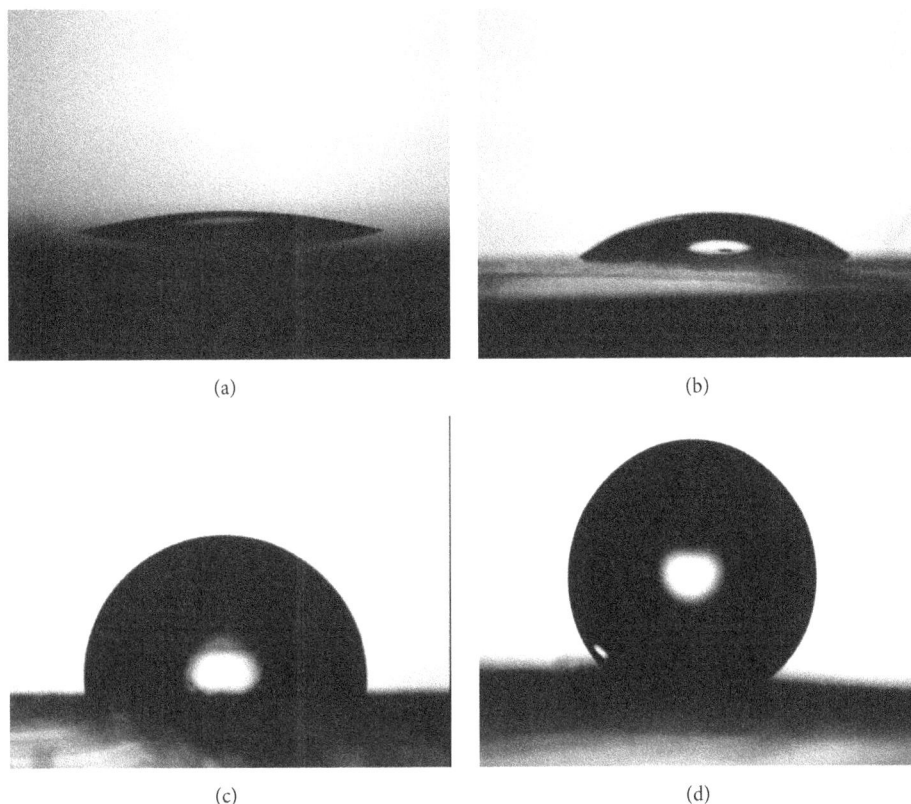

(a)

(b)

(c)

(d)

FIGURE 1: 5 μl drop of water (a) on glass substrate (treated with Piranha solution), (b) on nontreated glass substrate, (c) on polystyrene (PS) substrate, and (d) on hydrophobic TEFLON substrate.

multifractal aggregates of salt crystals, or nanoparticles [24–27]. In addition, the dried drop may develop crack patterns [8, 28–43], which can also be induced by external fields [35, 44]. It is important to realize that the shape of the droplet plays a crucial role in generating these patterns. Unless the drop is too large, its shape may be roughly approximated by a section of a sphere or a spherical cap. In some cases, when a crust or skin forms on the free surface [45], buckling instability may develop [46–49]. Nonuniform evaporation over different regions of the drop surface generates convection currents that determine mass transfer. Temperature and concentration gradients also develop [50–53], leading to surface tension gradients, resulting in thermal Marangoni flow [54–59] or/and solutal Marangoni flow [60]. Obviously, drying out of a large flat film of fluid would lead to a different situation. Figure 1 presents sessile droplets on hydrophilic and hydrophobic surfaces. A thin section of a sphere, where the height at the center is small compared to the lateral dimension, represents a sessile drop that wets the surface well (Figure 1(a)). Here, the angle of contact is very small. On the other hand, a section much larger than a hemisphere represents a strongly hydrophobic contact (Figure 1(d)).

Another possible geometry of a drop in contact with a solid surface is the *pendant drop*, which is suspended from a support *above* it, like a rain drop hanging from a leaf. The pendant drop has also been studied [69], but we do not discuss it further in the present article.

To eliminate the effect of substrate, one may turn to the pendant drop hung by a thin needle or nonwetting drop supported by a superhydrophobic surface. In this case, though its shape was quasi-spherical, the boundary condition for evaporation was, however, influenced by the contact points, thus in turn influencing the evaporation flux. Via levitation techniques, for instance, electrostatic, magnetic, and acoustic levitation, the contact of solid substrate can be completely avoided; thus it somewhat can provide a more uniform evaporation flux along the drop surface. But for most of levitation techniques, the levitation force to balance gravity is surface force, not body force; therefore, the natural convection cannot be suppressed. To further study evaporation without the influence of natural convection, experiments under microgravity, that is, levitation experiments in space, are expected.

Recent experiments on levitated droplets [70] represent the extreme case of a spherical drop with no contact to a substrate. This situation is relevant in space research. In the present review, we aim to discuss the whole range of contact angles and how it affects the morphology of the dried residue. Interplay of related factors such as the chemistry of the materials, interface tensions involved, and physical properties such as elastic or viscoelastic moduli and ambient drying conditions all need to be taken into consideration [71]. There can also be external perturbations such as electric/magnetic fields, heating/cooling of the substrate, or mechanical perturbations. We try to discuss these factors and their contribution as far as possible within the brief span of this review.

We excluded from our consideration some specific topics such as evaporation of sessile droplets on inclined (including vertical), patterned (textured), dissolvable, and porous substrates. To obtain some information about these topics, we can refer the reader to [72–76]. Besides, we omitted all effects connected with artificially inhomogeneous evaporation such as obstacles [77], masks [78], modulated gas phase convection [79], and infrared heating [80], which can force or impede evaporation at particular parts of the free surface of the drop.

The medical science application of the droplet problem utilizes characteristics of the droplet patterns found in dried body fluids, such as blood, serum, tear drops, saliva, and cervical fluid, for identifying diseases in patients. The technique offers a simple and inexpensive method in pathological diagnostics, which has been in use for several decades [3, 81].

A more recent technological innovation is the use of dried droplets with conducting inclusions such as silver nanoparticles or carbon nanotubes (CNT) to create a nearly invisible but connected conducting network on a transparent surface [27, 82, 83]. The demand for such transparent conductors is extremely high in photovoltaic devices of everyday use. The droplet method has a potential to fabricate such surfaces easily and inexpensively. There are other applications related to, for example, ink-jet printing [84] and high throughput drug screening [60].

2. Experimental Studies

In this section, we give an overview of experimental studies on evaporating droplets. We have tried to classify the large body of earlier work into subsections to facilitate the discussion. However, separating the experiments into clear-cut nonoverlapping sections is not always possible. We hope the reader will bear with us in this attempt to present an array of a colourful mass of tangled threads, sorted out as far as possible.

2.1. Evaporation of a Pure Fluid. A sessile drop of a pure liquid, such as water or methanol, when placed on a solid substrate evaporates completely without leaving any residue. The points of interest here are the spreading rate of the drop, the equilibrium shape reached, and how it evolves with time up to complete evaporation. These observations help in understanding the more complicated situations that follow. Obviously, the nature of fluid and ambient are important as well as the properties of the substrate and interaction between fluid and substrate [85]. So we have to consider the substrate here as well, but we leave situations where the substrate (or its absence) has a crucial role to Section 2.2.1.

A liquid drop on a solid surface evolves spontaneously until it reaches its equilibrium with minimum surface area consistent with Laplace's equation (see, e.g., [86]). The shape of the drop on a surface depends on the surface energy of the substrate and the liquid drop, as well as the interface energies between different phases. An additional energy is needed to create an interface between the liquid and solid; this is called

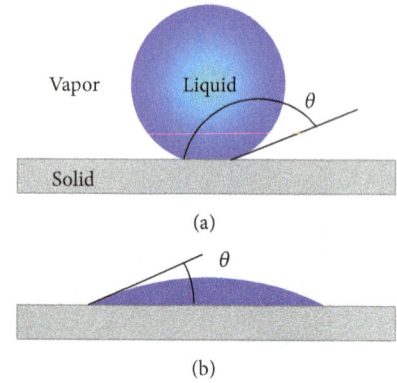

FIGURE 2: Contact angle between a drop and a substrate: (a) nonwettable substrate and (b) a partially wetted substrate.

the interface free energy. In equilibrium, a liquid drop satisfies Young's equation (see, e.g., [86]).

$$\gamma_s = \gamma_{sl} + \gamma_l \cos\theta, \qquad (1)$$

where γ_s, γ_{sl}, and γ_l represent the energy of the surface, solid-liquid interface, and the liquid-vapor interface, respectively. θ is the contact angle between liquid and the surface (Figure 2).

The effects of vapor adsorption and spreading pressure are not considered here. Further, the drop is considered to be much smaller than the capillary length l_c, which is defined as

$$l_c = \sqrt{\frac{\gamma_s}{\rho g}}; \qquad (2)$$

here, ρ is the density of the liquid and g is the acceleration due to gravity [1, 86]. Conventionally, the term "drop" is used when the volume of the liquid drop is more than 100 μl and for lower volumes the term "droplet" is used [87].

Assuming the drop to be a spherical cap, the drop boundary on the substrate, that is, the three-phase contact line (TPCL), where solid, liquid, and vapor meet, has the shape of a perfect circle, for an ideal smooth surface. In this article, we do not elaborate on rough or prepatterned surfaces, where there may be interesting deviations [88, 89]. We also leave out from our discussion sessile drops describing works on soft or liquid surfaces.

As the liquid evaporates with time, the drop can shrink in two ways [90] (Figure 3):

(1) The TPCL remains "pinned" to the substrate, leaving the circular solid-liquid contact surface constant; this is referred to as the *"Constant Contact Radius (CCR) mode"* of drying. Obviously, the angle of contact will decrease with time in this case.

(2) Alternatively, the *radius* of the TPLC can decrease, while the angle of contact remains constant. This is called the *"Constant Contact Angle (CCA) mode."*

These are two extreme cases but many real experiments show mixed behavior, where both the contact angle and the contact radius vary (so-called stick-and-slip mode; see, e.g., [91]).

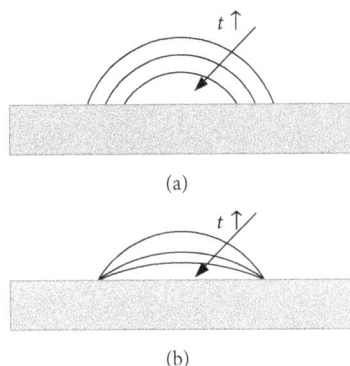

FIGURE 3: (a) Constant Contact Angle (CCA) mode and (b) Constant Contact Radius (CCR) mode.

CCR is the favored mode for drying on high energy surfaces. For example, in case of Piranha treated glass substrate, the contact angle of a water droplet is less than 15° and it is pinned during drying [92, 93]. On the other hand, in the case of low energy surfaces, for example, a TEFLON coated substrate, the contact angle is more than 150° and the mode of drying is CCA. However, things get complicated in the case of liquid drying on a PMMA or polystyrene substrate. The equilibrium contact angles are 70° and 90°, respectively, but the drying mode switches from CCR to CCA mode with time (Figure 4) [94]. Yu et al. [95] explain the switching from CCR to CCA mode from a thermodynamic point of view. Since we are not discussing rough or patterned surfaces, we refer the interested reader to relevant literature in this area. Modifications of Young's equation required for such problems are discussed in [96–99]. The question of contact angle hysteresis (CAH) also needs to be addressed here [100]. The effect of CAH on pattern formation is discussed in [101].

2.2. Evaporation of a Mixture.

In Section 2.1, we discussed how the shape and size of an evaporating drop of pure liquid change during drying. However, once the liquid evaporates completely, no trace of the drop is left, if the substrate is hard and inert. In this section, we come to more interesting observations. What happens when a drop of a solid-liquid suspension or a salt solution evaporates? What pattern does the solid residue leave on the substrate? Study of this phenomenon as a physics problem started with Deegan et al.'s [22] work on the "coffee stain." Noticeably, at the same time, similar independent researches have been published by Parisse and Allain [90, 102]. Moreover, pattern formation in desiccated drops of biological fluids has been known in medical community even several decades earlier [103, 104].

2.2.1. Evaporation of a Suspension of Micro- or Nanoparticles.

It is a common observation that when a spilt drop of coffee dries, it leaves a dark ring along the periphery of the droplet, the so-called *coffee-stain* pattern. Initially, the drop looked uniform, so why do the microscopic grains of coffee crowd along the edge during drying, leaving the central portion nearly clean?

Deegan et al. [22, 23] explained that this happens for a pinned boundary (CCR mode). This causes "capillary flow" internal currents from the center towards the TPCL, carrying the suspended solid particles to the boundary, where they get deposited in the form of a ring.

The necessary conditions for formation of the coffee stain are (i) evaporation and (ii) the fact that the drop should be pinned to the substrate during evaporation (Figure 5). Note that an alternative mechanism for coffee-ring deposition based on convection [105] and on active role of free surface [106] has been proposed.

Different patterns of drying are noted when the above conditions are not satisfied. If the TPCL does not remain pinned to the substrate but slips and sticks during drying, a series of concentric rings with varying radius are formed [91, 107, 108].

There is another important effect that competes with the capillary flow causing "coffee-ring effect" and leads to deposition of the suspended particles near the center of the drop. On the other hand, in some cases, the whole solid is deposited near the center of the drop [13, 23, 109]. This is more prominent when the angle of contact is large and is named the *Marangoni effect* [54, 110–112]. It arises due to a gradient in surface tension. There are two sources of such a gradient. First of all, a gradient in surface tension can be produced by a temperature gradient created by different rates of evaporation along the surface of the drop (thermal Marangoni effect). Since evaporation absorbs latent heat, regions near the TPCL cool more than the surface near the top of the drop. This leads to a convection current from the center of the drop surface to the periphery then to center of the bottom of the drop and again to the center of the surface. The temperature gradient in turn produces a gradient in surface tension. This drives the suspended particles along with the fluid in circulating paths from the top of the drop downward towards the center (Figure 6). Here they may get adsorbed on the surface or move towards the periphery and recirculate with the vortex formed. Notice that direction of the Marangoni flow depends on the relative thermal conductivities of the substrate and liquid, $k_R \equiv k_S/k_L$, reversing direction at a critical contact angle over the range $1.45 < k_R < 2$ [85]. It is important to note that a *single* circular flow or vortex is not the only possibility. It has been demonstrated that Marangoni convection induced by thermal conduction in the drop and the substrate can result in multiple vortices, depending on the ratio of substrate to fluid thermal conductivities, the substrate thickness, and the contact angle [113]. The Marangoni effect has been clearly demonstrated in organic liquids. Hu and Larson [111] show that, during evaporation, PMMA microparticles suspended in octane collect near the center of the drop. In water, however, the coffee-ring is observed. Another origin of a Marangoni flow is concentration gradient. Thermal and solutal Marangoni effects may compete with one another [60].

Marangoni effect should always be present in clean fluids but is suppressed in water if surfactants are present. According to Hu and Larson [111], it is difficult to avoid presence of trace surfactants even in "pure" water, so the coffee ring dominates.

(a)

(b)

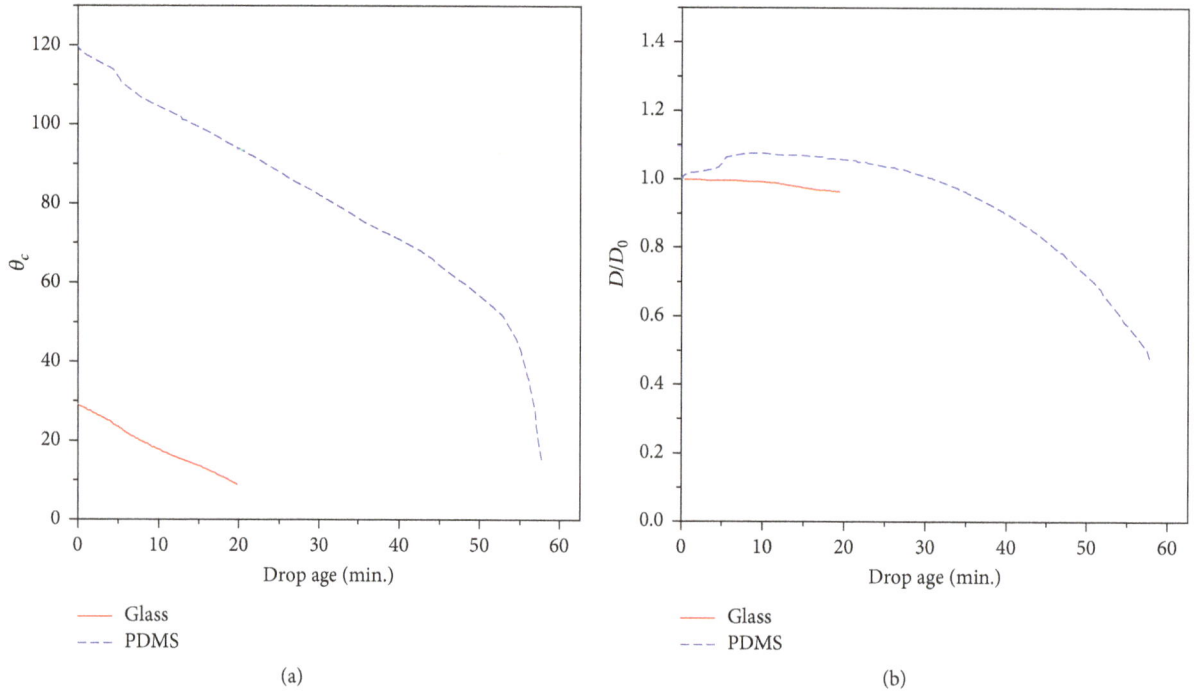

FIGURE 4: (a) Contact angle variations for a water droplet nontreated glass substrate (red solid curve) and PDMS substrate (blue-dashed curve); (b) normalized diameter of same drops on glass substrate (red solid curve) and PDMS substrate (blue-dashed curve).

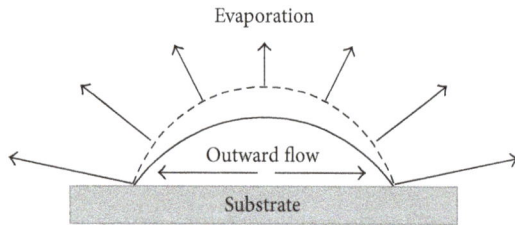

FIGURE 5: Evaporation and TPCL pinning as the origin of outward flow inside evaporating sessile droplet [22]. Initial position of the free surface is shown as the dashed line, whereas its new position due to evaporation is shown as the solid line.

FIGURE 6: Thermal Marangoni effect is produced by inhomogeneous cooling of the free surface due to evaporation. The temperature gradient in turn causes a surface tension gradient.

The competition between the capillary effect leading to the coffee ring and Marangoni effect is decided by several parameters including the thermal conductivities of the substrate and fluid [85, 113–117] and a Marangoni number (thermal) (Ma) has been defined to characterize this [54, 118]:

$$\text{Ma} = -\frac{\partial \gamma}{\partial T} \frac{H \Delta T}{\mu \kappa}. \qquad (3)$$

Here, ΔT is the temperature gradient, μ is the viscosity of the fluid, κ is the thermal diffusivity, the surface tension of the fluid is γ, and H is the height of the drop. So, hydrophobicity or high contact angle of the liquid on the substrate is needed for getting higher H to induce stronger Marangoni flow inside the droplet.

Once the factors deciding the dominance of capillary flow or Marangoni effect have been identified, it becomes possible to control the deposition pattern and choose whether a thin

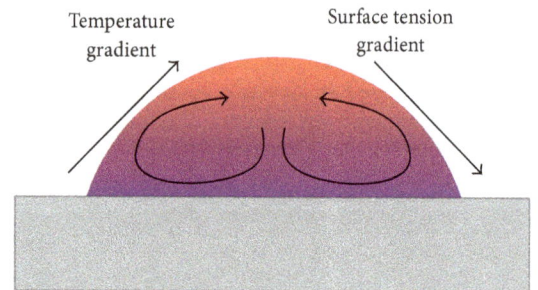

ring or a uniformly covered circular patch or a small spot at the center of the drop is desired. Besides physical and chemical properties of the fluid and substrate, manipulating the temperature distribution using heated/cooled substrates or microheaters has also been used to tailor the deposition pattern [119–121] for various applications. The effect of particle shape has also been found to play a role in the final desiccation pattern; ellipsoid-shaped particles suppress the coffee-ring effect as shown in several works [109, 117].

2.2.2. Evaporation of a Salt Solution. In Section 2.2.1, we considered a fluid drop containing suspended solid particles of nm to μm size, which do not interact with each other. Now we discuss the case of a *dissolved salt* in a suitable liquid. So here we have dissociated ions uniformly distributed in a liquid drop. Unlike the colloid particles in Section 2.2.1,

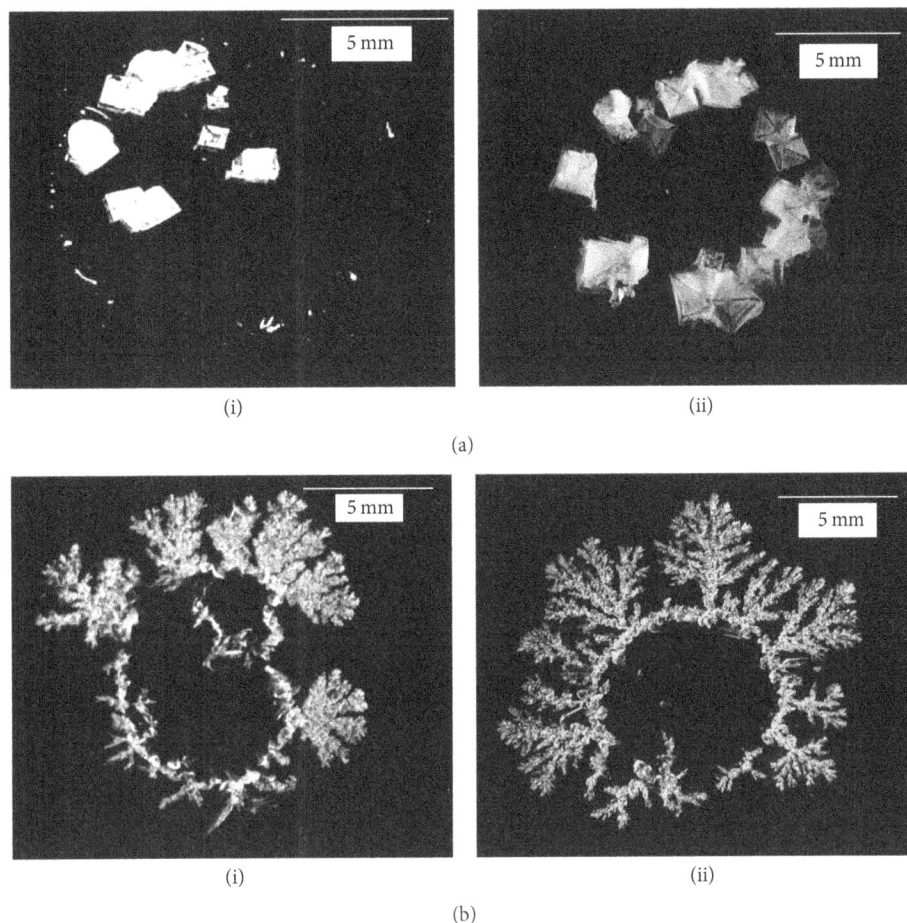

FIGURE 7: (a) and (b) are dried drops (50 μl) of NaCl and Na$_2$SO$_4$ solutions on Piranha treated glass substrates. (i) and (ii) represent different concentrations (0.05 M and 0.1 M, resp.) of salts in the solutions.

the ions tend to form *crystals* as the drop dries and salt concentration increases.

Dutta Choudhury et al. [24, 26] evaporated droplets of aqueous NaCl solution of different concentrations on a hydrophilic substrate. They found that the capillary flow dominates here and the salt collects along the pinned TPCL in the form of a ring of cubic crystals. The size and structure of the crystals depend on salt concentration and evaporation rate. For large concentrations, there is a tendency to form *hopper crystals*. Hopper crystals [122, 123] are somewhat like empty boxes and form when growth at the corners and edges is faster compared to the faces (Figure 7(a)). Droplets of aqueous copper sulphate solutions also display the coffee-ring effect, with crystals forming along the periphery of the drop [124–126]. However, Shahidzadeh et al. [124] did not find coffee rings on drying aqueous NaCl. Rather they found that the salt accumulates near the center of the dried drop. In their experiments, the crystals formed initially at the periphery but were later pushed towards the center. The conflicting findings of Dutta Choudhury et al. [24] and Shahidzadeh et al. may be due to finer details of the experiment. It is possible that fine imperfections/roughness on the surface in the case of the former group pinned the NaCl crystals

at the drop periphery, without allowing them to be pushed inward.

Other interesting forms of crystallization have been reported as well. Sodium sulphate crystallizes in two forms—thenardite and mirabilite. At high relative humidity (RH), a drop of aqueous sodium sulphate shows on drying tree-like growth of mirabilite on the substrate *outside the drop* in addition to crystal formation within the drop boundary (Figure 7(b)). This peculiar growth has been explained as follows: crystals formed along the TPCL release water of crystallization which seeps out from the drying droplet and the dendritic trees grow from these. At low RH, thenardite crystals grow within the drop boundary [125, 127].

2.2.3. Evaporation of a Complex Fluid Drop Containing Salt.
If the host liquid is a complex non-Newtonian fluid and salt is added to it, the solid residue after evaporation exhibits most beautiful patterns. Such patterns on dried biological fluids have long been studied [103, 104, 128–137], particularly as a tool for diagnosis of certain diseases [3, 62, 138–141]. Before discussing the salt added complex fluid, we should examine what happens in absence of the salt.

If a drop of gelatinized starch or gelatin or an aqueous solution of a polymer like polyethylene oxide is allowed to dry, usually a uniform circular film is produced. Depending on the nature of the fluid and substrate, the film either sticks to the substrate or can be cleanly pulled off [142].

Gelatin, starch, or a similar medium which increases the viscosity of the solution by orders of magnitude can change pattern formation of salts drastically. The flow fields within the drop get suppressed and salt crystals no longer show their normal morphology. The role of gelatin in tuning crystal growth has been known for a long time [26]. Goto et al. [143] demonstrate how tuning concentration of a drying gelatin and NaCl drop and manipulating concentration gradients create unique morphologies such as orthogonal or oblique lattices and curving patterns. They show that nonequilibrium growth conditions lead to various dendritic patterns.

If salts are present in the droplet, they usually crystallize during drying. Morphology of the salt crystals is very sensitive to the kind of salt, concentration, and type of colloidal particles, as well as the rate of evaporation [24–26, 29, 142, 144–148]. This sensitiveness allows using the morphology of salt crystals as an indicator, for example, to diagnose different diseases by drying drops of biological fluids [139, 149].

On addition of a salt, the salt crystallizes in different forms, often several different crystallization modes appear as drying proceeds, and video recordings observed under an optical microscope are fascinating to watch (see videos; a drop of aqueous gelatin solution with a little NaCl salt is allowed to dry; large crystals of salt form, followed by multifractal dendrites in-between and a drop of gelatin containing sodium sulfate forms patterns on drying; concentric rings and dendrites growing from them can be seen). Sodium chloride in gelatinized potato starch solution forms dendrites or hopper crystals of different morphology, depending on experimental conditions [24]. In gelatin, two distinct modes of pattern formation were observed [26, 142]. Formation of initial faceted rectilinear crystals of macroscopic dimensions (of ~mm) size was followed by a fine dendritic network observable only under a microscope. The faceted crystals appear to consist of NaCl only, while the dendrites may be a composite; further analysis is needed here. The dendritic pattern consists of fine self-similar branches meeting at right angles and has been shown to be multifractal [26]. Other salts also form interesting patterns when dried in a gelatin drop. Sodium sulphate forms a series of rings, which are grouped in bands, and dendritic crystals grow from the rings [126]. Copper sulphate forms feather-like patterns, with anisotropy evident from images under crossed polarizers [126]. These patterns also reveal fractal characteristics [150]. Effect of albumin concentrations on sodium chloride crystallization from drying drops of albumin-salt solutions has been experimentally investigated [147].

The processes occurring during the desiccation of the sessile colloidal droplets and morphology of the resulting precipitate depend on many different factors, for example, the nature and shape of the colloidal particles [109] and their initial volume fraction [147], the presence of admixtures (e.g., surfactants) in the solution [50, 107, 151, 152], ionic strength and pH of the solution [28], the properties of substrate

(thermal conductivity, whether hydrophilic/hydrophobic) [24, 85, 153, 154], and evaporation mode [32, 36, 155–157].

Surfactants have a strong effect on Marangoni flow as already discussed. A particle-surfactant mixture thus can modulate the capillary flow [18, 107, 152]. Colloidal sulphur with salt self-assembles into a wide variety of patterns, depending on conditions [158]. Different sodium salts are formed in situ, using different acids—monobasic, dibasic, and tribasic. For HCl, the typical NaCl aggregation pattern was observed (similar to [26]). For dibasic sulphuric acid, morphology similar to the two crystalline forms of sodium sulphate was observed and tribasic, citric acid produced a uniform deposition without any definite pattern.

Classification of possible desiccation modes may be done using two characteristic times, namely, the drying time, t_d, and the gelation time, t_g [28]. There are three different modes of colloidal sessile droplet desiccation [28]:

(1) $t_g \gg t_d$, where t_g is the gelation time and t_d is the desiccation time. The gelled phase occurs near the droplet edge and moves inward, while the central area of the droplet remains liquid.

(2) $t_g \approx t_d$. The gelled skin covers the free droplet surfaces. This thin shell cannot prevent evaporation of the solvent. The buckling instability occurs [48].

(3) $t_g \ll t_d$. The phase transition from sol to gel in the whole bulk of the droplet is almost instantaneous. The gelled droplet loses solvent via evaporation very slowly.

When $t_g \gg t_d$, the desiccation process can be divided into several stages (see, e.g., [63, 159, 160]):

(1) Initial single-phase liquid stage: the whole droplet is a sol. The outward flow carries suspended particles to the droplet edge until the volume fraction of the suspended particles, Φ, reaches the critical value, Φ_g. Note that particle-enriched region is extremely narrow, whereas the particle volume fraction in the central area of the droplet is almost constant along its radius. This stage was simulated in [161–163] and in [159].

(2) Intermediate two-phase stage: a gelled ring appears near the droplet edge and grows towards the droplet center. The volume fraction of the colloidal particles is constant inside the *foot*, that is, the outer gelled band, Φ_g, and almost constant in the sol, Φ, except for a rather narrow area near the phase front. This stage was simulated in [164–166] and in [159, 167].

(3) Final single-phase solid stage: the gelled deposit loses the remaining moisture very slowly. Some real fluids of interest (e.g., biological fluids) can contain both suspended particles and dissolved substances. In this case, the dendritic crystals can occur in the central area of a sample [25, 29, 60]. Finally, the desiccation crack patterns appear [8, 28–31, 33, 36, 38].

2.2.4. Biological Fluids. Most biological fluids can be regarded as a complex fluid containing salts. But we devote a

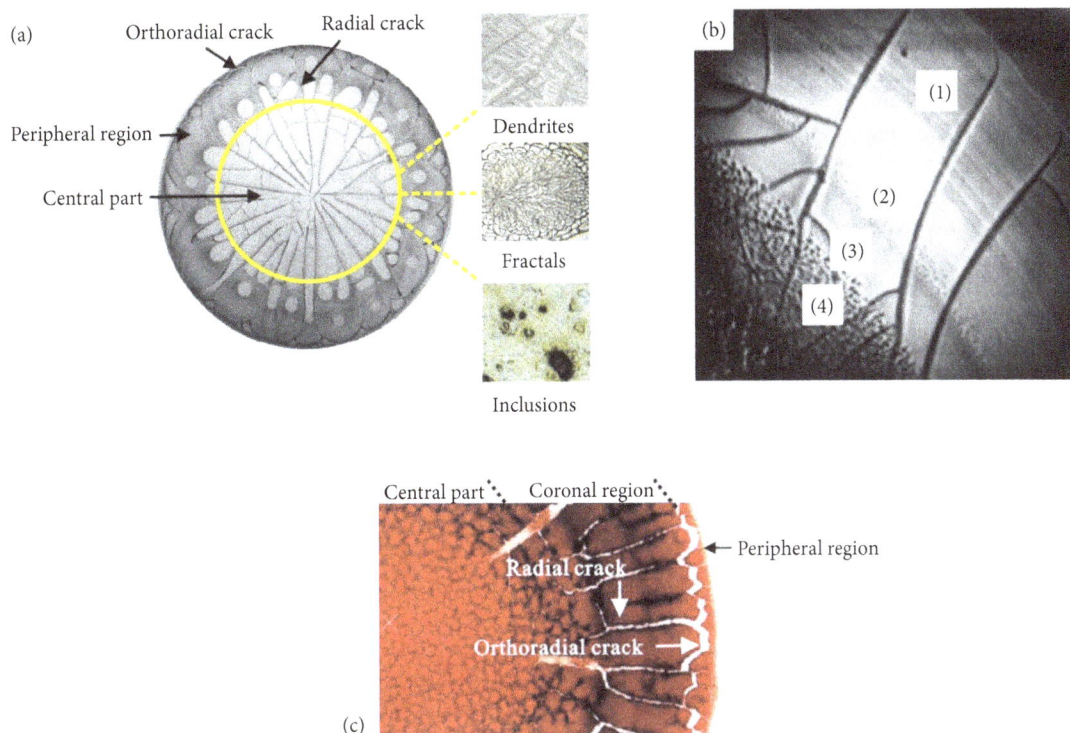

FIGURE 8: Desiccation patterns in the sessile drop of (a) blood plasma from a healthy adult (adopted from [61] with the permission of Springer and from [62] with the permission of Hindawi); (b) morphological details of four regional patterns in the sessile drop of BSA saline solution: (1) homogeneous protein film, (2) protein precipitates, (3) protein gel, and (4) salt crystal [63] (reproduced with permission of Elsevier); (c) whole blood droplet drying on a glass substrate (adopted from [64] with the permission of Cambridge University Press). The whole figure is reproduced from [65].

separate section to these liquids, because the complexity in composition and the bioactive nature of the constituents of body fluids make the evaporation of such liquid drops quite different from laboratory prepared micro or nanofluid drops.

Although pattern formation can be observed during drying of both inorganic and organic colloids, the case of biological fluids is increasingly attracting attention from the scientific community in recent years [64, 168, 169] as this study promises to be a simple and cost-effective technique for diagnosis [20]. We shall discuss further details of application in medicine in Section 4. In this section, we will review some of the experimental observations of drying on drops of human blood serum, whole blood, and other biofluids such as urine, saliva, and tears.

While performing the experiments seems very simple (table top experiments producing interesting patterns can be done even at home), understanding the physics behind the pattern formation phenomena turns out to be extremely complicated and involves a number of interrelated processes of different nature [10, 12]. During desiccation of biological fluids, a sequence of various physical and physicochemical processes can be observed [63, 170]. For example, redistribution of the components occurs. Protein molecules are carried by flows to the edge of the droplet and accumulate to form a gel. The salt is distributed over the whole area of the droplet almost uniformly. After complete drying of the droplet, a protein precipitate remains on the substrate in the form of a

ring; the width of the ring depends on concentrations of the protein and the salt [3, 171, 172]. Salt crystals can form fractal (dendritic) structures [24, 29, 142, 173, 174]. In the later stages of drying, a sample may crack [8, 28, 29]; the characteristic pattern of the cracks also helps in diagnosing diseases from which the subject may be suffering [175].

Blood serum mainly contains 90% (by mass) of water, 6% of macromolecular proteins, 1% of inorganic electrolytes, and other minor components [20]. Although its composition is complicated, blood serum behaves like a Newtonian fluid. Desiccation patterns of a sessile drop of blood serum drying on a solid substrate are generally characterized by two distinguishable regions: a peripheral region and a central part. Cracking patterns in orthoradial and radial directions are observed throughout the whole sessile drop, and crystal patterns with different morphologies accumulate in the central part [61], as shown in Figure 8(a).

Three major factors influence the morphologies of these serum patterns: concentration of inorganic salts, concentration of macromolecular proteins, and the wettability of the droplet on the substrate. Inorganic salts are essential for formation of crystal patterns in the central part [50]. High concentration of salts promotes the aggregation of macromolecular proteins, thus changing the morphologies of crystal patterns [176]. The wettability of the substrate plays an important role in determining the apparent contact angle of the sessile drop [177]. This may further influence the TPCL motion and the mass transportation during drying and in turn the desiccation patterns.

Whole human blood behaves like a non-Newtonian fluid unlike blood serum. It is composed of plasma (55% by volume) and cellular components (45% by volume) (i.e., red blood cells (RBCs), white blood cells (WBCs), and platelets); RBCs, WBCs, and platelets represent 97%, 2%, and 1% of the total volume of these cellular components, respectively. Desiccation patterns in the dried sessile drop of whole human blood are significantly different from those of blood plasma without cellular components [64, 153]. Patterns in the blood droplet from healthy adults dried on glass substrates consist of three distinct zones with different characteristic cracking patterns (Figure 8(b)), namely, a fine peripheral region adhering to the substrates, a coronal region with regularly ordered radial cracks and large-sized deposit plaques, and a central part with disordered chaotic cracks and small-sized deposit plaques [64]. Desiccation patterns of the blood droplet are significantly influenced by the external drying conditions, such as the RH and the wettability of the blood droplet on the substrates [36, 38, 157].

Characteristic desiccation patterns are also formed in the sessile drops of other biofluids (e.g., urine, saliva, and tear fluids) [11, 178, 179]. Yakhno et al. investigated the drying of sessile drops of urine and saliva from the healthy adults and divided it into three stages: the redistribution of materials leads to the continuous flattening of the droplet; the deposited macromolecular proteins aggregate to form the gel matrix; the inorganic salts induce phase transition of macromolecular proteins to form desiccation patterns [11]. The desiccation patterns of the tear droplet are characterized by a thin amorphous film in the peripheral region, with fern-like patterns in the central part. The thin amorphous film in the peripheral region has crack patterns as observed by SEM. The fern-like patterns in the central part are composed of cubic crystals and dendritic patterns adjacent to them. Energy dispersive X-ray analysis (EDXA) results revealed that dendritic patterns were predominantly made up of sodium and chloride, while cubic crystals were potassium and chloride [180].

Despite the application of the phenomena for practical purposes and considerable progress in understanding of the phenomena [10, 12], the theoretical description of the pattern formation in desiccating biological fluids is still incomplete. The physical, biophysical, biochemical, biological, and physicochemical processes occurring in the dehydration of biological fluids remain largely to be clarified.

Analysis based on a visual comparison of the structures formed by drying a liquid drop [3, 4, 81, 181, 182] has significant drawbacks. Conclusions are liable to be subjective, without techniques for defining quantitative parameters to characterize the structures. Computer pattern recognition may be tried to eliminate this shortcoming [36, 61, 183].

Other samples of biological origin which form liquid crystal phases have shown interesting patterns on drying droplets. For example, DNA [184] forms a ring-like deposit with zigzag patterns [185]. Cetyltrimethylammonium Bromide drops with salt were also found to form concentric rings and crystalline aggregates near the center [186].

2.3. Evaporation of Drops under Levitation. So far we discussed sessile droplets that sit on a solid substrate during drying. Even for the superhydrophobic case with contact angles approaching 180°, the dried pattern left on the substrate is two-dimensional or quasi-two-dimensional (when the deposits pile up forming rings or aggregates). The aspect ratio of the deposit, that is, the maximum height, divided by the drop radius is much less than 1. Is it possible for a drying drop to reduce to a spherical shell after evaporation? It turns out that this can be achieved by drying a levitated drop floating in air with zero contact to any surface. Suspending the drop by properly adjusted acoustic fields is the most convenient technique for this study. Radiative heating leads to evaporation of the solvent.

Tijerino et al. [66] evaporated droplets with nanosilica suspensions under different conditions of acoustic amplitudes and solid concentration to obtain residues with varied shapes like rings, bowls, and spheroids (Figure 9). Wulsten et al. [70] used acoustic levitation to investigate the effect of a solvent containing different polymers and the drug itraconazole on drying. The morphology of the residue was found to depend on the drug, while the polymer determined the drying rate. Single-phase droplets [70, 187] and binary or multiphase droplet [188, 189] have been studied using similar techniques.

Although liquid drops can be levitated by various levitation techniques ranging from electrostatic to diamagnetic [190], the acoustic technique is more suitable for evaporation experiments as the electromagnetic properties of the materials are not relevant here. The drying patterns of colloidal droplets depend on the particle-particle interactions and are significantly affected by the substrates as well. With the increase of substrate contact angle, the final residue pattern can be coffee-ring-like stains (Figure 10(a)) on hydrophilic substrate [37] and bowl-shaped residue (Figure 10(b)) on a hydrophobic substrate ($\Theta \sim 90°$) [67]. Under acoustic levitation, a ring-shaped residue has been obtained (Figure 10(c)) [68], which exhibits a geometrical similarity with the initial dog-bone shape of the acoustically levitated droplet [191]. These results highlight the important role played by the drop profile in drying pattern formation, and acoustic levitation provides the possibility to study drop evaporation at zero-contact condition for different initial shapes.

2.4. Desiccation Crack Patterns on Drying Drops. As a colloidal drop dries, viscosity increases and a sol-gel transition may occur. Stress accumulates nonuniformly due to the droplet shape and inhomogeneity of the fluid. This often leads to formation of cracks with characteristic patterns. In addition to the *crack patterns*, other quasi-3D effects are of interest such as wrinkling, buckling, and skin formation.

Desiccation crack patterns have been intensively investigated both experimentally and theoretically [28–32, 34, 36–41, 192]. State of the art may be found in the recently published book [8]. We discuss here some studies on crack formation in drying drops [16, 24, 29, 193].

The final drying pattern and crack nucleation vary with the kinetics of the evaporation rate. During solvent evaporation, curvature of the solvent-air meniscus is responsible for

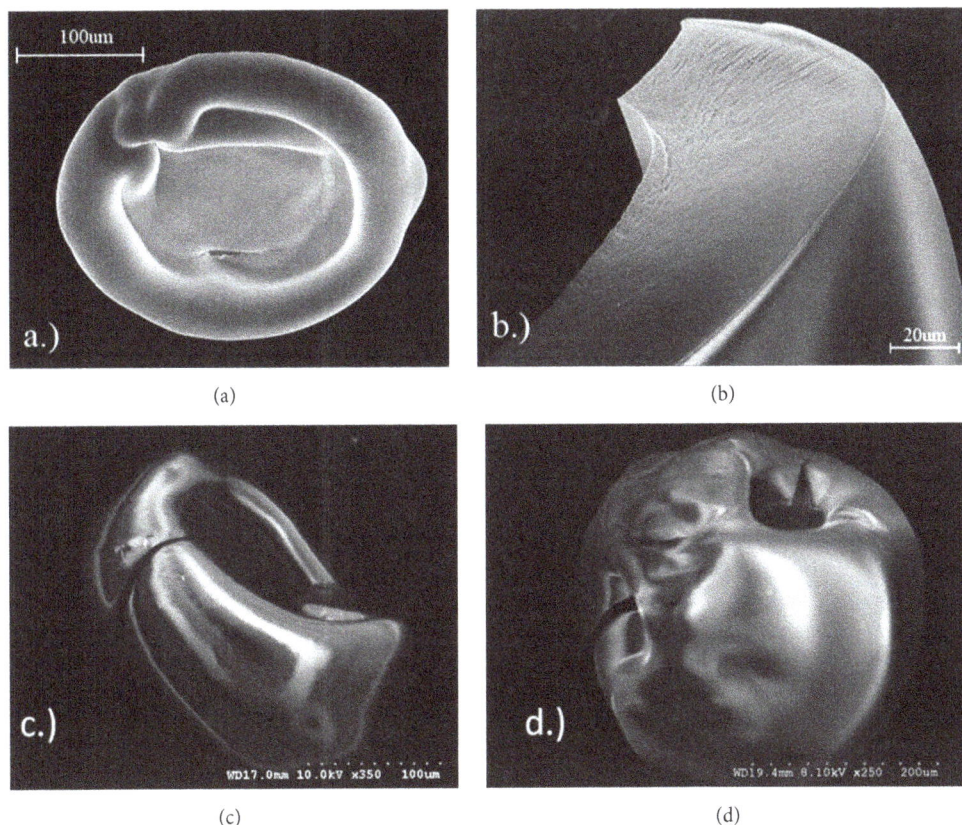

FIGURE 9: Different residues obtained from levitating droplets (reproduced from [66], with the permission of AIP Publishing).

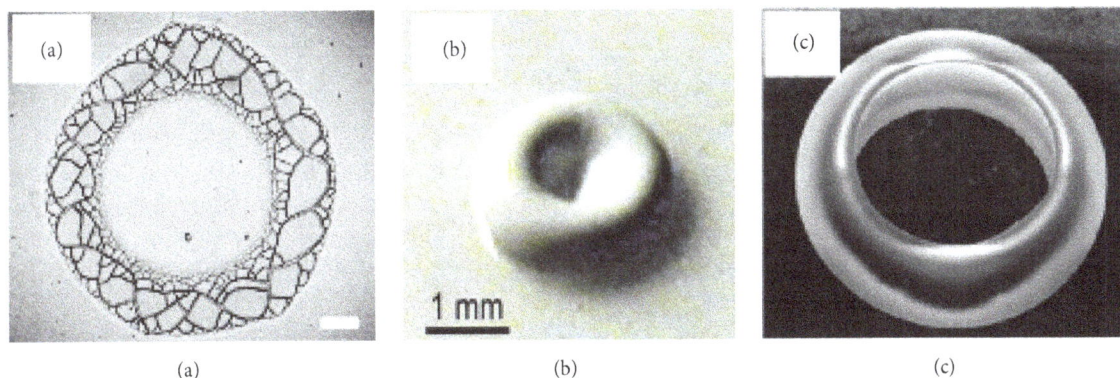

FIGURE 10: Different drying patterns obtained from varied conditions. (a) Coffee ring-like stain on hydrophilic substrate, reprinted with permission from [37]; (b) bowl-shaped relic on hydrophobic surface, reprinted with permission from [67]; (c) ring-shaped residue obtained from acoustic levitation, reprinted with permission from [68].

a capillary pressure in the liquid phase. The capillary pressure induces shrinkage of the porous matrix, which is constrained by the adhesion of the deposit to the glass substrate. As tensile stresses build up, the internal stresses become too great and fractures appear to release mechanical energy. The differences in pattern formation arise due to the competition between the drying process and the adhesion of the matrix on the substrate.

Annarelli et al. [29] worked on the evaporation, gelling, and the cracking behavior of a deposited drop of protein solution, bovine serum albumin. They observed that the cracks appearing at the gelling edge were regularly spaced and were a result of the competition between evaporation-induced evolution and relaxation-induced evolution. When the crack evolution is only evaporation-induced, the mean crack spacing is proportional to the layer thickness. However, in the case of a drop of bovine serum albumin, the evolution of cracks has been described in relation to the change with time of the average shrinkage stress. In this case, the mean crack spacing was observed to be inversely proportional to the deposit thickness. This is unexpected as normally crack spacing increases with thickness.

Brutin and his group worked on the pattern formation of desiccating droplets of human blood from which the coagulation protein had been removed [7, 64]. They studied the dynamics of the process of evaporation of a blood droplet using top-view visualization and the drop mass evolution during the drying process. Brutin et al. [64] showed that there are two distinct regimes of evaporation during the drying of whole blood. The first regime is driven by convection, diffusion, and gelation, while the second regime is only diffusive in nature. A diffusion model of the drying process allows a prediction of the transition between these two regimes of evaporation. Concentration of the solid mass in the drop was important and fracture occurred at a critical mass concentration of solid in a drying drop of blood. They showed that the final crack patterns formed on drying droplets of blood collected from a healthy person, anemic person, and hyperlipidemic person are quite different. But drawing conclusions for definite diagnosis is not so straightforward as the crack patterns are strongly affected by external conditions such as the ambient relative humidity and the nature of the substrate.

Brutin et al. [64] conclude that the final drying pattern and crack nucleation vary with the kinetics of the evaporation rate. The transfer of water to air is limited by diffusion and is controlled by the relative humidity in the surrounding air. The drying process of a sessile drop of blood is characterized by an evolution of the solution into a gel saturated with solvent. When the gel is formed, the new porous matrix formed by the aggregation of particles continues to dry by evaporation of the solvent, which causes the gel to consolidate.

Carle and Brutin [154] studied the influence of surface functional groups and substrate surface energy on the formation of crack patterns and on the dry-out shape in drying a water-based droplet of nanofluid. They have also studied desiccation of blood droplets [153] on different substrates such as glass and glass coated with gold or aluminium. They measured the rate of heat transfer from the substrate to the fluid drop. They show that wettability of the substrate by the fluid is the decisive factor, which can account for the differences in the morphology of the desiccated blood drop on different surfaces, rather than the thermal diffusivity which determines rate of heat transfer from the substrate to the drop. On metallic surfaces, where the drop is nearly hemispherical and a glassy skin forms on the fluid-air interface, there are hardly any cracks. On a glass surface, on the other hand, where the drop is more or less flat, an intricate pattern of cracks form. Figure 11 shows spiral cracks typical of albumin; these can be easily observed by drying egg white [8].

The study of desiccation crack patterns on clay-gel droplets dried in a static electric field [35] led to some interesting results. The number of cracks formed and the time of first appearance of the cracks could be related to the field strength through exponential relations. In a further set of observations, the field was applied for a very short finite duration and then switched off. Now the time required for crack appearance after switching off could be related to the

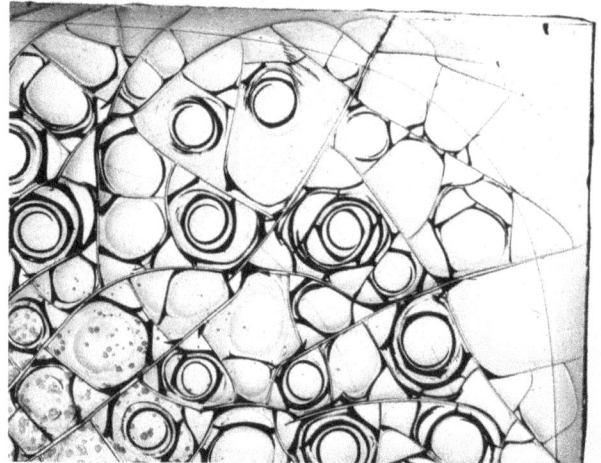

FIGURE 11: Crack patterns in a dried sample of albumin (courtesy of N.A. Koltovoi).

field strength and exposure time, when all quantities were appropriately scaled.

In a follow-up of this work [194], where platinum electrodes were used to avoid chemical interactions during drying, it was shown that the clay drop behaves like a leaky capacitor and could be modelled using methods of generalized calculus [195].

Effect of the drop constituents on *crack speed* has also been studied. Zhang et al. show that a colloidal solution of polytetrafluoroethylene particles cracks radially as it dries [196]. Crack speed varies as a power law with thickness and surfactants such as sodium dodecyl sulfate can be used to tune the cracking.

2.5. Desiccation under Perturbation. While there are several works on desiccating droplets done by different groups, there are very few studies on drying droplets in the presence of a perturbation. The contact angle of a conductive aqueous laden drop with organic or inorganic solutes or ambient oils changes with the application of alternating current (AC) voltage during drying. Banpurkar et al. [197] studied the above effects in experiments to demonstrate the potential of electrowetting-based tensiometry. Contact angle (θ) decreases with increasing amplitude (V_{AC}) of AC voltage following the linear relation of $\cos\theta$ with V_{AC}. They applied low frequency AC voltage and obtained interfacial tensions from $5\,\text{mJ/m}^2$ to $72\,\text{mJ/m}^2$, in close agreement with the macroscopic tensiometry for drop volumes between 20 and 2000 nL. Vancauwenberghe et al. [198] reviewed the effect of an electric field on a sessile drop. They observed that an external electric field can change the contact angle and shape of a droplet. The electric field also affects the evaporation rate during drying. The contact angle is not always an increasing function of the magnitude of the applied electric field but may be a decreasing function for some liquid droplets as well.

Studies have also been done on the effect of external perturbations during drying. The effect of electric fields has been studied on cracks formed in drops of drying clay gels

[35, 194]. Khatun et al. [35] and Hazra et al. [194] investigated desiccation cracks on drying droplets of aqueous Laponite solution in the presence of a static electric field (DC). The electric field had cylindrical geometry, with the peripheral electrode being an aluminium wire [35] or platinum wire [194] bent into a circular form with diameter of ~1.8 cm. A drop of Laponite gel was deposited inside this wire loop. Another aluminium/platinum wire with its tip touching the lower substrate through the centre of the drop acted as the central electrode. Typical cracks had radial symmetry and were found to emerge always from the positive electrode. The cracks formed even when the field was applied for a few seconds and then switched off; this was interpreted as a *memory effect*.

Sanyal et al. [199] experimentally investigated nanoparticle aggregation and structure alteration of evaporating sessile colloidal droplets when subjected to low frequency vibrations of the substrate. Low frequencies perturbed the droplet when the corresponding vibrational wavelength was comparable to the size. For forcing frequencies in the resonance band of lowest allowable mode, the change in the overall morphology of the deposit structure from natural droplet drying was pronounced, with a sharper wedge at the periphery. Recirculation and subsequent outer flow near the droplet edges created higher particle concentration, leading to faster growth of the peripheral wedge. For frequencies away from resonance, the internal flow was mostly uniform, leading to less enhanced wedge structure, similar in appearance to the case of natural evaporation. They demonstrated that, by using forced vibration, desired control of particle deposition could be achieved for various applications.

2.6. Interacting Droplets. We briefly mention another recent interesting line of research here. There are studies showing that a drying drop influences the pattern formation on another drying drop, provided that the distance between them does not exceed a certain limit. This is because the presence of another drop enhances the RH on the side nearer the other drop. So evaporation at the near side is less than that at the far side when two drops are placed side by side. Pradhan and Panigrahi [200] give an experimental as well as a simulation study of this problem.

Cira et al. [201] showed experimentally that two-component droplets of well-chosen miscible liquids such as propylene glycol and water deposited on clean glass at distances of up to several radii apart moved towards each other. This occurred over a wide range of concentrations and even when both droplets had the same concentration. The droplets increased speed as they approached each other. These long-range interactions were preserved even across a break in the glass slide. They explained that the two neighbouring droplets each lie in a gradient of water vapor produced by the other. This gradient causes a local increase in RH and thus decreased evaporation of the thin film on the adjacent portions of the droplets, breaking symmetry. The decreased evaporation leads to an increased water fraction in the thin film, hence increasing the interface tension between liquid and vapor for the film, denoted by $\gamma_{LV,film}$ locally. Asymmetric $\gamma_{LV,film}$ around the droplet causes a net force that drives the

droplets towards each other. Cira et al. [201] proposed a mathematical model to obtain the net force acting on each droplet as follows:

$$F = 2\gamma_{LV,droplets} m R \int_0^\pi \frac{(1 - RH_{room}) R \cos\psi}{\sqrt{d^2 + R^2 + 2Rd\cos\psi}} d\psi, \quad (4)$$

where m is the slope of a plot of apparent contact angle versus the relative humidity, RH; d is the distance between the droplets; ψ is a parameter of integration; RH_{room} is the ambient humidity far from the droplets. This net force causes droplet motion and is balanced by a viscous drag.

If droplets of different surface tension but equal concentration coalesce upon contact, fluid is directly exchanged between the droplets. This exchange of fluid leads to a surface tension gradient and Marangoni flow across both droplets, where the droplet of lower surface tension chases the droplet of higher surface tension, which in turn flees away. They explained the observed phenomena of droplets of sufficiently different concentrations exhibiting a "chasing phase" indicative of a repulsive force that comes into play as the gradient of the vapor pressure decreases $\gamma_{LV,film}$ around the droplets causing them to move away.

Using their understanding of "self-fuelled surface tension driven fluidic machines," the authors have explored several applications in food coloring, glass slides, and permanent Sharpie markers.

3. Theoretical Description and Modelling

3.1. Models of Evaporation. Any model of mass and heat transfer inside an evaporating drop and any model of deposit pattern formation are based on models of evaporation from free surface of the drop. The functional form of the evaporation rate depends on the rate-limiting step, which can be either the transfer rate across the liquid-vapor interface or the diffusive relaxation of the saturated vapor layer immediately above the free surface of the drop [202]. Hence, two main approaches should be mentioned.

The first one is based on the assumption that evaporation from free surface of the drop is steady-state and diffusion-limited. Analysis of this assumption validity can be performed in detail by Popov [202]. In the case of diffusion-limiting quasi-steady process, the vapor density above the liquid-vapor interface obeys the Laplace equation. When droplet shape is governed by surface tension, it can be treated as a spherical cap (see Section 1) and Laplace equation can be solved analytically. This solution has one essential drawback; namely, vapor flux is singular at TPCL. To suppress this physically senseless singularity, a correcting factor may be introduced [203–205].

The second one is based on the assumption that the rate-limiting factor is heat transfer [206]. In this case, a singularity is missing in analytical formula for the vapor rate except for the case of highly volatile droplet.

3.2. 2D Models of Mass Transfer. Modelling of the processes occurring during the drying of colloidal droplet solutions is very complicated, because these processes are extraordinarily varied and complex [63]. The authors have different views about the driving mechanisms that lead to the formation of the solid phase [22, 23, 90, 91, 202, 207]. For example, [208] considered competition of convection and sedimentation, but [159] considered competition of convection and diffusion. Numerous models were proposed during the last two decades. Several models describe some particular processes occurring during the colloidal droplet desiccation (e.g., capillary flow and mass transport processes) [23, 90, 105, 167, 202, 207–215]. Generally, models are developed for systems with low concentrations of the colloidal particles.

Two very different situations are possible when a colloidal sessile droplet desiccates. In the first case, the particles inside a droplet can interact with each other only mechanically (impacts). In this case, the deposit forms a porous medium. Such a medium prevents neither bulk flow inside it nor evaporation from its surface. Moreover, such a porous medium can enhance evaporation from its surface due to drainage effect [209]. In the second case, the colloidal particles can form strong interparticle bonds. In this case, hydrodynamic flows, particle diffusion, and solvent evaporation are restricted. The proposed theoretical models mainly deal with the first situation [23, 90, 105, 167, 202, 207–214]. Only a few models treat the deposit as impenetrable for flows and preventing evaporation [159, 165–167]. Nevertheless, the simulation of desiccated colloidal droplets with phase transition is extremely important for high-throughput drug screening [60], biostabilization [216], identification of fluids [183], and medical tests [139, 217–219]. The models in [159, 165–167] utilize sets of rather complicated partial differential equations (PDE).

Other approaches for simulating pattern formation in drying drops have also been tried. A simple Monte Carlo algorithm for evaporation and pattern formation has been developed by Dutta Choudhury et al. to reproduce the formation of faceted salt crystals and dendritic aggregates in drying droplets of aqueous gelatin containing NaCl [142]. The pattern formation can be correlated to the topological concept of the Euler number [220].

Several models describing desiccated sessile colloidal droplets have been reported recently [117, 159, 165, 167, 221]. They are based on the lubrication approximation [222]. This approach has several serious shortcomings [223] as enumerated below:

(1) Only thin films can be considered; all quantities are supposed to be dependent only on one radial coordinate.

(2) In fact, a two-phase system is considered as one-phase system; the gel is assumed to be a liquid with very high viscosity; the hydrodynamic equations are written for the whole droplet desiccation.

(3) The mathematical expression for evaporation flux above the free surface is speculative rather than being supported by experiments. To the best of our knowledge, measurements of the vapor flux above a system with sol-gel phase transition are not published yet.

(4) Knowledge of the effect of particle concentration on viscosity is needed for calculations. This dependence can be obtained from experiments with rather large volumes of colloid. Viscosity of a small droplet with a large free surface and large contact area with a substrate can deviate from this in a rather complex manner.

(5) It is assumed that all the molecules that get to the edge of the droplet pass into the solid phase. Generally, this assumption can be wrong in the presence of convection of any nature in a droplet. An inward flux of particles due to diffusion may also exist.

To overcome the limitations of the listed models, three-dimensional (3D) models need to be developed and utilized.

3.3. Modelling Flow in 3 Dimensions. A number of papers devoted to 3D models of processes inside evaporating droplets were published during the last few years. Mostly, the articles consider droplets of pure liquids and simulate flows within them [114, 224–229]. The analytical solutions of the Laplace equation which describe the velocity field inside evaporating droplets of a nonviscous liquid were obtained for the contact angle of 90° by Tarasevich [226] and for a case of arbitrary contact angle by Masoud and Felske [230].

Flow inside the boundary line of an evaporating liquid for any contact angle was found using Stokes approach [231]. Numerical calculations of the velocity field within evaporating droplets were performed using finite element method [54, 110, 225]. Presence of dissolved substances or suspended particles inside the droplets and deposit formation were not taken into account in these models.

3.4. 3D Models of Mass Transfer. 3D models describing the processes inside the particle-laden droplets were developed using both the continuum and discrete approaches. Development of discrete models was initiated by the requirements of modelling of evaporation-driven self-assembly (EDSA) or evaporation-induced self-assembly (EISA) [214, 232–239]. Additional references can be found in [240]. Recently published models considered the Brownian motion of particles inside the droplets. For instance, in the work of Petsi et al. [213], the Brownian motion of the particles is superimposed on the hydrodynamic flow calculated previously [231]. A continuum approach has also been applied in some works [59, 208, 209].

Conflicting results from experiments imply that available models may be too simplistic and more realistic theories need to be developed. For example, some experimental data indicate that transfer of the suspended particles to the edge of the drying droplets is possible only when the Marangoni effect is suppressed [111]. However, other experimental studies consider the Marangoni effect to be the driving force for the formation of a new phase on the edge of a drop. It appears therefore that the theoretical models are incomplete, or too drastic approximations/assumptions have been made while formulating them.

There is also some confusion due to nonuniform terminology used by different research groups. For example, [209] reports *depinning* for large contact angle and no depinning for small contact angle. But it refers to receding of the fluid from the solid deposited at the TPCL as "depinning." But many other researchers use the term "depinning" as the inward motion of the TPLC as a whole, leaving no deposit behind.

Another such instance is the fact that the direction of flow can be opposite to a direction that is predicted by calculations for the pure solvent [241]. Independent experiments confirmed that the flows in pure liquids and in liquids with admixtures go in different directions [242, 243]. In the multicomponent liquids of biological origin, the thermocapillary and solutocapillary effects can eliminate each other [60]. Calculations of various research groups have shown that during evaporation of the droplet of a pure liquid there are circular flows caused by the Marangoni stress. The flow is directed along the droplets base to its edge and along its surface towards to the center of the drop [54, 114, 225, 228, 244]. At the same time, experiments conducted with biological fluids exhibit opposite direction of flow [241–243]. According to Kistovich et al., Marangoni flow cannot generally occur during drying of the droplets of biological fluids; they suggest that the observed circular currents are caused by buoyant convection [105].

Most of the earlier models for simulating evaporation of droplets have been developed for single solvent droplets. Recently, a finite element model has been formulated by Diddens et al. [245, 246], which explains results of interesting experimental phenomena such as self-wrapping of ouzo drops on a superamphiphobic substrate [247].

Evaporation of a multicomponent drop is interesting because the solvent contains different liquids with different volatility, so the composition changes during evaporation. For example, in a water-glycerol mixture, where evaporation of glycerol can be neglected compared to water, a large contact angle is shown to lead to a reversed Marangoni flow. If the relative humidity is high, when water content of the drop is reduced due to evaporation, water vapor from the surroundings condenses onto the drop. The problem now becomes nonlinear and chaotic vortices form [246]. This has been observed experimentally as well [248].

On the other hand, for a water-ethanol mixture, where both components evaporate fast, the substrate cools rapidly and thermal transport has to be taken into account. Chaotic behavior is predicted by the model [246] and has also been observed experimentally [249–251]. There are, in addition, models focusing on a particular aspect of desiccation such as skin formation [252, 253].

3.5. Crystal Growth

The sensitivity of crystal morphology on various parameters impedes modelling because a lot of different effects have to be taken into account. In fact, all used models should be treated as semiempirical. The models often utilize the lattice approach [142, 254–257] and diffusion equation [25, 258]. Adequacy of some models [254–256] has been questioned [258]. Mainly, dendritic crystal growth can be observed at the final stages of drop desiccation. Both nonequilibrium growth and presence of impurities may produce dendritic shape of crystals [259]; these effects can

be reproduced in a simple model [260]. The semiempirical Monte Carlo approach by Dutta Choudhury et al. [142] qualitatively reproduces the crossover in faceted and dendritic crystal growth and shows the relevance of statistical methods in this problem through the Euler number [220].

The phase-field method [261] looks extremely promising for modelling crystal growth in desiccated colloidal droplets with salt admixtures, but it requires a lot of additional information, which is difficult to obtain experimentally.

4. Applications and Perspectives

Initially, the negative effect of the coffee ring was of concern to scientists and engineers, since it precludes uniform deposition in processes such as ink-jet printing. So methods to reverse it were in demand. We have discussed in Section 2.2.1 that this is possible by enhancing Marangoni flow or imposing temperature gradients. Another method is to use electrowetting [96, 262, 263]. This can be done by applying voltage though the drop by the technique known as eMALDI, introduced by Eral et al. [96]. Varying voltages give different deposition of salt patterns on substrates [262].

In recent times, however, the picture has changed; increasingly various patterns formed by evaporating drops are being put to good use, instead of being considered a hindrance. We briefly mention below some fields where the *nonuniform distribution* of solute in the drying drop has been helpful.

4.1. Application in Functional Materials

The group led by Shimoni et al. [264] created a connected network of coffee rings by deposition of conducting micro/nanoparticles on a transparent substrate. This produces a transparent conductor that is extremely important in today's technology for various photovoltaic devices. The droplet technique provides a much cheaper alternative to indium tin oxide or indium tin fluoride coated glass, traditionally used as transparent conducting material. Drying a drop of some specific solvents with CNT (carbon nanotubes) on a cooled substrate similarly produces a polygonal connected, self-assembled network, which is transparent as well as conducting [27]. Moreover, the coffee-ring effect is used for the separation of two different sized particles [265, 266].

4.2. Biomedical Application

The utility of studying patterns of dried biological fluids has been well known in medical diagnosis [138]. The structures observed after drying biological fluids on a horizontal impenetrable substrate attracted the attention of researchers as early as the 1950s [103, 104, 128–132, 134–137]. In the 1980s–1990s, doctors of the former Soviet Union began to use the appearance of structures formed by drying droplets of biological fluids for the diagnosis of various diseases [3, 4, 81, 181, 267]. Unfortunately, very few of these articles were published in English [62, 138, 140, 149, 182, 217].

Many constituents of biological fluids crystallize on drying. The presence of these crystals, their morphology, size, and abundance are of great help in pathological investigations. Denisov describes crystallization patterns observed in saliva of patients with gastrointestinal diseases [268]. He shows that box-shaped, cross-shaped, and dendritic crystals with multiple-level branching are observed in various samples. The shapes can be classified in a phase diagram with nonoverlapping groups to identify problems such as peptic ulcer, chronic gastritis, and other such diseases in patients.

The potential for diagnosis by using the evaporation patterns of whole blood or blood serum lies in the fact that blood composition may vary due to diseases, which in turn results in changes in the evaporation patterns. Researchers had used dried human serum patterns to diagnose metastatic carcinoma [269] and also found that various interesting patterns could be used to reveal different pathological information [61]. The dried drop patterns of blood serum were also suggested to be useful for disease diagnosis because some featured morphologies of blood serum patterns could be used to acquire information about the health state of human organisms [138]. However, the use of whole blood patterns for medical diagnosis was rarely reported [42, 153], and more systematic experiments are expected.

The coffee-ring effect has been used as low-resource diagnostics for detection of the malarial biomarker *Plasmodium falciparum* [270] and also as a biomarker elsewhere [138, 271–273]. Blood drop patterns are used extensively for diagnostic purposes [20]. *Crack patterns* in dried drops of biofluids are also used to extract valuable information in medical diagnosis. Dried droplets of blood and blood serum show characteristic crack patterns for patients suffering from anemia, hyperlipidemia, and other disorders [7, 21]. A related field where droplet patterns are of importance is forensics related to crime investigation [7]. More such areas are biopreservation [216, 274] and high-throughput drug screening, where pattern formation in the drying sample is not desirable [60].

Some other applications have come up too; patterns on dried droplets may be used for quality analysis of food grains [275], as well as alcoholic drinks [276], fast identification of fluid and substrate chemistry based on automatic pattern recognition of stains [183], assessment of quality of products [275], and Raman spectroscopy [180, 277–286].

4.3. Droplet Levitation. The idea behind drying a droplet under levitation is to eliminate the effect of gravity during drying and observe desiccation under *no contact* condition. However, the droplet levitation technology is a promising candidate for generating novel applications. For example, the technique of manipulating levitated drops used on liquid marbles [287] may be applied to insert desired components in dried shell structures [66] for drug delivery or other applications.

There is, however, one concern with evaporation under levitation. This is that although one can avoid substrate contact and maintain the evaporating drops in a quasi-spherical shape, the levitation techniques, for instance, acoustic levitation, often lead to an additional boundary layer to the levitated drop, which may influence the evaporation process. In addition, the levitation force could also influence the morphology of dried residues. To investigate the mass transfer and evaporation-driven assembly in a truly undisturbed manner, drop evaporation in a space station is highly desirable. This will be of great help in understanding the emergence of crust, formation of cavity, buckling of crust, and elucidating the effect of gravity on these processes.

5. Conclusion

There obviously remains much more work to be done in this interesting and useful area of research. At this juncture, some tasks can be specially emphasized:

(i) Obtaining new experimental data critically needed for the design and development of adequate models

(ii) The development of 3D models describing the redistribution of the components, the movement of the phase front, and the evolution of the profile of the drying colloidal droplets with salt admixtures. In these systems, phase transition from sol to gel is concentration-driven. The thermal phase transition from liquid to vapor also takes place in this system. This phase transition leads to a movement of the liquid-vapor phase boundary (i.e., the droplet volume decreases and droplet profile changes)

(iii) Considering additional effects that may be crucial to understanding the processes of pattern formation but have not yet been included in the models (e.g., variations of the viscosity of a colloid with time and concentration of salts and changes of the vapor flux above the free surface of the droplets when the phase boundary (sol-gel) is moving). Time-varying interactions between the components of the droplet (e.g., particles and ions) also need to be considered

(iv) Studying the effect of external fields such as electromagnetic and acoustic fields on the droplet evaporation process

(v) Analyzing the final pattern through tools such as fractal and multifractal characterization

To conclude, the simple but effective process of drying a fluid drop and observing it under a microscope (preferably with video recording) is rapidly developing into a new and exciting field of research. Several reviews [15–17, 20, 288, 289] and books [5–8] based on pattern formation during desiccation published within the short span of just four years confirm the intensely growing interest in this subject. Exploiting the full potential of this topic in basic science research and applications needs involvement and interaction between scientists and engineers from disciplines of physics, chemistry, biology, medicine, and other related fields.

Conflicts of Interest

The authors declare that there are no conflicts of interest regarding the publication of this paper.

Acknowledgments

Yuri Yu. Tarasevich acknowledges the funding from the Ministry of Education and Science of the Russian Federation

(Project no. 3.959.2017/4.6). Duyang Zang acknowledges the National Natural Science Foundation of China (Grant no. U1732129). Moutushi Dutta Choudhury acknowledges SERB, India, for providing financial support through NPDF (PDF/2016/001151/PMS).

References

[1] P. G. de Gennes, F. Brochard-Wyart, and D. Quéré, *Capillarity and Wetting Phenomena*, Springer, New York, NY, USA, 2004.

[2] D. J. Wedlock, *Controlled Particle, Droplet and Bubble Formation*, Butterworth-Heinemann, 1994.

[3] V. N. Shabalin and S. N. Shatohina, *Morphology of Human Biological Fluids*, Khrizostom, Moscow, Russia, (Russian) 2001.

[4] E. Rapis, *Protein and Life (Self-Assembling and Symmetry of Protein Nanostructures)*, Philobiblion, Jerusalem, Israel; Milta-PKPGIT, Moscow, Rusiia, (Russian) 2003.

[5] Z. Lin, Ed., *Evaporative Self-Assembly of Ordered Complex Structures*, World Scientific Publishing Company, 2010.

[6] P. Innocenzi, L. Malfatti, and P. Falcaro, *Water Droplets to Nanotechnology*, The Royal Society of Chemistry, 2013.

[7] D. Brutin, Ed., *Droplet Wetting and Evaporation*, Academic Press, Oxford, UK, 2015.

[8] L. Goehring, A. Nakahara, T. Dutta, S. Kitsunezaki, and S. Tarafdar, *Desiccation Cracks and Their Patterns: Formation and Modelling in Science and Nature, Statistical Physics of Fracture and Breakdown*, Wiley, 2015.

[9] S. K. Aggarwal and F. Peng, "A review of droplet dynamics and vaporization modeling for engineering calculations," *Journal of Engineering for Gas Turbines and Power*, vol. 117, no. 3, pp. 453–461, 1995.

[10] Yu. Y. Tarasevich, "Mechanisms and models of the dehydration self-organization in biological fluids," *Physics-Uspekhi*, vol. 47, no. 7, pp. 717–728, 2004.

[11] T. A. Yakhno, V. G. Yakhno, A. G. Sanin, O. A. Sanina, and A. S. Pelyushenko, "Protein and salt: Spatiotemporal dynamics of events in a drying drop," *Technical Physics*, vol. 49, no. 8, pp. 1055–1063, 2004.

[12] T. A. Yakhno and V. G. Yakhno, "Structural evolution of drying drops of biological fluids," *Technical Physics*, vol. 54, no. 8, pp. 1219–1227, 2009.

[13] Y. Choi, J. Han, and C. Kim, "Pattern formation in drying of particle-laden sessile drops of polymer solutions on solid substrates," *Korean Journal of Chemical Engineering*, vol. 28, no. 11, pp. 2130–2136, 2011.

[14] H. Y. Erbil, "Evaporation of pure liquid sessile and spherical suspended drops: A review," *Advances in Colloid and Interface Science*, vol. 170, no. 1-2, pp. 67–86, 2012.

[15] K. Sefiane, "Patterns from drying drops," *Advances in Colloid and Interface Science*, vol. 206, pp. 372–381, 2014.

[16] A. F. Routh, "Drying of thin colloidal films," *Reports on Progress in Physics*, vol. 76, no. 4, Article ID 046603, 2013.

[17] R. G. Larson, "Transport and deposition patterns in drying sessile droplets," *AIChE Journal*, vol. 60, no. 5, pp. 1538–1571, 2014.

[18] N. M. Kovalchuk, A. Trybala, and V. M. Starov, "Evaporation of sessile droplets," *Current Opinion in Colloid & Interface Science*, vol. 19, no. 4, pp. 336–342, 2014.

[19] M. Anyfantakis and D. Baigl, "Manipulating the coffee-ring effect: interactions at work," *ChemPhysChem*, vol. 16, no. 13, pp. 2726–2734, 2015.

[20] R. Chen, L. Zhang, D. Zang, and W. Shen, "Blood drop patterns: Formation and applications," *Advances in Colloid and Interface Science*, vol. 231, pp. 1–14, 2016.

[21] D. Brutin and V. Starov, "Recent advances in droplet wetting and evaporation," *Chemical Society Reviews*, vol. 47, no. 2, pp. 558–585, 2018.

[22] R. D. Deegan, O. Bakajin, T. F. Dupont, G. Huber, S. R. Nagel, and T. A. Witten, "Capillary flow as the cause of ring stains from dried liquid drops," *Nature*, vol. 389, no. 6653, pp. 827–829, 1997.

[23] R. D. Deegan, O. Bakajin, T. F. Dupont, G. Huber, S. R. Nagel, and T. A. Witten, "Contact line deposits in an evaporating drop," *Physical Review E: Statistical, Nonlinear, and Soft Matter Physics*, vol. 62, no. 1, pp. 756–765, 2000.

[24] M. Dutta Choudhury, T. Dutta, and S. Tarafdar, "Pattern formation in droplets of starch gels containing NaCl dried on different surfaces," *Colloids and Surfaces A: Physicochemical and Engineering Aspects*, vol. 432, pp. 110–118, 2013.

[25] T. Dutta, A. Giri, M. D. Choudhury, and S. Tarafdar, "Experiment and simulation of multifractal growth of crystalline NaCl aggregates in aqueous gelatin medium," *Colloids and Surfaces A: Physicochemical and Engineering Aspects*, vol. 432, pp. 127–131, 2013.

[26] A. Giri, M. Dutta Choudhury, T. Dutta, and S. Tarafdar, "Multifractal growth of crystalline NaCl aggregates in a gelatin medium," *Crystal Growth and Design*, vol. 13, no. 1, pp. 341–345, 2013.

[27] B. Roy, S. Karmakar, and S. Tarafdar, "Self assembled transparent conducting network of multi-walled carbon nanotubes formed by evaporation," *Materials Letters*, vol. 207, pp. 86–88, 2017.

[28] L. Pauchard, F. Parisse, and C. Allain, "Influence of salt content on crack patterns formed through colloidal suspension desiccation," *Physical Review E: Statistical Physics, Plasmas, Fluids, and Related Interdisciplinary Topics*, vol. 59, no. 3, pp. 3737–3740, 1999.

[29] C. C. Annarelli, J. Fornazero, J. Bert, and J. Colombani, "Crack patterns in drying protein solution drops," *European Physical Journal E*, vol. 5, no. 5, pp. 599–603, 2001.

[30] K.-T. Leung, L. Józsa, M. Ravasz, and Z. Néda, "Pattern formation: spiral cracks without twisting," *Nature*, vol. 410, no. 6825, p. 166, 2001.

[31] Z. Néda, K.-T. Leung, L. Józsa, and M. Ravasz, "Spiral cracks in drying precipitates," *Physical Review Letters*, vol. 88, no. 9, Article ID 095502, 2002.

[32] B. D. Caddock and D. Hull, "Influence of humidity on the cracking patterns formed during the drying of sol-gel drops," *Journal of Materials Science*, vol. 37, no. 4, pp. 825–834, 2002.

[33] E. Golbraikh, E. G. Rapis, and S. S. Moiseev, "On the crack pattern formation in a freely drying protein film," *Technical Physics*, vol. 48, no. 10, pp. 1333–1337, 2003.

[34] G. Jing and J. Ma, "Formation of circular crack pattern in deposition self-assembled by drying nanoparticle suspension," *The Journal of Physical Chemistry B*, vol. 116, no. 21, pp. 6225–6231, 2012.

[35] T. Khatun, T. Dutta, and S. Tarafdar, "Crack formation under an electric field in droplets of laponite gel: Memory effect and scaling relations," *Langmuir*, vol. 29, no. 50, pp. 15535–15542, 2013.

[36] W. Bou Zeid, J. Vicente, and D. Brutin, "Influence of evaporation rate on cracks' formation of a drying drop of whole blood," *Colloids and Surfaces A: Physicochemical and Engineering Aspects*, vol. 432, pp. 139–146, 2013.

[37] Y. Zhang, Z. Liu, D. Zang, Y. Qian, and K.-J. Lin, "Pattern transition and sluggish cracking of colloidal droplet deposition with polymer additives," *Science China Physics, Mechanics & Astronomy*, vol. 56, no. 9, pp. 1712–1718, 2013.

[38] B. Sobac and D. Brutin, "Desiccation of a sessile drop of blood: Cracks, folds formation and delamination," *Colloids and Surfaces A: Physicochemical and Engineering Aspects*, vol. 448, no. 1, pp. 34–44, 2014.

[39] F. Giorgiutti-Dauphiné and L. Pauchard, "Striped patterns induced by delamination of drying colloidal films," *Soft Matter*, vol. 11, no. 7, pp. 1397–1402, 2015.

[40] U. U. Ghosh, M. Chakraborty, A. B. Bhandari, S. Chakraborty, and S. DasGupta, "Effect of surface wettability on crack dynamics and morphology of colloidal films," *Langmuir*, vol. 31, no. 22, pp. 6001–6010, 2015.

[41] J. Y. Kim, K. Cho, S.-A. Ryu, S. Y. Kim, and B. M. Weon, "Crack formation and prevention in colloidal drops," *Scientific Reports*, vol. 5, Article ID 13166, 2015.

[42] R. Chen, L. Zhang, D. Zang, and W. Shen, "Understanding desiccation patterns of blood sessile drops," *Journal of Materials Chemistry B*, vol. 5, no. 45, pp. 8991–8998, 2017.

[43] H. Lama, M. G. Basavaraj, and D. K. Satapathy, "Tailoring crack morphology in coffee-ring deposits via substrate heating," *Soft Matter*, vol. 13, no. 32, pp. 5445–5452, 2017.

[44] Y. Men, W. Wang, P. Xiao et al., "Controlled evaporative self-assembly of Fe_3O_4 nanoparticles assisted by an external magnetic field," *RSC Advances*, vol. 5, pp. 31519–31524, 2015.

[45] Y. Li, Q. Yang, M. Li, and Y. Song, "Rate-dependent interface capture beyond the coffee-ring effect," *Scientific Reports*, vol. 6, Article ID 24628, 2016.

[46] L. Pauchard and C. Allain, "Buckling instability induced by polymer solution drying," *EPL (Europhysics Letters)*, vol. 62, no. 6, pp. 897–903, 2003.

[47] L. Pauchard and C. Allain, "Stable and unstable surface evolution during the drying of a polymer solution drop," *Physical Review E: Statistical, Nonlinear, and Soft Matter Physics*, vol. 68, no. 5, Article ID 052801, 2003.

[48] L. Pauchard and C. Allain, "Mechanical instability induced by complex liquid desiccation," *Comptes Rendus Physique*, vol. 4, no. 2, pp. 231–239, 2003.

[49] Y. Gorand, L. Pauchard, G. Calligari, J. P. Hulin, and C. Allain, "Mechanical instability induced by the desiccation of sessile drops," *Langmuir*, vol. 20, no. 12, pp. 5138–5140, 2004.

[50] T. A. Yakhno, V. V. Kazakov, O. A. Sanina, A. G. Sanin, and V. G. Yakhno, "Drops of biological fluids drying on a hard substrate: Variation of the morphology, weight, temperature, and mechanical properties," *Technical Physics*, vol. 55, no. 7, pp. 929–935, 2010.

[51] F. Girard, M. Antoni, and K. Sefiane, "Infrared thermography investigation of an evaporating sessile water droplet on heated substrates," *Langmuir*, vol. 26, no. 7, pp. 4576–4580, 2010.

[52] T. A. Yakhno, O. A. Sanina, M. G. Volovik, A. G. Sanin, and V. G. Yakhno, "Thermographic investigation of the temperature field dynamics at the liquid-air interface in drops of water solutions drying on a glass substrate," *Technical Physics*, vol. 57, no. 7, pp. 915–922, 2012.

[53] Y. Tsoumpas, S. Dehaeck, A. Rednikov, and P. Colinet, "Effect of Marangoni flows on the shape of thin sessile droplets evaporating into air," *Langmuir*, vol. 31, no. 49, pp. 13334–13340, 2015.

[54] H. Hu and R. G. Larson, "Analysis of the effects of Marangoni stresses on the microflow in an evaporating sessile droplet," *Langmuir*, vol. 21, no. 9, pp. 3972–3980, 2005.

[55] D. J. Harris and J. A. Lewis, "Marangoni effects on evaporative lithographic patterning of colloidal films," *Langmuir*, vol. 24, no. 8, pp. 3681–3685, 2008.

[56] H. Bodiguel and J. Leng, "Imaging the drying of a colloidal suspension," *Soft Matter*, vol. 6, no. 21, pp. 5451–5460, 2010.

[57] X. Xu, J. Luo, and D. Guo, "Radial-velocity profile along the surface of evaporating liquid droplets," *Soft Matter*, vol. 8, no. 21, pp. 5797–5803, 2012.

[58] Y.-C. Hu, Q. Zhou, H.-M. Ye, Y.-F. Wang, and L.-S. Cui, "Peculiar surface profile of poly(ethylene oxide) film with ring-like nucleation distribution induced by Marangoni flow effect," *Colloids and Surfaces A: Physicochemical and Engineering Aspects*, vol. 428, pp. 39–46, 2013.

[59] G. Son, "Numerical simulation of particle-laden droplet evaporation with the Marangoni effect," *The European Physical Journal Special Topics*, vol. 224, no. 2, pp. 401–413, 2015.

[60] P. Takhistov and H.-C. Chang, "Complex stain morphologies," *Industrial & Engineering Chemistry Research*, vol. 41, no. 25, pp. 6256–6269, 2002.

[61] M. E. Buzoverya, Yu. P. Shcherbak, and I. V. Shishpor, "Quantitative estimation of the microstructural inhomogeneity of biological fluid facies," *Technical Physics*, vol. 59, no. 10, pp. 1550–1555, 2014.

[62] S. N. Shatokhina, V. N. Shabalin, M. E. Buzoverya, and V. T. Punin, "Bio-liquid morphological analysis," *TheScientificWorldJOURNAL*, vol. 4, pp. 657–661, 2004.

[63] T. Yakhno, "Salt-induced protein phase transitions in drying drops," *Journal of Colloid and Interface Science*, vol. 318, no. 2, pp. 225–230, 2008.

[64] D. Brutin, B. Sobac, B. Loquet, and J. Sampol, "Pattern formation in drying drops of blood," *Journal of Fluid Mechanics*, vol. 667, pp. 85–95, 2011.

[65] R. Chen, L. Zhang, D. Zang, and w. Shen, "Wetting and drying of colloidal droplets: physics and pattern formation," in *Advances in Colloid Science*, InTech, 2016.

[66] E. Tijerino, S. Basu, and R. Kumar, "Nanoparticle agglomeration in an evaporating levitated droplet for different acoustic amplitudes," *Journal of Applied Physics*, vol. 113, no. 3, Article ID 034307, 2013.

[67] L. Chen and J. R. G. Evans, "Drying of colloidal droplets on superhydrophobic surfaces," *Journal of Colloid and Interface Science*, vol. 351, no. 1, pp. 283–287, 2010.

[68] S. Basu, E. Tijerino, and R. Kumar, "Insight into morphology changes of nanoparticle laden droplets in acoustic field," *Applied Physics Letters*, vol. 102, no. 14, Article ID 141602, 2013.

[69] R. G. Picknett and R. Bexon, "The evaporation of sessile or pendant drops in still air," *Journal of Colloid and Interface Science*, vol. 61, no. 2, pp. 336–350, 1977.

[70] E. Wulsten, F. Kiekens, F. van Dycke, J. Voorspoels, and G. Lee, "Levitated single-droplet drying: Case study with itraconazole dried in binary organic solvent mixtures," *International Journal of Pharmaceutics*, vol. 378, no. 1-2, pp. 116–121, 2009.

[71] R. Mukherjee, R. Pangule, A. Sharma, and G. Tomar, "Contact instability of elastic bilayers: Miniaturization of instability patterns," *Advanced Functional Materials*, vol. 17, no. 14, pp. 2356–2364, 2007.

[72] U. Thiele and E. Knobloch, "Thin liquid films on a slightly inclined heated plate," *Physica D: Nonlinear Phenomena*, vol. 190, no. 3-4, pp. 213–248, 2004.

[73] D. I. Yu, H. J. Kwak, S. W. Doh et al., "Wetting and evaporation phenomena of water droplets on textured surfaces," *International Journal of Heat and Mass Transfer*, vol. 90, pp. 191–200, 2015.

[74] N. Anantharaju, M. Panchagnula, and S. Neti, "Evaporating drops on patterned surfaces: Transition from pinned to moving triple line," *Journal of Colloid and Interface Science*, vol. 337, no. 1, pp. 176–182, 2009.

[75] S. K. Singh, S. Khandekar, D. Pratap, and S. A. Ramakrishna, "Wetting dynamics and evaporation of sessile droplets on nano-porous alumina surfaces," *Colloids and Surfaces A: Physicochemical and Engineering Aspects*, vol. 432, pp. 71–81, 2013.

[76] M. Gonuguntla and A. Sharma, "Polymer patterns in evaporating droplets on dissolving substrates," *Langmuir*, vol. 20, no. 8, pp. 3456–3463, 2004.

[77] C. Parneix, P. Vandoolaeghe, V. S. Nikolayev, D. Quéré, J. Li, and B. Cabane, "Dips and rims in dried colloidal films," *Physical Review Letters*, vol. 105, no. 26, Article ID 266103, 2010.

[78] D. J. Harris, H. Hu, J. C. Conrad, and J. A. Lewis, "Patterning colloidal films via evaporative lithography," *Physical Review Letters*, vol. 98, no. 14, Article ID 148301, 2007.

[79] H. M. J. M. Wedershoven, K. R. M. Deuss, C. Fantin, J. C. H. Zeegers, and A. A. Darhuber, "Active control of evaporative solution deposition by means of modulated gas phase convection," *International Journal of Heat and Mass Transfer*, vol. 117, pp. 303–312, 2018.

[80] A. Georgiadis, F. N. Muhamad, A. Utgenannt, and J. L. Keddie, "Aesthetically textured, hard latex coatings by fast IR-assisted evaporative lithography," *Progress in Organic Coatings*, vol. 76, no. 12, pp. 1786–1791, 2013.

[81] L. V. Savina, *Crystalloscopic Structures of Blood Serum of Healthy People and Patients*, Sov. Kuban, Krasnodar, Russia, 1999 (Russian).

[82] J. Gao, K. Kempa, M. Giersig, E. M. Akinoglu, B. Han, and R. Li, "Physics of transparent conductors," *Advances in Physics*, vol. 65, no. 6, pp. 553–617, 2016.

[83] A. Kumar and G. U. Kulkarni, "Evaluating conducting network based transparent electrodes from geometrical considerations," *Journal of Applied Physics*, vol. 119, no. 1, Article ID 015102, 2016.

[84] Z. Zhan, J. An, Y. Wei, V. T. Tran, and H. Du, "Inkjet-printed optoelectronics," *Nanoscale*, vol. 9, no. 3, pp. 965–993, 2017.

[85] W. D. Ristenpart, P. G. Kim, C. Domingues, J. Wan, and H. A. Stone, "Influence of substrate conductivity on circulation reversal in evaporating drops," *Physical Review Letters*, vol. 99, no. 23, Article ID 234502, 2007.

[86] L. D. Landau and E. M. Lifshitz, "Surface phenomena," in *Fluid Mechanics*, chapter 7, pp. 230–244, Pergamon Press, 1st edition, 1959.

[87] W. A. Sirignano, "Introduction," in *Fluid Dynamics and Transport of Droplets and Sprays*, pp. 1–6, Cambridge University Press, Cambridge, UK, 1999.

[88] R. Mukherjee, D. Bandyopadhyay, and A. Sharma, "Control of morphology in pattern directed dewetting of thin polymer films," *Soft Matter*, vol. 4, no. 10, pp. 2086–2097, 2008.

[89] R. Mukherjee, M. Gonuguntla, and A. Sharma, "Meso-patterning of thin polymer films by controlled dewetting: From nano-droplet arrays to membranes," *Journal of Nanoscience and Nanotechnology*, vol. 7, no. 6, pp. 2069–2075, 2007.

[90] F. Parisse and C. Allain, "Shape changes of colloidal suspension droplets during drying," *Journal de Physique II*, vol. 6, no. 7, pp. 1111–1119, 1996.

[91] R. D. Deegan, "Pattern formation in drying drops," *Physical Review E: Statistical Physics, Plasmas, Fluids, and Related Interdisciplinary Topics*, vol. 61, no. 1, pp. 475–485, 2000.

[92] A. Lazauskas and V. Grigaliūnas, "Float glass surface preparation methods for improved chromium film adhesive bonding," *Medziagotyra*, vol. 18, no. 2, pp. 181–186, 2012.

[93] J. Canning, I. Petermann, and K. Cook, "Surface treatment of silicate based glass: base Piranha treatment versus 193 nm laser processing," in *3rd Asia Pacific Optical Sensors Conference*, J. Canning and G. Peng, Eds., vol. 8351 of *Proceedings of SPIE*, Sydney, Australia, February 2012.

[94] Y. Ma, X. Cao, X. Feng, and H. Zou, "Fabrication of super-hydrophobic film from PMMA with intrinsic water contact angle below 90°," *Polymer Journal*, vol. 48, no. 26, pp. 7455–7460, 2007.

[95] D. I. Yu, H. J. Kwak, S. W. Doh et al., "Dynamics of contact line depinning during droplet evaporation based on thermodynamics," *Langmuir*, vol. 31, no. 6, pp. 1950–1957, 2015.

[96] H. B. Eral, D. Mampallil Augustine, M. H. G. Duits, and F. Mugele, "Suppressing the coffee stain effect: How to control colloidal self-assembly in evaporating drops using electrowetting," *Soft Matter*, vol. 7, no. 10, pp. 4954–4958, 2011.

[97] H. B. Eral, D. van den Ende, and F. Mugele, "Say goodbye to coffee stains," *Physics World*, vol. 25, no. 4, pp. 33–37, 2012.

[98] D. Mampallil, H. B. Eral, D. van den Ende, and F. Mugele, "Control of evaporating complex fluids through electrowetting," *Soft Matter*, vol. 8, no. 41, pp. 10614–10617, 2012.

[99] F.-C. Wang and H.-A. Wu, "Pinning and depinning mechanism of the contact line during evaporation of nano-droplets sessile on textured surfaces," *Soft Matter*, vol. 9, no. 24, pp. 5703–5709, 2013.

[100] A. Zigelman and O. Manor, "A model for pattern deposition from an evaporating solution subject to contact angle hysteresis and finite solubility," *Soft Matter*, vol. 12, no. 26, pp. 5693–5707, 2016.

[101] C. W. Extrand and Y. Kumagai, "Contact angles and hysteresis on soft surfaces," *Journal of Colloid and Interface Science*, vol. 184, no. 1, pp. 191–200, 1996.

[102] F. Parisse and C. Allain, "Drying of colloidal suspension droplets: Experimental study and profile renormalization," *Langmuir*, vol. 13, no. 14, pp. 3598–3602, 1997.

[103] C. Koch, "Feinbau und Entstehungsweise von Kristallstrukturen in getrockneten Tropfen hochmolekularsalzhaltiger Flüssigkeiten," *Kolloid-Zeitschrift*, vol. 138, no. 2, pp. 81–86, 1954.

[104] A. Solé, "Die rhythmischen Kristallisationen im Influenzstagramm," *Kolloid-Zeitschrift*, vol. 137, no. 1, pp. 15–19, 1954.

[105] A. V. Kistovich, Yu. D. Chashechkin, and V. V. Shabalin, "Formation mechanism of a circumferential roller in a drying biofluid drop," *Technical Physics*, vol. 55, no. 4, pp. 473–478, 2010.

[106] S. Jafari Kang, V. Vandadi, J. D. Felske, and H. Masoud, "Alternative mechanism for coffee-ring deposition based on active role of free surface," *Physical Review E: Statistical, Nonlinear, and Soft Matter Physics*, vol. 94, no. 6, Article ID 063104, 2016.

[107] M. Anyfantakis, Z. Geng, M. Morel, S. Rudiuk, and D. Baigl, "Modulation of the coffee-ring effect in particle/surfactant mixtures: The importance of particle-interface interactions," *Langmuir*, vol. 31, no. 14, pp. 4113–4120, 2015.

[108] W. Zhang, T. Yu, L. Liao, and Z. Cao, "Ring formation from a drying sessile colloidal droplet," *AIP Advances*, vol. 3, no. 10, Article ID 102109, 2013.

[109] P. J. Yunker, T. Still, M. A. Lohr, and A. G. Yodh, "Suppression of the coffee-ring effect by shape-dependent capillary interactions," *Nature*, vol. 476, no. 7360, pp. 308–311, 2011.

[110] H. Hu and R. G. Larson, "Analysis of the microfluid flow in an evaporating sessile droplet," *Langmuir*, vol. 21, no. 9, pp. 3963–3971, 2005.

[111] H. Hu and R. G. Larson, "Marangoni effect reverses coffee-ring depositions," *The Journal of Physical Chemistry B*, vol. 110, no. 14, pp. 7090–7094, 2006.

[112] H. Kim, F. Boulogne, E. Um, I. Jacobi, E. Button, and H. A. Stone, "Controlled uniform coating from the interplay of Marangoni flows and surface-adsorbed macromolecules," *Physical Review Letters*, vol. 116, no. 12, Article ID 124501, 2016.

[113] L. Yu. Barash, "Marangoni convection in an evaporating droplet: analytical and numerical descriptions," *International Journal of Heat and Mass Transfer*, vol. 102, pp. 445–454, 2016.

[114] L. Yu. Barash, T. P. Bigioni, V. M. Vinokur, and L. N. Shchur, "Evaporation and fluid dynamics of a sessile drop of capillary size," *Physical Review E: Statistical, Nonlinear, and Soft Matter Physics*, vol. 79, no. 4, Article ID 046301, 2009.

[115] L. Yu. Barash, "Dependence of fluid flows in an evaporating sessile droplet on the characteristics of the substrate," *International Journal of Heat and Mass Transfer*, vol. 84, pp. 419–426, 2015.

[116] Y. H. Chen, W. N. Hu, J. Wang, F. J. Hong, and P. Cheng, "Transient effects and mass convection in sessile droplet evaporation: The role of liquid and substrate thermophysical properties," *International Journal of Heat and Mass Transfer*, vol. 108, pp. 2072–2087, 2017.

[117] A. D. Eales, A. F. Routh, N. Dartnell, and G. Simon, "Evaporation of pinned droplets containing polymer—an examination of the important groups controlling final shape," *AIChE Journal*, vol. 61, no. 5, pp. 1759–1767, 2015.

[118] D. Schwabe and A. Scharmann, "Some evidence for the existence and magnitude of a critical marangoni number for the onset of oscillatory flow in crystal growth melts," *Journal of Crystal Growth*, vol. 46, no. 1, pp. 125–131, 1979.

[119] S. Chavan, H. Cha, D. Orejon et al., "Heat transfer through a condensate droplet on hydrophobic and nanostructured superhydrophobic surfaces," *Langmuir*, vol. 32, no. 31, pp. 7774–7787, 2016.

[120] X. Zhong and F. Duan, "Disk to dual ring deposition transformation in evaporating nanofluid droplets from substrate cooling to heating," *Physical Chemistry Chemical Physics*, vol. 18, no. 30, pp. 20664–20671, 2016.

[121] N. D. Patil, P. G. Bange, R. Bhardwaj, and A. Sharma, "Effects of substrate heating and wettability on evaporation dynamics and deposition patterns for a sessile water droplet containing colloidal particles," *Langmuir*, vol. 32, no. 45, pp. 11958–11972, 2016.

[122] J.-N. Tisserant, G. Wicht, O. F. Göbel et al., "Growth and alignment of thin film organic single crystals from dewetting patterns," *ACS Nano*, vol. 7, no. 6, pp. 5506–5513, 2013.

[123] S. Wolfram, *A New Kind of Science*, Wolfram Media, Champaign, Ill, USA, 2002.

[124] N. Shahidzadeh, M. F. L. Schut, J. Desarnaud, M. Prat, and D. Bonn, "Salt stains from evaporating droplets," *Scientific Reports*, vol. 5, Article ID 10335, 2015.

[125] N. Shahidzadeh-Bonn, S. Rafai, D. Bonn, and G. Wegdam, "Salt crystallization during evaporation: Impact of interfacial properties," *Langmuir*, vol. 24, no. 16, pp. 8599–8605, 2008.

[126] S. Bhattacharyya, B. Roy, and M. Dutta Choudhury, "Pattern formation in drying drops of colloidal copper sulphate solution on glass surface," *Journal of Surface Science and Technology*, vol. 32, pp. 79–84, 2017.

[127] C. Rodriguez-Navarro and E. Doehne, "Salt weathering: influence of evaporation rate, supersaturation and crystallization pattern," *Earth Surface Processes and Landforms*, vol. 24, no. 2-3, pp. 191–209, 1999.

[128] C. Koch, "Über Austrocknungssprünge," *Colloid and Polymer Science*, vol. 145, no. 1, pp. 7–14, 1956.

[129] C. Koch, "Periodische Erscheinungen an trocknenden Tropfen hochmolekularer Flüssigkeiten," *Kolloid-Zeitschrift*, vol. 150, no. 1, pp. 80–83, 1957.

[130] C. Koch, "Über Doppelbildungen (Dubletten) und gleichartige Erscheinungen in trockenen Tropfen hochmolekularer Flüssigkeiten (Dextran, Polyvinylpyrrolidon)," *Kolloid-Zeitschrift*, vol. 151, no. 1, pp. 62–66, 1957.

[131] C. Koch, "Über Polyvinylpyrrolidonkristallverbände," *Kolloid-Zeitschrift*, vol. 151, no. 2, pp. 122–126, 1957.

[132] C. Koch, "Kristallstrukturen und verwandte erscheinungen in getrockneten tropfen hochmolekular-salzhaltiger Flüssigkeiten (Kammerwasser)," *Documenta Ophthalmologica*, vol. 11, no. 1, pp. 182–189, 1957.

[133] A. Solé, "Vergleichende Stagoskopie des Blutserums," *Österreichische Zoologische Zeitschrift*, vol. 5, pp. 366–376, 1954.

[134] A. Solé, "Die Rhythmik im Influenzstagogramm," *Kolloid-Zeitschrift*, vol. 143, no. 2, pp. 73–83, 1955.

[135] A. Solé, "Untersuchung über die Bewegung der Teilchen im Stagogramm und Influenzstagogramm," *Kolloid-Zeitschrift*, vol. 151, no. 1, pp. 55–62, 1957.

[136] A. Solé, "Stagoskopische Untersuchungen über die Rhythmik einiger Aminosäuren sowie anderer organischer Verbindungen," *Kolloid-Zeitschrift*, vol. 151, no. 2, pp. 126–136, 1957.

[137] A. Solé, *Stagoskopie: Einfuhrung in Methodik, Theorie und Praxis fur Arzte und Naturforscher*, Franz Deuticke, Wien, Austria, 1960.

[138] V. N. Shabalin and S. N. Shatokhina, "Diagnostic markers in the structures of human biological liquids," *Singapore Medical Journal*, vol. 48, no. 5, pp. 440–446, 2007.

[139] A. K. Martusevich, Y. Zimin, and A. Bochkareva, "Morphology of dried blood serum specimens of viral hepatitis," *Hepatitis Monthly*, vol. 7, no. 4, pp. 207–210, 2007.

[140] T. A. Yakhno, A. A. Sanin, R. G. Ilyazov et al., "Drying drop technology as a possible tool for detection leukemia and tuberculosis in cattle," *Journal of Biomedical Science and Engineering*, vol. 8, no. 1, pp. 1–23, 2015.

[141] M. T. A. Rajabi and M. Sharifzadeh, "'Coffee ring effect' in ophthalmology: 'anionic dye deposition' hypothesis explaining normal lid margin staining," *Medicine*, vol. 95, no. 14, p. e3137, 2016.

[142] M. Dutta Choudhury, T. Dutta, and S. Tarafdar, "Growth kinetics of NaCl crystals in a drying drop of gelatin: Transition from faceted to dendritic growth," *Soft Matter*, vol. 11, no. 35, pp. 6938–6947, 2015.

[143] M. Goto, Y. Oaki, and H. Imai, "Dendritic growth of NaCl crystals in a gel matrix: Variation of branching and control of bending," *Crystal Growth and Design*, vol. 16, no. 8, pp. 4278–4284, 2016.

[144] C. C. Annarelli, L. Reyes, J. Fornazero, J. Bert, R. Cohen, and A. W. Coleman, "Ion and molecular recognition effects on the crystallization of bovine serum albumin-salt mixtures," *Crystal Engineering*, vol. 3, no. 3, pp. 173–194, 2000.

[145] T. Basu, M. M. Goswami, T. R. Middya, and S. Tarafdar, "Morphology and ion-conductivity of gelatin-LiClO$_4$ films: factional diffusion analysis," *The Journal of Physical Chemistry B*, vol. 116, no. 36, pp. 11362–11369, 2012.

[146] B. Roy, M. D. Choudhuri, T. Dutta, and S. Tarafdar, "Multi-scale patterns formed by sodium sulphate in a drying droplet of gelatin," *Applied Surface Science*, vol. 357, pp. 1000–1006, 2015.

[147] T. A. Yakhno, "Sodium chloride crystallization from drying drops of albumin–salt solutions with different albumin concentrations," *Technical Physics*, vol. 60, no. 11, pp. 1601–1608, 2015.

[148] G. Glibitskiy, D. Glibitskiy, O. Gorobchenko et al., "Textures on the surface of BSA films with different concentrations of sodium halides and water state in solution," *Nanoscale Research Letters*, vol. 10, no. 1, p. 155, 2015.

[149] A. K. Martusevich and N. F. Kamakin, "Crystallography of biological fluid as a method for evaluating its physicochemical characteristics," *Bulletin of Experimental Biology and Medicine*, vol. 143, no. 3, pp. 385–388, 2007.

[150] S. Bhattacharyya, B. Roy, and S. Tarafdar, "Aggregation of crystalline copper sulphate salt in a gelatin medium," *Fractals*, vol. 25, no. 5, Article ID 1750038, 2017.

[151] T. A. Yahno, V. G. Yahno, A. G. Sanin, A. S. Pelyushenko, O. B. Shaposhnikova, and A. S. Chernov, "The phenomenon of drying drops and the possibility of its practical use," *Nonlinear World*, vol. 5, pp. 54–65, 2007 (Russian).

[152] T. Still, P. J. Yunker, and A. G. Yodh, "Surfactant-induced Marangoni eddies alter the coffee-rings of evaporating colloidal drops," *Langmuir*, vol. 28, no. 11, pp. 4984–4988, 2012.

[153] D. Brutin, B. Sobac, and C. Nicloux, "Influence of substrate nature on the evaporation of a sessile drop of blood," *Journal of Heat Transfer*, vol. 134, no. 6, Article ID 061101, 2012.

[154] F. Carle and D. Brutin, "How surface functional groups influence fracturation in nanofluid droplet dry-outs," *Langmuir*, vol. 29, no. 32, pp. 9962–9966, 2013.

[155] V. H. Chhasatia, A. S. Joshi, and Y. Sun, "Effect of relative humidity on contact angle and particle deposition morphology of an evaporating colloidal drop," *Applied Physics Letters*, vol. 97, no. 23, Article ID 231909, 2010.

[156] W. Bou-Zeid and D. Brutin, "Effect of relative humidity on the spreading dynamics of sessile drops of blood," *Colloids and Surfaces A: Physicochemical and Engineering Aspects*, vol. 456, no. 1, pp. 273–285, 2014.

[157] W. Bou Zeid and D. Brutin, "Influence of relative humidity on spreading, pattern formation and adhesion of a drying drop of whole blood," *Colloids and Surfaces A: Physicochemical and Engineering Aspects*, vol. 430, pp. 1–7, 2013.

[158] S. Paria, R. Ghosh Chaudhuri, and N. N. Jason, "Self-assembly of colloidal sulfur particles on a glass surface from evaporating sessile drops: Influence of different salts," *New Journal of Chemistry*, vol. 38, no. 12, pp. 5943–5951, 2014.

[159] T. Okuzono, M. Kobayashi, and M. Doi, "Final shape of a drying thin film," *Physical Review E: Statistical, Nonlinear, and Soft Matter Physics*, vol. 80, no. 2, Article ID 021603, 2009.

[160] Y. Jung, T. Kajiya, T. Yamaue, and M. Doi, "Film formation kinetics in the drying process of polymer solution enclosed by bank," *Japanese Journal of Applied Physics*, vol. 48, no. 3, p. 031502, 2009.

[161] Yu. Yu. Tarasevich and D. M. Pravoslavnova, "Drying of a multi-component solution drop on a solid substrate: qualitative analysis," *Technical Physics*, vol. 52, no. 2, pp. 159–163, 2007.

[162] Yu. Yu. Tarasevich and D. M. Pravoslavnova, "Segregation in desiccated sessile drops of biological fluids," *The European Physical Journal E*, vol. 22, no. 4, pp. 311–314, 2007.

[163] Yu. Yu. Tarasevich, O. P. Isakova, V. V. Kondukhov, and A. V. Savitskaya, "Effect of evaporation conditions on the spatial redistribution of components in an evaporating liquid drop on a horizontal solid substrate," *Technical Physics*, vol. 55, no. 5, pp. 636–644, 2010.

[164] I. V. Vodolazskaya, Yu. Yu. Tarasevich, and O. P. Isakova, "The model of phase boundary motion in drying sessile drop of colloidal solution," *Nonlinear World*, vol. 8, pp. 142–150, 2010 (Russian).

[165] Yu. Yu. Tarasevich, I. V. Vodolazskaya, and O. P. Isakova, "Desiccating colloidal sessile drop: dynamics of shape and concentration," *Colloid & Polymer Science*, vol. 289, no. 9, pp. 1015–1023, 2011.

[166] I. V. Vodolazskaya and Yu. Yu. Tarasevich, "The model of drying sessile drop of colloidal solution," *Modern Physics Letters B*, vol. 25, no. 15, pp. 1303–1310, 2011.

[167] K. Ozawa, E. Nishitani, and M. Doi, "Modeling of the drying process of liquid droplet to form thin film," *Japanese Journal of Applied Physics*, vol. 44, no. 6, pp. 4229–4234, 2005.

[168] B. Sobac and D. Brutin, "Structural and evaporative evolutions in desiccating sessile drops of blood," *Physical Review E: Statistical, Nonlinear, and Soft Matter Physics*, vol. 84, no. 1, Article ID 011603, 2011.

[169] K. Sefiane, "On the formation of regular patterns from drying droplets and their potential use for bio-medical applications," *Journal of Bionic Engineering*, vol. 7, supplement 1, pp. S82–S93, 2010.

[170] T. A. Yakhno, A. G. Sanin, O. A. Sanina, and V. G. Yakhno, "Dynamics of mechanical properties of drying drops of biological liquids as a reflection of the features of self-organization of their components from nano- to microlevel," *Biophysics*, vol. 56, no. 6, pp. 1005–1010, 2011.

[171] V. E. Prokhorov, "Some features of the biological liquid droplet revealed from its dehydrated phase structure analysis," in *Proceedings of the 15th International Conference "Fluxes and Structures in Fluids: Physics of Geospheres"*, abstracts 2, p. 187, 2009.

[172] Y. J. P. Carreón, J. González-Gutiérrez, M. I. Pérez-Camacho, and H. Mercado-Uribe, "Patterns produced by dried droplets of protein binary mixtures suspended in water," *Colloids and Surfaces B: Biointerfaces*, vol. 161, pp. 103–110, 2018.

[173] Yu. Yu. Tarasevich and A. K. Ayupova, "Effect of diffusion on the separation of components in a biological fluid upon wedge-shaped dehydration," *Technical Physics*, vol. 48, no. 5, pp. 535–540, 2003.

[174] H. M. Gorr, J. M. Zueger, D. R. McAdams, and J. A. Barnard, "Salt-induced pattern formation in evaporating droplets of lysozyme solutions," *Colloids and Surfaces B: Biointerfaces*, vol. 103, pp. 59–66, 2013.

[175] T. A. Yakhno, V. G. Yakhno, and A. V. Sokolov, "Shaping processes in drying drops of serum in norm and pathology," *Biophysics*, vol. 50, no. 4, pp. 638–645, 2005.

[176] G. Chen and G. J. Mohamed, "Complex protein patterns formation via salt-induced self-assembly and droplet evaporation," *European Physical Journal E*, vol. 33, no. 1, pp. 19–26, 2010.

[177] K. A. Esmonde-White, F. W. L. Esmonde-White, M. D. Morris, and B. J. Roessler, "Characterization of biofluids prepared by sessile drop formation," *The Analyst*, vol. 139, no. 11, pp. 2734–2741, 2014.

[178] R. López Solís, L. T. Castro, D. S. Toro, M. Srur, and H. T. Araya, "Microdesiccates produced from normal human tears display four distinctive morphological components," *Biological Research*, vol. 46, no. 3, pp. 299–305, 2013.

[179] L. Traipe-Castro, D. Salinas-Toro, D. López et al., "Dynamics of tear fluid desiccation on a glass surface: a contribution to tear quality assessment," *Biological Research*, vol. 47, no. 1, article 25, 2014.

[180] E. I. Pearce and A. Tomlinson, "Spatial location studies on the chemical composition of human tear ferns," *Ophthalmic and Physiological Optics*, vol. 20, no. 4, pp. 306–313, 2000.

[181] E. G. Rapis, "Formation of an ordered structure upon drying of the protein film," *Soviet Technical Physics Letters*, vol. 14, p. 679, 1988.

[182] T. A. Yakhno, O. A. Sedova, A. G. Sanin, and A. S. Pelyushenko, "On the existence of regular structures in liquid human blood serum (plasma) and phase transitions in the course of its drying," *Technical Physics*, vol. 48, no. 4, pp. 399–403, 2003.

[183] N. Kim, Z. Li, C. Hurth, F. Zenhausern, S.-F. Chang, and D. Attinger, "Identification of fluid and substrate chemistry based on automatic pattern recognition of stains," *Analytical Methods*, vol. 4, no. 1, pp. 50–57, 2012.

[184] I. I. Smalyukh, O. V. Zribi, J. C. Butler, O. D. Lavrentovich, and G. C. L. Wong, "Structure and dynamics of liquid crystalline pattern formation in drying droplets of DNA," *Physical Review Letters*, vol. 96, no. 17, Article ID 177801, 2006.

[185] H. Maeda, "Observation of spatially rhythmic patterns from evaporating collagen solution droplets," *Langmuir*, vol. 16, no. 26, pp. 9977–9982, 2000.

[186] B. Roy, S. Karmakar, A. Giri, and S. Tarafdar, "Pattern formation of drying lyotropic liquid crystalline droplet," *RSC Advances*, vol. 6, no. 113, pp. 112695–112703, 2016.

[187] R. Mondragon, L. Hernandez, J. E. Julia et al., "Study of the drying behavior of high load multiphase droplets in an acoustic levitator at high temperature conditions," *Chemical Engineering Science*, vol. 66, no. 12, pp. 2734–2744, 2011.

[188] A. L. Yarin, G. Brenn, and D. Rensink, "Evaporation of acoustically levitated droplets of binary liquid mixtures," *International Journal of Heat and Fluid Flow*, vol. 23, no. 4, pp. 471–486, 2002.

[189] G. Brenn, L. J. Deviprasath, F. Durst, and C. Fink, "Evaporation of acoustically levitated multi-component liquid droplets," *International Journal of Heat and Mass Transfer*, vol. 50, no. 25–26, pp. 5073–5086, 2007.

[190] E. H. Brandt, "Levitation in physics," *Science*, vol. 243, no. 4889, pp. 349–355, 1989.

[191] C. P. Lee, A. V. Anilkumar, and T. G. Wang, "Static shape and instability of an acoustically levitated liquid drop," *Physics of Fluids*, vol. 3, no. 11, pp. 2497–2515, 1991.

[192] F. Giorgiutti-Dauphiné and L. Pauchard, "Elapsed time for crack formation during drying," *European Physical Journal E*, vol. 37, no. 5, article 39, 2014.

[193] N. Tsapis, E. R. Dufresne, S. S. Sinha et al., "Onset of buckling in drying droplets of colloidal suspensions," *Physical Review Letters*, vol. 94, no. 1, Article ID 018302, 2005.

[194] S. Hazra, T. Dutta, S. Das, and S. Tarafdar, "Memory of electric field in laponite and how it affects crack formation: modeling through generalized calculus," *Langmuir*, vol. 33, no. 34, pp. 8468–8475, 2017.

[195] S. Das, *Functional Fractional Calculus*, Springer, New Delhi, India, 2011.

[196] Y. Zhang, Y. Qian, Z. Liu, Z. Li, and D. Zang, "Surface wrinkling and cracking dynamics in the drying of colloidal droplets," *The European Physical Journal E*, vol. 37, no. 9, article 84, 7 pages, 2014.

[197] A. G. Banpurkar, K. P. Nichols, and F. Mugele, "Electrowetting-based microdrop tensiometer," *Langmuir*, vol. 24, no. 19, pp. 10549–10551, 2008.

[198] V. Vancauwenberghe, P. Di Marco, and D. Brutin, "Wetting and evaporation of a sessile drop under an external electrical field: A review," *Colloids and Surfaces A: Physicochemical and Engineering Aspects*, vol. 432, pp. 50–56, 2013.

[199] A. Sanyal, S. Basu, S. Chowdhuri, P. Kabi, and S. Chaudhuri, "Precision control of drying using rhythmic dancing of sessile nanoparticle laden droplets," *Applied Physics Letters*, vol. 104, no. 16, Article ID 163108, 2014.

[200] T. K. Pradhan and P. K. Panigrahi, "Deposition pattern of interacting droplets," *Colloids and Surfaces A: Physicochemical and Engineering Aspects*, vol. 482, pp. 562–567, 2015.

[201] N. J. Cira, A. Benusiglio, and M. Prakash, "Vapour-mediated sensing and motility in two-component droplets," *Nature*, vol. 519, no. 7544, pp. 446–450, 2015.

[202] Y. O. Popov, "Evaporative deposition patterns: Spatial dimensions of the deposit," *Physical Review E: Statistical, Nonlinear, and Soft Matter Physics*, vol. 71, no. 3, Article ID 036313, 2005.

[203] M. Cachile, O. Bénichou, and A. M. Cazabat, "Evaporating droplets of completely wetting liquids," *Langmuir*, vol. 18, no. 21, pp. 7985–7990, 2002.

[204] M. Cachile, O. Bénichou, C. Poulard, and A. M. Cazabat, "Evaporating droplets," *Langmuir*, vol. 18, no. 21, pp. 8070–8078, 2002.

[205] C. Poulard, O. Bénichou, and A. M. Cazabat, "Freely receding evaporating droplets," *Langmuir*, vol. 19, no. 21, pp. 8828–8834, 2003.

[206] J. P. Burelbach, S. G. Bankoff, and S. H. Davis, "Nonlinear stability of evaporating/condensing liquid films," *Journal of Fluid Mechanics*, vol. 195, pp. 463–494, 1988.

[207] B. J. Fischer, "Particle convection in an evaporating colloidal droplet," *Langmuir*, vol. 18, no. 1, pp. 60–67, 2002.

[208] E. Widjaja and M. T. Harris, "Particle deposition study during sessile drop evaporation," *AIChE Journal*, vol. 54, no. 9, pp. 2250–2260, 2008.

[209] R. Bhardwaj, X. Fang, and D. Attinger, "Pattern formation during the evaporation of a colloidal nanoliter drop: A numerical and experimental study," *New Journal of Physics*, vol. 11, Article ID 075020, 2009.

[210] R. Zheng, "A study of the evaporative deposition process: pipes and truncated transport dynamics," *The European Physical Journal E: Soft Matter and Biological Physics*, vol. 29, no. 2, pp. 205–218, 2009.

[211] T. A. Witten, "Robust fadeout profile of an evaporation stain," *EPL (Europhysics Letters)*, vol. 86, no. 6, Article ID 64002, 2009.

[212] R. V. Craster, O. K. Matar, and K. Sefiane, "Pinning, retraction and terracing of evaporating droplets containing nanoparticles," *Langmuir*, vol. 25, no. 6, pp. 3601–3609, 2009.

[213] A. J. Petsi, A. N. Kalarakis, and V. N. Burganos, "Deposition of Brownian particles during evaporation of two-dimensional sessile droplets," *Chemical Engineering Science*, vol. 65, no. 10, pp. 2978–2989, 2010.

[214] H.-S. Kim, S. S. Park, and F. Hagelberg, "Computational approach to drying a nanoparticle-suspended liquid droplet," *Journal of Nanoparticle Research*, vol. 13, no. 1, pp. 59–68, 2011.

[215] A. Zigelman and O. Manor, "The deposition of colloidal particles from a sessile drop of a volatile suspension subject to particle adsorption and coagulation," *Journal of Colloid and Interface Science*, vol. 509, pp. 195–208, 2018.

[216] V. Ragoonanan and A. Aksan, "Heterogeneity in desiccated solutions: Implications for biostabilization," *Biophysical Journal*, vol. 94, no. 6, pp. 2212–2227, 2008.

[217] V. N. Shabalin and S. N. Shatokhina, "Autogenous rhythms and self-organization of biological fluids," *Bulletin of Experimental Biology and Medicine*, vol. 122, no. 10, pp. 967–973, 1996.

[218] T. Yakhno, V. Yakhno, A. Sanin, O. Sanina, and A. Pelyushenko, "Dynamics of phase transitions in drying drops as an information parameter of liquid structure," *Nonlinear Dynamics*, vol. 39, no. 4, pp. 369–374, 2005.

[219] A. A. Killeen, N. Ossina, R. C. McGlennen et al., "Protein self-organization patterns in dried serum reveal changes in B-cell disorders," *Molecular Diagnosis & Therapy*, vol. 10, no. 6, pp. 371–380, 2006.

[220] T. Khatun, T. Dutta, and S. Tarafdar, ""Islands in sea" and "lakes in Mainland" phases and related transitions simulated on a square lattice," *The European Physical Journal B*, vol. 90, no. 11, p. 213, 2017.

[221] A. D. Eales, N. Dartnell, S. Goddard, and A. F. Routh, "The impact of trough geometry on film shape. A theoretical study of droplets containing polymer, for P-OLED display applications," *Journal of Colloid and Interface Science*, vol. 458, pp. 53–61, 2015.

[222] D. M. Anderson and S. H. Davis, "The spreading of volatile liquid droplets on heated surfaces," *Physics of Fluids*, vol. 7, no. 2, pp. 248–265, 1995.

[223] N. I. Lebovka, V. A. Gigiberiya, O. S. Lytvyn, Yu. Yu. Tarasevich, I. V. Vodolazskaya, and O. P. Bondarenko, "Drying of sessile droplets of laponite-based aqueous nanofluids," *Colloids and Surfaces A: Physicochemical and Engineering Aspects*, vol. 462, pp. 52–63, 2014.

[224] H. Hu and R. G. Larson, "Evaporation of a sessile droplet on a substrate," *The Journal of Physical Chemistry B*, vol. 106, no. 6, pp. 1334–1344, 2002.

[225] R. Mollaret, K. Sefiane, J. R. E. Christy, and D. Veyret, "Experimental and numerical investigation of the evaporation into air of a drop on a heated surface," *Chemical Engineering Research and Design*, vol. 82, no. 4, pp. 471–480, 2004.

[226] Yu. Yu. Tarasevich, "Simple analytical model of capillary flow in an evaporating sessile drop," *Physical Review E: Statistical, Nonlinear, and Soft Matter Physics*, vol. 71, no. 2, Article ID 027301, 2005.

[227] E. Widjaja and M. T. Harris, "Numerical study of vapor phase-diffusion driven sessile drop evaporation," *Computers & Chemical Engineering*, vol. 32, no. 10, pp. 2169–2178, 2008.

[228] G. J. Dunn, S. K. Wilson, B. R. Duffy, S. David, and K. Sefiane, "A mathematical model for the evaporation of a thin sessile liquid droplet: Comparison between experiment and theory," *Colloids and Surfaces A: Physicochemical and Engineering Aspects*, vol. 323, no. 1-3, pp. 50–55, 2008, Bubble and Drop Interfaces—Selected papers from the International Workshop on Bubble and Drop Interfaces, 25–28 March 2007, Granada, Spain.

[229] L. Yu. Barash, "Influence of gravitational forces and fluid flows on the shape of surfaces of a viscous fluid of capillary size," *Physical Review E: Statistical, Nonlinear, and Soft Matter Physics*, vol. 79, no. 2, Article ID 025302, 2009.

[230] H. Masoud and J. D. Felske, "Analytical solution for inviscid flow inside an evaporating sessile drop," *Physical Review E: Statistical, Nonlinear, and Soft Matter Physics*, vol. 79, no. 1, Article ID 016301, 2009.

[231] A. J. Petsi and V. N. Burganos, "Stokes flow inside an evaporating liquid line for any contact angle," *Physical Review E: Statistical, Nonlinear, and Soft Matter Physics*, vol. 78, no. 3, Article ID 036324, 2008.

[232] C.-L. Hsu, S.-M. Chu, K. Wood, and Y.-R. Yang, "Percolation in two-dimensional nanoparticle films from colloidal self-assembly," *Physica Status Solidi (a)*, vol. 204, no. 6, pp. 1856–1862, 2007.

[233] W. Chen, J. Koplik, and I. Kretzschmar, "Molecular dynamics simulations of the evaporation of particle-laden droplets," *Physical Review E: Statistical, Nonlinear, and Soft Matter Physics*, vol. 87, no. 5, Article ID 052404, 2013.

[234] A. Crivoi and F. Duan, "Fingering structures inside the coffee-ring pattern," *Colloids and Surfaces A: Physicochemical and Engineering Aspects*, vol. 432, pp. 119–126, 2013.

[235] P. Lebedev-Stepanov and K. Vlasov, "Simulation of self-assembly in an evaporating droplet of colloidal solution by dissipative particle dynamics," *Colloids and Surfaces A: Physicochemical and Engineering Aspects*, vol. 432, pp. 132–138, 2013.

[236] N. I. Lebovka, S. Khrapatiy, R. Melnyk, and M. Vygornitskii, "Effects of hydrodynamic retardation and interparticle interactions on the self-assembly in a drying droplet containing suspended solid particles," *Physical Review E: Statistical, Nonlinear, and Soft Matter Physics*, vol. 89, no. 5, Article ID 052307, 2014.

[237] A. Crivoi and F. Duan, "Three-dimensional Monte Carlo model of the coffee-ring effect in evaporating colloidal droplets," *Scientific Reports*, vol. 4, article 4310, pp. 1–6, 2014.

[238] M. Fujita, O. Koike, and Y. Yamaguchi, "Direct simulation of drying colloidal suspension on substrate using immersed free surface model," *Journal of Computational Physics*, vol. 281, pp. 421–448, 2015.

[239] H. Hwang and G. Son, "A level-set method for the direct numerical simulation of particle motion in droplet evaporation," *Numerical Heat Transfer, Part B: Fundamentals*, vol. 68, no. 6, pp. 479–494, 2015.

[240] M. Fujita and Y. Yamaguchi, "Mesoscale modeling for self-organization of colloidal systems," *Current Opinion in Colloid & Interface Science* , vol. 15, no. 1-2, pp. 8–12, 2010.

[241] M. G. Zaleskiĭ, V. L. Emanuél', and M. V. Krasnova, "Physical and chemical characteristics of structurization of a biological fluid drop as examplified by the 'LITOS-system' diagnosticum," *Klinicheskaia Laboratornaia Diagnostika*, no. 8, pp. 20–24, 2004 (Russian).

[242] R. N. Bardakov, Yu. D. Chashechkin, and V. V. Shabalin, "Hydrodynamics of a drying multicomponent liquid droplet," *Journal of Fluid Dynamics*, vol. 45, no. 5, pp. 803–816, 2010.

[243] Y. D. Chashechkin and R. N. Bardakov, "Formation of texture in residue of a drying drop of a multicomponent fluid," *Doklady Physics*, vol. 55, no. 2, pp. 68–72, 2010.

[244] F. Girard, M. Antoni, A. Steinchen, and S. Faure, "Numerical study of the evaporating dynamics of a sessile water droplet," *Microgravity Science and Technology*, vol. 18, no. 3-4, pp. 42–46, 2006.

[245] C. Diddens, J. G. M. Kuerten, C. W. M. van der Geld, and H. M. A. Wijshoff, "Modeling the evaporation of sessile multi-component droplets," *Journal of Colloid and Interface Science*, vol. 487, pp. 426–436, 2017.

[246] C. Diddens, "Detailed finite element method modeling of evaporating multi-component droplets," *Journal of Computational Physics*, vol. 340, pp. 670–687, 2017.

[247] H. Tan, C. Diddens, M. Versluis, H.-J. Butt, D. Lohse, and X. Zhang, "Self-wrapping of an ouzo drop induced by evaporation on a superamphiphobic surface," *Soft Matter*, vol. 13, no. 15, pp. 2749–2759, 2017.

[248] S. Shin, I. Jacobi, and H. A. Stone, "Bénard-Marangoni instability driven by moisture absorption," *EPL (Europhysics Letters)*, vol. 113, no. 2, Article ID 24002, 2016.

[249] J. R. E. Christy, Y. Hamamoto, and K. Sefiane, "Flow transition within an evaporating binary mixture sessile drop," *Physical Review Letters*, vol. 106, no. 20, Article ID 205701, 2011.

[250] R. Bennacer and K. Sefiane, "Vortices, dissipation and flow transition in volatile binary drops," *Journal of Fluid Mechanics*, vol. 749, no. 5, pp. 649–665, 2014.

[251] X. Zhong and F. Duan, "Flow regime and deposition pattern of evaporating binary mixture droplet suspended with particles," *The European Physical Journal E*, vol. 39, no. 2, article 18, pp. 1–6, 2016.

[252] T. Okuzono, K. Ozawa, and M. Doi, "Simple model of skin formation caused by solvent evaporation in polymer solutions," *Physical Review Letters*, vol. 97, no. 13, Article ID 136103, 2006.

[253] M. G. Hennessy, G. L. Ferretti, J. T. Cabral, and O. K. Matar, "A minimal model for solvent evaporation and absorption in thin films," *Journal of Colloid and Interface Science*, vol. 488, pp. 61–71, 2017.

[254] L. M. Martyushev, V. D. Seleznev, and S. A. Skopinov, "Reentrant kinetic phase transitions during dendritic growth of crystals in a two-dimensional medium with phase separation," *Technical Physics Letters*, vol. 23, no. 7, pp. 495–497, 1997.

[255] L. M. Martiouchev, V. D. Seleznev, and S. A. Skopinov, "Computer simulation of nonequilibrium growth of crystals in a two-dimensional medium with a phase-separating impurity," *Journal of Statistical Physics*, vol. 90, no. 5-6, pp. 1413–1427, 1998.

[256] L. M. Martyushev and V. D. Seleznev, "Self-similarity in the kinetic growth regime of a crystal in a phase-separating medium," *Technical Physics Letters*, vol. 25, no. 10, pp. 833–835, 1999.

[257] A. Crivoi and F. Duan, "Evaporation-induced formation of fractal-like structures from nanofluids," *Physical Chemistry Chemical Physics*, vol. 14, no. 4, pp. 1449–1454, 2012.

[258] Yu. Yu. Tarasevich, "Computer simulation of crystal growth from solution," *Technical Physics*, vol. 46, no. 5, pp. 627–629, 2001.

[259] J. S. Langer, "Instabilities and pattern formation in crystal growth," *Reviews of Modern Physics*, vol. 52, no. 1, pp. 1–28, 1980.

[260] Yu. Yu. Tarasevich, V. O. Konstantinov, and A. K. Ayupova, "Simulation of dendritic growth of salt crystals in biological fluids," in *Izvestija vuzov. Severo-Kavkazskij Region*, Estestvennye nauki Special Issue. Math. Mod., pp. 147–149, (Russian) 2001.

[261] J. A. Warren and W. J. Boettinger, "Prediction of dendritic growth and microsegregation patterns in a binary alloy using the phase-field method," *Acta Metallurgica et Materialia*, vol. 43, no. 2, pp. 689–703, 1995.

[262] J. Zhang, M. K. Borg, K. Ritos, and J. M. Reese, "Electrowetting controls the deposit patterns of evaporated salt water nanodroplets," *Langmuir*, vol. 32, no. 6, pp. 1542–1549, 2016.

[263] X. Li, K. L. Maki, and M. J. Schertzer, "A low cost fabrication method for electrowetting assisted desiccation of colloidal droplets," in *Symposia: Fluid Measurement and Instrumentation; Fluid Dynamics of Wind Energy; Renewable and Sustainable Energy Conversion; Energy and Process Engineering; Microfluidics and Nanofluidics; Development and Applications in Computational Fluid Dynamics; DNSLES and Hybrid RANSLES Methods*, vol. 1B, ASME, 2017.

[264] A. Shimoni, S. Azoubel, and S. Magdassi, "Inkjet printing of flexible high-performance carbon nanotube transparent conductive films by "coffee ring effect"," *Nanoscale*, vol. 6, no. 19, pp. 11084–11089, 2014.

[265] N. R. Devlin, K. Loehr, and M. T. Harris, "The separation of two different sized particles in an evaporating droplet," *AIChE Journal*, vol. 61, no. 10, pp. 3547–3556, 2015.

[266] T.-S. Wong, T.-H. Chen, X. Shen, and C.-M. Ho, "Nanochromatography driven by the coffee ring effect," *Analytical Chemistry*, vol. 83, no. 6, pp. 1871–1873, 2011.

[267] A. V. Vorobiev, A. K. Martusevich, and S. P. Peretyagin, *Crystallogenesis Biological Fluids and Substrates in Assessing the State of the Organism*, FGU "NNIITO Rosmedtechnology", Nizhny Novgorod, Russia, (Russian) 2008.

[268] I. G. Denisov, Y. V. Grinkova, A. A. Lazarides, and S. G. Sligar, "Directed self-assembly of monodisperse phospholipid bilayer nanodiscs with controlled size," *Journal of the American Chemical Society*, vol. 126, no. 11, pp. 3477–3487, 2004.

[269] E. Rapis, "A change in the physical state of a nonequilibrium blood plasma protein film in patients with carcinoma," *Technical Physics*, vol. 47, no. 4, pp. 510–512, 2002.

[270] C. P. Gulka, J. D. Swartz, J. R. Trantum et al., "Coffee rings as low-resource diagnostics: Detection of the malaria biomarker plasmodium falciparum histidine-rich protein-II using a surface-coupled ring of Ni(II)NTA gold-plated polystyrene particles," *ACS Applied Materials & Interfaces*, vol. 6, no. 9, pp. 6257–6263, 2014.

[271] J. R. Trantum, D. W. Wright, and F. R. Haselton, "Biomarker-mediated disruption of coffee-ring formation as a low resource diagnostic indicator," *Langmuir*, vol. 28, no. 4, pp. 2187–2193, 2012.

[272] J. R. Trantum, M. L. Baglia, Z. E. Eagleton, R. L. Mernaugh, and F. R. Haselton, "Biosensor design based on Marangoni flow in an evaporating drop," *Lab on a Chip*, vol. 14, no. 2, pp. 315–324, 2014.

[273] L. Bahmani, M. Neysari, and M. Maleki, "The study of drying and pattern formation of whole human blood drops and the effect of thalassaemia and neonatal jaundice on the patterns," *Colloids and Surfaces A: Physicochemical and Engineering Aspects*, vol. 513, pp. 66–75, 2017.

[274] R. Less, K. L. M. Boylan, A. P. N. Skubitz, and A. Aksan, "Isothermal vitrification methodology development for non-cryogenic storage of archival human sera," *Cryobiology*, vol. 66, no. 2, pp. 176–185, 2013.

[275] M. O. Kokornaczyk, G. Dinelli, I. Marotti, S. Benedettelli, D. Nani, and L. Betti, "Self-organized crystallization patterns from evaporating droplets of common wheat grain leakages as a potential tool for quality analysis," *TheScientificWorldJOURNAL*, vol. 11, pp. 1712–1725, 2011.

[276] J. González-Gutiérrez, R. Pérez-Isidoro, and J. C. Ruiz-Suárez, "A technique based on droplet evaporation to recognize alcoholic drinks," *Review of Scientific Instruments*, vol. 88, no. 7, Article ID 074101, 2017.

[277] K. A. Esmonde-White, S. V. Le Clair, B. J. Roessler, and M. D. Morris, "Effect of conformation and drop properties on surface-enhanced Raman spectroscopy of dried biopolymer drops," *Applied Spectroscopy*, vol. 62, no. 5, pp. 503–511, 2008.

[278] K. A. Esmonde-White, G. S. Mandair, F. Raaii, B. J. Roessler, and M. D. Morris, "Raman spectroscopy of dried synovial fluid droplets as a rapid diagnostic for knee joint damage," in *Biomedical Optical Spectroscopy, 68530Y*, vol. 6853 of *Proceedings of SPIE*, San Jose, Calif, USA, January 2008.

[279] K. A. Esmonde-White, G. S. Mandair, F. Raaii et al., "Raman spectroscopy of synovial fluid as a tool for diagnosing osteoarthritis," *Journal of Biomedical Optics*, vol. 14, no. 3, Article ID 034013, 2009.

[280] K. A. Esmonde-White, G. S. Mandair, F. W. L. Esmonde-White, F. Raaii, B. J. Roessler, and M. D. Morris, "Osteoarthritis screening using raman spectroscopy of dried humansynovial fluid drops," in *Optics in Bone Biology and Diagnostics, 71660J*, vol. 7166 of *Proceedings of SPIE*, San Jose, Calif, USA, January 2009.

[281] J. Filik and N. Stone, "Drop coating deposition Raman spectroscopy of protein mixtures," *Analyst*, vol. 132, no. 6, pp. 544–550, 2007.

[282] J. Filik and N. Stone, "Raman point mapping of tear ferning patterns," in *Biomedical Optical Spectroscopy, 685309*, vol. 6853 of *Proceedings of SPIE*, San Jose, Calif, USA, January 2008.

[283] J. Filik and N. Stone, "Investigation into the protein composition of human tear fluid using centrifugal filters and drop coating deposition Raman spectroscopy," *Journal of Raman Spectroscopy*, vol. 40, no. 2, pp. 218–224, 2009.

[284] D. Zhang, K. Vangala, D. Jiang, S. Zou, and T. Pechan, "Drop coating deposition raman spectroscopy of fluorescein isothiocyanate labeled protein," *Applied Spectroscopy*, vol. 64, no. 10, pp. 1078–1085, 2010.

[285] N. C. Dingari, G. L. Horowitz, J. W. Kang, R. R. Dasari, and I. Barman, "Raman spectroscopy provides a powerful diagnostic tool for accurate determination of albumin glycation," *PLoS ONE*, vol. 7, no. 2, Article ID e32406, 2012.

[286] E. Kočišová, A. Vodáková, and M. Procházka, "DCDR spectroscopy as efficient tool for liposome studies: aspect of preparation procedure parameters," *Spectroscopy: An International Journal*, vol. 27, no. 5-6, pp. 349–353, 2012.

[287] D. Zang, J. Li, Z. Chen, Z. Zhai, X. Geng, and B. P. Binks, "Switchable opening and closing of a liquid marble via ultrasonic levitation," *Langmuir*, vol. 31, no. 42, pp. 11502–11507, 2015.

[288] X. Zhong, A. Crivoi, and F. Duan, "Sessile nanofluid droplet drying," *Advances in Colloid and Interface Science*, vol. 217, pp. 13–30, 2015.

[289] C. Sadek, P. Schuck, Y. Fallourd, N. Pradeau, C. Le Floch-Fouéré, and R. Jeantet, "Drying of a single droplet to investigate process–structure–function relationships: a review," *Dairy Science & Technology*, vol. 95, no. 6, pp. 771–794, 2015.

14

Liquid Crystals as Phase Change Materials for Thermal Stabilization

Eva Klemenčič [ID][1] **and Mitja Slavinec** [ID][1,2]

[1]*Faculty of Natural Sciences and Mathematics, University of Maribor, Maribor, Slovenia*
[2]*Academic Scientific Union of Pomurje (PAZU), Murska Sobota, Slovenia*

Correspondence should be addressed to Mitja Slavinec; mitja.slavinec@um.si

Academic Editor: Charles Rosenblatt

Thermal stabilization exploiting phase change materials (PCMs) is studied theoretically and numerically. Using the heat source approach in numerical simulations, we focus on phase change temperature as a key factor in improving thermal stabilization. Our focus is to analyze possible mechanisms to tune the phase change temperature. We use thermotropic liquid crystals (LCs) as PCMs in a demonstrative system. Using the Landau-de Gennes mesoscopic approach, we show that an external electric field or appropriate nanoparticles (NPs) dispersed in LCs can be exploited to manipulate the phase change temperature.

1. Introduction

In order to achieve and maintain an optimum ambient temperature, one must regulate heat flow using different heating or cooling systems, which are usually large energy consumers. Recently, several studies [1–5] focused on the development of innovative materials to passively reduce temperature fluctuations. Phase change materials (PCMs) are suitable candidates, as they undergo a phase change during which they exchange latent heat with the surroundings.

The main function of PCMs is saving excess heat that is accumulated at higher temperature conditions and releasing it at a lower temperature. Therefore, the efficiency of a system using PCMs is strongly correlated with phase change temperature. For example, if one wants to decrease room temperature oscillation, it would be optimal to have a PCM with phase change around the desired room temperature. If the room temperature exceeds the temperature of the PCM composite and reaches the phase change temperature, the PCM undergoes a phase change. During the phase change, the PCM accumulates the excess heat in form of latent heat. The opposite happens as the temperature lowers towards the phase change temperature and the latent heat is released to the surroundings.

In this paper, we focus on analyzing possible mechanisms to tune the phase change temperature in order to obtain optimal thermal stabilization. We used a nematic liquid crystal (LC) system as an experimental testing ground. LCs are experimentally suitable due to their liquid character, softness, large susceptibility to relatively weak external perturbations, and optical anisotropy [6, 7]. We considered thermotropic LC in which, by decreasing the temperature from the isotropic (I) phase, the nematic (N) phase is obtained via first-order continuous symmetry breaking phase transition [8–10]. In the uniaxial bulk NLC phase, rod-like molecules exhibit long-range orientational ordering commonly presented by the nematic director $\vec{n}$. Fluctuations about $\vec{n}$ describe the uniaxial order parameter S. Note that $|\vec{n}| = 1$ and states $\pm\vec{n}$ are equivalent due to the so-called head-to-tail invariance. At the mesoscopic scale, we described nematic orientational ordering by the nematic tensor order parameter $\mathbf{Q}$ [11]. Our main goal was to identify different mechanisms to control and manipulate the phase change temperature. In general, the presence of nanoparticles (NPs), external electric fields, and other perturbations gives rise to elastic distortions [12–16] that can result in the temperature shift of the phase transition.

First we introduce a numerical model to analyze thermal stabilization using PCMs. The model is based on the heat

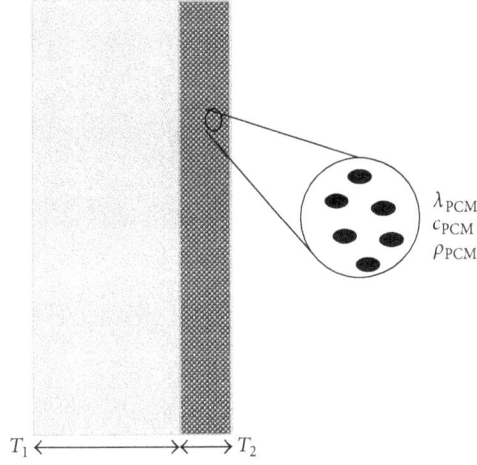

FIGURE 1: Schematic representation of two layers, one integrating PCMs.

source approach and finite difference method. Next, we present a theoretical Landau-de Gennes model to examine the impact of the external electric field and NPs on the phase change temperature of NLC system. We present and discuss results in Section 3. In the final section, we conclude our findings.

2. Modeling

2.1. Thermal Stabilization with PCM. In this section, we describe the use of PCMs for the improvement of thermal stabilization. We considered one-dimensional heat transfer through the planar surface of a two-layered composite (Figure 1) with one layer integrating homogeneously distributed PCMs. Our numerical model is based on the heat source approach [17–20], in which latent heat is modeled as an additional heat source. In terms of enthalpy (h), the transient heat transfer equation in one dimension is expressed as

$$\rho \frac{\partial h}{\partial t} = \lambda \frac{\partial T^2}{\partial x^2}, \tag{1}$$

where ρ is density and λ thermal conductivity of each layer.

Assuming high porosity of the composite layer, the expansion of PCMs during the phase transition does not change the total volume. Additionally, we considered different physical properties of the solid (s) and liquid (l) phase of PCMs.

In order to simplify their numerical simulations, other studies [17, 19, 20] determined the temperature range where phase transition occurs. In this study, we assumed that phase transition is instant at one specific critical temperature T_c. We introduced the factor r that equals zero when the PCM is in the solid phase ($T < T_c$) and one when the PCM is in the liquid phase ($T > T_c$).

Using a model of effective physical quantities [18] we defined the effective specific heat capacity (c_e), thermal conductivity (λ_e), and density (ρ_e) as follows:

$$c_e = \alpha \left(c_{PCM}^{(l)} r + c_{PCM}^{(s)} (1 - r) \right) + (1 - \alpha) c_A, \tag{2a}$$

$$\lambda_e = \alpha \left(\lambda_{PCM}^{(l)} r + \lambda_{PCM}^{(s)} (1 - r) \right) + (1 - \alpha) \lambda_A, \tag{2b}$$

$$\rho_e = (1 - \alpha) \rho_A + \alpha \left(\rho_{PCM}^{(l)} r + \rho_{PCM}^{(s)} (1 - r) \right), \tag{2c}$$

where α represents the relative amount of PCM integrated in the composite layer and subscripts A and B correspond to two different layers of studied composite (see Figure 1).

We derived the heat transfer equation for the composite incorporating PCMs in terms of the enthalpy $h_e = c_e T + \alpha r Q_L$ as follows:

$$\rho_e c_e \frac{\partial T}{\partial t} + \rho_e \alpha Q_L \frac{\partial r}{\partial t} = \lambda_e \frac{\partial T^2}{\partial x^2}, \tag{3}$$

where Q_L is the latent heat of the integrated PCM. For numerical simulations, we used the finite difference method.

2.2. Phase Change Temperature Shift. Next, we theoretically analyzed possible mechanisms to tune the phase change temperature. We considered the NLC rectangular cell, where lateral confinement walls impose strong homogeneous tangential anchoring (Figure 2). The characteristic confinement size R is assumed to be larger than the cell thickness h.

Using the nematic tensor order parameter [6], we described LC orientational ordering as follows:

$$\mathbf{Q} = \sum_{i=1}^{3} s_i \vec{e}_i \otimes \vec{e}_i, \tag{4}$$

where $\vec{e}_i$ are eigenvectors, s_i are eigenvalues of $\mathbf{Q}$, and the symbol $\otimes$ represents tensor product. In case of uniaxial ordering, it holds as follows [12]:

$$\mathbf{Q} = S \left(\vec{n} \otimes \vec{n} - \frac{1}{3} \mathbf{I} \right), \tag{5}$$

where $\mathbf{I}$ is the identity matrix.

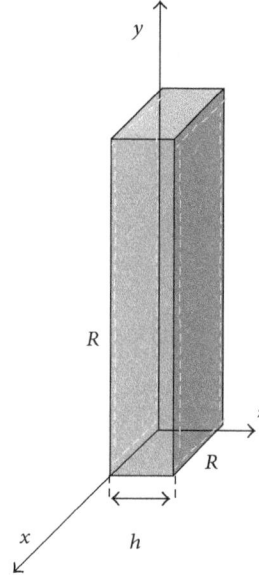

FIGURE 2: NLC cell of thickness h along the z-Cartesian coordinate with conflicting strong homogeneous tangential anchoring.

Next, we expressed the free energy density as a sum of the condensation nematic term f_n, elastic term f_e, external field term f_f, and surface term f_s [12, 13]:

$$f_n = \frac{A_0 (T - T_*)}{2} \mathrm{Tr} \mathbf{Q}^2 - \frac{B}{3} \mathrm{Tr} \mathbf{Q}^3 + \frac{C}{4} \left(\mathrm{Tr} \mathbf{Q}^2 \right)^2, \quad (6a)$$

$$f_e = \frac{L}{2} |\nabla \mathbf{Q}|^2, \quad (6b)$$

$$f_f = \frac{\varepsilon_0 \Delta \varepsilon}{2} \vec{E} \cdot \mathbf{Q} \vec{E}, \quad (6c)$$

$$f_s = \frac{w}{2} \mathrm{Tr} \left(\mathbf{Q} - \mathbf{Q_s} \right), \quad (6d)$$

where A_0, B, C are material-dependent constants and T_* is the isotropic supercooling temperature. The elastic term evaluates deviations from a spatially homogeneous orientational ordering in terms of the elastic constant L. The external field term describes the contribution of an external electric field $\vec{E}$, where $\Delta \varepsilon$ stands for the electric field anisotropy constant and ε_0 represents the electrical permittivity constant. Since we imposed homogeneous tangential anchoring at the lateral surfaces with the anchoring strength w, $\mathbf{Q_s}$ describes the degree of orientational ordering enforced by the confining surface.

The I-N phase change temperature is determined by material-dependent constants:

$$T_{\mathrm{IN}} = T_* + \frac{B^2}{(4A_0 C)}. \quad (7)$$

Next, assuming uniaxial ordering described by (5), we imposed a distortion to orientational ordering on a characteristic length scale R. On average, it holds as follows (see (6b)) [21, 22]:

$$\overline{f_e} \sim \frac{L}{2} \frac{S^2}{R^2}, \quad (8)$$

where the overbar $\overline{(\cdots)}$ indicates spatial averaging.

Then, we expressed the average contributions of the condensation nematic term and elastic term to the average free energy density $\overline{f}$ as follows [23–25]:

$$\overline{f} \sim \frac{A_0 (T - T_*)}{3} \overline{S}^2 - \frac{B}{9} \overline{S}^3 + \frac{C}{9} \overline{S}^4 + \frac{L}{2} \frac{\overline{S}^2}{R^2}$$

$$\equiv \frac{A_0 \left(T - T_*^{\mathrm{eff}} \right)}{3} \overline{S}^2 - \frac{B}{9} \overline{S}^3 + \frac{C}{9} \overline{S}^4, \quad (9)$$

where

$$T_*^{\mathrm{eff}} = T_* - \frac{3}{2} \frac{L}{A_0 R^2}. \quad (10)$$

Considering L, A_0, and T_* are material-dependent coefficients, one could vary the effective phase change temperature of the NLC system by varying the typical distortion length R. We obtained the phase change temperature shift $\Delta T = T_{\mathrm{IN}} - T_{\mathrm{IN}}(R)$ expressed as follows (see (7)) [26, 27]:

$$\Delta T_{\mathrm{IN}} (R) = -\frac{3}{2} \frac{L}{A_0 R^2}. \quad (11)$$

Since the temperature shift is connected to R, we focused on imposing different distortions that can be described by a linear deformation imposed characteristic length R.

Below, we present two experimentally attainable possibilities influencing the phase change temperature of NLC: the presence of an external electric field and NPs. Since the presence of NPs and the external electric field could induce biaxiality, we introduce the degree of biaxiality as follows [28]:

$$\beta^2 = 1 - \frac{6 \left(\mathrm{Tr} \mathbf{Q}^3 \right)^2}{\left(\mathrm{Tr} \mathbf{Q}^2 \right)^3}, \quad (12)$$

which is zero for uniaxial states and one when the state exhibits a maximum degree of biaxiality.

For further convenience, we introduce three characteristic lengths: the external field coherence length $\xi_f = \sqrt{LS/(\varepsilon_0 \Delta \varepsilon E^2)}$, the surface anchoring length $d_e = L/w$, and the bare biaxial order parameter correlation length $\xi_b = \sqrt{LC}/B$. In addition, we introduce parametrization and scaling as described in the following [12]:

$$\mathbf{Q}(x, y) = (q_3 + q_1) \vec{e}_x \otimes \vec{e}_x + (q_3 - q_1) \vec{e}_y \otimes \vec{e}_y$$

$$+ q_2 \left(\vec{e}_x \otimes \vec{e}_y + \vec{e}_y \otimes \vec{e}_x \right) - 2q_3 \vec{e}_z \quad (13)$$

$$\otimes \vec{e}_z,$$

where q_1, q_2, and q_3 are x and y functions and $\vec{e}_3 = \vec{e}_z$ is an eigenvector of $\mathbf{Q}$.

In this scaling, we define the scaled temperature as $t = (T - T_*)/(T_{**} - T_*)$. Here, the superheating temperature of the nematic phase is $T_{**} = T_* + B^2/24A_0C$. For numerical convenience we introduce $\tau = 1 + \sqrt{1 - t}$. With this in mind, we express the equilibrium uniaxial order parameter as $S_{eq} = S_{**}\tau$ where $S_{**} = B/4C = S_{eq}(T_{**})$.

Finally, we obtain bulk Euler-Lagrange equations for variation parameters q_1, q_2, and q_3 by minimization of the free energy density [29]:

$$\Delta_{\perp} q_1 \left(\frac{\xi_b^{(0)}}{R} \right)^2 - \frac{\tau}{6} q_1 + 2q_2 q_1 - \frac{q_1}{2} \left(3q_2^2 + q_1^2 + q_3^2 \right)$$

$$+ \frac{1}{4} \left(\frac{\xi_b^{(0)}}{\xi_f} \right)^2 \left(\left(\vec{e}_x \cdot \vec{e} \right)^2 - \left(\vec{e}_y \cdot \vec{e} \right)^2 \right) = 0, \quad (14a)$$

$$\Delta_{\perp} q_2 \left(\frac{\xi_b^{(0)}}{R} \right)^2 - \frac{\tau}{6} q_2 + \frac{1}{3} \left(q_1^2 + q_3^2 - 3q_2^2 \right)$$

$$- \frac{q_2}{2} \left(3q_2^2 + q_1^2 + q_3^2 \right) + \frac{1}{12} \left(\frac{\xi_b^{(0)}}{\xi_f} \right)^2 = 0, \quad (14b)$$

$$\Delta_{\perp} q_3 \left(\frac{\xi_b^{(0)}}{h} \right)^2 - \frac{\tau}{6} q_3 + 2q_2 q_3 - \frac{q_3}{2} \left(3q_2^2 + q_1^2 + q_3^2 \right)$$

$$+ \frac{1}{2} \left(\frac{\xi_b^{(0)}}{\xi_f} \right)^2 \left(\vec{e}_x \cdot \vec{e} \right) \left(\vec{e}_y \cdot \vec{e} \right) = 0, \quad (14c)$$

where $\Delta_{\perp} = \partial^2/\partial x^2 + \partial^2/\partial y^2$. Euler-Lagrange equations are solved using the standard relaxation method numerically.

3. Results and Discussion

In this section, we first analyze the numerical results of a one-dimensional heat transfer through composite wall with a focus on varying the phase change temperature. In our simulations, the composite wall (labeled GPCM) consisted of two layers: 20,0 cm thick panel at the outside and 3,0 CM thick gypsum wallboard with PCMs at the inside. The latter is also true in real building constructions, as the purpose of PCM is to stabilize the interior temperature. The outside temperature fluctuated according to typical day/night temperatures in the summertime (from 15°C to 35°C) and the room temperature (T_{room}) was set to a constant value, in our case to 21°C. We examined cases for three different phase change temperatures (T_c) of PCM and obtained different scenarios for the optimal thermal stabilization depending on the initial temperature (T_i) of the system.

Figure 3 depicts the incoming heat flux (q_{in}) time dependency for three different cases. First, we set the initial temperature of the composite wall to the average value of the outside temperature ($T_i > T_{room}$). It is evident that the presence of PCMs affects q_{in} as soon as the phase change temperature T_c is reached. At this point, q_n is close to zero, or zero for $T_c = T_{room}$ (black solid curve), until the latent heat storage capacity becomes full. By varying the phase change temperature, we show that the time period of a steady incoming heat flux is the longest when T_c is slightly above the room temperature (red dotted curve). On the other hand, in order to maintain the constant room temperature in this case, the incoming heat flux should be compensated using different cooling devices. The zero incoming heat flux is reachable only by setting the phase change temperature equal to the room temperature. Negative values of q_{in}, for the case when $T_c < T_{room}$ (blue dashed curve), correspond to a decrease of a room temperature below the desired value. Next, we analyzed cases for $T_i < T_{room}$ (Figure 3(b)) and $T_i = T_{room}$ (Figure 3(c)). Figures 3(b) and 3(c) confirm that the incoming heat flux can be entirely or partly stored in the form of the latent heat at the T_c. It is also evident that temperature oscillations in all three cases are reduced noticeably, especially for $T_c = T_{room}$ and $T_c > T_{room}$. The numerical obtained results show that thermal stabilization could be more efficient by using a small variation of the phase change temperature. Nevertheless, it is evident that the incoming heat flux fluctuates the most for the composite with no PCMs (black dotted curve).

To better understand the efficiency of PCM composites, we numerically analyzed thermal stabilization for three different composites used in real-life applications. Figure 4 shows a comparison of an alternative composite GPCM (black solid curve) with composites labeled BS1 (blue dashed curve) and BS2 (red dotted curve). BS1 and BS2 both consist of classic building materials: brick wall of thickness 50,0 cm for BS1 and 20,0 cm for BS2 and 3,0 cm thick insulation layer of Styrofoam placed at the outside of a wall. We simulated typical outside temperature fluctuations for the summertime, as described above. In this case the phase change temperature of GPCM is set to the room temperature. It is noticeable that thicker wall of BS1 lowers temperature oscillations, in comparison with BS2. This is expected due to larger thermal capacity. Furthermore, it is also evident that the alternative composite GPCM, of equal thickness as BS2, improves thermal stability even more. The incoming heat flux after 48 hours of simulations is close to 0% for GPCM and around 5% for BS1. Therefore, we can conclude that both systems BS1 and GPCM optimize the thermal stability, but by using alternative composites with PCMs construction walls can be approximately 2 times thinner.

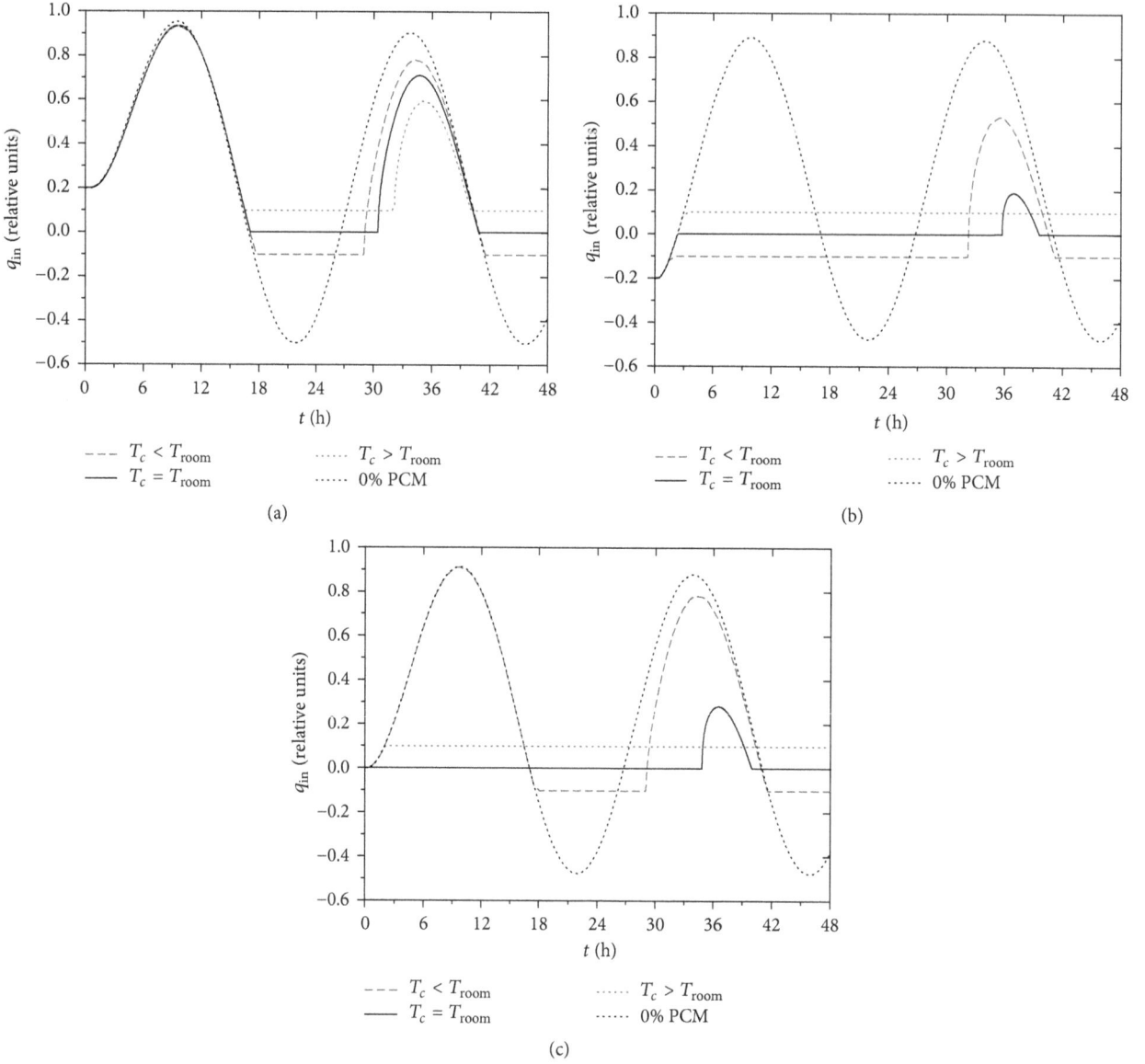

FIGURE 3: The time dependency of the incoming heat flux (relative units) for three values of phase change temperature $T_c = T_{room}$ (black solid curve), $T_c > T_{room}$ (red dotted curve), and $T_c < T_{room}$ (blue dashed curve). For comparison, the composite without PCM (black dotted curve) is included. Initial temperature is set to (a) $T_i > T_{room}$, (b) $T_i < T_{room}$, and (c) $T_i = T_{room}$.

Note that in all simulations we considered temperature oscillations between 15°C and 35°C for total time of 48 hours. In order to obtain more realistic results, additional simulations using different temperature ranges and longer simulation time are welcomed. According to our numerical outcomes (Figures 3 and 4), we assume similar thermal stabilization, with the main difference in the direction of the heat flow, when simulating low temperatures. In this case, we expect that the lowest temperature oscillations would be obtained for $T_c = T_{room}$ and $T_c < T_{room}$. Since even small variation in T_c affects the thermal stability, it is reasonable to explore possible mechanisms to develop PCM composites with tuneable phase change temperature.

We analyzed the impact of external electric fields and NPs on the phase change temperature T_c of NLC. We considered

square-shaped NP that enforces strong tangential boundary conditions at its surface in the presence and absence of an external electric field. In our study, we set the lateral confinement size R below 1 micron, comparable to ξ_b. Figure 5 demonstrates typical nematic configurations in the absence and presence of an external electric field. Figure 5(a) depicts the diagonal structure, which is stable for cases when $R \gg \xi_b$ and $E = 0$. At the corners of the cell, opposing conditions give rise to defects. In the center of each defect, a negative uniaxial order along $\vec{e}_z$ is established and is surrounded by a rim of a maximal degree of biaxiality. By increasing the external electric field strength, we obtained qualitatively different configurations for $R/\xi_f = 10$ (Figure 5(b)) and for $R/\xi_f = 100$ (Figure 5(c)). In the latter case, the external electric field triggered a surface order-reconstruction type

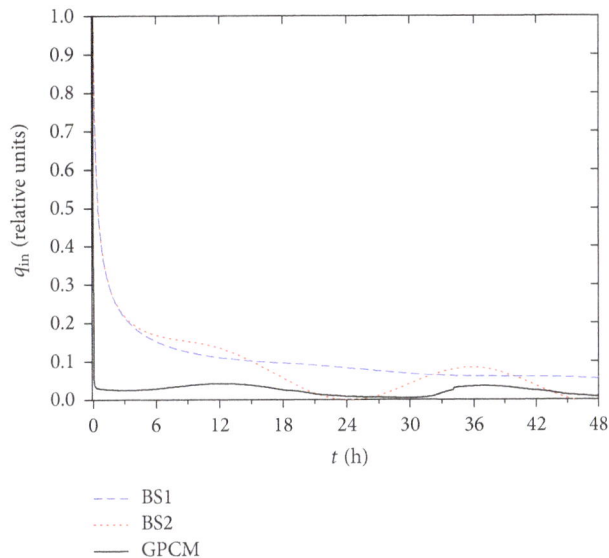

FIGURE 4: The time dependency of the incoming heat flux (relative units) for three real-life composites BS1 (blue dashed curve), BS2 (red dotted curve), and GPCM (black solid curve).

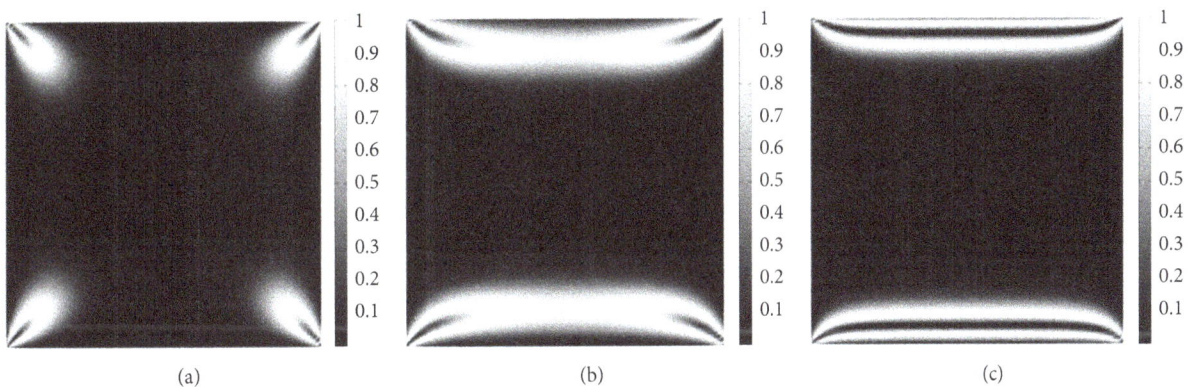

FIGURE 5: The degree of biaxiality β^2 in the absence of NPs for $R/\xi_b = 7$. Uniaxial state is presented in black; a maximum degree of biaxiality is presented in white. The diagonal structure is formed (a) in the absence of an external electric field, (b) for $R/\xi_f = 10$, and (c) for $R/\xi_f = 100$.

of structural transition. Therefore, by application of strong enough external field, one obtains the qualitatively different configuration of NLC, which would affect the phase transition temperature according to (11).

Next, we studied the impact of square-shaped NP in the absence of an external electric field (Figure 6). We placed NP at four different positions of the NLC cell: (i) at the center, (ii) at the bottom boundary, (iii) at the left boundary, and (iv) at the left bottom edge. The NP acts as a source of elastic distortions that result in locally induced biaxiality.

We obtained an equal configuration (but rotated) for two positions of NP: close to the left boundary and close to the bottom boundary (Figures 6(b) and 6(c)). This shows that the presence of NPs in an NLC cell affects the typical distortion length R. Regarding (10) and (11), we conclude that NPs affect effective phase change temperatures. Note that in general several NPs could be introduced which increases complexity and richness of phenomena.

Additionally, we studied the combined impact of NPs and external electric fields. The external electric field broke the

symmetry of the system (see Figure 7) and enabled an order-reconstruction type transition at the bottom plate.

4. Conclusions

In this study, we numerically assessed the impact of PCMs on thermal stabilization. Numerical simulations based on the heat source method confirm that PCMs integrated in the composite material reduced temperature oscillations and therefore improved thermal stabilization. We focused on different cases by varying the phase change temperature around room temperature and showed that the efficiency of the thermal stabilization depends on both the phase change temperature and the desired room temperature. Therefore, it is reasonable to develop tuneable PCM composites. Our main goal was to find and analyze possible mechanisms to control and manipulate phase change temperature. We considered NLC cells as PCMs, as they are commonly used in theoretical and experimental testing.

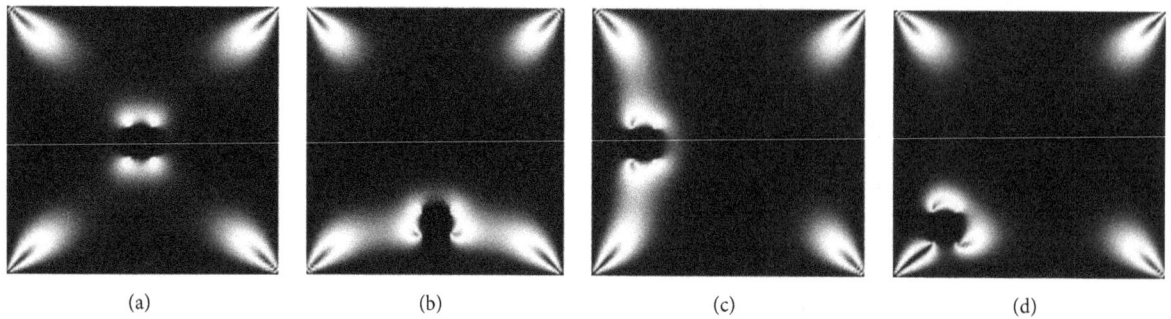

FIGURE 6: Absence of external electric field, $R/\xi_b = 7$. Qualitatively different configurations are obtained for NP placed at the (a) center, (b) bottom boundary, (c) left boundary, and (d) left bottom edge. The color scheme is the same as in Figure 5.

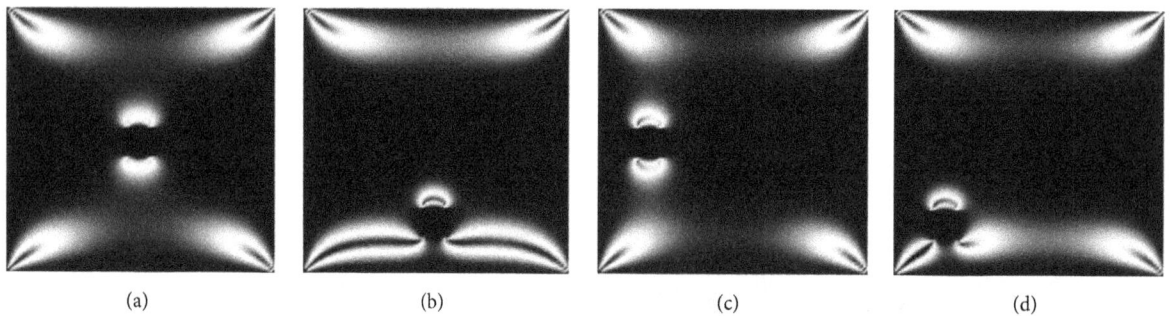

FIGURE 7: Configurations for combined impact of the external electric field $R/\xi_f = 50$ and four different positions of NPs (a) at the center, (b) bottom boundary, (c) left boundary, and (d) left bottom edge, for $R/\xi_b = 7$. The color scheme is the same as in Figure 5.

In our theoretical section, we showed that phase change temperature can be shifted by varying the typical distortion length R. We then examined the impact of external electric fields and NPs on the NLC configuration and consequently on the phase change temperature. We demonstrated that we could shift the phase change temperature by changing the external electric field in the absence of NPs. When an external electric field is absent, an NP can also effectively change the typical characteristic length of the NLC system.

LCs can be used as PCMs to improve thermal stabilization. Furthermore, one can even tune the phase change temperature with relatively simple mechanisms. One of the main disadvantages of using LCs for thermal stabilization is relatively high cost, corresponding to the relatively small amount of latent heat. Nevertheless, LCs have potential for future applications as PCMs, especially in the space industry.

Conflicts of Interest

The authors declare that there are no conflicts of interest regarding the publication of this paper.

References

[1] D. Zhou, C. Y. Zhao, and Y. Tian, "Review on thermal energy storage with phase change materials (PCMs) in building applications," *Applied Energy*, vol. 92, pp. 593–605, 2012.

[2] V. V. Tyagi and D. Buddhi, "PCM thermal storage in buildings: a state of art," *Renewable & Sustainable Energy Reviews*, vol. 11, no. 6, pp. 1146–1166, 2007.

[3] D. Li, L. Yang, and J. Lam, "Zero energy buildings and sustainable development implications —A review," *Energy*, vol. 51, pp. 1–10, 2013.

[4] N. Bhikhoo, A. Hashemi, and H. Cruickshank, "Improving thermal comfort of low-income housing in Thailand through passive design strategies," *Sustainability*, vol. 9, no. 8, article no. 1440, 2017.

[5] M. Casini, *Smart Buildings: Advanced Materials and Nanotechnology to Improve Energy-Efficiency and Environmental Performance*, Woodhead Publishing (Elsevier), Amsterdam, The Netherlands, 2016.

[6] M. Kleman and O. D. Lavrentovich, *Soft Matter Physics*, Springer-Verlag, Berlin, Germany, 2002.

[7] H. de J Wim, *Liquid crystal elastomers: Materials and Applications*, Springer-Verlag, Berlin, Germany, 2012.

[8] V. Popa-Nita, "Statics and kinetics at the nematic-isotropic interface in porous media," *The European Physical Journal B*, vol. 12, no. 83, 1999.

[9] D. E. Feldman, "Quasi-long-range order in nematics confined in random porous media," *Physical Review Letters*, vol. 84, no. 21, pp. 4886–4889, 2000.

[10] A. Aharony and E. Pytte, "Infinite susceptibility phase in random uniaxial anisotropy magnets," *Physical Review Letters*, vol. 45, no. 19, pp. 1583–1586, 1980.

[11] N. J. Mottram and C. Newton, "Introduction to Q-tensor theory," Research Report no. 10, University of Strathclyde

Mathematics, Glasgow, UK, 2004.

[12] S. Kralj and A. Majumdar, "Order reconstruction patterns in nematic liquid crystal wells," *Proceedings of the Royal Society of London*, vol. 470, no. 2169, 2014.

[13] M. Ambrožič, S. Kralj, and E. G. Virga, "Defect-enhanced nematic surface order reconstruction," *Physical Review E: Statistical, Nonlinear, and Soft Matter Physics*, vol. 75, no. 3, Article ID 031708, 2007.

[14] N. Schopohl and T. J. Sluckin, "Defect core structure in nematic liquid crystals," *Physical Review Letters*, vol. 59, no. 22, pp. 2582–2584, 1987.

[15] N. D. Mermin, "The topological theory of defects in ordered media," *Reviews of Modern Physics*, vol. 51, no. 3, pp. 591–648, 1979.

[16] S. Kralj, Z. Bradač, and V. Popa-Nita, "The influence of nanoparticles on the phase and structural ordering for nematic liquid crystals," *Journal of Physics: Condensed Matter*, vol. 20, no. 24, Article ID 244112, 2008.

[17] Y. Dutil, D. R. Rouse, N. B. Salah, S. Lassue, and L. Zalewski, "A review on phase-change materials: Mathematical modeling and simulations," *Renewable Sustainable Energy Reviews*, vol. 15, pp. 112–130, 2011.

[18] A. M. Borreguero, M. Luz Sánchez, J. L. Valverde, M. Carmona, and J. F. Rodríguez, "Thermal testing and numerical simulation of gypsum wallboards incorporated with different PCMs content," *Applied Energy*, vol. 88, no. 3, pp. 930–937, 2011.

[19] S. N. AL-Saadi and Z. Zhai, "Modelling phase change materials embedded in building enclosure: a review," *Renewable and Sustainable Energy Reviews*, vol. 21, pp. 659–673, 2013.

[20] A. Guiavarch, D. Bruneau, and B. Peuportier, "Evaluation of thermal effect of pcm wallboards by coupling simplified phase change model with design tool," *Journal of Building Construction and Planning Research*, vol. 02, no. 01, pp. 12–29, 2014.

[21] G. Cordoyiannis, A. Zidanšek, G. Lahajnar et al., "Influence of confinement in controlled-pore glass on the layer spacing of smectic- A liquid crystals," *Physical Review E: Statistical, Nonlinear, and Soft Matter Physics*, vol. 79, no. 5, Article ID 051703, 2009.

[22] Z. Bradač, S. Kralj, and S. Žumer, "Early stage domain coarsening of the isotropic-nematic phase transition," *The Journal of Chemical Physics*, vol. 135, no. 2, p. 024506, 2011.

[23] V. Popa-Nita and S. Kralj, "Liquid crystal-carbon nanotubes mixtures," *The Journal of Chemical Physics*, vol. 132, no. 2, Article ID 024902, 2010.

[24] R. Repnik, A. Ranjkesh, V. Simonka, M. Ambrozic, Z. Bradac, and S. Kralj, "Symmetry breaking in nematic liquid crystals: Analogy with cosmology and magnetism," *Journal of Physics: Condensed Matter*, vol. 25, no. 40, Article ID 404201, 2013.

[25] A. Ranjkesh, M. Ambrožič, S. Kralj, and T. J. Sluckin, "Computational studies of history dependence in nematic liquid crystals in random environments," *Physical Review E: Statistical, Nonlinear, and Soft Matter Physics*, vol. 89, no. 2, Article ID 022504, 2014.

[26] S. Kralj, G. Cordoyiannis, A. Zidanšek et al., "Presmectic wetting and supercritical-like phase behavior of octylcyanobiphenyl liquid crystal confined to controlled-pore glass matrices," *The Journal of Chemical Physics*, vol. 127, no. 15, Article ID 154905, 2007.

[27] V. Popa-Nita and S. Kralj, "Random anisotropy nematic model: Nematic-non-nematic mixture," *Physical Review E: Statistical, Nonlinear, and Soft Matter Physics*, vol. 73, no. 4, Article ID 041705, 2006.

[28] M. Ambrožič, F. Bisi, and E. G. Virga, "Director reorientation and order reconstruction: competing mechanisms in a nematic cell," *Continuum Mechanics and Thermodynamics*, vol. 20, no. 4, pp. 193–218, 2008.

[29] M. Slavinec, E. Klemenčič, M. Ambrožič, and M. Krašna, "Impact of nanoparticles on nematic ordering in square wells," *Advances in Condensed Matter Physics*, vol. 2015, Article ID 532745, 11 pages, 2015.

15

Effect of Direct Current on Solid-Liquid Interfacial Tension and Wetting Behavior of Ga–In–Sn Alloy Melt on Cu Substrate

Limin Zhang[ID],[1] Ning Li,[1] Hui Xing,[1] Rong Zhang[ID],[1] and Kaikai Song[2]

[1]Shaanxi Key Laboratory of Condensed Matter Structures and Properties and Key Laboratory of Space Applied Physics and Chemistry, Ministry of Education, School of Science, Northwestern Polytechnical University, 127 West Youyi Road, Xi'an, Shaanxi 710072, China
[2]School of Mechanical, Electrical & Information Engineering, Shandong University at Weihai, 180 Wenhua Xilu, Weihai, Shandong 264209, China

Correspondence should be addressed to Rong Zhang; xbwl01@mail.nwpu.edu.cn

Academic Editor: Liyuan Zhang

The effect of direct current (DC) on the wetting behavior of Cu substrate by liquid Ga–25In–13Sn alloy at room temperature is investigated using a sessile drop method. It is found that there is a critical value for current intensity, below which the decrease of contact angle with increasing current intensity is approximately linear and above which contact angle tends to a stable value from drop shape. Current polarity is a negligible factor in the observed trend. Additionally, the observed change in contact angles is translated into the corresponding change in solid-liquid interfacial tension using the equation of state for liquid interfacial tensions. The solid-liquid interfacial tension decreases under DC. DC-induced promotion of solute diffusion coefficient is likely to play an important role in determining the wettability and solid-liquid interfacial tension under DC.

1. Introduction

During the solidification process of materials, the application of electric current including direct current, alternating current (AC), and electric current pulse (ECP) has developed into a promising technique to modify the solidifying structure in the past decades due to their high efficiency and cleanliness [1–3]. Consequently, much of the fundamental research has been performed towards understanding the mechanism behind the modification of the solidification structure by electric current [4–7]. However, the exact modification mechanism is not fully understood and it is still controversial.

To date, the main reasonable hypotheses proposed to explain the modification of the solidification structure are the increase in the solid-liquid interfacial tension [8], the reduction in the nucleation activation energy [9], the suppression of grain growth due to Joule heating [10], current crowding due to the differing electrical conductivities of solid and liquid [11], electromigration [12], and melt flow caused by Lorentz force [13], but all these taken together do not provide a consistent picture of the solidification structure evolution

under electric current. This is mainly due to the lack of conclusive evidence provided by theoretical and experimental studies in support of the proposed hypotheses, especially in the solid-liquid interfacial tension and its related wettability of solid phase by liquid phase which play an important role in determining the kinetics of crystal nucleation and growth [14]. To better understand the intrinsic mechanism of applied electric current to the solidification structure of alloys, a knowledge of the correlation between current parameter and the solid-liquid interfacial tension during solidification process of alloys is required. However, very little experimental investigations have been performed on this issue.

To our knowledge, although as yet no measuring method is applicable for the solid-liquid interfacial tension when DC passes through the solid-liquid interface, recent studies on effect of an applied DC on the solid-liquid interfacial reactions in metal-metal system have given an implication for the determination of the solid-liquid interfacial tension and the wettability. Xu et al. [15] found that the application of DC could improve the wettability of molten Bi on Cu substrate using a DC-coupled sessile drop method, and current polarity

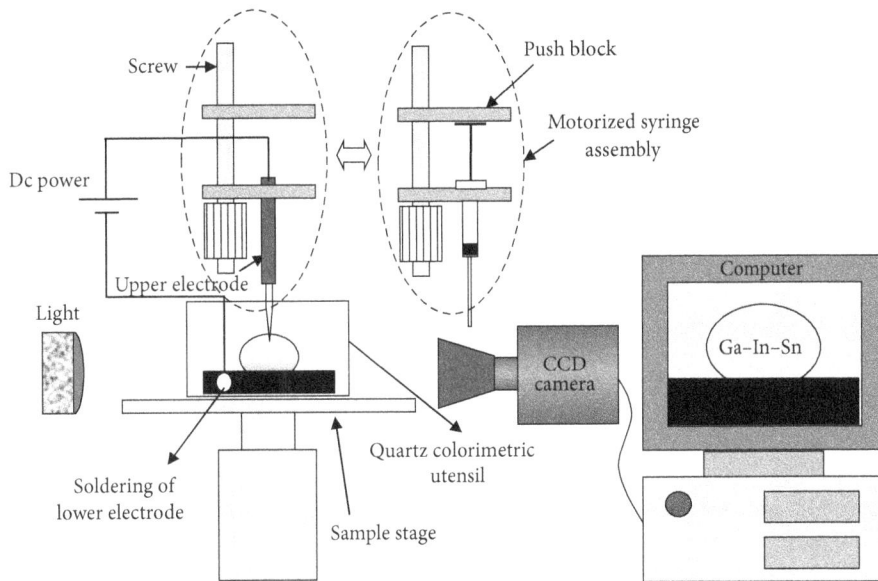

FIGURE 1: Schematic illustration of direct current-sessile drop apparatus.

did not give rise to significant difference on the wettability. Here, it should be pointed out that there are some drawbacks to their designed experiment. For example, the dissolution of the Cu electrode in liquid Sn at the setting temperature needs to be taken into account, and thus the wetting result is blemished. Gu et al. [16, 17] solved the above problem by selecting the graphite electrode to replace the Cu electrode. It was observed that similar results were obtained for the wettability of molten Sn on Cu substrate or Fe substrate. Moreover, the applied DC promoted the dissolution of the Cu substrate in molten Sn which was enhanced with increasing current intensity. It prevented the determination of solid-liquid interfacial tension from contact angles through Young's equation.

The aim of the present work is to obtain quantitative data on the wettability of Cu by molten alloy using a DC-coupled sessile drop method and to compare the solid-liquid interfacial tension under different current conditions. For the work, Ga–25In–13Sn ternary near-eutectic alloy is chosen as droplet at room temperature, because this alloy is liquid due to its low melting point. In this situation, the dissolution of the upper electrode in liquid alloy and the solid-liquid interfacial reaction can be neglected. This makes it possible to reveal the influence of DC on the solid-liquid interfacial tension and the wettability for the explanation of the solidification structure evolution under electric current. In addition, the potential mechanism of DC on the change of solid-liquid interfacial tension and wettability is also discussed in detail.

2. Experimental Procedure

The Ga–25In–13Sn (all percentages are wt-% unless otherwise stated) ternary near-eutectic alloy was prepared using high-purity metals of Ga (>99.999%), In (>99.99%), and Sn (>99.99%) in a resistance furnace. After allowing time for melt homogenization, the alloy was taken out and aspirated into a microinjector. Pure Cu plates (>99.9%) in a size of

$15 \times 15 \times 4 \text{ mm}^3$ were used as substrates. They were mechanically polished to a mirror surface with roughness of a few nanometers using diamond pastes and then ultrasonically cleaned in acetone. The measurements of contact angle and surface tension under DC were performed using a DC-coupled sessile drop apparatus, as schematically shown in Figure 1. As can be seen, it was comprised of two main devices: a high speed (frame rate = 200 Hz) contact angle instrument (Powereach, China) and the electric transmission mission device consisting of a DC power and a Cu wire of Φ 1.8 mm with a sharp-pointed head used as the upper electrode and Cu substrate, on the side of which the Cu wire was held, used as the bottom electrode. Cu substrate was put at the bottom of quartz colorimetric utensil and horizontally placed on the sample stage. Subsequently, liquid drop of Ga–In–Sn alloy with a volume of 40 μl was injected on the Cu substrate surface with microinjector. After that the upper electrode, which replaced the position of microinjector, was fixed to the motorized syringe assembly. The sample stage can move in both horizontal and vertical direction to achieve a sufficient contact of the upper electrode with liquid drop.

In DC-coupled sessile drop experiments, the sharp-pointed upper electrode was initially moved down to be inserted into the liquid drop. This moment was defined as the starting time of the wettability (i.e., $t = 0$ s), and then let the liquid drop stand for 5 minutes without DC. After that, DC of 1 A with a predetermined polarity was continuously applied for 5 minutes at room temperature, and then the 2D projection of 3D drop was recorded by a high speed framing camera. Repeating the above operations, each time current intensity added 1 A until current intensity increased to 11 A. Due to the existence of relaxation time for the steady-state contact angle, the wetting experiment without DC was performed at room temperature and image was taken by the camera every 5 minutes, one which lasted until 65 minutes. Contact angles were calculated from the

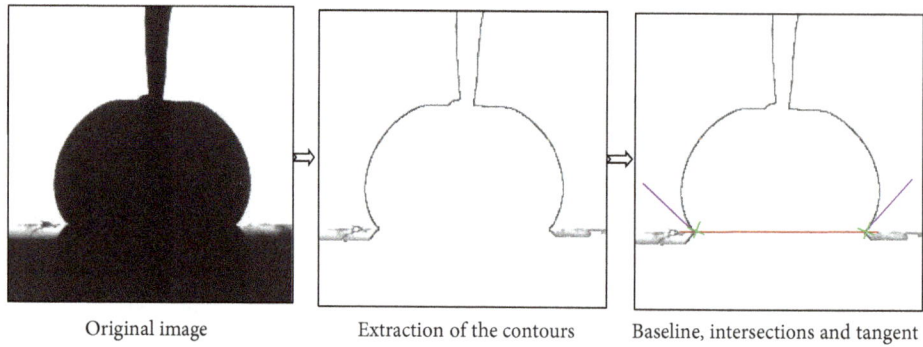

| Original image | Extraction of the contours | Baseline, intersections and tangent |

FIGURE 2: Measurement of contact angle using image-processing techniques.

captured images using drop shape analysis. First the drop profile and the surface of Cu substrate were extracted using image-processing techniques, and then the intersection of the baseline with the tangent was obtained. The different steps in the processing technique are presented in Figure 2.

3. Results and Discussion

In order to conveniently describe current polarity, the direction of DC is defined as a positive one with the Cu substrate connected to the cathode, while the corresponding negative DC is one with the Cu substrate connected to the anode. Typical liquid drop shapes of Ga–In–Sn alloy on Cu substrate under different conditions are illustrated in Figure 3. Figures 3(a)–3(c) represent the metal drop spreading on Cu substrate at different time without DC; the corresponding drop contours are overlapped in Figure 3(d). It can be seen that wetting degree of Cu substrate by Ga–In–Sn alloy drop is poor and does not vary with time in the absence of DC. It suggests that metal drop spreading on Cu substrate forms the quasi-equilibrium shape within 5 minutes. Since the spreading rate was accelerated by DC according to [15, 16], the time required to form the approximate equilibrium shape during application of DC should be less than 5 minutes. In the present experiment, the duration of DC is 5 minutes in order to ensure sufficient spreading of Ga–In–Sn alloy on Cu substrate. Effect of positive DC on the wetting behavior of liquid drop of Ga–In–Sn alloy on Cu substrate is depicted in Figures 3(e)–3(h), while the corresponding results under the condition of a negative DC are shown in Figures 3(i)–3(l). The wettability is still poor when DC is applied regardless of current polarity. However, the application of DC of +6 A deforms the drop shape compared with that without DC, and there is no increase in the deformation of drop when current intensity increases to +11 A, as shown in Figure 3(h). Similar deformation of drop is obtained upon DC reversal (Figure 3(l)).

Since wettability is usually characterized by contact angle of liquid on solid surface in three-phase equilibrium, contact angle measurements are necessary to profoundly understand effect of DC on the wettability of Cu substrate by Ga–In–Sn alloy melt. The plot of contact angle as a function of current intensity at room temperature is shown in Figure 4. As a result, DC reduces the contact angle regardless of current polarity. The decrease of contact angle with increasing current

intensity is approximately linear when the applied DC is not more than 6 A in current intensity, and there is a critical value for current intensity, above which contact angle remains almost constant. Moreover, current polarity seems to have no significant effect on contact angle.

According to Young's equation (1), the contact angle of liquid drop on solid surface is determined by the force balance between the interfacial tensions at the solid-liquid-gas interface:

$$\gamma_{lg} \cos\theta = \gamma_{sg} - \gamma_{sl}, \tag{1}$$

where θ is the Young contact angle, γ_{lg} is the liquid-gas surface tension, γ_{sg} is the solid-gas surface tension, and γ_{sl} is the solid-liquid interfacial tension. It should be pointed out that (1) is only applicable to thermodynamically meaningful contact angles [18]. In that case, it suggests that the observed decrease in contact angles is a manifestation of the effect of DC on interfacial tensions. It is very probable that the solid-gas surface tension is unaffected by DC due to the lack of mobility of atom of the solid. Therefore, it is believed that the observed decrease in contact angles is a consequence of a change in liquid-gas and solid-liquid interfacial tensions.

Taking the solid-gas surface tension as a constant, effect of DC on contact angle can be translated into the corresponding effects in terms of the liquid-gas and solid-liquid interfacial tensions using the equation of state [19]:

$$\gamma_{lg} = \frac{2\gamma_{sg}}{\left(\sqrt{1 + \sin^2\theta} + \cos\theta\right)},$$

$$\gamma_{sl} = \frac{\gamma_{sg}\left(\sqrt{1 + \sin^2\theta} - \cos\theta\right)}{\left(\sqrt{1 + \sin^2\theta} + \cos\theta\right)}. \tag{2}$$

Equation (2) describes the dependence of liquid-gas and solid-liquid interfacial tensions on contact angle, respectively. They can be employed to qualitatively estimate the change of liquid-gas and solid-liquid interfacial tensions under DC compared with those without DC. Figure 5 shows variation of the calculated solid-liquid and liquid-gas interfacial tensions ratio with current intensity and polarity. Obviously, with increasing current density, the ratio of calculated solid-liquid interfacial tension (γ_{sl}^e) under DC to that (γ_{sl0}) in the absence of DC decreases at first and tends to a constant when

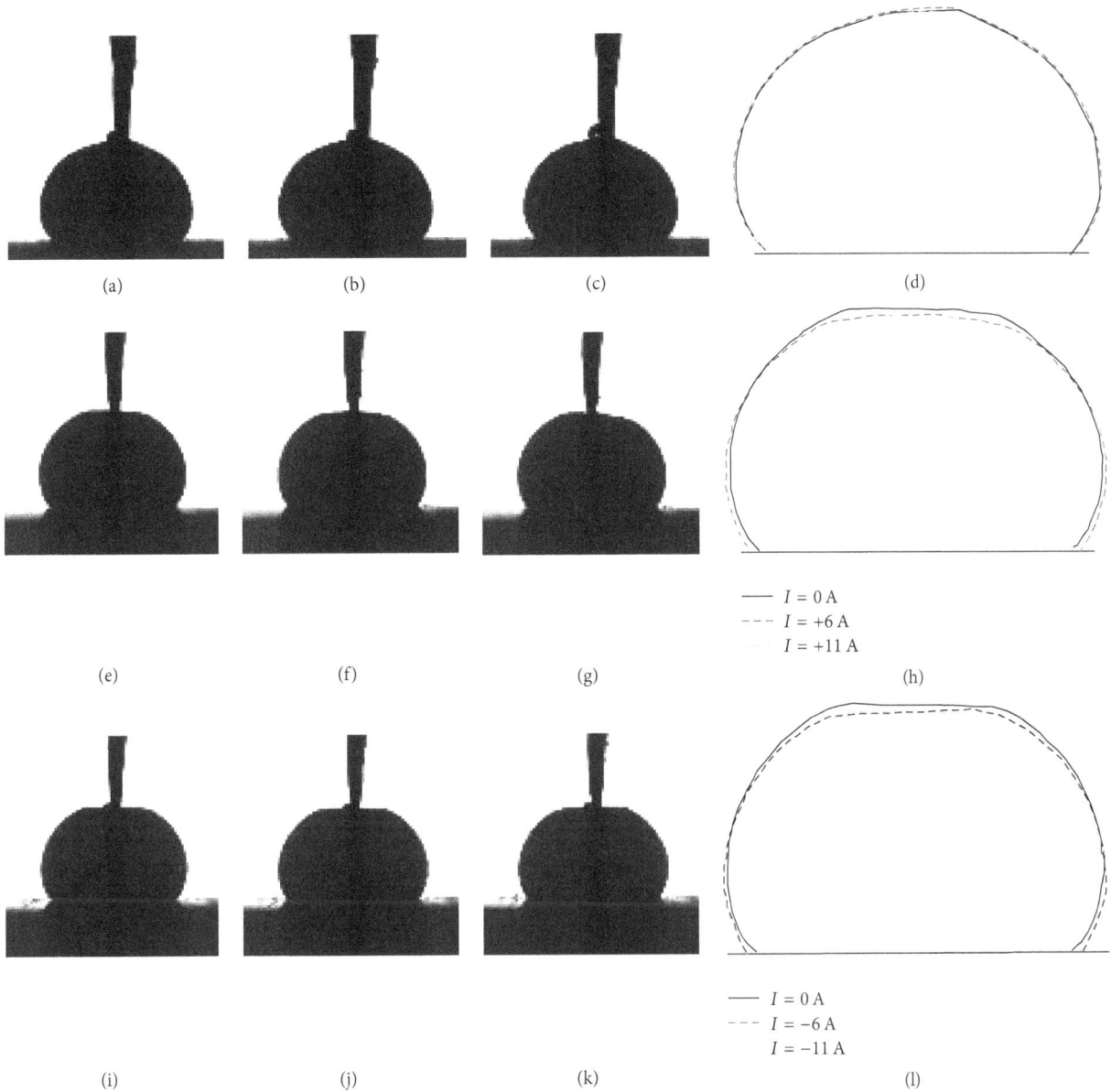

FIGURE 3: Typical liquid Ga–In–Sn drop shapes on Cu substrate under different conditions. Without DC after different standing time of (a) 5 min, (b) 35 min, and (c) 65 min, the three aforementioned drop shapes are amplified and overlapped in Figure (d). With positive DC (e) 0 A, (f) 6 A, and (g) 11 A, the three aforementioned drop shapes are amplified and overlapped in Figure (h). With negative DC (i) 0 A, (j) 6 A, and (k) 11 A, the three aforementioned drop shapes are amplified and overlapped in Figure (l).

current intensity exceeds 6 A regardless of current polarity (Figure 5(a)). Similar tendency is observed for the ratio of calculated liquid-gas interfacial tension (γ_{lg}^{e}) under DC to that (γ_{lg0}) in the absence of DC. It indicates that the observed change in contact angle is a consequence of the decrease in the liquid-gas and solid-liquid interfacial tensions.

Since the measurement of contact angle is performed using a DC-coupled sessile drop method at room temperature, the mass transfer and interfacial action at the triple-phase region are negligible. In this case, Young's equation and its extended formulas are applicable to determinate liquid-gas and solid-liquid interfacial tensions from contact angle in the present study. As indicated above, the observed decrease in

contact angle results from the decrease of liquid interfacial tensions. The potential mechanism causing the above change in liquid interfacial tensions refers to a series of current effects such as Joule heating, electromigration, electric potential energy, and convection caused by Lorentz force. Therefore, the contributions of these effects to liquid interfacial tensions should be seriously considered.

When electric current passes through a conductor, Gibbs free energy increases by adding an extra term (δG_{e}) at the same temperature [20, 21]. It is speculated that solute atoms in alloys will be excited to a higher energy state under DC and affect the solute diffusion process. An explanation to the influence of DC on the diffusion activation energy (Q) of

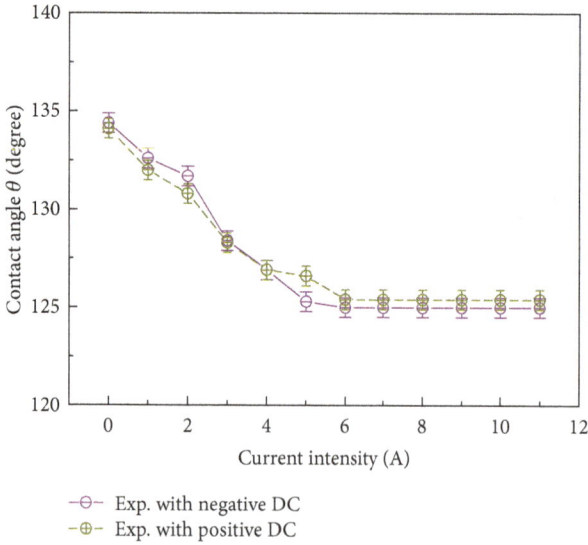

FIGURE 4: Contact angle versus current intensity and polarity.

solute is schematically illustrated in Figure 6(a). Generally, the diffusion activation energy, sometimes called the diffusion barrier, is defined as the minimum energy required to start atom migration. Position 1 and position 2 represent the initial location and target location of atom migration, respectively. In the absence of DC, the diffusion barrier (ΔG) is given by $\Delta G = G_2 - G_1$. The diffusion barrier (ΔG_e) decreases under DC due to the contribution of an extra term (δG_e). It suggests that DC reduces the diffusion activation energy of solute and then increases its diffusion coefficient (D) at the same temperature in terms of the relationship between D, Q, and absolute temperature (T) [22]. Similarly, Zhao and coworkers have found that the calculated diffusion coefficient for solid Ni in liquid Al increases with increasing current density under DC and approaches a relatively stable value at a certain critical current density, regardless of current polarity [23]. It is concluded that DC plays an important role in promoting the diffusion process of metal atoms.

Additionally, it should be noted that the physical parameters of liquid alloys, liquid-gas interfacial tension (γ_{lg0}), viscosity (η), and solute diffusion coefficient (D) are intensively correlative. Therefore, the relationship between γ_{lg0}, η, and D can be given with some empirical models [22]:

$$\eta = \frac{16}{15}\sqrt{\frac{M}{kT}} \cdot \gamma_{lg0},$$
$$D = \frac{kT}{6\pi r \eta},$$
(3)

where M is the absolute atomic mass, k is the Boltzmann constant, and r is the characteristic radius of solute atom. According to (3), the liquid-gas interfacial tension is inversely proportional to the solute diffusion coefficient, implying that the DC-induced promotion of solute diffusion coefficient obviously decreases the liquid-gas interfacial tension of liquid Ga–In–Sn alloy regardless of current polarity, which is in agreement with the calculated results. As indicated above,

the changing trends for solid-liquid and liquid-gas interfacial tensions are similar regarding current intensity dependence. Thus, it is likely that the reduction of diffusion activation energy is the main factor for the reduction in contact angle and solid-liquid interfacial tension under DC.

Since previous studies have proved that Joule heating, electromigration, and convection caused by Lorentz force play significant roles in determining the dissolutive wetting process of molten metal on solid substrate at higher temperature [15, 16], it is worth studying how to affect the wettability of Cu by liquid Ga–In–Sn alloy and its related solid-liquid interfacial tension at room temperature by the above-mentioned current effects. Here, current density as an important parameter in assessing current effects is essential to be determined and its average values are estimated as 10–110 Acm^{-2} in our experiment. In prior work it was observed that effect of Joule heating effect on temperature was negligible for Sn–Bi alloy under DC of 50 Acm^{-2}. The change of temperature (ΔT) under DC is proportional to current density squared (j^2) and electrical resistivity of materials (ρ_e, approximately 7.0×10^{-7} $\Omega \cdot$m for Sn–Bi alloy at 500 K [24] and 2.9×10^{-7} $\Omega \cdot$m for Ga–In–Sn alloy at room temperature [25]) [26]. As a consequence, the obvious change of temperature occurs only when current density exceeds about 120 Acm^{-2} in this study. Thus, effect of Joule heating in liquid drop on the wettability should be negligible with the present DC range of 1–11 A. Generally, current densities of the order of 10^2 Acm^{-2} are required to produce a substantial electromigration, which enhances with increasing current density [12]. Composition fluctuation at the solid-liquid interface as the direct expression of electromigration [4] will affect the solid-liquid interfacial tension and depends on current polarity. Therefore, the difference of the solid-liquid interfacial tension occurs upon DC reversal and increases with increasing current density, which is inconsistent with our experimental results. It suggests that the contribution of electromigration is not expected to be significant to decrease the solid-liquid interfacial tension. As has been reported before, Lorentz force caused by the interaction between the applied DC and its own induced magnetic field gives rise to significant convection in liquid phase [23]. In our case, flow field in the drop, which is similar to that reported in [27], provides a driving force for composition homogenization and weakens composition fluctuation at the solid-liquid interface due to electromigration. According to drop shape, the liquid near the upper electrode has a higher current density, which is estimated as 40–440 Acm^{-2}. It implies that temperature gradient as a result of differences in Joule heating associated with differences in cross-sectional area of drop should be considered when current intensity of the applied DC exceeds 3 A, which can lead to Marangoni convection in drop, as schematically shown in Figure 6(b). Note that there is no significant difference between the direction of Marangoni convection and that of convection induced by Lorentz force. Marangoni convection also plays a role in restraining composition fluctuation at the solid-liquid interface. Suppression of electromigration by convection explains the fact that there is no distinction in the wettability upon DC reversal.

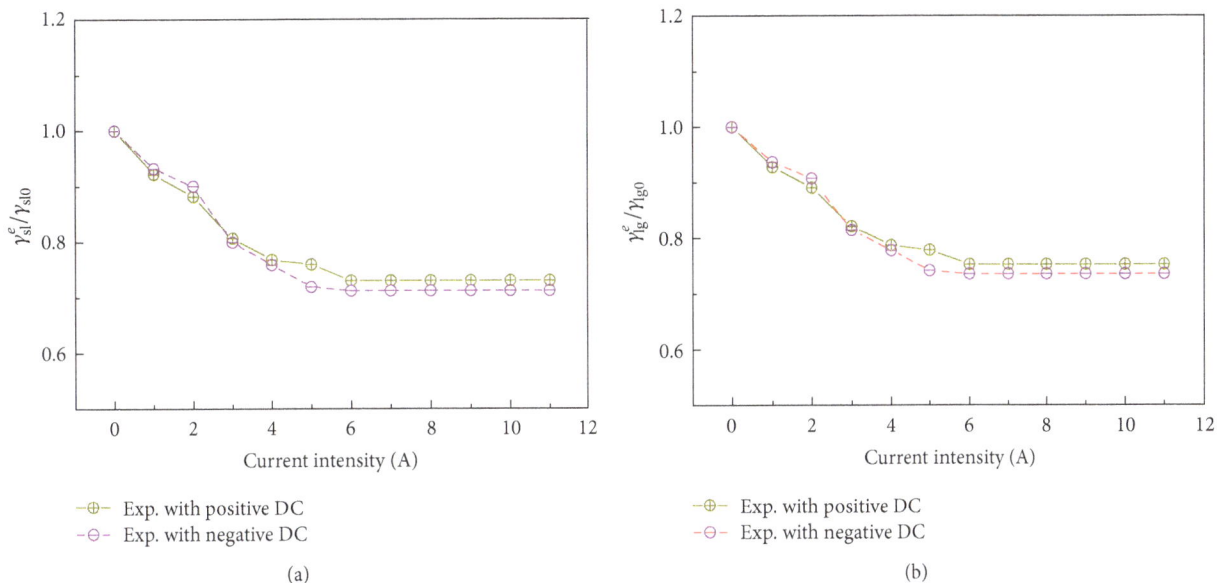

FIGURE 5: (a) Variation of solid-liquid interfacial tension ratio ($\gamma_{sl}^{e}/\gamma_{sl0}$) with current intensity and polarity. (b) Variation of liquid-gas surface tension ($\gamma_{lg}^{e}/\gamma_{lg0}$) with current intensity and polarity.

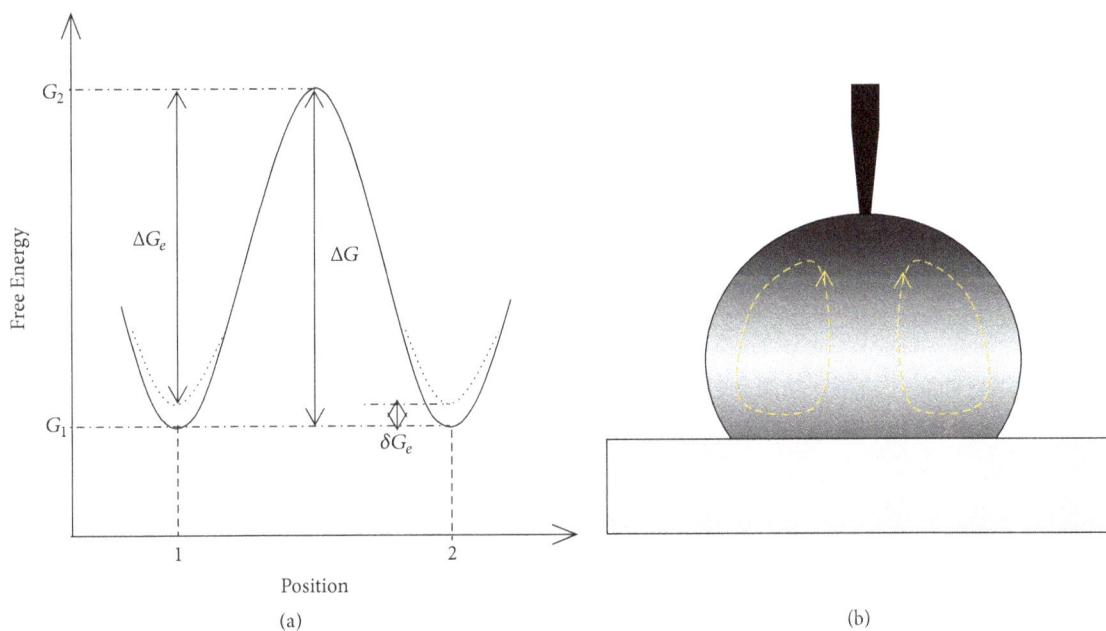

FIGURE 6: (a) Schematic explanation for the effect of DC on the diffusion activation energy. (b) Schematic of Marangoni convection in the Ga–In–Sn/Cu system under DC.

4. Conclusions

We have investigated the wetting behavior of Ga–In–Sn alloy melt on Cu substrate and solid-liquid interfacial tension under DC at room temperature. The application of DC has a pronounced effect on the wettability. There is a critical value for current intensity, below which the steady-state contact angle decreases with increasing current intensity and above which contact angle remains almost constant. In addition, the direction of DC does not play a role in the

observed effects under the same current intensity. Using the equation of state that describes the dependence of solid-liquid interfacial tension on contact angle, the change of solid-liquid interfacial tension under DC can be qualitatively determined. The application of DC leads to significant reduction of solid-liquid interfacial tension. It is speculated that the reduction of contact angle and its corresponding solid-liquid interfacial tension is mainly attributed to the promotion of solute diffusion coefficient under DC. Moreover, Joule heating, Marangoni convection induced by nonuniform distribution of temperature filed, electromigration, and convection caused by Lorentz force do not have a noticeable effect.

Conflicts of Interest

The authors declare that there are no conflicts of interest regarding the publication of this paper.

Acknowledgments

This work is supported by the National Natural Science Foundation of China (Grant nos. 51704242 and 51701160), the Young Scholars Program of Shandong University (Weihai), and the NPU Foundation for Fundamental Research in China (JC201272).

References

[1] T. Wang, J. Xu, T. Xiao et al., "Evolution of dendrite morphology of a binary alloy under an applied electric current: An in situ observation," *Physical Review E: Statistical, Nonlinear, and Soft Matter Physics*, vol. 81, no. 4, Article ID 042601, 2010.

[2] S. R. Coriell, G. B. McFadden, A. A. Wheeler, and D. T. J. Hurle, "The effect of an electric field on the morphological stability of the crystal-melt interface of a binary alloy. II. Joule heating and thermoelectric effects," *Journal of Crystal Growth*, vol. 94, no. 2, pp. 334–346, 1989.

[3] M. Nakada, Y. Shiohara, and M. C. Flemings, "Modification of Solidification Structures by Pulse Electric Discharging," *ISIJ International*, vol. 30, no. 1, pp. 27–33, 1990.

[4] L. Zhang, N. Li, R. Zhang et al., "Effect of Direct Current on Microstructure Evolution of Directionally Solidified Sn-70 wt pct Bi Alloy at Different Pulling Rates," *Metallurgical and Materials Transactions A: Physical Metallurgy and Materials Science*, vol. 46, no. 9, pp. 4174–4182, 2015.

[5] X. Liao, Q. Zhai, J. Luo, W. Chen, and Y. Gong, "Refining mechanism of the electric current pulse on the solidification structure of pure aluminum," *Acta Materialia*, vol. 55, no. 9, pp. 3103–3109, 2007.

[6] D. Räbiger, Y. Zhang, V. Galindo, S. Franke, B. Willers, and S. Eckert, "The relevance of melt convection to grain refinement in Al-Si alloys solidified under the impact of electric currents," *Acta Materialia*, vol. 79, pp. 327–338, 2014.

[7] H. Conrad, "Influence of an electric or magnetic field on the liquid-solid transformation in materials and on the microstructure of the solid," *Materials Science and Engineering: A Structural Materials: Properties, Microstructure and Processing*, vol. 287, no. 2, pp. 205–212, 2000.

[8] X. F. Zhang, W. J. Lu, and R. S. Qin, "Removal of MnS inclusions in molten steel using electropulsing," *Scripta Materialia*, vol. 69, no. 6, pp. 453–456, 2013.

[9] L. Jianming, L. Shengli, L. Jin, and L. Hantong, "Modification of solidification structure by pulse electric discharging," *Scripta Materialia*, vol. 31, no. 12, pp. 1691–1694, 1994.

[10] J. Zhu, T. Wang, F. Cao, W. Huang, H. Fu, and Z. Chen, "Real time observation of equiaxed growth of Sn-Pb alloy under an applied direct current by synchrotron microradiography," *Materials Letters*, vol. 89, pp. 137–139, 2012.

[11] C. Song, Y. Guo, Y. Zhang et al., "Effect of currents on the microstructure of directionally solidified Al4.5 wt% Cu alloy," *Journal of Crystal Growth*, vol. 324, no. 1, pp. 235–242, 2011.

[12] L. Zhang, N. Li, H. Xing et al., "Microstructure evolution of directionally solidified Sn-Bi alloy under different medium-density direct current," *Journal of Crystal Growth*, vol. 430, pp. 80–86, 2015.

[13] L. Zhang, H. Liu, N. Li et al., "The relevance of forced melt flow to grain refinement in pure aluminum under a low-frequency alternating current pulse," *Journal of Materials Research*, vol. 31, no. 3, pp. 396–404, 2016.

[14] K. Mondal and B. S. Murty, "On the prediction of solid-liquid interfacial energy of glass forming liquids from homogeneous nucleation theory," *Materials Science and Engineering: A Structural Materials: Properties, Microstructure and Processing*, vol. 454-455, pp. 654–661, 2007.

[15] Q.-G. Xu, X.-B. Liu, and H.-F. Zhang, "Effect of direct electric current on wetting behavior of molten Bi on Cu substrate," *Transactions of Nonferrous Metals Society of China*, vol. 20, no. 8, pp. 1452–1457, 2010.

[16] Y. Gu, P. Shen, N.-N. Yang, and K.-Z. Cao, "Effects of direct current on the wetting behavior and interfacial morphology between molten Sn and Cu substrate," *Journal of Alloys and Compounds*, vol. 586, pp. 80–86, 2014.

[17] P. Shen, Y. Gu, N.-N. Yang, R.-P. Zheng, and L.-H. Ren, "Influences of electric current on the wettability and interfacial microstructure in Sn/Fe system," *Applied Surface Science*, vol. 328, pp. 380–386, 2015.

[18] D. Y. Kwok and A. W. Neumann, "Contact angle measurement and contact angle interpretation," *Advances in Colloid and Interface Science*, vol. 81, no. 3, pp. 167–249, 1999.

[19] D. Y. Zhu, P. Q. Pin, X. B. Luo et al., "Novel characterization of wetting properties and the calculation of liquid–solid interface tension (I)," *Science Technology and Engineering*, vol. 7, no. 13, pp. 3057–3062, 2007.

[20] R. S. Qin, E. I. Samuel, and A. Bhowmik, "Electropulse-induced cementite nanoparticle formation in deformed pearlitic steels," *Journal of Materials Science*, vol. 46, no. 9, pp. 2838–2842, 2011.

[21] X. L. Wang, J. D. Guo, Y. M. Wang, X. Y. Wu, and B. Q. Wang, "Segregation of lead in Cu-Zn alloy under electric current pulses," *Applied Physics Letters*, vol. 89, no. 6, Article ID 061910, 2006.

[22] I. Egry, "On the relation between surface tension and viscosity for liquid metals," *Scripta Materialia*, vol. 28, no. 10, pp. 1273–1276, 1993.

[23] J. F. Zhao, C. Unuvar, U. Anselmi-Tamburini, and Z. A. Munir, "Kinetics of current-enhanced dissolution of nickel in liquid aluminum," *Acta Materialia*, vol. 55, no. 16, pp. 5592–5600, 2007.

[24] X.-F. Li, F.-Q. Zu, H.-F. Ding, J. Yu, L.-J. Liu, and Y. Xi, "High-temperature liquid-liquid structure transition in liquid Sn-Bi alloys: Experimental evidence by electrical resistivity method," *Physics Letters A*, vol. 354, no. 4, pp. 325–329, 2006.

[25] Y. Plevachuk, V. Sklyarchuk, S. Eckert, G. Gerbeth, and R. Novakovic, "Thermophysical properties of the liquid Ga-In-Sn eutectic alloy," *Journal of Chemical & Engineering Data*, vol. 59, no. 3, pp. 757–763, 2014.

[26] L. M. Zhang, N. Li, R. Zhang, H. Xing, K. Song, and J. Y. Wang, "Effect of medium-density direct current on dendrites of directionally solidified Pb–50Sn alloy," *Materials Science and Technology (United Kingdom)*, vol. 32, no. 18, pp. 1877–1885, 2016.

[27] P. A. Nikrityuk, K. Eckert, R. Grundmann, and Y. S. Yang, "An impact of a low voltage steady electrical current on the solidification of a binary metal alloy: A numerical study," *Steel Research International*, vol. 78, no. 5, pp. 402–408, 2007.

16

Threading Dislocations Piercing the Free Surface of an Anisotropic Hexagonal Crystal

Salem Neily,[1] **Sami Dhouibi,**[1] **and Roland Bonnet** ⓘ[2]

[1]*Laboratoire de Physique de la Matière Condensée et Nanosciences LR 11 ES 40, Faculté des Sciences, Université de Monastir, rue de l'Environnement, 5019 Monastir, Tunisia*
[2]*Université Grenoble Alpes, CNRS, SIMaP, 38000 Grenoble, France*

Correspondence should be addressed to Roland Bonnet; ralbonnet@sfr.fr

Academic Editor: Sergei Sergeenkov

Inclined threading dislocations (TDs) piercing the oriented free surface of a crystal are currently observed after growth of oriented thin films on substrates. Up to date the unique way to treat their anisotropic elastic properties nearby the free surface region is to use the integral formalism, which assumes no dislocation core size and needs numerical double integrations. In a first stage of the work, a new and alternative approach to the integral formalism is developed using double Fourier series and the concept of a finite core size, which is often observed in high-resolution transmission electron microscopy. In a second stage, the integral formalism and the Fourier series approaches are applied to the important case of a TD piercing the basal free surface of a hexagonal crystal. For this particular geometry, easy-to-use expressions are derived and compared to a third approach previously known for a plate-like crystal. Finally, the numerical interest and the convergence of these approaches are tested using the basal free surface of the GaN compound, in particular for TDs with Burgers vectors **c** and (**a** + **c**).

1. Introduction

Epitaxial growth of semiconductor materials offers opportunities in combining and modifying structural and electronic parameters [1–5]. Unfortunately, the thin crystalline film deposited on the substrate often contains high densities of dislocations piercing the free surface, the so-called threading dislocations (TDs), which affect adversely device properties. At high temperature, a reduction of the density of some TDs can occur via fusion or annihilation reactions. It is therefore of importance to study their intrinsic elastic properties. The aim of the present work is first to investigate the elastic field of an inclined TD piercing the free surface of any anisotropic, semi-infinite crystal. The inclination angle with respect to the normal is denoted θ. For isotropic crystals, exact solutions are well known for $\theta < \pi/2$ [6, 7] and $\theta = \pi/2$ [8, 9]. We present a new, alternative approach to the TD elastic field to that developed previously by Lothe et al. [10] who uses the integral formalism and requires in general a set of double integrations.

This new approach is developed in terms of double Fourier series and can take account of a finite dislocation core size, which is not possible with the integral formalism. These two approaches are then applied to the important case of a hexagonal crystal (parameters a and c) limited by a basal plane. The particular geometry of a TD parallel to the **c** axis enables simplified expressions to be derived. These latter are compared to a third approach developed previously for a plate of finite thickness [11] from stress potentials [12]. Since wurtzite gallium nitride (GaN) has attractive optoelectronic applications, such as light emitting diode and laser diodes [1, 3–5], this compound is used in the present work to perform numerical applications.

2. Methods

2.1. The Fourier Calculation Approach. A description of the elastic field of inclined dislocations lying in an isotropic plate was recently presented from a Fourier calculation method

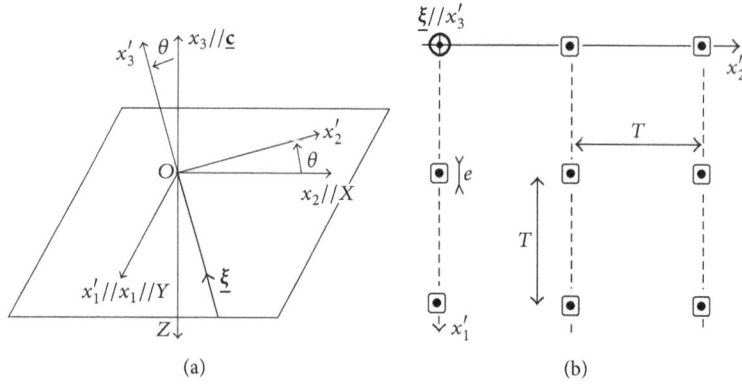

FIGURE 1: Cartesian frames and symbols attached to (a) an isolated inclined threading dislocation (TD) oriented by $\boldsymbol{\xi}//Ox'_3$ in the plane Ox_2x_3 and (b), two infinite sets of periodically distributed TDs oriented along $\boldsymbol{\xi}$. Each of these TDs has a cylindrical core, the section of which is a small square.

[20, 21]. Below, this method is generalized to treat any anisotropic semi-infinite crystal. For the calculation, we use two Cartesian coordinate systems with the same origin O, Figure 1(a). The system $Ox_1x_2x_3$ is linked to the free surface and oriented with the axis Ox_3 directed outside matter. The direction Ox_1 is any direction taken in the surface. The second system denoted $Ox'_1x'_2x'_3$, is attached to a TD placed and oriented along Ox'_3. This latter system is obtained from $Ox_1x_2x_3$ by a rotation θ around Ox_1. The TD terminates in O, is oriented by the unit vector $\boldsymbol{\xi}//Ox'_3$, and has the Burgers vector $\mathbf{b}$.

The first step of the calculation is to consider a biperiodic series of largely spaced and identical emerging TDs. Its stress field can be written as the sum ($j, k = 1$–3)

$$\sigma_{jk} = \sigma_{jk}{}^{\infty} + \sigma_{jk}{}^{\text{rel}}, \tag{1}$$

in which $\sigma_{jk}{}^{\infty}$ is the stress field generated in an infinite crystal and $\sigma_{jk}{}^{\text{rel}}$ is the unknown relaxation stress field that cancels the forces applied along the plane chosen for the surface. To fulfil (1) the idea is to use partial results obtained in [20] for $\sigma_{jk}{}^{\infty}$ and in [22] for $\sigma_{jk}{}^{\text{rel}}$.

Let us first consider the calculation of $\sigma_{jk}{}^{\infty}$.

In the plane $x'_3 = 0$, Figure 1(b), the theoretical dislocation lines form a square pattern of black points. These lines are distant by the period T along Ox'_1 and Ox'_2. Each dislocation line is oriented parallel to Ox'_3 and is associated with a finite core. According to some high-resolution transmission electron microscopy observations and numerical simulations, for example, [23–25], a TD core can have a variety of atomic structures. These latter have certainly an influence on the elastic properties very near the theoretical dislocation line [25]. However, beyond a few $|\mathbf{b}|$ values and thanks to the St. Venant principle, elasticity can be applied without accounting for true limiting boundary conditions around the core. In the present work, for simplicity, the core section is chosen as a square, limited by an infinite prism with a section e^2, limited by portions of planes parallel to planes $x'_1 = x'_2 = \pm e/2$. Inside the core the dislocation density is taken as $\mathbf{b}/e^2$, while outside the prism, the dislocation density is zero. Each prism

can therefore be described as an extended core dislocation. Around the axis Ox'_3, when T increases and the core size e decreases, the elastic field tends gradually to that of a classical Volterra dislocation. Along the axes Ox_1 and Ox_2, the periodicity of the defects is T and $T/\cos\theta$.

According to [20], the displacement $u_k{}^{\infty}$ and stress fields $\sigma_{jk}{}^{\infty}$ of the infinite dislocation set can be written in the $Ox_1x_2x_3$ frame in the form of double Fourier series. The $\sigma_{jk}{}^{\infty}$ stress field can be expressed as (integers $m, n \neq 0$)

$$\sigma_{jk}{}^{\infty} = \Sigma_{m,n} F_{jk}{}^{(m,n)} \exp\left[2i\pi\left(G_1x_1 + G_2x_2 + G_3x_3\right)\right], \tag{2}$$

in which the preexponential terms $F_{jk}{}^{(m,n)}$ can be calculated explicitly with

$$G_1 = \frac{m}{T},$$

$$G_2 = \left(\frac{n}{T}\right)\cos\theta, \tag{3}$$

$$G_3 = \left(-\frac{n}{T}\right)\sin\theta.$$

The terms $F_{jk}{}^{(m,n)}$ depend on the elastic constants C_{ijkl} taken in $Ox_1x_2x_3$, on the Burgers vector $\mathbf{b}$, on the inclination angle θ, and on the geometrical parameters e and T. However, since their expressions are much too long to be written extensively, they are not given in the present text.

Let us now consider the calculation of $\sigma_{jk}{}^{\text{rel}}$.

Since this stress field should also be periodic, it can be expressed formally as a double Fourier series. As shown in [22], the formal forms of the associated displacement field can be written as (m and $n \neq 0$)

$$u_k{}^{\text{rel}} = \Sigma_{m,n}\Sigma_{\alpha=1-3}P_\alpha\lambda_{\alpha k}\exp\left[2i\pi p_\alpha x_3\right]$$
$$\cdot \exp\left[2i\pi\left(G_1x_1 + G_2x_2\right)\right], \tag{4}$$

in which the constants P_α are three unknown complex constants attached to each harmonic term. They depend on the order (m, n) of each harmonic term and on the limiting boundary conditions of the problem. The constants p_α and $\lambda_{\alpha k}$ are found from the nontrivial solutions of the equations

$$\left(\psi_{jk} + p_\alpha \phi_{jk} + p_\alpha^2 C_{j3k3}\right)\lambda_{\alpha k} = 0. \tag{5}$$

For $\alpha = 1\text{--}3$, the constants p_α are complex chosen with a negative complex part. The corresponding constants $\lambda_{\alpha k}$ are normalized solutions of the same equation. In addition

$$\psi_{jk} = C_{j1k1}G_1^2 + C_{j2k2}G_2^2 + \left(C_{j1k2} + C_{j2k1}\right)G_1 G_2,$$
$$\phi_{jk} = \left(C_{j1k3} + C_{j3k1}\right)G_1 + \left(C_{j3k2} + C_{j2k3}\right)G_2. \tag{6}$$

For convergence towards $x_3 < 0$, only the negative complex parts of the constants p_α are retained. The stress field is finally calculated from derivation of (4) and application of the Hooke law:

$$\sigma_{jk}^{\text{rel}} = 2i\pi \Sigma_{m,n} \Sigma_{\alpha=1\text{--}3} P_\alpha L_{\alpha jk} \exp\left[2i\pi p_\alpha x_3\right]$$
$$\cdot \exp\left[2i\pi\left(G_1 x_1 + G_2 x_2\right)\right], \tag{7}$$

The constants $L_{\alpha jk}$ are

$$L_{\alpha jk} = \lambda_{\alpha q}\left(G_1 C_{jkq1} + G_2 C_{jkq1} + p_\alpha C_{jkq3}\right). \tag{8}$$

The unknown constants P_α are found from the following condition: the addition of the stress fields (2) and (7) are such that zero stresses act on the plane $x_3 = 0$, that is, ($k = 1\text{--}3$),

$$\left(\sigma_{3k}^\infty + \sigma_{3k}^{\text{rel}}\right) = 0 \quad (\text{for } x_3 = 0). \tag{9}$$

For each k value, this expression represents in fact a double Fourier series, which must be identical to zero. The only solution is that all its terms (m, n) are zero. This leads, for each term (m, n), to a system of three linear equations, the solution of which gives the values of the three complex unknowns P_α. From the knowledge of the P_α values associated with each term (m, n), the total displacement field is obtained by the addition ($\mathbf{u}^\infty + \mathbf{u}^{\text{rel}}$), in which $\mathbf{u}^\infty$ is calculated from expression (5) in [20] and $\mathbf{u}^{\text{rel}}$ from expression (4).

The total stress field is finally obtained from the addition of expressions (2) and (7).

As a consequence, three parameters are required to evaluate the elastic field around an isolated TD: the period T should much larger than $|\mathbf{b}|$, while the size e could be estimated to a few $|\mathbf{b}|$. Of course, a sufficiently large number of terms (m, n) must be retained to obtain a sufficiently precise solution.

For the case of a hexagonal crystal, $\theta = 0$ and $Ox_3 // \mathbf{c}$, simplified expressions can be obtained, briefly listed below. The axis Ox_1 can be chosen anywhere in the basal plane.

Constants p_α are the following:

$$p_1 = -i\left\{\left[\frac{(C_{11} - C_{12})}{(2C_{44})}\right]\left(G_1^2 + G_2^2\right)\right\}^{1/2},$$
$$p_2 = -i\left\{\left[\frac{(\beta_1 - \beta_2)}{(2C_{33}C_{44})}\right]\left(G_1^2 + G_2^2\right)\right\}^{1/2}, \tag{10}$$
$$p_3 = -i\left\{-\left[\frac{(\beta_2 + \beta_1)}{(2C_{33}C_{44})}\right]\left(G_1^2 + G_2^2\right)\right\}^{1/2}.$$

Compact expressions are then obtained for the P_α constants because the terms $L_{\alpha j1}$ and $L_{\alpha j2}$ are not involved in the calculations. As a result, for brevity, denoting $L_{\alpha j3}$ ($\alpha = 1\text{--}3$, $j = 1\text{--}3$) as $L_{\alpha j}$ and F_{j3} as F_j, respectively, the P_α constants become

$$P_1 = \left(\frac{i}{\beta_3}\right)\left[F_3\left(L_{22}L_{31} - L_{21}L_{32}\right)\right.$$
$$\left. - F_2\left(L_{23}L_{31} - L_{21}L_{33}\right) - F_1\left(L_{22}L_{33} - L_{23}L_{32}\right)\right],$$

$$P_2 = \left(\frac{i}{\beta_3}\right)\left[F_3\left(L_{11}L_{32} - L_{12}L_{31}\right)\right. \tag{11}$$
$$\left. + F_2\left(L_{13}L_{31} - L_{11}L_{33}\right) + F_1\left(L_{12}L_{33} - L_{13}L_{32}\right)\right],$$

$$P_3 = \left(\frac{i}{\beta_3}\right)\left[F_3\left(L_{12}L_{21} - L_{11}L_{22}\right)\right.$$
$$\left. + F_2\left(L_{13}L_{21} - L_{11}L_{23}\right) + F_1\left(L_{12}L_{23} - L_{13}L_{22}\right)\right]$$

with

$$\beta_1 = \left[\left(C_{13}^2 - C_{11}C_{33}\right)\left(\left(C_{13} + 2C_{44}\right)^2 - C_{11}C_{33}\right)\right]^{1/2},$$
$$\beta_2 = C_{13}\left(C_{13} + 2C_{44}\right) - C_{11}C_{33},$$
$$\beta_3 = 2\pi\left[L_{13}\left(L_{22}L_{31} - L_{21}L_{32}\right)\right. \tag{12}$$
$$\left. - L_{12}\left(L_{23}L_{31} - L_{21}L_{33}\right) + L_{11}\left(L_{23}L_{32} - L_{22}L_{32}\right)\right].$$

Similar expressions are found for the particular cases $m = 0$ or $n = 0$.

The elastic field described by the above Fourier approach can be compared with that calculated with the integral formalism [10]. After a brief presentation of this latter approach, we present simplified expressions for a hexagonal crystal (parameters a and c) that has a basal free surface.

2.2. The Integral Formalism Approach. In the integral formalism, the relaxation stress field of a TD dislocation is calculated for convenience in the frame $OXYZ$ chosen in [10] Figure 1(a). The components of the Burgers vector are denoted $\mathbf{b} = (b_X, b_Y, b_Z)$. It is generated by a plane fan-shaped distribution of straight infinitesimal dislocations placed in the plane $x_3 = 0$ of the unlimited crystal. This distribution is located in the plane $Z = 0$ and has the Burgers vector density $\mathbf{b}(\psi) = \{b_1(\psi), b_2(\psi), b_3(\psi)\}$. As in [10], ψ is the oriented polar angle of an infinitesimal dislocation linked to the plane OXY. The field σ_{jk}^{rel} is obtained from double integrations: the first one is needed to compute the energy matrix $B(\psi)$ related to each dislocation, while the second one is required to sum the contributions of all the planar set of dislocations.

For the particular orientation $\theta = 0$, we obtain compact expressions for the image force dF that applies on a TD segment $d\lambda$ in the plane $Y = 0$ and for the relaxation stress σ_{jk}^{rel}. The integral (24) given in [10] can be greatly simplified. If λ is a distance along the TD taken from the emerging point and $d\lambda$ a length element, we obtain the simple expression

$$dF = \frac{b_X b_Z C_{13}\left(C_{11} - C_{12}\right)d\lambda}{\left(4\pi C_{11}\lambda\right)}. \tag{13}$$

This expression shows clearly that $dF \neq 0$ for mixed dislocations, while $dF = 0$ for pure screw ($b_X = 0$) or prismatic edge dislocations ($b_Z = 0$). This explains the trend, for mixed TDs, to have directions systematically misoriented with respect to the $\mathbf{c}$ axis, for example, [2].

When $\theta = 0$ the density $\mathbf{b}(\psi)$ becomes simple to express because the calculation shows that the energy coefficient matrix B attached to an infinitesimal dislocation placed in the plane $Z = 0$ is no more depending on its orientation ψ. This constant matrix can be evaluated from expressions (73, 74) in [26] or other similar procedures indicated in [13, 27]. In passing, some misprints in [26] are indicated in Appendix A.

The Burgers vector of an infinitesimal dislocation has the following simple form:

$$
\mathbf{b}(\psi) = \left(-\frac{1}{\pi}\right)
$$
$$
\cdot \begin{pmatrix} \left(\dfrac{b_Z C_{44}}{B_{11}}\right) \cos(\psi) \\[2mm] \left(\dfrac{b_Z C_{44}}{B_{22}}\right) \sin(\psi) \\[2mm] \dfrac{[b_X \cos(\psi) + b_Y \sin(\psi)] C_{13}(C_{11} - C_{12})}{(C_{11} B_{33})} \end{pmatrix}, \tag{14}
$$

in which B_{11}, B_{22} and B_{33} are the nonzero elements of the energy coefficient matrix B.

The relaxed displacement field $\mathbf{u}^{\text{rel}}$ in a point defined by its cylindrical coordinates (R_1, ψ_1, θ_1) is then calculated from the addition of all the contributions of the infinitesimal dislocations with Burgers vector $\mathbf{b}(\psi)d\psi$, in the plane $Z = 0$.

As an example, for the edge case with $\mathbf{b}(b_X, 0, 0)$, the nonzero components of the stress tensor are

$$
\sigma_{31}^{\text{rel}}
$$
$$
= \frac{b_X(C_{12} - C_{11})C_{13}C_{44}}{2\pi^3 C_{11} B_{33}} \int_\Delta \frac{[\cos(\psi)]^2}{\widehat{\rho}_1(\psi)\sin(\psi - \psi_1)} d\psi, \tag{15}
$$
$$
\sigma_{32}^{\text{rel}}
$$
$$
= \frac{b_X(C_{12} - C_{11})C_{13}C_{44}}{2\pi^3 C_{11} B_{33}} \int_\Delta \frac{\sin(\psi)\cos(\psi)}{\widehat{\rho}_1(\psi)\sin(\psi - \psi_1)} d\psi,
$$

in which [10] (see Appendix B)

$$
\widehat{\rho}_1(\psi) = R_1\left[\cos(\theta_1)^2 + \sin(\theta_1)^2 \sin(\psi - \psi_1)^2\right]^{1/2}. \tag{16}
$$

For the screw case with $\mathbf{b}(0, 0, b_Z)$, the stress tensor is

$$
\left(\sigma^{\text{rel}}\right) = \begin{pmatrix} \int_\Delta 2H(\psi)\cos(\psi)^2 d\psi & \int_\Delta 2H(\psi)\sin(2\psi) d\psi & 0 \\[2mm] \int_\Delta 2H(\psi)\sin(2\psi) d\psi & \int_\Delta 2H(\psi)\sin(\psi)^2 d\psi & 0 \\[2mm] 0 & 0 & \int_\Delta \left[\dfrac{2C_{13}}{(C_{11}+C_{12})}\right]H(\psi) d\psi \end{pmatrix} \tag{17}
$$

in which

$$
H(\psi) = \frac{b_Z(C_{12} - C_{11})(C_{12} + C_{11})C_{44}}{8\pi^3 C_{11} B_{11} B_{22}}
$$
$$
\cdot \frac{(B_{11} + B_{22} + (B_{22} - B_{11})\cos(2\psi))}{\widehat{\rho}_1(\psi)\sin(\psi - \psi_1)}. \tag{18}
$$

The integration domain Δ is indicated in [10] with a crossed integration symbol. Therefore, it is not performed on π exactly. For numerical applications, to avoid zero denominators in (13)–(16), an angular domain Δ running from $(\psi_1 + \varepsilon)$ to $(\psi_1 + \pi - \varepsilon)$ should be used with ε tending to zero. Our experience is that convergence is ensured with ε equal to 0.0001, a value sufficiently small to obtain a good precision on the stresses nearby the TD core.

2.3. The Stress Potential Approach. The displacement field regarding a mixed dislocation with $\boldsymbol{\xi}//\mathbf{c}(\theta = 0)$ was already expressed for a plate-like hexagonal crystal in [14] but unfortunately this author did not consider a semi-infinite crystal. In the present work, the solution is found in searching for the limit of the displacement field $\mathbf{u}^{\text{rel}}$ around the upper emerging point when the plate thickness tends to infinity. The

calculation is carried out in the $OXYZ$ frame, which gives the following result:

(i) Edge dislocation $\mathbf{b}//OX$: the field $\mathbf{u}^{\text{rel}}$ is derived from two stress potentials ϕ_1 and ϕ_2 mentioned in [14] but calculated at the limit of an infinite thickness:

$$
\mathbf{u}^{\text{rel}}
$$
$$
= \left[\frac{\partial}{\partial X}(\phi_1 + \phi_2), \frac{\partial}{\partial Y}(\phi_1 + \phi_2), \frac{\partial}{\partial Z}(k_1\phi_1 + k_2\phi_2)\right] \tag{19}
$$

with

$$
\phi_1 = \left(\frac{b_X C_{13}(C_{11} - C_{12})\nu_2}{2\pi C_{11} C_{44}(1 + k_1)(\nu_2 - \nu_1)}\right)\left(\frac{Y}{r}\right)
$$
$$
\cdot \int_{\lambda_{\min}}^{\lambda_{\max}} \frac{J_1(\lambda r)}{\lambda^2} \exp(-\nu_1 Z\lambda) d\lambda, \tag{20}
$$

in which r is the polar radius, while the parameters ν_1, ν_2, ν_3, k_1, k_2 are given in Appendix C. The potential ϕ_2 is obtained from ϕ_1 by interchanging ν_1 and ν_2 and interchanging k_1 and k_2. $J_1(\lambda r)$ is the Bessel function of the first kind for the integer order 1. In expression (17), since the integration domain cannot be exactly π, $\lambda_{\min}$ should is taken close to 0, while $\lambda_{\max}$

TABLE 1: Elastic constants of GaN. In anisotropic elasticity, the equilibrium angle θ_{eq} of the inclined (**a** + **c**) TD is small, weakly depending on the existence of a free surface (columns 8 and 9). It is not the case in isotropic elasticity (columns 10 and 11). Symbols A, S and I means Anisotropy, Surface and Isotropy.

(GPa) Refs.	C_{11}	C_{12}	C_{13}	C_{33}	C_{44}	μ (GPa), ν	θ_{eq} AS	θ_{eq} A∞	θ_{eq} IS	θ_{eq} I∞
[13]	367	135	103	405	95	113, 0.27	2.6°	3.8°	4.8°	11.6°
[14]	388	154	84	458	85	112, 0.28	2.5°	3.6°	4.9°	11.6°
[15]	329	109	80	357	91	107, 0.25	2.4°	3.6°	4.5°	11.6°
[16]	390	145	106	398	105	120, 0.26	2.8°	4.4°	4.8°	11.6°
[17]	374	106	70	379	101	121, 0.22	2.0°	3.0°	4.2°	11.6°
[18]	359	129	92	389	98	113, 0.26	2.4°	3.7°	4.7°	11.6°
[19]	334	132	99	372	86	100, 0.28	2.6°	3.8°	4.9°	11.6°

should be large enough. A few numerical applications show that values $(\lambda_{min}, \lambda_{max}) = (0.0001, 100)$ give similar values to those obtained with the Fourier approach and the integral formalism.

(ii) Screw dislocation **b**//OZ: only the tangential component is nonzero:

$$u_\psi^{rel} = \left[\frac{b_Z}{(2\pi\nu_3)}\right] \int_{\lambda_{min}}^{\lambda_{max}} \left[\frac{J_1(\lambda r)}{\lambda}\right] \exp\left(-\nu_3 Z\lambda\right) d\lambda. \quad (21)$$

Expression (21) only depends on C_{44} and C_{66} via ν_3.

3. Application to GaN

Gallium nitride is chosen to perform numerical applications because of its technological interest for electronic devices that can work under high power density and at relatively high temperature [4, 5]. The growth of GaN can be stabilized by epitaxial growth on bulky GaN substrates or a foreign substrate, for example, (111)Si, SiC, or sapphire [2, 4, 5, 15]. An investigation on GaN thin films is of particular interest because of its TDs that have anisotropic electrical properties regarding cathodoluminescence and photoluminescence. These properties are sensitive to the Burgers vectors and also to the TD directions [5].

The thickness of the GaN layer is assumed to be large enough to neglect the elastic interaction of the TD with the GaN/substrate interface. Different anisotropic elastic constants Cij of GaN are reported in several experimental or theoretical works. These constants are taken from [16–19, 28–30] and are indicated in Table 1. GaN has the structure of wurtzite (space group P6$_3$mc) with lattice parameters a = 0.319 nm and c = 0.519 nm [15].

Since the isotropic approximation is sometimes used in literature, column 7 indicates the calculated Young modulus μ and the Poisson ratio ν relative to each Cij set. μ and ν are found from an averaging of the elastic properties of GaN according to the method proposed in [31]. In this reference, a way is proposed to define an anisotropy factor coefficient A for any crystal symmetry. In the present work, using the Cij constants of [16–19, 28–30], the A values are found between 1.27 and 1.78. These two values correspond to the Cij constants mentioned in [19] and [17], respectively.

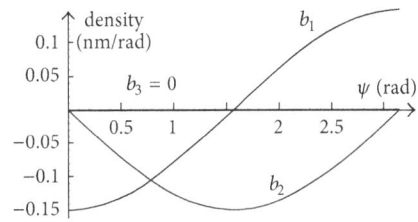

FIGURE 2: Case of the **c** screw TD oriented along the normal OZ. Evolution of the dislocation density $\mathbf{b}(b_1, b_2, b_3)$ in the basal plane $Z = 0$ of GaN.

Previous transmission electron microscopy investigations, for example, [2, 3, 5, 15], show that a large fraction of TDs cross the free surface with directions $\boldsymbol{\xi}$ parallel or not far from the **c** axis. They can be of edge-type, screw-type, or mixed-type. The local elastic equilibrium nearby the emerging point of a (**a** + **c**)-type TD can be treated from the image force theorem due to Lothe et al. [10]. For instance, taking OX//-**a** and a dislocation direction $\boldsymbol{\xi}$ in the prismatic plane (**a**, **c**), a systematic equilibrium angle θ_{eq} can be obtained. In Table 1, this angle is mentioned for each set of Cij constants given in [16–19, 28–30] including isotropic approximation. The comparison of columns 8 and 9 shows a small anisotropic effect since θ_{eq} changes from averages equal to 2.5° (free surface, column 8) to 3.7° (no free surface, column 9). On the contrary, with the isotropic approximation, columns 10 and 11 indicate a strong effect of the free surface since θ_{eq} increases from averages equal to 4.7° (free surface) to 11.6° (no free surface). These last results demonstrate that the isotropic approximation is not reliable for GaN.

Other TDs in GaN are sometimes directed along the **c** axis, while other TDs have lines inclined noticeably with respect to **c**. For the screw case ($\theta = 0$) and **b** = **c**, the dislocation density $\mathbf{b}(\psi)$ is given by the expression (14). Taking the Cij constants indicated in [16], Figure 2 illustrates the sinusoidal shape of the two nonzero components of the dislocation density $\mathbf{b}(\psi)$, that is, $b_1(\psi)$ and $b_2(\psi)$. Our experience is that, for $\theta = 0$, the stress potential approach is more efficient to calculate the stress relaxation field than the integral formalism or the Fourier series.

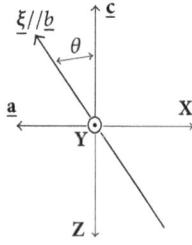

FIGURE 3: Orientation ξ of the $(\mathbf{a} + \mathbf{c})$ screw TD. The inclination angle θ is 31.5°.

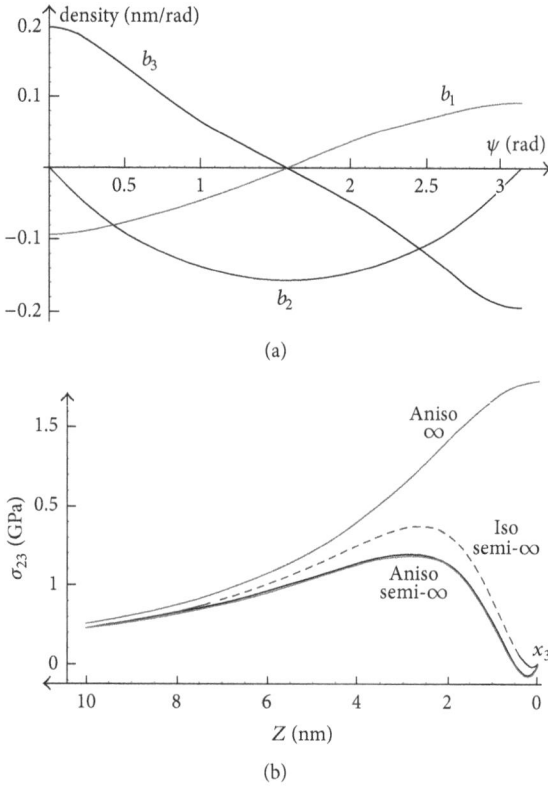

(a)

(b)

FIGURE 4: Case of the $(\mathbf{a} + \mathbf{c})$ screw TD. (a) Change with ψ of the three components of the dislocation density $\mathbf{b}(b_1, b_2, b_3)$ in the plane $Z = 0$ and (b) evolution of the σ_{23} stress component along the straight line $(X = 0, Y = 2 \text{ nm})$ according to different assumptions.

For $\theta \neq 0$, the Fourier approach and the integral formalism have similar performances. The numerical applications below, which use the elastic constants mentioned in line 2 Table 1, correspond to the geometry depicted in Figure 3. The TD is an inclined $(\mathbf{a} + \mathbf{c})$ screw dislocation placed in the $Y = 0$ plane. Figure 4(a) describes the changes of the dislocation density in the plane $Z = 0$. The component $b3$ is nonzero for this screw dislocation, Figure 4, conversely to the orientation $\theta = 0$.

Figure 4(b) exhibits, as an example, the evolution of the σ_{23} stress component along a straight line parallel to OZ and cutting the surface $Z = 0$ at the point $(X, Y) = (0, 2 \text{ nm})$. This inclined TD line is chosen because it gives some ideas about the results obtained from different approaches. All obtained

curves converge far from the free surface but give different values close to the emerging point of the TD. The lower curve denoted "Aniso-∞" seems a bold line but is in fact the quasi-superposition of two curves, one calculated from the integral formalism and the other calculated from the Fourier series approach. For this latter approach, the calculation parameters were the following: period $T = 30$ nm, core size $e = |\mathbf{b}|$. The squared geometry of the core has therefore no practical impact on the values of the stress at points placed about 2 or 3 nm from the dislocation axis. The number of harmonic terms (m, n) chosen for summing the Fourier series was tested from visual convergence with the curve obtained from the integral approach. For points at 2 or 3 nm from the core, the number of harmonic terms should include a large number of terms for a good convergence (m and n can reach about a few hundreds). The same kind of difficulty is encountered with the integral formalism when it is applied to points close to the core. In the calculations, we adopted the integration domain $(\pi - \varepsilon)$ with $\varepsilon = 0.0001$ rad. Table 1 shows that, with the other sets of Cij constants, the amplitudes of these curves change hardly.

The dashed curve denoted "Iso semi-∞" is calculated from the Yoffe-Shaibani-Hazzledine solution [6, 7] and the upper curve denoted "Aniso-∞" assumes an infinite anisotropic crystal. The part of this latter curve, towards the negative Z, is not represented but is symmetrical with respect to $x_3 = 0$. The bold line denoted "Aniso-∞" shows that both approaches presented in Sections 2 and 3 give the same result for the σ_{23} stress and includes in particular the slight negative values nearby $x_3 = 0$, that is, the region close to the dislocation core (a few nanometers). In this region, the stress is slightly negative, which is not the case in the isotropic approximation. This negative value nearby $x_3 = 0$ is clearly due to the proximity to the TD core and not a calculation artefact generated by a lack of precision. Anisotropy effect is therefore far to be negligible nearby the free surface and the dislocation core.

4. Conclusions

In this work, the elastic field of an inclined TD has been investigated using different approaches. The first one, based on Fourier series and the concept of extended dislocation core, is developed in the present work from an appropriate combination of partial results obtained in [20, 22]. It proves to be a valuable alternative method to a well-known approach, the so-called integral formalism [10]. The computation times required by both approaches are found roughly similar and depend on the wanted precision. There is a strong interest in using these two independent approaches to test the ability of elasticity to give the same values around the theoretical core. Both approaches converge beyond one nanometer from the theoretical core, as verified for some TDs in GaN. A third approach using stress potentials and assuming $\theta = 0$ is also numerically tested for the first time (at the author's knowledge).

When the anisotropic constants tend to those of an isotropic crystal, we verified numerically that the three above approaches converge to the same stress field, namely, that obtained from the Yoffe-Shaibani-Hazzledine expressions [6, 7]. Correction of misprints should be introduced in [7]; see [32].

For a dislocation normal to the basal free surface of a hexagonal crystal ($\theta = 0$), some general expressions of the integral formalism are simplified. They were tested positively in regions far from the free surface and close to the emerging core region of a TD. The numerical results exhibited in the four last columns of Table 1 and the curves in Figure 4(b) show clearly that the neglect of both elastic anisotropy and the presence of a free surface can be a confusion source in the interpretation of the angular equilibrium of a TD.

Of course, it is expected that the atomic structure of the dislocation core could influence the elastic field at distance from the core lower than a few Burgers vector moduli. According to our experience the assumption of a square section of the dislocation core has no sensible influence on the stress field beyond a few nanometers from the dislocation line provided that the core size e is less than the modulus of the Burgers vector.

Appendix

A. Misprints in [26]

The following misprints can be noticed in Appendix C of [26].

(i) A_1 should read

$$A_1 = \frac{(C_{11} + C_{12})}{(4C_{11})}. \tag{A.1}$$

(ii) In expression σ_{33}, replace the sign + in the first numerator by the sign −.

(iii) In expression σ_{12}, the second denominator should be squared.

B. Misprint in [10]

In [10], the first expression in (14) should be corrected. All the trigonometric terms should be squared.

C. Useful Expressions Given in [11]

The parameters ν_1, ν_2, ν_3, k_1, k_2 are defined in [14]

$$\nu_1 = \left[\frac{\left(\overline{C}_{13} - C_{13} \right)\left(\overline{C}_{13} + C_{13} + 2C_{44} \right)}{\left(4C_{33}C_{44} \right)} \right]^{1/2}$$
$$+ \left[\frac{\left(\overline{C}_{13} + C_{13} \right)\left(\overline{C}_{13} - C_{13} - 2C_{44} \right)}{\left(4C_{33}C_{44} \right)} \right]^{1/2} \tag{C.1}$$

with

$$\overline{C}_{13} = \left(\frac{C_{33}}{C_{11}} \right)^{1/2}. \tag{C.2}$$

ν_2 is obtained from expression (C.1), in which the sign + at the end of the first line is changed into −. The other parameters are

$$\nu_3 = \left(\frac{C_{66}}{C_{44}} \right)^{1/2},$$
$$k_1 = \frac{\left[\left(C_{11}/\nu_1{}^2 \right) - C_{44} \right]^{1/2}}{\left(C_{11} + C_{44} \right)}. \tag{C.3}$$

k_2 is obtained from k_1 by changing ν_1 into ν_2.

Conflicts of Interest

The authors declare that there are no conflicts of interest regarding the publication of the paper.

References

[1] C. Romanitan, R. Gavrila, and M. Danila, "Comparative study of threading dislocations in GaN epitaxial layers by nondestructive methods," *Materials Science in Semiconductor Processing*, vol. 57, pp. 32–38, 2017.

[2] S. K. Mathis, A. E. Romanov, L. F. Chen, G. E. Beltz, W. Pompe, and J. S. Speck, "Modeling of threading dislocation reduction in growing GaN layers," *Journal of Crystal Growth*, vol. 231, no. 3, pp. 371–390, 2001.

[3] H. Morkoç, *Handbook of Nitride Semiconductors and Devices, Materials Properties, Physics and Growth*, vol. 2, Wiley-VCH Verlag GmbH & Co. KGaA, 2008, Chapter 1.

[4] H. Iwata, H. Kobayashi, T. Kamiya et al., "Annealing effect on threading dislocations in a GaN grown on Si substrate," *Journal of Crystal Growth*, vol. 468, pp. 835–838, 2017.

[5] Y. Yao, Y. Ishikawa, M. Sudo, Y. Sugawara, and D. Yokoe, "Characterization of threading dislocations in GaN (0001) substrates by photoluminescence imaging, cathodoluminescence mapping and etch pits," *Journal of Crystal Growth*, vol. 468, pp. 484–488, 2017.

[6] E. H. Yoffe, "A dislocation at a free surface," *Philosophical Magazine*, vol. 6, pp. 1147–1155, 1961.

[7] S. J. Shaibani and P. M. Hazzledine, "The displacement and stress fields of a general dislocation close to a free surface of an isotropic solid," *Philosophical Magazine*, vol. 44, no. 3, pp. 657–665, 1981.

[8] J. D. Eshelby and A. N. Stroh, "Dislocations in thin plates," *Philosophical Magazine*, vol. 42, pp. 1401–1405, 1951.

[9] K. Honda, "Dislocation walls consisting of double arrays in white tin single crystals," *Japanese Journal of Applied Physics*, vol. 18, no. 2, pp. 215–224, 1979.

[10] J. Lothe, V. L. Indenbom, and V. A. Chamrov, "Elastic field and self-force of dislocations emerging at the free surfaces of an anisotropic halfspace," *Physica Status Solidi (b)—Basic Solid State Physics*, vol. 111, no. 2, pp. 671–677, 1982.

[11] A. Y. Belov, "Fields of displacement and stresses of rectilinear dislocation in an anisotropic plate," *Kristallografiya*, vol. 3, pp. 550–558, 1987.

[12] S. G. Lekhnitskii, *Theory of Elasticity of an Anisotropic Elastic Body*, J. J. Brandstatter, Ed., Holden-day, San Francisco, Calif, USA, 1963, Librairie of Congress Catalog Card Number: 63-15647, chapter 4.

[13] J. P. Hirth and J. Lothe, *Theory of Dislocations*, John Wiley Sons, New York, NY, USA, 2nd edition, 1982.

[14] A. Y. Belov, "Dislocations emerging at planar boundaries," in *Modern Problems in Condensed Matter Sciences, Elastic Strain Fields and Dislocations Mobility*, V. I.. Indenbom and J. Lothe, Eds., vol. 31, pp. 432–438, North-Holland, Amsterdam, 1992.

[15] P. B. Hirsch, J. G. Lozano, S. Rhode et al., "The dissociation of the [a + c] dislocation in GaN," *Philosophical Magazine*, vol. 93, no. 28-30, pp. 3925–3938, 2013.

[16] A. F. Wright, "Elastic properties of zinc-blende and wurtzite AlN, GaN, and InN," *Journal of Applied Physics*, vol. 82, p. 2833, 1997.

[17] K. Kim, W. R. Lambrecht, and B. Segall, "Erratum: Elastic constants and related properties of tetrahedrally bonded BN, AlN, GaN, and InN [Physical Review. B 53, 16310]," *Physical Review B: Condensed Matter and Materials Physics*, vol. 56, no. 11, pp. 7018-7019, 1997.

[18] Z. Usman, C. Cao, G. Nabi et al., "First-principle electronic, elastic, and optical study of cubic gallium nitride," *The Journal of Physical Chemistry A*, vol. 115, no. 24, pp. 6622–6628, 2011.

[19] A. Polian, M. Grimsditch, and I. Grzegory, "Elastic constants of gallium nitride," *Journal of Applied Physics*, vol. 79, no. 6, pp. 3343-3344, 1996.

[20] K. Saito, R. O. Bozkurt, and T. Mura, "Dislocation stresses in a thin film due to the periodic distributions of dislocations," *Journal of Applied Physics*, vol. 43, no. 1, pp. 182–188, 1972.

[21] R. Bonnet, S. Neily, and S. Youssef, "Inclined dislocations in plates," *Physica Status Solidi (b)—Basic Solid State Physics*, vol. 251, no. 11, pp. 2307–2313, 2014.

[22] R. Bonnet, "Periodic elastic fields in anisotropic two-phase media. Application to interfacial dislocations," *Acta Metallurgica et Materialia*, vol. 29, no. 2, pp. 437–445, 1981.

[23] C. Kisielowski, B. Freitag, X. Xu, S. P. Beckman, and D. C. Chrzan, "Sub-angstrom imaging of dislocation core structures: How well are experiments comparable with theory?" *Philosophical Magazine*, vol. 86, no. 29-31, pp. 4575–4588, 2006.

[24] B. G. Mendis, Y. Mishin, C. S. Hartley, and K. J. Hemker, "Use of the Nye tensor in analyzing HREM images of bcc screw dislocations," *Philosophical Magazine*, vol. 86, no. 29-31, pp. 4607–4640, 2006.

[25] X. Hu and S. Wang, "Nonplanar core structure of the screw dislocations in tantalum from the improved Peierls–Nabarro theory," *Philosophical Magazine*, vol. 98, no. 6, pp. 484–516, 2017.

[26] L. J. Teutonico, "Dislocations in hexagonal crystals," *Materials Science and Engineering: A Structural Materials: Properties, Microstructure and Processing*, vol. 6, no. 1, pp. 27–47, 1970.

[27] M. Dupeux and R. Bonnet, "Stresses, displacements and energy calculations for interfacial dislocations in anisotropic two-phase media," *Acta Metallurgica et Materialia*, vol. 28, no. 6, pp. 721–728, 1980.

[28] Y. Takagi, M. Ahart, T. Azuhata, T. Sota, K. Suzuki, and S. Nakamura, "Brillouin scattering study in the GaN epitaxial layer," *Physica B: Condensed Matter*, vol. 219-220, no. 1-4, pp. 547–549, 1996.

[29] N. Nakamura, H. Ogi, and M. Hirao, "Elastic, anelastic, and piezoelectric coefficients of GaN," *Journal of Applied Physics*, vol. 111, no. 1, Article ID 013509, 2012.

[30] Y. Duan, J. Li, S. Li, and J. Xia, "Elasticity, band-gap bowing, and polarization of $Al_xGa_{1-x}N$ alloys," *Journal of Applied Physics*, vol. 103, no. 2, p. 023705, 2008.

[31] N. Fribourg-Blanc, M. Dupeux, G. Guenin, and R. Bonnet, "Détermination par extensométrie et mesures ultrasonores des six constantes elastiques du cristal Al_2Cu (π). Duscussion de l'anisotropie," *Journal of Applied Crystallography*, vol. 12, no. 2, pp. 151–155, 1979.

[32] R. Bonnet, S. B. Youssef, A. Boussaïd, C. H. Belgacem, and M. Fnaiech, "Free Surface Relaxation and Two-Beam TEM Imaging of Straight Dislocations in Thin Foils," *Defect and Diffusion Forum*, vol. 224-225, pp. 1–12, 2004.

17

Absence of the Rashba Splitting of Au(111) Surface Bands

I. N. Yakovkin ⓘ

Institute of Physics of National Academy of Sciences of Ukraine, Prospect Nauki 46, Kiev 03028, Ukraine

Correspondence should be addressed to I. N. Yakovkin; yakov@iop.kiev.ua

Academic Editor: Golam M. Bhuiyan

The electronic structure of Au(111) films is studied by means of relativistic DFT calculations. It is found that the twinning of the surface bands, observed in photoemission experiment, does not necessarily correspond to the spin-splitting of the surface states caused by the break of the inversion symmetry at the surface. The twinning of the bands of clean Au(111) films can be obtained within nonrelativistic or scalar-relativistic approximation, so that it is not a result of spin-orbit coupling. However, the spin-orbit coupling does not lead to the spin-splitting of the surface bands. This result is explained by Kramers' degeneracy, which means that the existence of a surface itself does not destroy the inversion symmetry of the system. The inversion symmetry of the Au(111) film can be broken, for example, by means of adsorption, and a hydrogen monolayer deposited on one face of the film indeed leads to the appearance of the spin-splitting of the bands.

1. Introduction

The twin surface bands of Au(111) with a parabolic dispersion, pertinent to nearly free electrons, were found in angle-resolved photoemission study by LaShell et al. [1] and explained as a result of the spin-splitting caused by spin-orbit coupling (SOC). Probably, this interpretation, in large part, was inspired by a rapidly growing interest to so-called topological insulators [2–5] and, in this regard, to the Rashba splitting [6–8] of surface bands crossing E_F. One of the most persuasive arguments in support of this explanation was the absence of the splitting at Γ (the center of the Brillouin zone), which seemingly indicated that the splitting of the surface bands for $\mathbf{k}_\parallel \neq 0$ was allowed due to an absence of inversion symmetry at the surface and thus the lifting of Kramers' degeneracy. (Recall that the inversion symmetry of a crystal $E(-\mathbf{k}, \uparrow) = E(\mathbf{k}, \uparrow)$ together with the time-reversal symmetry $E(-\mathbf{k}, \uparrow) = E(\mathbf{k}, \downarrow)$ leads to $E(\mathbf{k}, \uparrow) = E(\mathbf{k}, \downarrow)$, that is, to the spin degeneracy of the bands for all $\mathbf{k}$.)

However, as noted by Reinert [9], since these results were not reproduced over several years and the splitting of the bands was not observed by means of the scanning tunneling spectroscopy [10], this interpretation was later questioned, and different mechanisms, such as the herringbone reconstruction of the Au(111) surface [11], were proposed to explain the splitting of the surface bands. On the other hand, the Rashba-type splitting of the surface bands was later supported by tight-binding calculations [12] and photoemission studies (also for similar surface states of Ag(111) and Cu(111) [13]) and first-principles calculations for Au(111) slabs [14].

The theory of the Rashba effect [6] is based on the model of a 2D electron gas, which might be applicable for true surface states in semiconductor heterostructures, but obviously not for surface resonances in metals. Indeed, the degree of the localization of the surface resonances of Au(111) is relatively weak and this feature causes a significant interaction between the surface states of opposite surfaces even for rather thick (up to ~20 monolayers) Au(111) films [14]. For this reason, as was noted by Koroteev et al. [15], the appearance of the Rashba splitting of the Au(111) surface state "came as surprise". It should be noted also that the splitting, estimated within the 2D electron gas model, was found to be by several orders of magnitude smaller than that observed in experiment [16], so that it was possible to state only a qualitative agreement with the theory.

In the systems with asymmetry with respect to the axis normal to the surface, Rashba splitting leads to the spin polarization of the bands within the surface plane. By analogy, it was suggested that, in the case of the Au(111)

surface, the asymmetry is brought about by the surface potential, in particular by the surface barrier. Then, for the Au(111), the twin surface states must show an opposite spin polarization, and several studies were performed to reveal the spin polarization of the surface bands of Au(111) by means of spin- and angle-resolved photoelectron spectroscopy [16]. However, as it was mentioned in [16], the spin polarization of the initial state (photohole) is not necessarily that of the photoelectron, in particular when the spin-orbit coupling is strong (like in the case of Au with $Z = 79$). Therefore, the interpretation of spin-resolved photoemission spectra can become complicated due to the various spin polarization effects.

Hence, while the splitting of the surface bands of Au(111) has been well established in a number of photoemission experiments, the commonly recognized interpretation of the splitting of the surface bands as a result of the spin-orbit coupling steel needs a better justification. Then, the cornerstone suggestion that a surface itself can cause the break of the symmetry has led to the generally accepted concept of the nature of the splitting of surface states of metal surfaces (see [8] for a recent review). This concept, however, has obvious shortcomings. First, any surface cannot be separated from the bulk and therefore cannot have its own symmetry. In other words, any surface-induced lowering of symmetry means the lowering of the symmetry of the net system, so that the symmetry of the interior cannot be considered separately. Second, any real crystal or film always has an opposite surface, which restores the inversion symmetry in a general case. It should be noted in this regard that the model of a semi-infinite crystal seems improper for the study of the spin-splitting of surface bands caused by SOC. Third, if the existence of a surface were sufficient to initiate the break of inversion symmetry, the Rashba-type spin-splitting would be observable for surface bands of all clean surfaces of all metals, while in fact there are only a few reports of the splitting for clean metal surfaces (e.g., for Ag(111) [14], Bi(111) [15], Cu(111) [17], Cu(110) [18], and W(110) [19]; for review, see [7–9]), which, because of biased determination of the spin polarization and actual surface conditions, might be interpreted differently, or for surfaces covered by adsorbed layers, which could destroy the inversion symmetry.

In the present paper, we revisit the interpretation of the twinning surface bands of the Au(111) surface, which has become a basis of subsequent interpretation of the bands of other metal surfaces, by means of fully relativistic (that is, with account for spin-orbit coupling) DFT calculations for the repeated-slab model. The calculations were performed with ABINIT [20] set of programs using Troullier-Martins [21] pseudopotentials and plane-wave decomposition of the wave functions. The LDA exchange-correlation potential was in the Goedecker-Teter-Hutter (GTH) form [22]. The cutoff energy of 24 Ha and $6 \times 6 \times 1$ lattice of k-points provided the adopted 0.001 Ha convergence of the energy. In the course of the standard structural optimization, the cutoff was decreased to 20 Ha and calculations were carried out within semirelativistic approximation (i.e., without SOC).

2. Results and Discussion

Figure 1 shows the band structures calculated for a 13-layer Au(111) film in both the semirelativistic and full relativistic approximations. The surface bands (marked red in Figure 1) appear in a doublet form already in scalar calculations, which means that the splitting seen in Figure 1(a) stems not from spin-orbit coupling but is a result of possible interaction of electrons in the surface states of the terminating surfaces, as it was suggested in [14]. Indeed, the spin-orbit interaction only slightly increases the estimated binding energies of the bands (the relativistic surface bands are shifted down by ~0.08 eV) but does not initiate the spin-splitting, so that each surface band in Figure 1(b) still contains 2 electrons with opposite spins.

It was suggested [14] that the split of the surface bands stems from the interaction between surface states of opposite faces of the slab, and since this split is significant, the spin-splitting, being noticeably smaller, is not revealed. Then, it was proposed to increase the number of layers (and thus the thickness of the model slab) to diminish the interaction between the Au(111) surfaces.

Results of the calculations for various thicknesses of the films confirm these suggestions (Figure 2). In particular, with increasing the thickness of the slab, the splitting of the surface bands noticeably decreases (from 0.170 eV at Γ for 11-layer slab) and, perfectly consistent with results of calculations in [14], for the 23-layer film becomes negligible (0.015 eV).

To facilitate comparison and discussion of obtained results, it is convenient to present the band structure, calculated for the 23-layer Au(111) slab, in the same scale as adopted by Nicolay et al. [14] (Figure 3).

It is evident from comparison of Figures 3(a) and 3(b) that the account for spin-orbit interaction results in a well-pronounced splitting of the bands, which seemingly remain degenerated at Γ point, just like it must be for Rashba splitting. It should be emphasized that the relativistic bands (including not only surface bands but also the bands originated by the interior), shown in Figure 3(b), are almost perfectly consistent with the bands presented in [14] (in particular, the binding energies at Γ, of ~0.50 eV, are the same). It is just this splitting of the surface bands that was interpreted as a Rashba-type spin-splitting, thus apparently supporting the concept of the brake of the inversion symmetry at the surface and hence the lifting of the spin degeneracy of the bands.

In fact, however, the splitting of the Au(111) surface bands seen in Figure 3(b) is found to be not the Rashba splitting. This conclusion directly follows from a detailed analysis of the bands using actual numerical values obtained in the relativistic calculations. Specifically, each surface band of the doublet is twice degenerated; that is, it contains 2 states with opposite spins, so that there are not 2 but 4 bands which form 2 spin-degenerated bands. Furthermore, these bands are not perfectly degenerated at Γ point but split by 0.011 eV, which is close to the value obtained in scalar-relativistic calculations (a quite similar splitting at Γ also can be revealed from Figure 1 in [14] by zooming).

In other words, there is no spin-splitting of the surface bands despite an evident similarity with Rashba effect, which

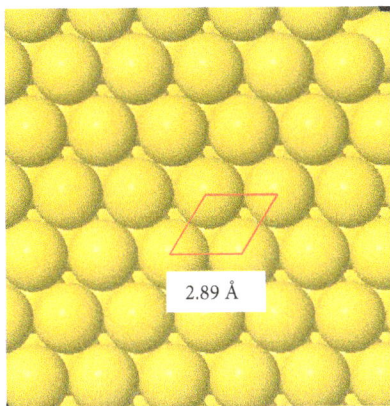

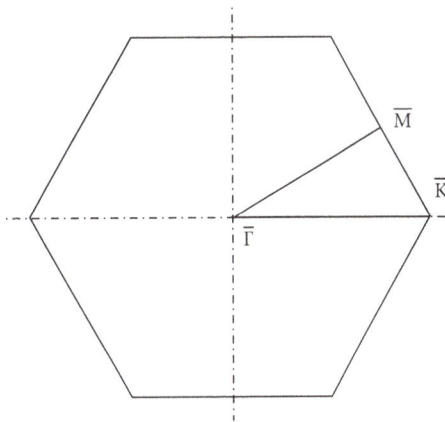

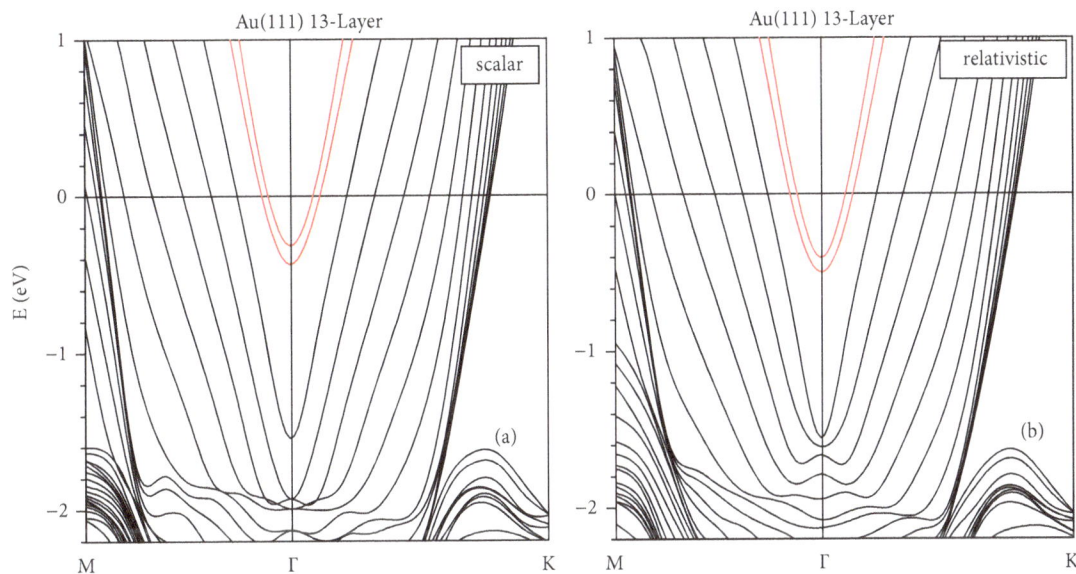

FIGURE 1: The band structure of Au(111) calculated for 13-layer film in the semirelativistic (a) and full relativistic (b) approximations. The surface bands are marked red.

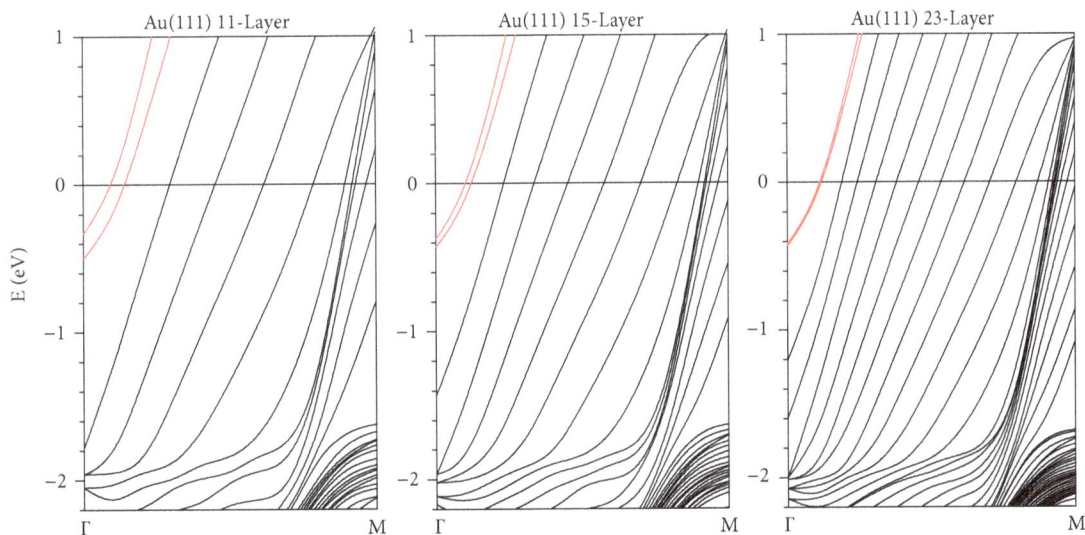

FIGURE 2: The band structure of Au(111) calculated for various thicknesses of films in semirelativistic approximation.

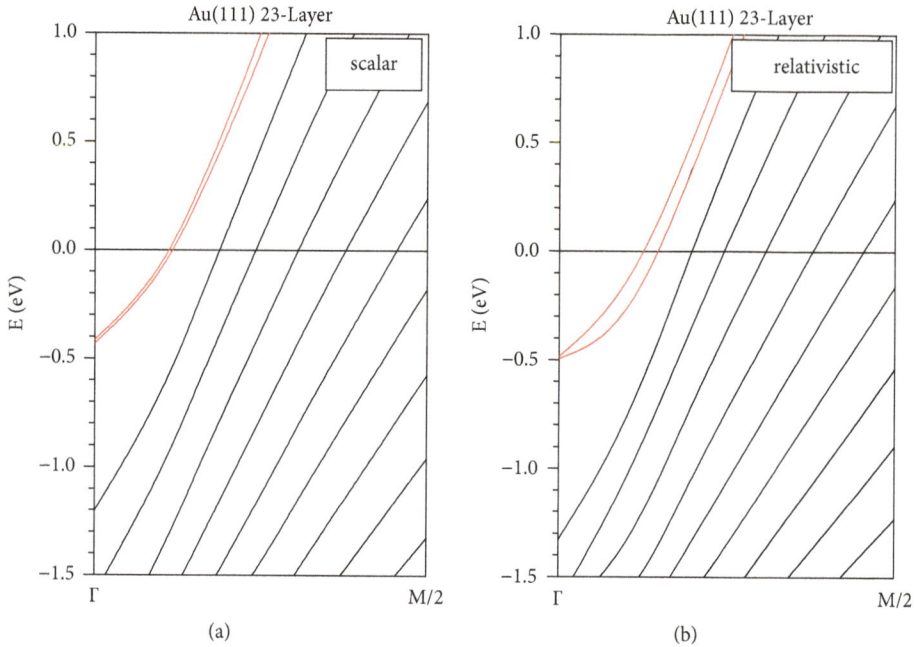

FIGURE 3: The band structure calculated for the 23-layer Au(111) slab in semirelativistic (a) and full relativistic (b) approximations.

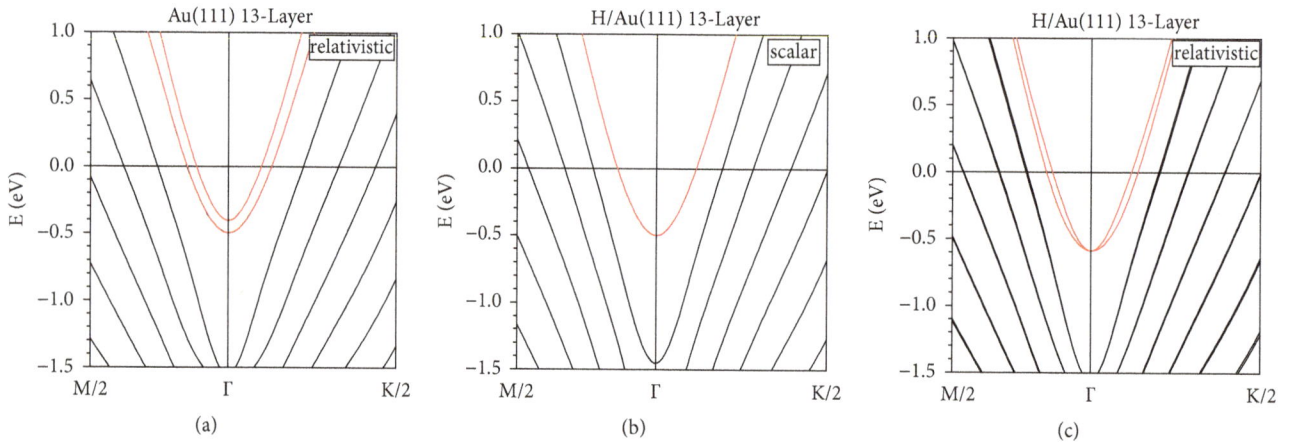

FIGURE 4: Band structures of the 13-layer Au(111) slab (a) and the slab with H overlayer (b, c), calculated in semirelativistic (b) and full relativistic (a, c) approximations.

here is apparent. The Rashba splitting could be switched on by breaking the inversion symmetry of the slab with a help of adsorbed layer, so we have calculated semirelativistic and relativistic band structures for Au(111) film with one surface covered by H (Figure 4) (similar method was adopted in [15], for Bi(111) films, to eliminate, as it was suggested, the interaction between surface states).

The twinning of surface bands of the 13-layer Au(111) film (Figure 4(a)) stems from the interaction between two faces of the slab [14], as noted above. Hydrogen adsorption on one surface of the slab leads to the coupling with the surface state and therefore only one of the parabolic surface bands, corresponding, as it is widely recognized [8–14], to the opposite (clean) Au(111) surface, remains in vicinity of E_F (Figure 4(b)). In contrast to the films with clean surfaces, the

slab with one surface covered by adsorbate (hydrogen in the present case) has no inversion symmetry. Consequently, the SOC coupling lifts the spin degeneracy and causes the spin-splitting of the bands (Figure 4(c)). At Γ point, the bands are exactly degenerated, as it should be according to Kramers' degeneracy at $\mathbf{k} = 0$.

3. Conclusion

Results of present calculations of the band structures of Au(111) films lead to conclusion that the twinning of the surface bands, observed in photoemission experiment, does not necessarily correspond to the spin-splitting of the surface states caused by the break of the inversion symmetry at the surface. In calculations for thin Au(111) films, the twinning can be obtained within nonrelativistic or scalar-relativistic approximation, but these surface bands are split in energy at Γ point. The energy gap between the bands decreases with increasing the film thickness and, for the 23-layer film, consistent with earlier findings [14], is essentially eliminated.

The account for spin-orbit interaction, in perfect agreement with calculations by Nicolay et al. [14], results in a well-pronounced splitting of the surface bands, which seemingly remain degenerated at Γ point. However, a detailed analysis of the bands, using actual numerical values obtained in the relativistic calculations, leads to conclusion that each surface band of the doublet is twice degenerated; that is, it contains two states with opposite spins. In other words, there is no spin-splitting of the surface bands despite an apparent similarity with Rashba effect. This result ultimately indicates that the existence of a surface itself is insufficient to remove the inversion symmetry of the system and thus to lift Kramers' degeneracy. The inversion symmetry can be eliminated by H overlayer adsorbed on one face of the Au(111) film, which leads to the appearance of the spin-splitting of the bands.

It should be stressed that these obtained results do not discard the possibility of creation of metal surfaces or adsorbed films having spin-split surface bands, but this effect ultimately requires the break of inversion symmetry, which can be accomplished either by interaction with the substrate or by adsorption, for example, of H monolayer.

Conflicts of Interest

The author declares that there are no conflicts of interest.

References

[1] S. LaShell, B. A. McDougall, and E. Jensen, "Spin splitting of an Au(111) surface state band observed with angle resolved photoelectron spectroscopy," *Physical Review Letters*, vol. 77, no. 16, pp. 3419–3422, 1996.

[2] S. Murakami, "Quantum spin hall effect and enhanced magnetic response by spin-orbit coupling," *Physical Review Letters*, vol. 97, Article ID 236805, 2006.

[3] L. Fu and C. L. Kane, "Topological insulators with inversion symmetry," *Physical Review B: Condensed Matter and Materials Physics*, vol. 76, no. 4, 2007.

[4] M. Z. Hasan and C. L. Kane, "Colloquium: topological insulators," *Reviews of Modern Physics*, vol. 82, no. 4, pp. 3045–3067, 2010.

[5] B. Yan and S.-C. Zhang, "Topological materials," *Reports on Progress in Physics*, vol. 75, Article ID 096501, 2012.

[6] Y. A. Bychkov and E. I. Rashba, "Properties of a 2D electron gas with lifted spectral degeneracy," *JETP Letters*, vol. 39, p. 78, 1984.

[7] G. Bihlmayer, Y. M. Koroteev, P. M. Echenique, E. V. Chulkov, and S. Blügel, "The Rashba-effect at metallic surfaces," *Surface Science*, vol. 600, no. 18, pp. 3888–3891, 2006.

[8] E. E. Krasovski, "Spin–orbit coupling at surfaces and 2D materials," *Journal of Physics: Condensed Matter*, vol. 27, Article ID 493001, 2015.

[9] F. Reinert, "Spin–orbit interaction in the photoemission spectra of noble metal surface states," *Journal of Physics: Condensed Matter*, vol. 15, no. 5, p. S693, 2003.

[10] D. Fujita, K. Amemiya, T. Yakabe, H. Nejoh, T. Sato, and M. Iwatsuki, "Observation of two-dimensional Fermi contour of a reconstructed Au(111) surface using Fourier transform scanning tunneling microscopy," *Surface Science*, vol. 423, pp. 160–168, 1999.

[11] S. Narasimhan and D. Vanderbilt, "Elastic stress domains and the herringbone reconstruction on Au(111)," *Physical Review Letters*, vol. 69, no. 10, pp. 1564–1567, 1992.

[12] L. Petersen and P. Hedegard, "A simple tight-binding model of spin–orbit splitting of sp-derived surface states," *Surface Science*, vol. 459, pp. 49–56, 2000.

[13] F. Reinert, G. Nicolay, S. Schmidt, D. Ehm, and S. Hüfner, " Direct measurements of the ," *Physical Review B: Condensed Matter and Materials Physics*, vol. 63, no. 11, 2001.

[14] G. Nicolay, F. Reinert, S. Hüfner, and P. Blaha, "Spin-orbit splitting of the L-gap surface state on Au(111) and Ag(111)," *Physical Review B: Condensed Matter and Materials Physics*, vol. 65, no. 3, Article ID 03340, 2001.

[15] Y. M. Koroteev, G. Bihlmayer, J. E. Gayone et al., "Strong spin-orbit splitting on Bi surfaces," *Physical Review Letters*, vol. 93, no. 4, pp. 046403-1, 2004.

[16] M. Muntwiler, M. Hoesch, V. N. Petrov, and M. Hengsberger, "Spin- and angle-resolved photoemission spectroscopy study of the Au(111) Shockley surface state," *Journal of Electron Spectroscopy and Related Phenomena*, vol. 137-140, pp. 119–123, 2004.

[17] A. Tamai, W. Meevasana, P. D. King et al., "Spin-orbit splitting of the Shockley surface state on Cu(111)," *Physical Review B: Condensed Matter and Materials Physics*, vol. 87, no. 7, Article ID 075113, 2013.

[18] J. Jiang, S. S. Tsirkin, K. Shimada et al., "Many-body interactions and Rashba splitting of the surface state on Cu(110)," *Physical Review B: Condensed Matter and Materials Physics*, vol. 89, no. 8, Article ID 085404, 2014.

[19] K. Miyamoto, A. Kimura, K. Kuroda et al., "Spin-Polarized Dirac-Cone-Like Surface State with," *Physical Review Letters*, vol. 108, no. 6, Article ID 066808, 2012.

[20] X. Gonze, J.-M. Beuken, R. Caracas et al., "First-principles computation of material properties: the ABINIT software project," *Computational Materials Science*, vol. 25, no. 3, pp. 478–492, 2002.

[21] N. Troullier and J. L. Martins, "Efficient pseudopotentials for plane-wave calculations," *Physical Review B: Condensed Matter and Materials Physics*, vol. 43, p. 1993, 1991.

[22] S. Goedecker and M. Teter, "Separable dual-space Gaussian pseudopotentials," *Physical Review B: Condensed Matter and Materials Physics*, vol. 54, no. 3, pp. 1703–1710, 1996.

18

Nanoparticle-Enabled Ion Trapping and Ion Generation in Liquid Crystals

Yuriy Garbovskiy ⓘ

UCCS BioFrontiers Center and Department of Physics, University of Colorado Colorado Springs, 1420 Austin Bluffs Parkway, Colorado Springs, CO 80919, USA

Correspondence should be addressed to Yuriy Garbovskiy; ygarbovs@uccs.edu

Academic Editor: Charles Rosenblatt

Nowadays, nanomaterials in liquid crystals and their possible applications in the design of tunable, responsive, and wearable devices are among the most promising research directions. In the majority of cases, all liquid crystal based devices have one thing in common; namely, they are driven by electric fields. This type of device driving can be altered by minor amounts of ions typically present in liquid crystal materials. Therefore, it is very important to understand how nanodopants can affect ions in liquid crystals. In this paper, a recently developed model of contaminated nanoparticles is applied to existing experimental data. The presented analysis unambiguously indicates that, in general, nanomaterials in liquid crystals can behave as a source of ions or as ion traps. Physical factors determining the type of the nanoparticle behaviour and their effects on the concentration of ions in liquid crystals are discussed.

1. Introduction

Thermotropic liquid crystals are widely used in the design of tunable electro-optical devices. They include liquid crystal displays (LCD) [1], tunable lenses [2], filters [3], wave plates [3], retarders [3], diffractive optical elements [4], optical shutters [5, 6], and smart windows [7], to name a few. In the majority of cases, all of them are driven by electric fields [8]. Ions, normally present in liquid crystals in small quantities, can alter the performance of liquid crystals [9–12]. Typically, liquid crystal devices such as LCD and tunable optical elements (filters, retarders, etc.) utilize the electric field effect when the applied electric field reorients liquid crystal molecules. This type of liquid crystal based applications considers ions a nuisance because of many negative side effects caused by ions in liquid crystal devices (image sticking, image flickering, reduced voltage holding ratio, overall slow response) [9–12]. There are also electro-optical devices relying on ions in liquid crystals (optical shutters and smart windows) [5–7]. That is why an understanding of possible sources of ion generation in liquid crystals is very important [13].

Ions in molecular liquid crystals can be generated in different ways: (i) the dissociation of neutral molecules in the bulk of liquid crystals (these dissociating species can be inherently present or added intentionally) [10–12, 14, 15]; (ii) ionic impurities as chemicals left over from the chemical synthesis [11, 16–18]; (iii) chemical degradation of liquid crystals [19]; (iv) ionic contaminants originated from the glue [20] and from the alignment layers [21–25] of the liquid crystal cell; (v)-(vii) ions generated by means of ionizing radiation [26, 27], through electrochemical reactions [28–30], and by relatively high electric fields [31–33].

Recently, the dispersion of nanomaterials in liquid crystals has emerged as a promising way to modify their properties and design novel materials suitable for many applications [[37, 38] and references therein]. From perspectives of ion generation in liquid crystals, a very important question is how can nanodopants affect the behaviour of ions in liquid crystals. Ion-related effects of nano-objects in liquid crystals were reported in many publications reviewed in a recent paper [[39] and references therein]. Different research groups reported that various types of nanomaterials (metal [39–43], semiconductor [35, 44], dielectric [34, 36, 45], ferroelectric [46–51] magnetic [52], and carbon-based [53–56]) changed the concentration of ions in liquid crystals in different ways. Despite the variety of existing experimental results on ions and nano-objects in liquid crystals, they can be broadly categorized into the following groups: (i) papers reporting the decrease in the concentration of ions in liquid crystals (the ion trapping regime); (ii) publications presenting the increase in the concentration of ions (the ion generation regime); (iii) the combination of both ion trapping and ion generation regimes (depending on the concentration of nanodopants in liquid crystals) [39]. A very important finding is that the same type of nanomaterials dispersed in different liquid crystals can result in different regimes (ion trapping or ion generation) [34, 39, 53]. An elementary model of these regimes was recently proposed and developed in a series of papers [57–61]. This model introduced the ionic contamination of nanomaterials as a key factor enabling the possibility of different regimes (ion trapping (or ion capturing regime), ion generation (or ion releasing regime), and no change regime) in liquid crystals doped with nanomaterials [57–61]. However, the origin of the ionic contamination of nanomaterials remains practically unexplored.

In this paper, the aforementioned model of contaminated nanoparticles in liquid crystals is applied to existing experimental results with the aim of shedding some light on the nature of ionic contamination of nanodopants and ion generation/ion trapping in liquid crystals doped with such nanomaterials.

2. Elementary Model

To simplify the discussion, consider liquid crystals containing some mobile ions and nanoparticles contaminated with the same type of fully ionized ionic species. These ionic species are characterized by their volume concentration $n^+ = n^- = n$. The discussion of contaminated nanomaterials is needed to account for the possibility of experimentally observed ion trapping and ion generation regimes [57–61]. If contaminated nanoparticles are dispersed in liquid crystals, interactions between ions and nanoparticles will result in the change of the total concentration of mobile ions in liquid crystal/nanoparticle colloids. In short, a fraction of ionic contaminants can leave the surface of nanoparticles thus enriching the liquid crystal host with ions. This process can be considered as nanoparticle-enabled ion generation in liquid crystals. The reverse process, namely, the trapping of mobile ions by the surface of nanoparticles, also takes place. The competition between these two processes will result in the steady state characterized by a constant concentration of mobile ions in liquid crystals doped with nanoparticles ($dn/dt = 0$). In the simplest case, the ion releasing (or ion generation) process can be associated with the desorption of ions from the surface of nanoparticles, while the ion trapping process can be described by the adsorption of ions onto the surface of nanoparticles. As a result, the concentration of mobile ions in this system is governed by the following rate equation [62]:

$$\frac{dn}{dt} = -k_a^{NP} n_{NP} A_{NP} \sigma_S^{NP} n \left(1 - \Theta_{NP}\right)$$
$$+ k_d^{NP} n_{NP} A_{NP} \sigma_S^{NP} \Theta_{NP} \quad (1)$$

The first term of (1) describes the adsorption of ions onto the surface of nanoparticles (ion trapping process), and the second term accounts for the ion desorption from the surface of nanoparticles (ion generation process). In (1), n is the concentration of mobile ions; t denotes time; A_{NP} is the surface area of a single nanoparticle (for simplicity, consider spherical nanoparticles characterized by their radius R_{NP}; in this case $A_{NP} = 4\pi R_{NP}^2$ and this parameter determines the dependence of the concentration of mobile ions in liquid crystal nanocolloids on the size of nanoparticles); n_{NP} is the volume concentration of nanoparticles (in many practical cases the weight concentration of nanoparticles ω_{NP}, which is related to the volume concentration as $n_{NP} \approx \omega_{NP}(\rho_{LC}/\rho_{NP})(1/V_{NP})$ (V_{NP} is the volume of a single nanoparticle, and $\rho_{LC}(\rho_{NP})$ is the density of liquid crystals (nanoparticles)), is preferred); σ_S^{NP} is the surface density of all adsorption sites on the surface of a single nanoparticle; Θ_{NP} is the fractional surface coverage of nanoparticles defined as $\Theta_{NP} = \sigma_{NP}/\sigma_S^{NP}$ (σ_{NP} is the surface density of adsorption sites on the surface of nanoparticles occupied by ions); k_a^{NP} is the adsorption rate constant; and k_d^{NP} is the desorption rate constant. The aforementioned physical parameters (σ_S^{NP}, k_a^{NP}, and k_d^{NP}) are material-dependent and, for a given system under study (liquid crystals-nanoparticles), are considered as constants.

The applicability and limitations of (1) to compute the concentration of mobile ions in liquid crystals doped with nanoparticles were recently discussed in papers [63–65]. It should be stressed that (1) is an approximation which is reasonably applicable to describe ions in molecular liquid crystals. In a general case, a more rigorous approach utilizing Poisson-Boltzmann equation should be considered [66–69].

In the steady-state regime ($dn/dt = 0$), typically achieved in experiments, (1) should be solved along with the conservation law of the total number of ions expressed by

$$n_0 + n_{NP} A_{NP} \sigma_S^{NP} \nu_{NP} = n + n_{NP} A_{NP} \sigma_S^{NP} \Theta_{NP} \quad (2)$$

where n_0 is the initial concentration of ions in liquid crystals and ν_{NP} is the contamination factor of nanoparticles. The contamination factor of nanoparticles accounts for their ionic contamination and equals the fraction of the adsorption sites on the surface of nanoparticles occupied by ionic contaminants prior to dispersing nanodopants in liquid crystals [70, 71].

The ionic contamination of nanoparticles is a key factor enabling the possibility of different regimes, namely, the ion trapping regime (Figure 1, solid and dashed-dotted curves), the ion generation regime (Figure 1, dashed curve), and no change regime (Figure 1, dotted curve).

The switching between these regimes is controlled by the aforementioned contamination factor of nanoparticles ν_{NP}. The ion trapping regime is observed if $\nu_{NP} < \nu_{NP}^C$ ($n_0 > n_C$), the ion generation regime is reached if $\nu_{NP} > \nu_{NP}^C$ ($n_0 < n_C$), and no change regime is achieved if $\nu_{NP} = \nu_{NP}^C$ ($n_0 = n_C$), where ν_{NP}^C is the critical contamination factor of nanoparticles defined as $\nu_{NP}^C = n_0 K_{NP}/(1 + n_0 K_{NP})$ and n_C is the critical concentration of ions written as $n_C = \nu_{NP}/K_{NP}(1 - \nu_{NP})$

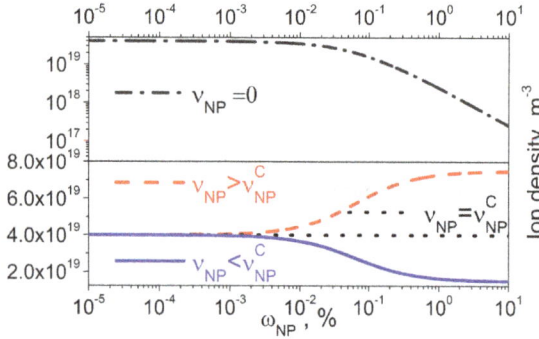

FIGURE 1: Liquid crystals doped with nanoparticles: the ion density as a function of the weight concentration of nanodopants, ω_{NP}. The case of 100% pure ($\nu_{NP} = 0$) nanoparticles in liquid crystals is represented by dashed-dotted curve (ion trapping regime). Different regimes of contaminated nanoparticles in liquid crystals are also shown: ion generation (or ion releasing) regime (dashed curve, $\nu_{NP} = 1.5 \times 10^{-3}$); ion trapping (or ion capturing) regime (solid curve, $\nu_{NP} = 3 \times 10^{-4}$); no change regime (dotted curve, $\nu_{NP} = \nu_{NP}^{C} = 8 \times 10^{-4}$). Physical parameters used in simulations: $\sigma_S^{NP} = 10^{18} m^{-2}$, $K_{NP} = k_a^{NP}/k_d^{NP} = 2 \times 10^{-23} m^3$, $n_0 = 4 \times 10^{19} m^{-3}$, $R_{NP} = 10 nm$, and $\rho_{NP}/\rho_{LC} = 3.9$.

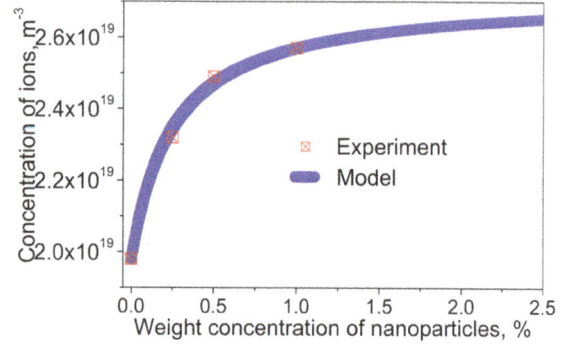

FIGURE 2: The concentration of ions in liquid crystals doped with TiO$_2$ nanoparticles as a function of the weight concentration of nanodopants. The presented model (solid curve) is in good agreement with reported experimental data points (squares) [34]. Physical parameters used in simulations: $\sigma_S^{NP} = 0.8 \times 10^{18} m^{-2}$, $K_{NP} = 1.6 \times 10^{-23} m^3$, $n_0 = 1.98 \times 10^{19} m^{-3}$, $\nu_{NP} = 4.35 \times 10^{-4}$, $R_{NP} = 25 nm$, and $\rho_{NP}/\rho_{LC} = 3.9$.

and $K_{NP} = k_a^{NP}/k_d^{NP}$ (Figure 1) [57]. Both ion trapping and ion generation regimes are more pronounced if higher concentrations of nanoparticles are used (Figure 1). It should be noted that the dispersion of 100% pure nanoparticles in liquid crystals results in the ion trapping regime only (Figure 1, dashed-dotted curve). Another very important feature of the behaviour of contaminated nanoparticles in liquid crystals is an existence of the saturation effect. Upon increasing the concentration of nanoparticles in liquid crystals, the concentration of mobile ions increases or decreases towards its saturation level given by the critical concentration $n_C = \nu_{NP}/K_{NP}(1 - \nu_{NP})$ (Figure 1, solid and dashed curves). On the contrary, the dispersion of 100% pure nanoparticles ($\nu_{NP} = 0$) in liquid crystals does not exhibit the saturation effect (Figure 1, dashed-doted curve).

3. Case Study: TiO$_2$ Nanoparticles in Liquid Crystals [34]

The effects of titanium dioxide nanoparticles on the concentration of mobile ions in nematic liquid crystals were recently reported by Shcherbinin and Konshina [34]. By dispersing TiO$_2$ nanoparticles (Plasmotherm, Moscow) in commercially available nematic liquid crystals (ZhK1282, NIOPIK, Moscow) the ion releasing regime was observed (Figure 2, squares) [34]. These results indicate that, in the considered case, nanoparticles in liquid crystals act as a source of ions. In other words, nanoparticles, upon their dispersion in liquid crystals, increase the total concentration of mobile ions (Figure 2). According to the model described in the previous section, this increase in the concentration of ions can be caused by the ionic contamination of nanoparticles. By applying (1)-(2) of the presented model, a very good

agreement between theoretical predictions and experimental results can be achieved (Figure 2). Interestingly, a minor level of the ionic contamination of nanoparticles ($\nu_{NP} = 4.35 \times 10^{-4} \ll 1$) is enough to increase the concentration of ions in liquid crystals (Figure 2). Given its small values, this uncontrolled ionic contamination of nanoparticles can easily happen during the preparation and handling of these nanomaterials.

By adding the ionic surfactant (cetyltrimethylammonium bromide, abbreviated as CTABr) to the same nematic liquid crystals (ZhK1282), the total concentration of ions was increased by nearly one order of magnitude [34]. Dispersing the same TiO$_2$ nanoparticles in these, enriched with ions, liquid crystals, the ion trapping regime was achieved [34]. To analyse these experimental results, we can consider two dominant types of ionic species in ZhK1282 liquid crystals. Pristine liquid crystals (prior to doping them with CTABr surfactants and nanoparticles) are characterized by dominant ionic species of the concentration n_1. Liquid crystals doped with surfactants are characterized by two dominant types of ionic species, namely, n_1 and n_2. In the regime of low surface coverage of nanoparticles ($\nu_{NP} \ll 1$ and $\Theta_{NP} \ll 1$), these two dominant types of ionic species can be treated independently. As a result, the dispersion of nanoparticles in liquid crystals increases the concentration of ions n_1 (Figure 2) and decreases the concentration of ionic species n_2 (Figure 3) (Table 1).

The ion trapping regime, corresponding to the effects of nanoparticles on the concentration of ions n_2, is shown in Figure 3. As can be seen from Figure 3, there is a very good agreement between theoretical predictions based on the model of contaminated nanoparticles ((1)-(2)) and reported experimental data. As expected, TiO$_2$ nanoparticles were not contaminated with ionic species n_2 prior to dispersing them in liquid crystals ($\nu_{NP} = 0$). As a result, the only possibility is an experimentally observed ion trapping regime (Figure 3).

TABLE 1: The concentration of ionic species in liquid crystals doped with TiO_2 nanoparticles [34].

Weight concentration of TiO_2, %	n_1, m^{-3} (LC1)	$n_1 + n_2$, m^{-3} (LC2)	n_2, m^{-3} (LC2)
0	1.98×10^{19}	1.345×10^{20}	1.147×10^{20}
0.25	2.32×10^{19}	1.0525×10^{20}	8.2×10^{19}
0.50	2.49×10^{19}	7.275×10^{19}	4.78×10^{19}
1.00	2.57×10^{19}	6.325×10^{19}	3.755×10^{19}

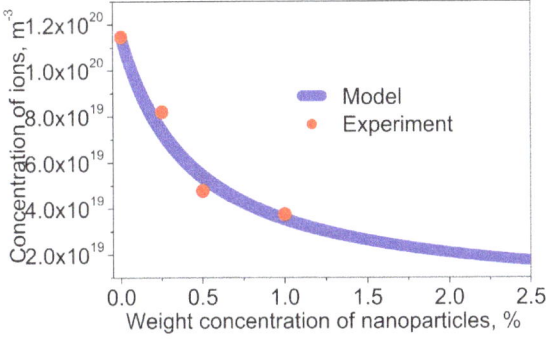

FIGURE 3: The concentration of ions in liquid crystals doped with TiO_2 nanoparticles as a function of the weight concentration of nanodopants. The presented model (solid curve) is in good agreement with reported experimental data points (circles) [34]. Physical parameters used in simulations: $\sigma_S^{NP} = 2 \times 10^{18} \, m^{-2}$, $K_{NP} = 3.65 \times 10^{-24} \, m^3$, $n_0 = 1.147 \times 10^{20} \, m^{-3}$, $\nu_{NP} = 0$, $R_{NP} = 25 \, nm$, and $\rho_{NP}/\rho_{LC} = 3.9$.

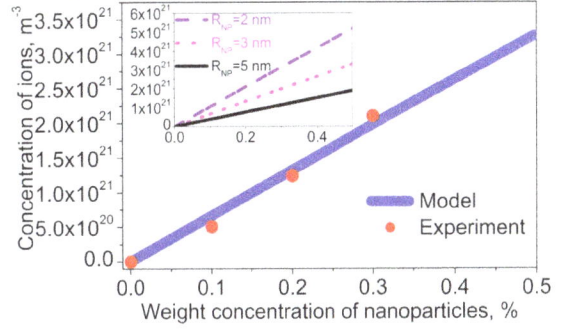

FIGURE 4: The concentration of "slow" ions in liquid crystals doped with CdSe/ZnS quantum dots as a function of the weight concentration of nanodopants. The presented model (solid curve) is in good agreement with reported experimental data points (circles) (Table 2 of paper [35]). An inset shows the same dependence calculated at different sizes of quantum dots. Physical parameters used in simulations: $\sigma_S^{NP} = 10^{18} \, m^{-2}$, $K_{NP} = 10^{-26} \, m^3$, $n_0 = 0 \, m^{-3}$, $\nu_{NP} = 3.379 \times 10^{-3}$, $R_{NP} = 3 \, nm$, and $\rho_{NP}/\rho_{LC} = 5.091$.

4. Case Study: CdSe/ZnS Core/Shell Nanoparticles in Liquid Crystals [35]

The behaviour of ionic impurities in nematic liquid crystals (ZhK1289, NIOPIK, Moscow) doped with CdSe/ZnS core/shell quantum dots was studied by Shcherbinin and Konshina [35]. Prior to dispersing them in liquid crystals, quantum dots were functionalized with trioctylphosphine oxide (TOPO) ligands. Shcherbinin and Konshina reported that functionalized quantum dots dispersed in liquid crystals enriched the liquid crystal host with a new type of ions. These ions ("slow" ions as called in paper [35]) were not present in liquid crystals prior to mixing them with nanodopants. Thus, functionalized quantum dots were considered a primary source of these ions. It was found that the concentration of "slow" ions was a linear function of the weight concentration of nanodopants in liquid crystals [35]. This linear dependence can also be obtained by means of (1)-(2) (steady-state regime) and assuming the following conditions: $n_{NP} K_{NP} A_{NP} \sigma_S^{NP} \ll 1$ and $K_{NP} n \ll 1$. In this case, the concentration of mobile ions in liquid crystals doped with nanoparticles can be written as

$$n \approx n_0 + n_{NP} A_{NP} \sigma_S^{NP} \nu_{NP} \qquad (3)$$

As can be seen, (3) is a linear function of the concentration of nanoparticles. Again, a good match between reported experimental data and the proposed model is achieved (Figure 4).

An inset (Figure 4) also shows the effect of the nanoparticle size on the concentration of ions in liquid crystals. In (1)–(3) this size effect is caused by the surface area of nanoparticles, A_{NP}. In the case of spherical nanodopants, $A_{NP} = 4\pi R_{NP}^2$. According to Figure 4 (inset), at the same concentration of nanoparticles, the ion releasing effect is greater if smaller nanoparticles are used. Additional details can also be found in recently published papers [63, 65].

5. Case Study: Cu_7PS_6 Nanoparticles in Liquid Crystals [36]

An increase in the electrical conductivity of 6CB nematic liquid crystals through doping them with Cu_7PS_6 nanoparticles was recently reported by Kovalchuk and coworkers [36]. An interesting feature of the observed effect was the saturation of the increase in the electrical conductivity at higher concentrations of nanoparticles [36].

The concentration of mobile ions n can be estimated through

$$\lambda = q\mu n \qquad (4)$$

where λ is the measured electrical conductivity, q is the charge of a single ion, and μ is its mobility [10–12]. As a result, the obtained values of the concentration of mobile ions in 6CB liquid crystals doped with Cu_7PS_6 are in a very good

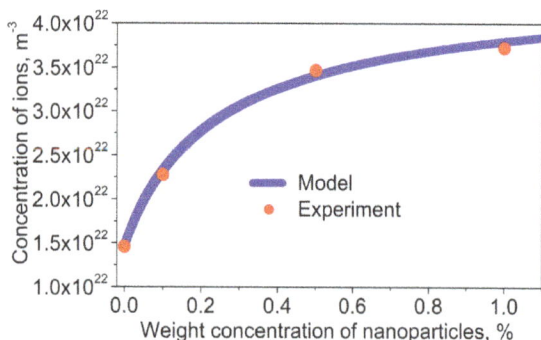

FIGURE 5: The concentration of ions in liquid crystals doped with Cu_7PS_6 nanoparticles as a function of the weight concentration of nanodopants. The presented model (solid curve) is in good agreement with reported experimental data points (circles) [36]. Physical parameters used in simulations: $\sigma_S^{NP} = 7 \times 10^{18} \, m^{-2}$, $K_{NP} = 10^{-23} \, m^3$, $n_0 = 1.466 \times 10^{22} \, m^{-3}$, $\nu_{NP} = 0.3075$, $R_{NP} = 58.5 \, nm$, $\rho_{NP}/\rho_{LC} = 4.907$, $q = 3.2 \times 10^{-19} \, C$, and $\mu = 10^{-10} \, m^2/V \times s$.

agreement with theoretical predictions based on (1)-(2) of the model of contaminated nanoparticles (Figure 5).

In the case shown in Figure 5, nanoparticles act as a source of ions in liquid crystals, thus increasing the electrical conductivity of the system under study. The concentration of ions in 6CB liquid crystals doped with Cu_7PS_6 nanoparticles as a function of the nanoparticle concentration exhibits a monotonous increase $n(\omega_{NP})$ towards the saturation level. In the framework of the proposed model, this saturation level is given by the critical concentration of ions, $n_C = \nu_{NP}/K_{NP}(1 - \nu_{NP})$. In the considered example, $\nu_{NP} = 0.35$, $K_{NP} = 10^{-23} \, m^3$, and thus $n_C = 4.44 \times 10^{22} \, m^{-3}$. Relatively high values of the contamination factor ν_{NP} can indicate that, in the case of Cu_7PS_6 nanoparticles, ionic contaminants are inherently present in these nanoparticles. In other words, the origin of these ions can be associated with the chemical structure of Cu_7PS_6 nanoparticles and the possibility of their self-dissociation rather than with ionic contaminants originated during their preparation and handling. The dissociation of Cu_7PS_6 nanoparticles resulting in the generation of copper ions can be reasonably expected for this type of materials [36].

6. Conclusions

According to existing experimental reports, nanoparticles in liquid crystals can behave either as ion traps or as ion generating objects. If nanodopants are 100% pure, the ion trapping regime is the only possible outcome of their dispersion in liquid crystals. On the contrary, liquid crystals doped with contaminated nanoparticles can exhibit three different regimes, namely, the ion trapping regime, ion generation regime, and no change regime. Both ion trapping and ion generation regimes are characterized by the saturation effect (Figure 1, solid and dashed curves, and Figures 2, 5). This effect, absent in the case of 100% pure nanoparticles in liquid crystals (Figure 1, dashed-dotted curve), is quantitatively described by the critical concentration of ions,

$n_C = \nu_{NP}/K_{NP}(1 - \nu_{NP})$. Thus, the ionic contamination of nanodopants is an important physical quantity, enabling the type of the achieved regime (ion trapping or ion generation) and leading to the saturation effect. In the framework of the model of contaminated nanomaterials, this ionic contamination is quantified by means of the dimensionless contamination factor ν_{NP} ($0 \leq \nu_{NP} \leq 1$). The predictions of this model are in very good agreement with reported experimental results (Figures 2–5). In addition, an analysis of experimental results using the model of contaminated nanoparticles can shed some light on the origin of ionic contaminants.

The ionic contamination of nanomaterials can originate from different sources during their production and handling. For example, it can be a minor fraction of ionic contaminants left over from the chemical synthesis. In the considered case study (Figure 4), trioctylphosphine oxide (TOPO) ligands are prone to uncontrolled contamination (including ionic contaminants) during the chemical synthesis [72]. As a result, once quantum dots functionalized with TOPO are dispersed in liquid crystals, ionic contaminants can be released in the bulk of the liquid crystal host. This assumption is also consistent with recent findings reported by Urbanski and Lagerwall [42]. In the case of uncontrolled ionic contamination of nanodopants, the values of the contamination factor are typically relatively low. Interestingly, even such small values of the contamination factor ($\nu_{NP} \ll 1$, on the order of 10^{-3} -10^{-4} or less, see Figures 1–4) are enough to alter the concentration of mobile ions in liquid crystals.

Ionic contaminants can also be inherently present in nanoparticles. In this case, the value of the contamination factor is relatively high ($\nu_{NP} \approx 0.1 - 1$). An example of this, inherent ionic "contamination," includes self-dissociating nanomaterials. Such nanomaterials can dissociate and generate ions in liquid crystals (Figure 5).

In general, contaminated nanoparticles should be considered as an important source of ion generation in liquid crystals. In addition, under certain conditions, they can act as ion traps. Further studies are needed to understand physical-chemical mechanisms governing these processes and identities of ionic contaminants. In the long run, this understanding will enable numerous applications of liquid crystals doped with nanomaterials.

Conflicts of Interest

The author declares that there are no conflicts of interest.

Acknowledgments

The author would like to acknowledge the support provided by the UCCS BioFrontiers Center at the University of Colorado.

References

[1] K. Naoyuki, Ed., *The Liquid Crystal Display Story. 50 Years of Liquid Crystal R&D that lead The Way to the Future*, Springer, Tokyo, Japan, 2014.

[2] Y. Lin, Y. Wang, and V. Reshetnyak, "Liquid crystal lenses with tunable focal length," *Liquid Crystals Reviews*, vol. 5, no. 2, pp. 111–143, 2017.

[3] I. Abdulhalim, "Non-display bio-optic applications of liquid crystals," *Liquid Crystals Today*, vol. 20, no. 2, pp. 44–60, 2011.

[4] L. De Sio, D. E. Roberts, Z. Liao et al., "Beam shaping diffractive wave plates [Invited]," *Applied Optics*, vol. 57, no. 1, pp. A118–A121, 2018.

[5] M. W. Geis, P. J. Bos, V. Liberman, and M. Rothschild, "Broadband optical switch based on liquid crystal dynamic scattering," *Optics Express*, vol. 24, no. 13, pp. 13812–13823, 2016.

[6] E. A. Konshina and D. P. Shcherbinin, "Study of dynamic light scattering in nematic liquid crystal and its optical, electrical and switching characteristics," *Liquid Crystals*, vol. 45, no. 2, pp. 292–302, 2017.

[7] R. Dabrowski, J. Dziaduszek, J. Bozetka et al., "Fluorinated smectics - New liquid crystalline medium for smart windows and memory displays," *Journal of Molecular Liquids*, 2017.

[8] D. K. Yang and S. T. Wu, *Liquid Crystal Devices*, John Wiley & Sons, Hoboken, Hoboken, NJ, USA, 2006.

[9] S. Naemura, "Electrical properties of liquid-crystal materials for display applications," *Materials Research Society - Proceedings*, vol. 559, pp. 263–274, 1999.

[10] V. G. Chigrinov, *Liquid Crystal Devices: Physics and Applications*, Artech House, Boston, MA, USA, 1999.

[11] L. M. Blinov, *Structure and Properties of Liquid Crystals*, Springer Netherlands, New York, NY, USA, 2010.

[12] K. Neyts and F. Beunis, *Handbook of Liquid Crystals: Physical Properties and Phase Behavior of Liquid Crystals*, vol. 2, Chapter 11, Ion transport in liquid crystals, Wiley-VCH, Germany, 2014.

[13] P. P. Korniychuk, A. M. Gabovich, K. Singer, A. I. Voitenko, and Y. A. Reznikov, "Transient and steady electric currents through a liquid crystal cell," *Liquid Crystals*, vol. 37, no. 9, pp. 1171–1181, 2010.

[14] R. Chang and J. M. Richardson, "The Anisotropic Electrical Conductivity of M.B.B.A. Containing Tetrabutyl-Ammonium Tetraphenyl-Boride," *Molecular Crystals and Liquid Crystals*, vol. 28, no. 1-2, pp. 189–200, 1974.

[15] M. I. Barnik, L. M. Blinov, M. F. Grebenkin, S. A. Pikin, and V. G. Chigrinov, "Electrohydrodynamic instability in nematic liquid crystals," *Sov Phys JETP*, vol. 42, no. 3, pp. 550–553, 1976.

[16] G. Briere, F. Gaspard, and R. Herino, "Ionic residual conduction in the isotropic phase of a nematic liquid crystal," *Chemical Physics Letters*, vol. 9, no. 4, pp. 285–288, 1971.

[17] S. Naemura and A. Saivada, "Ionic conduction in nematic and smectic a liquid crystals," *Molecular Crystals and Liquid Crystals*, vol. 400, pp. 79–96, 2003.

[18] H.-Y. Hung, C.-W. Lu, C.-Y. Lee, C.-S. Hsu, and Y.-Z. Hsieh, "Analysis of metal ion impurities in liquid crystals using high resolution inductively coupled plasma mass spectrometry," *Analytical Methods*, vol. 4, no. 11, pp. 3631–3637, 2012.

[19] M. Sierakowski, "Ionic interface-effects in electro-optical LC-cells," *Molecular Crystals and Liquid Crystals Science and Technology Section A: Molecular Crystals and Liquid Crystals*, vol. 375, pp. 659–677, 2002.

[20] S. Murakami and H. Naito, "Electrode and interface polarizations in nematic liquid crystal cells," *Japanese Journal of Applied Physics*, vol. 36, no. 4 A, pp. 2222–2225, 1997.

[21] H. Naito, Y. Yasuda, and A. Sugimura, "Desorption processes of adsorbed impurity ions on alignment layers in nematic liquid crystal cells," *Molecular Crystals and Liquid Crystals Science and Technology Section A: Molecular Crystals and Liquid Crystals*, vol. 301, pp. 85–90, 1997.

[22] M. Mizusaki, S. Enomoto, and Y. Hara, "Generation mechanism of residual direct current voltage for liquid crystal cells with polymer layers produced from monomers," *Liquid Crystals*, vol. 44, no. 4, pp. 609–617, 2017.

[23] R. Kravchuk, O. Koval'Chuk, and O. Yaroshchuk, "Filling initiated ion transport processes in liquid crystal cell," *Molecular Crystals and Liquid Crystals Science and Technology Section A: Molecular Crystals and Liquid Crystals*, vol. 384, no. I, pp. 111–119, 2002.

[24] E. A. Konshina and D. P. Shcherbinin, "Effect of granular silver films morphology on the molecules orientation and ion contamination of nematic liquid crystal," *Bulletin of the Moscow State Regional University (Physics and Mathematics)*, no. 4, pp. 103–113, 2017.

[25] Y. Garbovskiy, "Time-dependent electrical properties of liquid crystal cells: unravelling the origin of ion generation," *Liquid Crystals*, pp. 1–9, 2018.

[26] A. V. Kovalchuk, O. D. Lavrentovich, and V. A. Linev, "Electrical conductivity of γ-irradiated cholesteric liquid crystals," *Sov. Tech. Phys. Lett*, vol. 14, no. 5, pp. 381-382, 1988.

[27] H. Naito, K. Yoshida, M. Okuda, and A. Sugimura, "Transient current study of ultraviolet-light-soaked states in n-pentyl-p-n-cyanobiphenyl," *Japanese Journal of Applied Physics*, vol. 33, no. 10, pp. 5890-5891, 1994.

[28] S. Barret, F. Gaspard, R. Herino, and F. Mondon, "Dynamic scattering in nematic liquid crystals under dc conditions. I. Basic electrochemical analysis," *Journal of Applied Physics*, vol. 47, no. 6, pp. 2375–2377, 1976.

[29] S. Barret, F. Gaspard, R. Herino, and F. Mondon, "Dynamic scattering in nematic liquid crystals under dc conditions. II. Monitoring of electrode processes and lifetime investigation," *Journal of Applied Physics*, vol. 47, no. 6, pp. 2378–2381, 1976.

[30] H. S. Lim, J. D. Margerum, and A. Graube, "Electrochemical properties of dopants and the D-C dynamic scattering of a nematic liquid crystal," *J. Electrochem. Soc.: Solid State Science and technology*, vol. 124, no. 9, pp. 1389–1394, 1977.

[31] T. C. Chieu and K. H. Yang, "Transport properties of ions in ferroelectric liquid crystal cells," *Japanese Journal of Applied Physics*, vol. 28, no. 11 R, pp. 2240–2246, 1989.

[32] S. Murakami and H. Naito, "Charge injection and generation in nematic liquid crystal cells," *Japanese Journal of Applied Physics*, vol. 36, no. 2, pp. 773–776, 1997.

[33] H. De Vleeschouwer, A. Verschueren, F. Bougrioua et al., "Long-term ion transport in nematic liquid crystal displays," *Japanese Journal of Applied Physics*, vol. 40, no. 5 A, pp. 3272–3276, 2001.

[34] D. P. Shcherbinin and E. A. Konshina, "Impact of titanium dioxide nanoparticles on purification and contamination of nematic liquid crystals," *Beilstein Journal of Nanotechnology*, vol. 8, pp. 2766–2770, 2017.

[35] D. P. Shcherbinin and E. A. Konshina, "Ionic impurities in nematic liquid crystal doped with quantum dots CdSe/ZnS," *Liquid Crystals*, vol. 44, no. 4, pp. 648–655, 2017.

[36] O. V. Kovalchuk, I. P. Studenyak, V. Yu. Izai et al., "Saturation effect for dependence of the electrical conductivity of planar oriented liquid crystal 6CB on the concentration of Cu7PS6 nanoparticles, Semiconductor Physics," *Quantum Electronics & Optoelectronics*, vol. 20, no. 4, pp. 437–441, 2017.

[37] Y. A. Garbovskiy and A. V. Glushchenko, "Liquid crystalline colloids of nanoparticles: Preparation, properties, and applications," *Solid State Physics - Advances in Research and Applications*, vol. 62, pp. 1–74, 2010.

[38] J. P. F. Lagerwall and G. Scalia, *Liquid Crystals with Nano and Microparticles (Series in Soft Condensed Matter: Volume 7)*, vol. 7, World Scientific Publishing Co., Singapore, Singapore, 2016.

[39] Y. Garbovskiy and I. Glushchenko, "Nano-objects and ions in liquid crystals: Ion trapping effect and related phenomena," *Crystals*, vol. 5, no. 4, pp. 501–533, 2015.

[40] R. K. Shukla, X. Feng, S. Umadevi, T. Hegmann, and W. Haase, "Influence of different amount of functionalized bulky gold nanorods dopant on the electrooptical, dielectric and optical properties of the FLC host," *Chemical Physics Letters*, vol. 599, pp. 80–85, 2014.

[41] F. V. Podgornov, R. Wipf, B. Stühn, A. V. Ryzhkova, and W. Haase, "Low-frequency relaxation modes in ferroelectric liquid crystal/gold nanoparticle dispersion: impact of nanoparticle shape," *Liquid Crystals*, vol. 43, no. 11, pp. 1536–1547, 2016.

[42] M. Urbanski and J. P. F. Lagerwall, "Why organically functionalized nanoparticles increase the electrical conductivity of nematic liquid crystal dispersions," *Journal of Materials Chemistry C*, vol. 5, no. 34, pp. 8802–8809, 2017.

[43] F. V. Podgornov, M. Gavrilyak, A. Karaawi, V. Boronin, and W. Haase, "Mechanism of electrooptic switching time enhancement in ferroelectric liquid crystal/gold nanoparticles dispersion," *Liquid Crystals*, pp. 1–9, 2018.

[44] E. Konshina, D. Shcherbinin, and M. Kurochkina, "Comparison of the properties of nematic liquid crystals doped with TiO 2 and CdSe/ZnS nanoparticles," *Journal of Molecular Liquids*, 2017.

[45] A. Chandran, J. Prakash, J. Gangwar et al., "Low-voltage electro-optical memory device based on NiO nanorods dispersed in a ferroelectric liquid crystal," *RSC Advances*, vol. 6, no. 59, pp. 53873–53881, 2016.

[46] R. K. Shukla, C. M. Liebig, D. R. Evans, and W. Haase, "Electro-optical behaviour and dielectric dynamics of harvested ferroelectric LiNbO3 nanoparticle-doped ferroelectric liquid crystal nanocolloids," *RSC Advances*, vol. 4, no. 36, pp. 18529–18536, 2014.

[47] Y. Garbovskiy and I. Glushchenko, "Ion trapping by means of ferroelectric nanoparticles, and the quantification of this process in liquid crystals," *Applied Physics Letters*, vol. 107, no. 4, 2015.

[48] R. Basu and A. Garvey, "Effects of ferroelectric nanoparticles on ion transport in a liquid crystal," *Applied Physics Letters*, vol. 105, no. 15, p. 151905, 2014.

[49] Y.-C. Hsiao, S.-M. Huang, E.-R. Yeh, and W. Lee, "Temperature-dependent electrical and dielectric properties of nematic liquid crystals doped with ferroelectric particles," *Displays*, vol. 44, pp. 61–65, 2016.

[50] P. Kumar, S. Debnath, N. V. Rao, and A. Sinha, "Nanodoping: a route for enhancing electro-optic performance of bent core nematic system," *Journal of Physics: Condensed Matter*, vol. 30, no. 9, p. 095101, 2018.

[51] S. Al-Zangana, M. Turner, and I. Dierking, "A comparison between size dependent paraelectric and ferroelectric BaTiO3 nanoparticle doped nematic and ferroelectric liquid crystals," *Journal of Applied Physics*, vol. 121, no. 8, 2017.

[52] Khushboo, P. Sharma, P. Malik, and K. K. Raina, "Electro-optic, dielectric and optical studies of NiFe2O4-ferroelectric liquid crystal: a soft magnetoelectric material," *Liquid Crystals*, vol. 43, no. 11, pp. 1671–1681, 2016.

[53] S. Tomylko, O. Yaroshchuk, O. Kovalchuk, U. Maschke, and R. Yamaguchi, "Dielectric properties of nematic liquid crystal modified with diamond nanoparticles," *Ukrainian Journal of Physics*, vol. 57, no. 2, pp. 239–243, 2012.

[54] B.-R. Jian, C.-Y. Tang, and W. Lee, "Temperature-dependent electrical properties of dilute suspensions of carbon nanotubes in nematic liquid crystals," *Carbon*, vol. 49, no. 3, pp. 910–914, 2011.

[55] M.-J. Cho, H.-G. Park, H.-C. Jeong et al., "Superior fast switching of liquid crystal devices using graphene quantum dots," *Liquid Crystals*, vol. 41, no. 6, pp. 761–767, 2014.

[56] P.-C. Wu, L. N. Lisetski, and W. Lee, "Suppressed ionic effect and low-frequency texture transitions in a cholesteric liquid crystal doped with graphene nanoplatelets," *Optics Express*, vol. 23, no. 9, pp. 11195–11204, 2015.

[57] Y. Garbovskiy, "Switching between purification and contamination regimes governed by the ionic purity of nanoparticles dispersed in liquid crystals," *Applied Physics Letters*, vol. 108, no. 12, 2016.

[58] Y. Garbovskiy, "Impact of contaminated nanoparticles on the non-monotonous change in the concentration of mobile ions in liquid crystals," *Liquid Crystals*, vol. 43, no. 5, pp. 664–670, 2016.

[59] Y. Garbovskiy, "Electrical properties of liquid crystal nanocolloids analysed from perspectives of the ionic purity of nanodopants," *Liquid Crystals*, vol. 43, no. 5, pp. 648–653, 2016.

[60] Y. Garbovskiy, "Adsorption of ions onto nanosolids dispersed in liquid crystals: Towards understanding the ion trapping effect in nanocolloids," *Chemical Physics Letters*, vol. 651, pp. 144–147, 2016.

[61] Y. Garbovskiy, "Ion capturing/ion releasing films and nanoparticles in liquid crystal devices," *Applied Physics Letters*, vol. 110, no. 4, 2017.

[62] Y. Garbovskiy, "Kinetics of Ion-Capturing/Ion-Releasing Processes in Liquid Crystal Devices Utilizing Contaminated Nanoparticles and Alignment Films," *Nanomaterials*, vol. 8, no. 2, p. 59, 2018.

[63] Y. Garbovskiy, "Adsorption/desorption of ions in liquid crystal nanocolloids: the applicability of the Langmuir isotherm, impact of high electric fields and effects of the nanoparticle's size," *Liquid Crystals*, vol. 43, no. 6, pp. 853–860, 2016.

[64] Y. Garbovskiy, "The purification and contamination of liquid crystals by means of nanoparticles. The case of weakly ionized species," *Chemical Physics Letters*, vol. 658, pp. 331–335, 2016.

[65] Y. Garbovskiy, "Ions and size effects in nanoparticle/liquid crystal colloids sandwiched between two substrates. The case of two types of fully ionized species," *Chemical Physics Letters*, vol. 679, pp. 77–85, 2017.

[66] G. Barbero and L. R. Evangelista, *Adsorption Phenomena and Anchoring Energy in Nematic Liquid Crystals*, Taylor & Francis, Boca Raton, FL, USA, 2006.

[67] V. Steffen, L. Cardozo-Filho, E. A. Silva, L. R. Evangelista, R. Guirardello, and M. R. Mafra, "Equilibrium modeling of ion adsorption based on Poisson-Boltzmann equation," *Colloids and Surfaces A: Physicochemical and Engineering Aspects*, vol. 468, pp. 159–166, 2015.

[68] F. Batalioto, A. M. Figueiredo Neto, and G. Barbero, "Ion trapping on silica nanoparticles: Effect on the ζ-potential," *Journal of Applied Physics*, vol. 122, no. 16, 2017.

[69] V. Steffen, E. Silva, L. Evangelista, and L. Cardozo-Filho, "Debye–Hückel approximation for simplification of ions adsorption equilibrium model based on Poisson–Boltzmann equation," *Surfaces and Interfaces*, vol. 10, pp. 144–148, 2018.

[70] Y. Garbovskiy, "Nanoparticle enabled thermal control of ions in liquid crystals," *Liquid Crystals*, vol. 44, no. 6, pp. 948–955, 2017.

[71] Y. Garbovskiy, "Ions in liquid crystals doped with nanoparticles: conventional and counterintuitive temperature effects," *Liquid Crystals*, vol. 44, no. 9, pp. 1402–1408, 2017.

[72] F. Wang, R. Tang, J. L.-F. Kao, S. D. Dingman, and W. E. Buhro, "Spectroscopic identification of tri-n-octylphosphine oxide (TOPO) impurities and elucidation of their roles in cadmium selenide quantum-wire growth," *Journal of the American Chemical Society*, vol. 131, no. 13, pp. 4983–4994, 2009.

19

Electrically Tunable Diffraction Grating based on Liquid Crystals

Chuen-Lin Tien ⓘ,[1,2] Rong-Ji Lin,[2] Shu-Hui Su,[2] and Chi-Ting Horng ⓘ[3,4]

[1]Department of Electrical Engineering, Feng Chia University, Taichung 40724, Taiwan
[2]Ph.D. Program of Electrical and Communications Engineering, Feng Chia University, Taichung 40724, Taiwan
[3]Department of Ophthalmology, Fooyin University Hospital, Pingtung, Taiwan
[4]Department of Pharmacy, Tajen University, Pingtung, Taiwan

Correspondence should be addressed to Chi-Ting Horng; chitinghorng@gmail.com

Academic Editor: Daniel Ho

A periodic electric field is generated in the grating-like electrodes cell by an applied voltage and results in the reorientation of liquid crystals. The linearly polarized probe beam experienced periodic distribution of refractive index and formed a phase grating. He-Ne laser was used as the probe beam to detect the zeroth-order (o) and first-order (+1) diffraction intensities. The experimental results showed that the diffraction grating can be switched on by applying a small voltage. The optimal first-order diffraction efficiency is about 12%. The dependence of the first-order diffraction efficiency on the polarization of the probed beam is also discussed herein.

1. Introduction

Liquid crystal (LC) phase grating has numerous applications, such as laser beam steering, beam shaping, fiber-optic communications, highly efficient projection displays, wide-viewing and direct-view displays, and other modifications of light intensity or phase [1–9].

For nematic liquid crystals (NLCs) based phase grating, because the operation voltage changing the orientation of the molecular director is usually lower, depending on the dielectric anisotropy ($\Delta\varepsilon$) of LCs, an effective refractive index (Δn) of the medium is experienced by a propagating beam and has become an ideal candidate as an electrically controllable device due to the large optical birefringence [3, 10–15].

There are several ways to fabricate the liquid crystals diffraction gratings. Gibbons et al. have shown liquid crystal grating by exposing a dye-doped polymer layer to an ultraviolet interference pattern. Chen et al. have shown a hybrid liquid crystal configuration to create a double-rubbed polyimide layer for polarization-independent grating [16–20]. Honma et al. have demonstrated multidomain alignment regions liquid crystal gratings through a microrubbing technique [12, 13]. Wen et al. have shown a dual-domain

polarization grating which consists of right- and left-handed twisted regions created from scribing a polyimide layer [5].

In this paper, we propose a simple phase grating with tunable diffraction efficiency by applying a small voltage. The grating-like patterning of the electric field for the alignment of liquid crystals is easy to achieve and results in the periodic distribution of refractive index. The physical mechanisms for the first-order diffraction efficiency versus polarization angles of the probed beam are also discussed.

2. Experiment

We studied the liquid crystals phase grating in 25 μm-thick LC cell containing homogenously aligned E7. The cell is with grating-like electrode width $w = 15\ \mu$m. An LC mixture (E7) with $\Delta\varepsilon = 14.4$ (dielectric constants at 1kHz) and $\Delta n=0.218$ (refractive indices difference at λ=633 nm and T=20°C) was injected into the cell via capillary flow. Figure 1(a) shows the experimental setup for characterizing the liquid crystals phase grating. He-Ne laser ($\lambda = 633$ nm) was used as the probe beam. The transmission axis of the linear polarizer was set in the x-direction. An AC power supply (~1 KHz) applied voltage to the grating-like electrode cell and produced the periodic electric field on the cell and resulted in the periodic

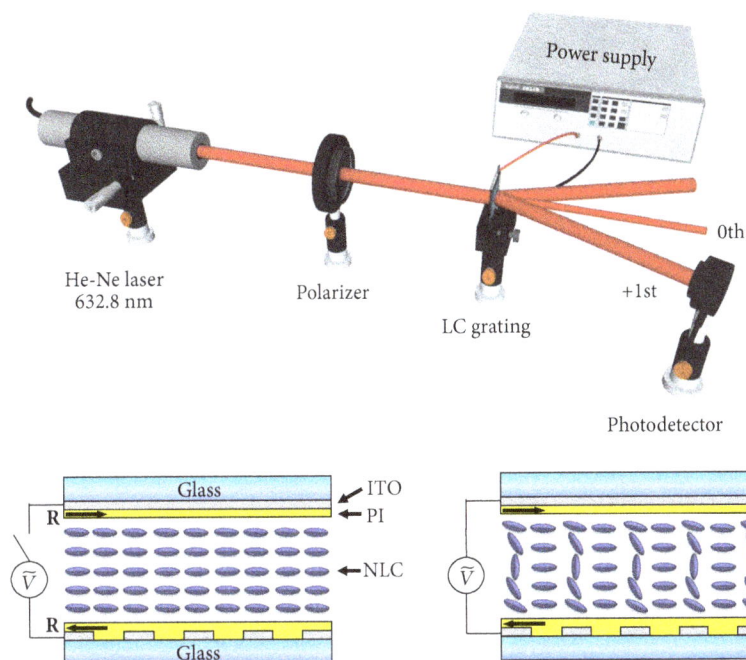

FIGURE 1: (a) Experimental setup for characterizing the performance of the phase grating. (b) The periodic distribution of refractive index of liquid crystals with an external *ac* voltage.

distribution of the refractive index of liquid crystals, as shown in Figure 1(b). The measured diffraction signals for the zeroth and the first order were detected by the photodetectors and recorded by the computer.

3. Results and Discussions

First, the diffraction intensities of the liquid crystal gratings in the zeroth and the first orders were recorded as a function of time, with an applied voltage of ~4.0V, as shown in Figure 2. The x- linearly polarized light from a He-Ne laser light source is normally incident on the sample; the zeroth-order diffraction rapidly decayed and the first-order diffraction was raised as the voltage was in "on" state. The diffraction intensities rise or decay to a stable level at t=250 ms. Next, the voltage was in "off" state at t=3250 ms; the zeroth-order diffraction was gradually increased and the first-order diffraction was gradually raised to an initial state at t=~3800 ms.

Figure 3(a) plots the first-order diffraction intensity versus voltage for a liquid crystal phase grating formed from the periodic grating-like electrodes cell. The threshold voltage occurred at 2.0 V. After 2.0 V, the diffraction intensity gradually grew. The shape of the first-order diffraction versus voltage curve was indicated; as the voltage is increased to 4.0 V, the liquid crystal director orientation may tend to be homeotropically aligned in the electrical field, and the linearly polarized probe beam in the x-direction experienced the difference in refractive index modulation between electrode and nonelectrode stripes zones, which was significant. From 4.0 V to 10.0 V, the diffraction intensity gradually decays. The phenomenon is speculated to be the result of the disordering of the alignment of liquid crystals in

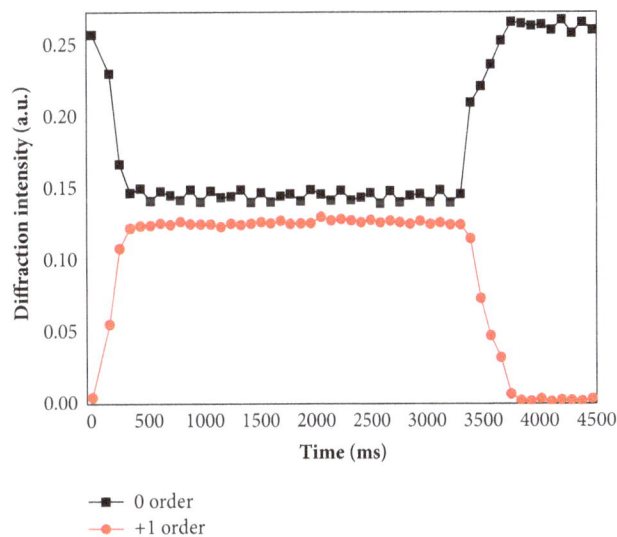

FIGURE 2: The diffraction intensities of the liquid crystal gratings in zeroth and the first orders were recorded as a function of time, using an applied voltage of ~4.0V.

the nonelectrode domains due to the edge effect of the electric field, reducing the difference in the refractive index modulation. After 10.0V, the first-order diffraction decays to a stable level. Figure 3(b) shows the images of the diffraction signals taken at 0, 2, 4, 10, and 20 V. At 0 V, only the zeroth-order diffraction signal exists; at 2 V, the first-order diffraction appears and the high order diffraction signals (4th-order diffraction) can be observed; at 4 V, the first-order diffraction becomes the brightest and the 6th-order diffraction signals can be observed; at 10.0 V and 20.0V, the

(a)   (b)

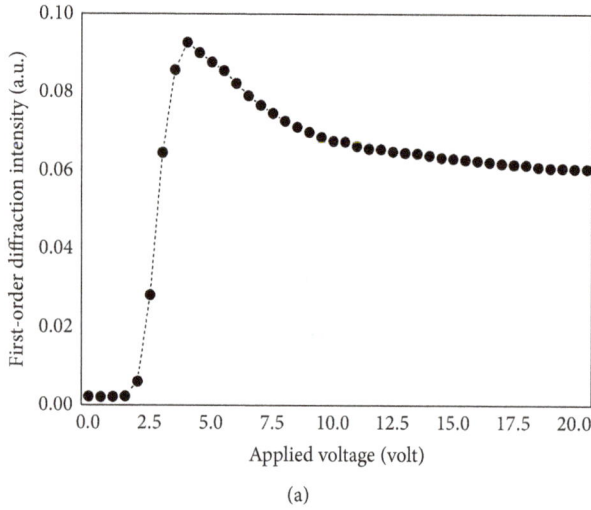

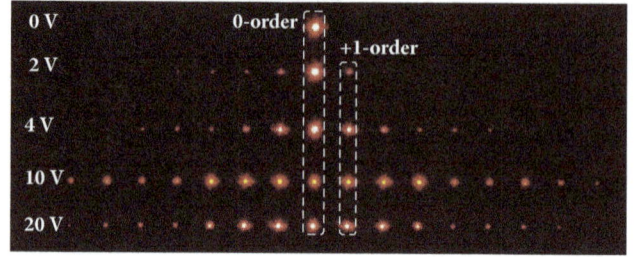

FIGURE 3: (a) The first-order diffraction intensity versus voltage for a liquid crystal phase grating formed from the periodic grating-like electrodes cell. (b) Diffraction patterns at different applied voltages: 0, 2, 4, 10, and 20 V, respectively.

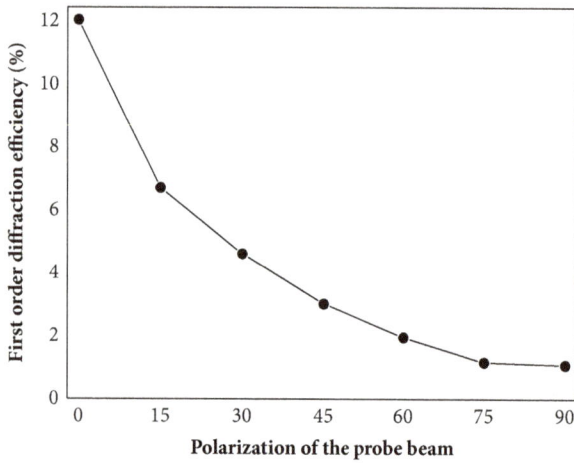

FIGURE 4: The diffraction efficiency of the first-order beams versus polarization of the probe beam with applied voltage ~4.0V.

first-order diffraction signal decays, and 8th-order diffraction signals can be observed.

Figure 4 shows the measured diffraction efficiency of the first-order beams with various polarization of the probe beam with applied voltage ~4.0V. The first-order diffraction efficiency η_1 is defined as the ratio between the intensity of 1st diffracted order and the total intensity at $V = 0$, described by

$$\eta_1 = \frac{I_1}{I_0} \qquad (1)$$

The polarization of the probed beam is adjusted by rotating a polarizer from $0°$ to $90°$ corresponding to the adjustment of the polarization of the probe beam from x polarized to y polarized. The intensity of the probe beam maintains a constant value as the polarization is adjusted. The polarization angles of the probed beam are $0°$, $15°$, $30°$, $45°$, $60°$, $75°$, and $90°$, respectively. When the polarization angle is at $0°$, the

probe beam experiences ordinary (n_o) and extraordinary (n_e) refractive indices in the regions with and without the grating-like electrode stripes of the liquid crystal grating, respectively. The modulation of the refractive indices $\Delta n = ne - no$ causes the phase grating effect in liquid crystals. The first diffraction efficiency η_1 is ~12.0%. In considering the polarization angle other than the x polarization ($0°$), the effective refractive index experienced by the probed beam is a function of the polarization angle, according to [21]

$$n_{\text{eff}}(\theta) = \left(\frac{n_o n_e}{n_o^2 \cos^2\theta} + n_e^2 \sin^2\theta \right)^{1/2} \qquad (2)$$

where θ is the angle between the x-axis and the polarization of the probe beam. As the polarization angle is adjusted from $0°$ to $90°$, the effective refractive index n_{eff} experienced by the beam in the regions without the electrode stripe is reduced; the refractive index is constant in the regions with the electrode stripe. The modulation of the refractive index influences the phase grating. Therefore, the decrease in the anisotropy of the refractive index weakens the phase grating effect. The phase grating effect gradually decays as the polarization angle is increased. When the probe beam is linearly polarized in the y direction ($90°$) it experiences almost the same ordinary refractive index in the regions with and without the electrode stripe. The first diffraction efficiency η_1 is ~1.1%. The results reveal that the phase grating effect is directly influenced by the polarization of the probe beam and can be switchable by adjusting the polarization of the probe beam.

4. Conclusions

In summary, the reorientation of liquid crystals can be obtained by applying external voltage in the grating-like electrodes cell; the linearly polarized probe beam experienced periodic distribution of refractive index, resulting in the liquid crystal phase grating. The zeroth-order and the first-order diffraction intensities were probed by a He-Ne laser. The diffraction grating can be switched on by applying a small voltage (~2.0 V). The optimal first-order diffraction efficiency is about 12%. The first-order diffraction efficiency can also be tuned from 12.0% to 1.1% by adjusting the polarization of the probe beam.

Conflicts of Interest

The authors declare that they have no conflicts of interest.

Acknowledgments

This work was supported by the Ministry of Science and Technology (MOST) of Taiwan under Grant MOST 106-2221-E-035-072-MY2. The authors also appreciate the Precision Instrument Support Center of Feng Chia University for providing the measurement facilities.

References

[1] D. J. Field, "Relations between the statistics of natural images and the response properties of cortical cells," *Journal of the Optical Society of America A: Optics, Image Science & Vision*, vol. 4, no. 12, p. 2379, 1987.

[2] H. Murai, "Electro-optic properties of liquid crystal phase gratings and their simulation using a homogeneous alignment model," *Liquid Crystals*, vol. 15, pp. 627–642, 1993.

[3] J. Chen, P. J. Bos, H. Vithana, and D. L. Johnson, "Electro-optically controlled liquid crystal diffraction grating," *Applied Physics Letters*, vol. 67, no. 18, pp. 2588–2590, 1995.

[4] J. N. Eakin, Y. Xie, R. A. Pelcovits, M. D. Radcliffe, and G. P. Crawford, "Zero voltage Freedericksz transition in periodically aligned liquid crystals," *Applied Physics Letters*, vol. 85, no. 10, pp. 1671–1673, 2004.

[5] B. Wen, R. G. Petschek, and C. Rosenblatt, "Nematic liquid-crystal polarization gratings by modification of surface alignment," *Applied Optics*, vol. 41, no. 7, p. 1246, 2002.

[6] L. Komitov, G. P. Bryan-Brown, E. L. Wood, and A. B. Smout, "Alignment of cholesteric liquid crystals using periodic anchoring," *Journal of Applied Physics*, vol. 86, no. 7, pp. 3508–3511, 1999.

[7] C. Rosenblatt, "Nanostructured surfaces: scientific and optical device applications," *Molecular Crystals and Liquid Crystals*, vol. 412, no. 1, pp. 117–134, 2004.

[8] T. Shioda, B. Wen, and C. Rosenblatt, "Step-wise Fréedericksz transition in a nematic liquid crystal," *Journal of Applied Physics*, vol. 94, no. 12, p. 7502, 2003.

[9] G. P. Sinha, C. Rosenblatt, and L. V. Mirantsev, "Disruption of surface-induced smectic order by periodic surface corrugations," *Physical Review E: Statistical, Nonlinear, and Soft Matter Physics*, vol. 65, no. 4, Article ID 041718, 2002.

[10] S. Varghese, G. P. Crawford, C. W. Bastiaansen, D. K. de Boer, and D. J. Broer, "Microrubbing technique to produce high pretilt multidomain liquid crystal alignment," *Applied Physics Letters*, vol. 85, no. 2, pp. 230–232, 2004.

[11] S. Varghese, S. Narayanankutty, C. W. M. Bastiaansen, G. P. Crawford, and D. J. Broer, "Patterned alignment of liquid crystals by μ-rubbing," *Advanced Material*, vol. 16, no. 18, pp. 1600–1605, 2004.

[12] M. Honma and T. Nose, "Liquid-Crystal Reflective Beam Deflector with Microscale Alignment Pattern," *Japanese Journal of Applied Physics*, vol. 43, no. 12, pp. 8151–8155, 2004.

[13] M. Honma, K. Yamamoto, and T. Nose, "Periodic reverse-twist nematic domains obtained by microrubbing patterns," *Journal of Applied Physics*, vol. 96, no. 10, pp. 5415–5419, 2004.

[14] D. J. Versteeg, C. W. M. Bastiaansen, and D. J. Broer, "Influence of laser writing of polyimides on the alignment of liquid crystals," *Journal of Applied Physics*, vol. 91, no. 7, pp. 4191–4195, 2002.

[15] M. Schadt, H. Seiberle, and A. Schuster, "Optical patterning of multi-domain liquid-crystal displays with wide viewing angles," *Nature*, vol. 381, pp. 212–215, 1996.

[16] W. M. Gibbons, P. J. Shannon, S. T. Sun, and B. J. Swetlin, "Surface-mediated alignment of nematic liquid crystals with polarized laser light," *Nature*, vol. 351, pp. 49-50, 1991.

[17] B. W. Lee and N. A. Clark, "Alignment of liquid crystals with patterned isotropic surfaces," *Science*, vol. 291, p. 2576, 2001.

[18] L. Nikolova and T. Todorov, "Diffraction efficiency and selectivity of polarization holographic recording," *Optica Acta: International Journal of Optics*, vol. 31, no. 5, p. 579, 1984.

[19] S. G. Cloutier, D. A. Peyrot, T. V. Galstian, and R. A. Lessard, "Measurement of permanent vectorial photoinduced anisotropy in azo-dye-doped photoresist using polarization holography," *Journal of Optics A: Pure and Applied Optics*, vol. 4, p. S228, 2002.

[20] P. S. Ramanujam, "Evanescent polarization holographic recording of sub-200-nm gratings in an azobenzene polyester," *Optics Letters*, vol. 28, p. 2375, 2003.

[21] C. Kuo, S. Huang, I. Jiang, and M. Tsai, "Multiguide directional coupler using switchable liquid-crystalline optical channels," *Journal of Applied Physics*, vol. 97, no. 10, p. 103113, 2005.

20

Voltage-Controllable Guided Propagation in Nematic Liquid Crystals

Hsin-Yu Yao[1] **and Shang-Min Yeh** (iD)[2]

[1]*Department of Ophthalmology, Kaohsiung Armed Forces General Hospital, Kaohsiung 802, Taiwan*
[2]*Department of Optometry, Central Taiwan University of Science and Technology, Taichung 40601, Taiwan*

Correspondence should be addressed to Shang-Min Yeh; optom.yap@gmail.com

Academic Editor: Jia-De Lin

Voltage-controllable guided channels are formed in a planar nematic liquid crystals cell. The director of liquid crystals can be aligned by applying external voltage, which results in a difference of the refractive index between two adjacent channels; therefore, the incidence beam can be coupled from one channel to another. First, we discussed the propagation of the beam and the self-focusing in a single channel; then we discussed the propagation of the beam and the coupling effect in the two channels. The results showed that the propagation of the beam can be selected in each channel by applying voltages in the two individual electrode channels.

1. Introduction

Optical waveguide elements play an important role in the applications of optical communication, optical signal processing, integrated optical circuits, and optical networking [1, 2]. Many devices have been designed for the split, combine, couple, and phase modulation of the optical signal systems [1, 3]. Controlling the path of the propagating beam is the primary aim of the optical switching system, and the optical signals can be transferred to different guided channels. Somekh *et al.* proposed the concept of optical waveguide arrays [3], and the potential for optical switching applications has attracted much attention. Haus et al. theoretically predicted that the optical signal switch can be achieved by the external control of the waveguide arrays [4, 5]. Christodoulides et al. predicted the existence of solitary waves in arrays, and such unique properties have developed all-optical signal processing [6]. Gia Russo et al. investigated guided light and found the directional characteristics in the layered crystalline media [7, 8]. Channin et al. studied the waveguide characteristics and discussed the voltage-addressable optical properties in a liquid-crystal medium [9, 10]. Aligned nematic liquid crystals are a good choice for large changes in optical properties, which can be easily driven by applying external voltage. Tsai et al. investigated a multiguided directional coupler based on planar-aligned nematic liquid crystals and discussed the dependence of the coupling effect on the external voltage, the polarization of the incident beam, and the temperature [11, 12].

This work discusses guided light in a single channel and two channels, where the propagation of the beam can be selected in each channel by applying different voltage in the individual electrode channels. In addition, self-focusing and the coupling effect are discussed.

2. Preparation of Sample and Experimental Setup

The nematic liquid crystal (NLC) in this experiment is E7 (n_e = 1.7462 and n_o = 1.5216 at 20°C for λ = 589 nm; nematic phase ranges = −10 to 60.5°C, from Merck). An empty cell is constructed with two indium-tin-oxide (ITO) coated glass plates. One of the two plates is etched with a two-stripe ITO pattern as the upper electrode, and the other plate is used as the grounding electrode. The spacing between the etched region and the nonetched region is 15 μm. These two plates are coated with polyimide film and rubbed parallel to the ITO electrode stripes (z-axis). The glass slides are separated by two

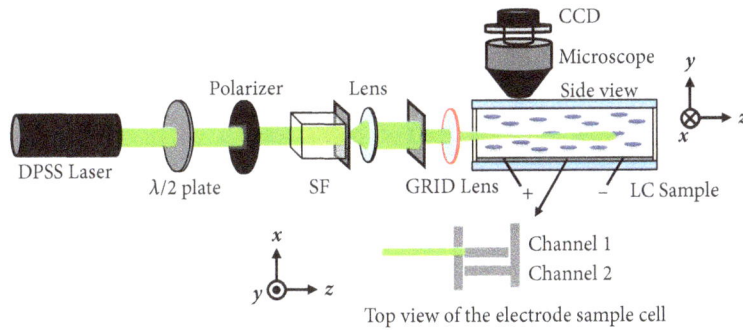

FIGURE 1: Experimental setup of voltage-controllable guided propagation in a nematic liquid crystals coupler.

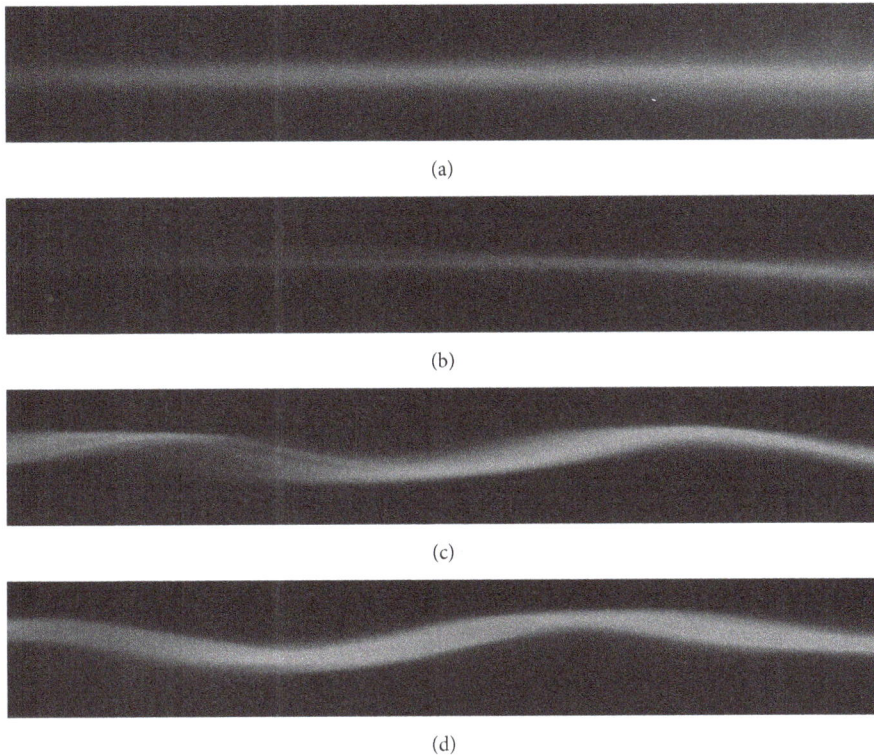

(a)

(b)

(c)

(d)

FIGURE 2: The guided light in a single channel by applying external voltages of (a) 0 V, (b) 1.5 V, (c) 2.2 V, and (d) 3.0 V.

25 μm thick plastic spacers, and the NLCs are injected into the empty cell to form a directional coupler.

Figure 1 presents voltage-controllable guided propagation in a nematic liquid crystals coupler. A linearly polarized (along the x-axis) beam of the 532 nm diode-pump solid-state laser (DPSS) impinges normally onto the side of the sample cell, which is focused on the cross-sectional region of the NLCs medium within the striped electrode of channel 1 using a gradium singlet lens (GRID lens). The focal point is around 3.0 mm from the surface of the side glass, and the focus spot has a diameter of about 3.2 μm. A half-wave plate ($\lambda/2$ WP, for 532 nm) and a polarizer are inserted between the DPSS laser and the spatial filter (SF) in order to change the direction of polarization and the intensity of the incident beam. By applying an external voltage on the sample cell, the NLC molecules are reoriented, which results in the distribution of the refractive index in the medium, and the

optical channels are formed under a single or two electrode stripes.

Guided propagation, self-focusing, and the coupling effect are discussed in three different situations: guided light propagation (I) in a single channel, (II) in two channels by applying equal external voltages in each of the two electrode stripes, and (III) in two channels by applying the different external voltages in each of the two electrode stripes.

3. Results and Discussions

Figure 2 presents a guided light in a single channel by applying various external voltages. A laser beam, which is linearly polarized in the x-direction, is introduced into the channel from the left and propagates along the z-axis. With the applied voltage of $V_{app} = 0$ V, the guided light diverges after propagating distance $z = 25$ μm, as shown in Figure 2(a).

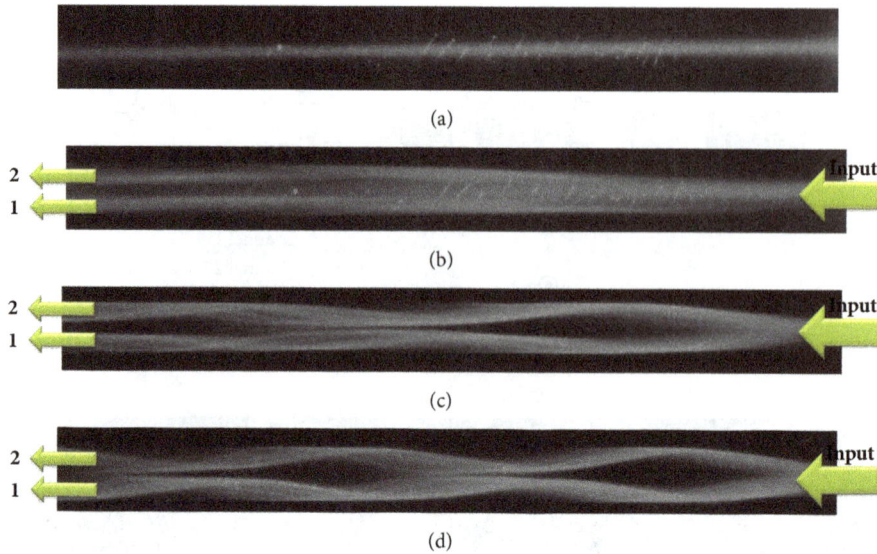

FIGURE 3: The guided light propagation by applying the same external voltages of (a) 0 V, (b) 1.5 V, (c) 2.2 V, and (d) 2.5 V in the two channels.

Initially, the liquid crystal molecules are horizontally aligned to the z-axis, while the x-direction polarized beam encounters the ordinary uniform refractive index distribution in the medium. With the increased applied voltage of $V_{app} = 1.5$ V, the guided light propagates in the channel, as shown in Figure 2(b). The liquid crystal molecules tend to align in the x-axis, while the x-direction polarized beam sees the extraordinary and ordinary refractive index of the liquid crystals in the guided channel and outside the channel, respectively, thus forming a waveguide-like structure. With the applied voltages of $V_{app} = 2.2$ V and 3.0 V, the self-focusing of the propagated beam is observed, as shown in Figures 2(c) and 2(d). The liquid crystal molecules around the channel can be reoriented due to the edge effect of the electric field, which forms an approximately gradient distribution of the refractive index around the channel in the medium.

Figure 3 presents the guided light propagation by applying the same external voltages of $V_{app} = 0$ V, 1.5 V, 2.2 V, and 2.5 V in the two channels. Initially, the incident beam is introduced into channel 1 with $V_{app} = 0$ V, as shown in Figure 3(a). With the applied voltage of $V_{app} = 1.5$ V, a part of the guided light gradually trends to channel 2, while part of the guided light still propagates in channel 1, as shown in Figure 3(b). Waveguide-like structures are formed, which results in total reflection in both channel 1 and channel 2. With $V_{app} = 2.2$ V and 2.5 V, each of the two beams travels back and forth in the vicinity of its channel, as shown in Figures 3(c) and 3(d). An approximate gradient distribution of the refractive index around the two channels is formed, which results in self-focusing in each channel.

Figure 4 presents the guided light propagation by applying fixed voltage V_{app} (1) = 2.2 V in channel 1 and applying voltage V_{app} (2) = 0 V, 1.5 V, 2.2 V, and 2.8 V in channel 2, which correspond to Figures 4(a)–4(d), respectively. Initially,

the incident beam is introduced into channel 1, and self-focusing is observed, as shown in Figure 4(a). At V_{app} (2) = 1.5 V (i.e., V_{app} (1) is larger than V_{app} (2)), a small part of the guided light gradually couples to channel 2, as shown in Figure 4(b). In the condition of V_{app} (2) = 2.2 V (i.e., V_{app} (1) is equal to V_{app} (2)), the coupling effect is clearly observed, and the propagation behavior seems to be the same in both channel 1 and channel 2, as shown in Figure 4(c). At V_{app} (2) = 2.8 V (i.e., V_{app} (1) is smaller than V_{app} (2)), a large part of the guided light gradually couples to channel 2, as shown in Figure 4(d). The results show that the propagation of the incident beam can be easily tuned in channel 1 or in channel 2 due to the distribution of the refractive index when applying external voltage.

4. Conclusions

In summary, voltage-controllable guided channels are formed in a planar nematic liquid crystals cell. Nematic liquid crystal molecules can be easily reoriented by applying external voltage, which results in the difference of the refractive index between two neighboring channels. Therefore, the incidence beam can be coupled from one channel to another.

First, we discussed the propagation properties of the beam in a single channel. The self-focusing of the light can be observed at $V_{app} = 2.2$ V and 3.0 V. Then we discussed the guided light propagation by applying the voltages in the two channels; the beam coupled from one channel to another can be observed. The results show that the propagation of the incident beam can be easily tuned in each of the two channels, due to the distribution formation of the refractive index by applying external voltage.

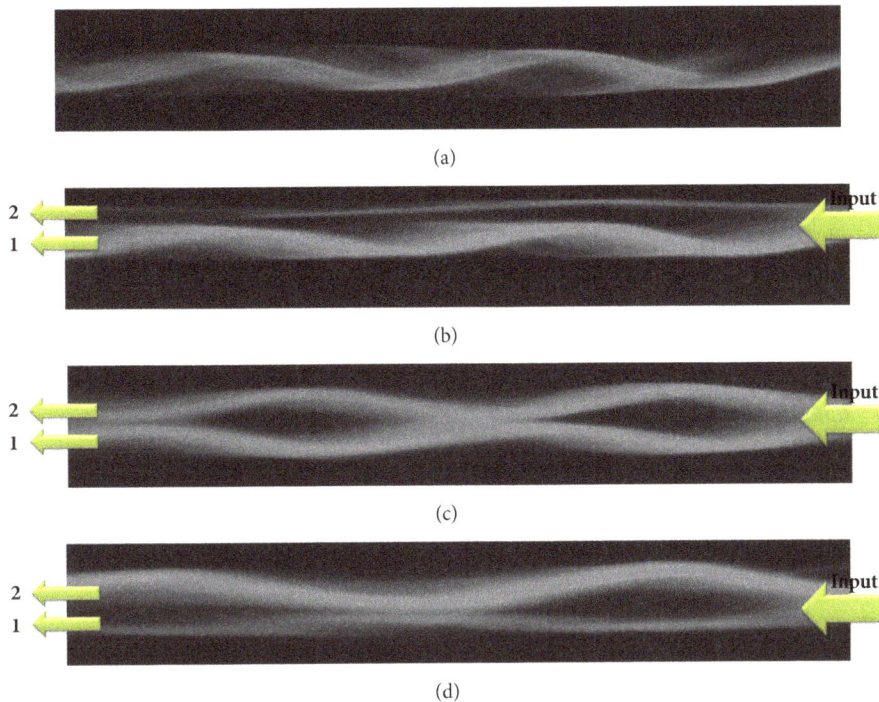

FIGURE 4: The guided light propagation by applying the fixed voltage of V_{app} (1) = 2.2 V in channel 1 and applying voltages of V_{app} (2) = 0 V, 1.5 V, 2.2 V, and 2.8 V in channel 2, which correspond to (a)–(d), respectively.

Conflicts of Interest

The authors declare that they have no conflicts of interest.

Acknowledgments

This study was financially supported by Clinical Research grants from Kaohsiung Armed Forces General Hospital, Taiwan (no. 106-17).

References

[1] B. E. A. Saleh and M. C. Teich, *Fundamentals of Photonics*, Wiley, 2nd edition, 2007.

[2] L. Wang, Y. Wang, and X. Zhang, "Embedded metallic focus grating for silicon nitride waveguide with enhanced coupling and directive radiation," *Optics Express*, vol. 20, no. 16, pp. 17509–17521, 2012.

[3] S. Somekh, E. Garmire, A. Yariv, H. L. Garvin, and R. G. Hunsperger, "Channel optical waveguide directional couplers," *Applied Physics Letters*, vol. 22, no. 1, pp. 46-47, 1973.

[4] S. F. Su, L. Jou, and J. Lenart, "A Review on Classification of Optical Switching Systems," *IEEE Communications Magazine*, vol. 24, no. 5, pp. 50–55, 1986.

[5] H. A. Haus and L. Molter-Orr, "Coupled Multiple Waveguide Systems," *IEEE Journal of Quantum Electronics*, vol. 19, no. 5, pp. 840–844, 1983.

[6] D. N. Christodoulides and R. I. Joseph, "Discrete self-focusing in nonlinear arrays of coupled waveguides," *Optics Expresss*, vol. 13, no. 9, pp. 794–796, 1988.

[7] D. P. Gia Russo and J. H. Harris, "Wave propagation in anisotropic thin-film optical waveguides," *Journal of the Optical Society of America*, vol. 63, no. 2, pp. 138–145, 1973.

[8] W. K. Burns and J. Warner, "Mode dispersion in uniaxial optical waveguides*," *Journal of the Optical Society of America*, vol. 64, no. 4, p. 441, 1974.

[9] D. J. Channin, "Optical waveguide modulation using nematic liquid crystal," *Applied Physics Letters*, vol. 22, no. 8, pp. 365-366, 1973.

[10] J. P. Sheridan, J. M. Schnur, and T. G. Giallorenzi, "Electro-optic switching in low-loss liquid-crystal waveguides," *Applied Physics Letters*, vol. 22, no. 11, pp. 560-561, 1973.

[11] M.-S. Tsai, C.-T. Kuo, S.-Y. Huang, C.-C. Shih, and I.-M. Jiang, "Voltage-controlled multiguide directional coupler formed in a planar nematic liquid crystal film," *Applied Physics Letters*, vol. 85, no. 6, pp. 855–857, 2004.

[12] C. Kuo, S. Huang, I. Jiang, and M. Tsai, "Multiguide directional coupler using switchable liquid-crystalline optical channels," *Journal of Applied Physics*, vol. 97, no. 10, pp. 103–113, 2005.

21

Enhanced Photoluminescence in Gold Nanoparticles Doped Homogeneous Planar Nematic Liquid Crystals

Chi-Huang Chang,[1,2] Rong-Ji Lin,[3,4] Chuen-Lin Tien (iD),[3,5] and Shang-Min Yeh (iD)[6]

[1]*Department of Ophthalmology, Chung Shan Medical University Hospital, Taichung 402, Taiwan*
[2]*Department of Medicine, Chung Shan Medical University, Taichung 402, Taiwan*
[3]*Ph.D. Program of Electrical and Communications Engineering, Feng Chia University, Taichung 40724, Taiwan*
[4]*Department of Optometry, Da-Yeh University, Changhua 515, Taiwan*
[5]*Department of Electrical Engineering, Feng Chia University, Taichung 40724, Taiwan*
[6]*Department of Optometry, Central Taiwan University of Science and Technology, Taichung 40601, Taiwan*

Correspondence should be addressed to Shang-Min Yeh; optom.yap@gmail.com

Academic Editor: Daniel Ho

This study reported the photoluminescence (PL) of gold nanoparticles (GNPs) doped planar nematic liquid crystals (NLCs) and observed around 64% enhancement in PL intensity with suitable doping amounts of GNPs in liquid crystals 5CB. The enhancement in PL intensity has been attributed to the increased surface area from GNPs, which results in increased emissions due to the increased scattering of excitation. The subsequent decay of PL intensity with doping more amounts of GNPs in liquid crystals 5CB was due to the aggregation of the GNPs, which resulted in decayed emissions due to the decay of the scattering of excitation. The concentration and the size of GNPs, as well as the orientation of the LCs' director, with respect to the excitation, which depend on the intensity of the PL, were also investigated.

1. Introduction

Research in the field of liquid crystals (LCs) has developed due to its imposing properties [1–4], especially, the physical and chemical properties, which are widely applied in liquid-crystal displays (LCDs). The main drawback of liquid-crystal displays (LCDs) is low brightness, which is due to the use of absorbing color filters and dichroic sheet polarizers. The luminescence of LCD is a possible method to improve the low brightness issue [5–7]. The problem of making a luminescent LCD is that light emissions in the visible region of the electromagnetic spectrum are lessened by using pure LC materials [8–10]. The enhanced luminescence of the LC materials may definitely realize the emissive LCDs [11–13]. Doping metal materials or metal nanoparticles (NPs) in LCs have attracted much attention due to the enhanced electro-optical properties of the doped LC materials. Palewska et al. investigated the influence of electric field on photoluminescence of lanthanide-doped nematic liquid crystals and obtained a highly resolved luminescence and luminescence

excitation spectra [14]. Kumar et al. reported the characterization and photoluminescence (PL) in gold nanoparticles doped ferroelectric liquid crystals and obtained enhancement in PL intensity [3]. Kuo et al. reported the enhancement of photoluminescence (PL) intensity of NLC doped with silver NPs [15]. Tanabe et al. reported the full-Color tunable photoluminescent ionic liquid crystals based on tripodal pyridinium, pyrimidinium, and quinolinium salts [16]. Lu et al. reported the electrically switchable photoluminescence of fluorescent-molecule-dispersed liquid crystals [17]. This study investigates photoluminescence (PL) in GNPs doped homogeneous planar NLCs and obtains about 64% enhancement in PL intensity with doping suitable amounts of GNPs in liquid crystals 5CB.

2. Preparation of Sample and Experimental Setup

The materials adopted in this work are nematic LCs (NLCs), 5CB (from Merck), and GNPs with diameters of 13 nm,

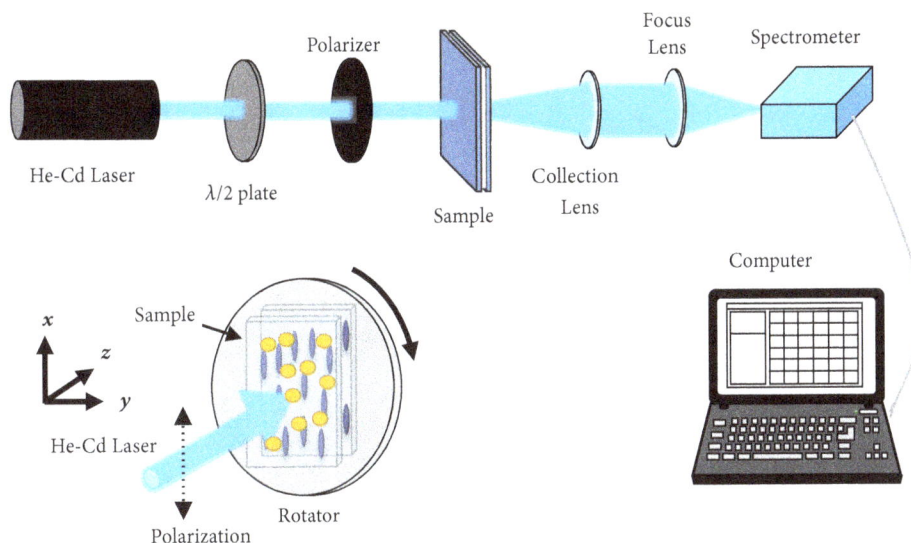

FIGURE 1: Experimental setup for measuring the PL spectra of the enhanced photoluminescence in GNPs doped homogeneous planar NLCs.

32 nm, and 56 nm (the concentrations of GNPs are $5*10^9$ particles/ml, $9*10^9$ particles/ml, $1.3*10^{10}$ particles/ml, $1.8*10^{10}$ particles/ml, and $2.5*10^{10}$ particles/ml). These materials are uniformly mixed and capillarity injected into the sample cell, which is assembled from two indium-tin-oxide-coated glass slides separated by two $5.4\mu m$-thick plastic spacers. The homogeneous planar alignment of the NLCs is accomplished unidirectionally by rubbing polyimide on the inner surfaces of the two glass substrates.

Figure 1 depicts the experimental setup for the investigation of the enhanced photoluminescence in GNPs doped homogeneous planar NLCs, where the CW helium-cadmium laser (wavelength: 325 nm, power: 1 mW) is focused on the sample cell. A half-wave plate ($\lambda/2$ for 325 nm) and a polarizer are placed in front of the sample cell to maintain the excitation power of 1 mw, the polarization of the excitation beam is fixed parallel to the x-axis, and the director of the NLCs is rotated every 15 degrees from x-axis (0 degree) to y-axis (90 degree). The spectra of photoluminescence (PL) are measured with a spectrometer and analyzed by computer.

3. Results and Discussions

Figure 2 shows the experimental absorption and fluorescence emission spectra of the pure NLCs cell (without doping GNPs). The absorption spectrum covered from 288 to 340 nm and the photoluminescence spectrum was recorded from 350 nm to 500 nm, respectively. The maxima of the absorption and fluorescence emissions are about 321 and 394 nm, respectively. The inset of Figure 2 shows the image of the fluorescence emission under the pumped helium-cadmium laser, which is operated at 325 nm.

Figure 3 shows the photoluminescence spectra of GNPs doped homogeneous planar NLCs, where the concentrations of GNPs are $5*10^9$ particles/ml, $9*10^9$ particles/ml, $1.3*10^{10}$ particles/ml, $1.8*10^{10}$ particles/ml, and $2.5*10^{10}$ particles/ml, respectively; and the gold nanoparticle size is 13 nm in

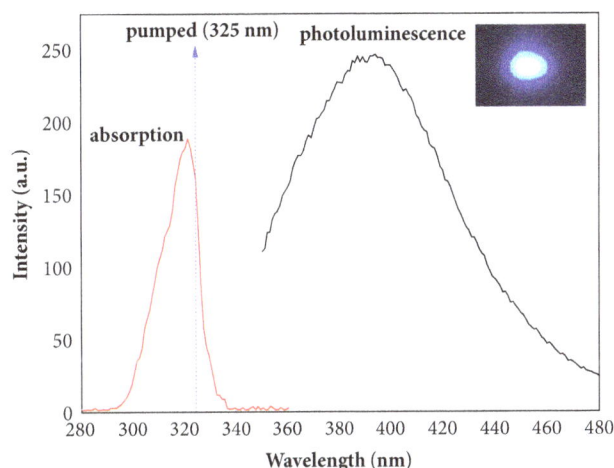

FIGURE 2: The experimental absorption and fluorescence emission spectra of the pure NLCs cell and the image of the fluorescence emission under the pumped helium-cadmium laser operated at 325 nm.

diameter. Figure 3(a) shows that the PL intensity gradually increased with the increased amount of GNPs per ml in NLCs. In the pure NLCs, the peak value of the PL intensity is around 250.4; the peak values of PL intensities are 303.5, 324.4, and 372.8, respectively, which correspond to doping GNPs with the concentrations of $5*10^9$ particles/ml, $9*10^9$ particles/ml, and $1.3*10^{10}$ particles/ml. The ratio between the maximum and minimum of peak intensity is ~1.49, meaning 49% enhancement in PL intensity with suitable concentrations of $1.3*10^{10}$ particles/ml.

Figure 3(b) shows that PL intensity gradually decayed with the increased amount of GNPs per ml in NLCs. The peak values of PL intensity are 372.8, 319.3, and 275.0, respectively, which correspond to doping GNPs with concentrations of $1.3*10^{10}$ particles/ml, $1.8*10^{10}$ particles/ml, and $2.5*10^{10}$ particles/ml.

(a)

(b)

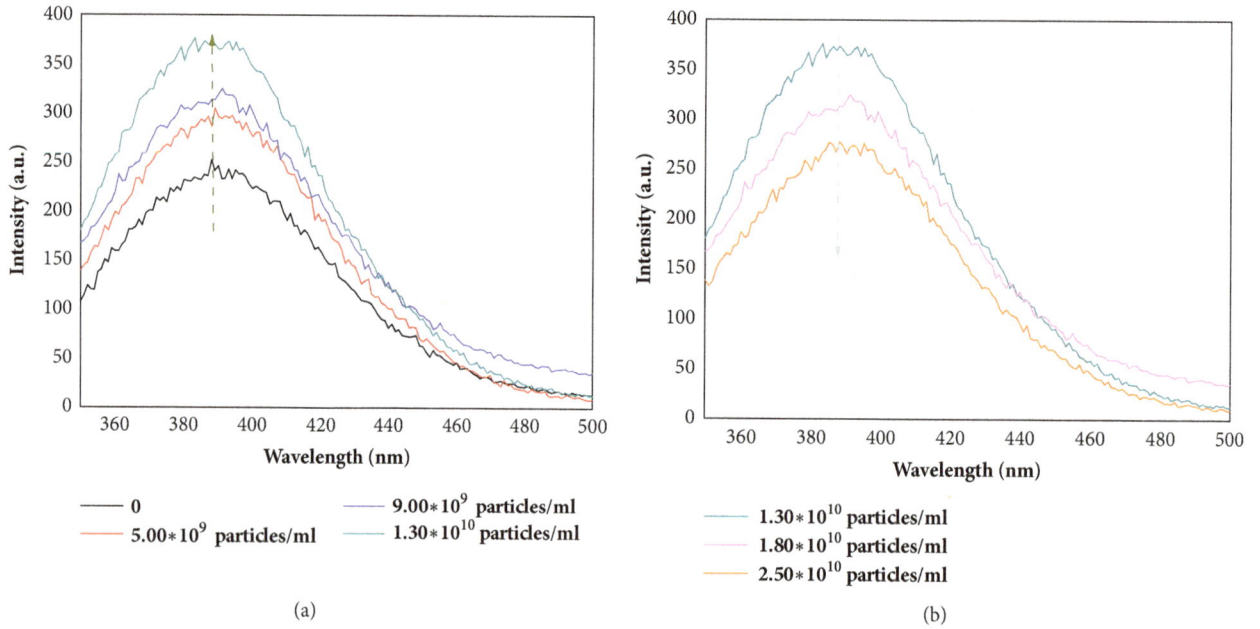

FIGURE 3: The photoluminescence spectra of GNPs doped homogeneous planar NLCs, where the concentrations of GNPs are (a) 0, $5*10^9$, $9*10^9$, and $1.3*10^{10}$ *particles/ml*, respectively, and (b) $1.3*10^{10}$, $1.8*10^{10}$, and $2.5*10^{10}$ *particles/ml*, respectively, and the gold nanoparticle size is 13 nm in diameter.

The enhancement of photoluminescence intensity is attributed to the increased surface area from the GNPs, which strengthens the multiple reflections of the excitation beam and results in the local surface plasmon resonance effect. This effect becomes more obvious with suitable doping concentrations of GNPs in the NLCs (NLCs). The decay of PL intensity is attributed to the aggregation of the GNPs, which cause the surface area to decrease, resulting in the reduction of the energy transfer effect between NLCs and GNPs, and this effect leads to the decrease of PL intensity.

Figure 4 shows the peak intensity of PL emissions versus concentrations of GNPs with diameter sizes of 13 nm, 32 nm, and 56 nm. The peak intensity of the PL emission gradually increases in the range of concentration from 0 (pure NLCs) to $1.3*10^{10}$ *particles/ml*. The fluorescence is enhanced by the scattering of the excitation beam due to the increased surface area of GNPs. At the same concentration of gold nanoparticles (lower than $1.3*10^{10}$ *particles/ml*), the peak intensity of PL emission is stronger for 13 nm gold particles. This is because the smaller nanoparticles have larger surface areas, which causes the multiple reflections and scattering of the excitation beam. As the concentration of GNPs is larger than $1.3*10^{10}$ *particles/ml*, the peak intensity of the PL emission for a diameter size of 13 nm gradually decreases; instead, it increases the PL emissions for 32 nm and 56 nm. The smaller nanoparticles are aggregated, which reduces the surface areas, and aggregation may disrupt the ordering state of the liquid-crystal alignment. The increasing PL emissions for 32 nm and 56 nm diameters are due the gradually increased surface areas, which result in multiple reflections and scattering of the excitation beam. For GNPs of 32 nm, the peak intensity of the PL emission gradually decays as the concentration of GNPs becomes larger than $1.8*10^{10}$ *particles/ml*. This is

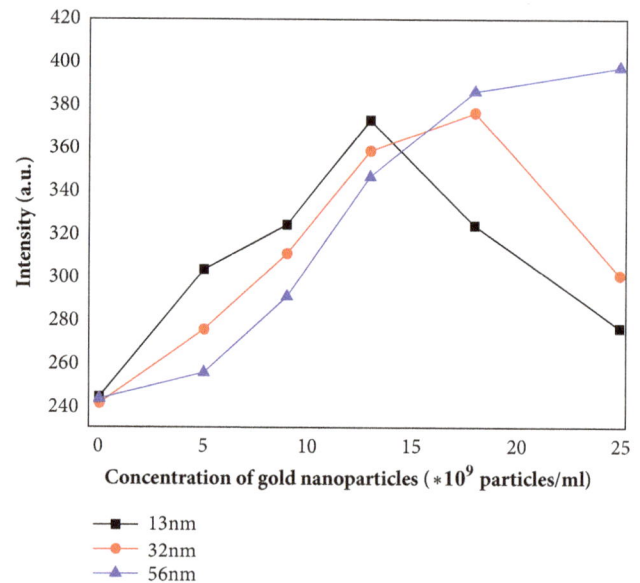

FIGURE 4: The peak intensity of PL emission versus concentrations of GNPs with diameter sizes of 13 nm, 32 nm, and 56 nm.

because the aggregation of the nanoparticles reduces PL intensity. For GNPs of 32 nm, the peak intensity gradually increases with the increased concentration of GNPs; thus, decreased PL emissions can be expected, as the concentration of nanoparticles is sufficient.

Figure 5 shows the peak intensity of the PL emission of pure NLCs and doping GNPs versus the rotational degree of NLCs' director, where the concentration of GNPs is $1.3*10^{10}$ *particles/ml*. The polarization of the excitation beam is fixed parallel to the x-axis, and the director of the NLCs is rotated

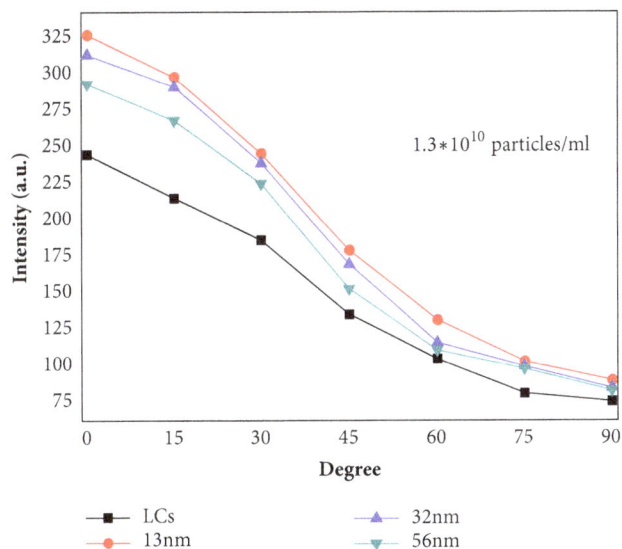

FIGURE 5: The peak intensity of PL emissions of pure NLCs and doping GNPs versus the rotational degree of NLCs' director, where the concentration of GNPs is $1.3*10^{10}$ particles/ml.

every 15 degrees from the x-axis (0 degree) to the y-axis (90 degree). All peak intensities for gold nanoparticles doping NLCs are stronger than pure NLCs', as the rotational degree is smaller than 45°. As the rotational degree gradually increases to 90°, all peak intensities decay to a similar value. The ratio between the absorption of NLCs in the x- and y-axis, denoted as $A_{0°}/A_{90°}$, is around 3.21, and the ratio between the PL intensity in the x- and y-axis, denoted as $PL_{0°}/PL_{90°}$, is around 3.32. These results show that the absorption of the liquid crystal critically dominates PL intensity. However, the $A_{0°}/A_{90°}$ is decreased; instead, the $PL_{0°}/PL_{90°}$ is increased by adding gold nanoparticles in the NLCs. This result shows that the doping of GNPs causes the reduction of the ordering state of NLCs, resulting in decreased absorption. As the added GNPs provide more scattering surface area, it results in the enhancement of PL intensity.

4. Conclusions

This work investigated the photoluminescence (PL) of the GNPs doped planar NLCs. The PL intensity gradually increased with the increased amount of GNPs per ml in NLCs. The results show 64% enhancement in PL intensity with doping suitable amounts of GNPs in liquid crystals 5CB. The enhanced PL intensity is attributed to the increased surface area from the GNPs, resulting in increased emissions due to the increase of the scattering of excitation. The subsequent decay of PL intensity was attributed to the aggregation of the GNPs, as it caused decreased surface area, resulting in the reduction of the energy transfer effect between NLCs and GNPs. The size effect of GNPs was discussed, and the results show that the peak intensity of PL emission was stronger for the 13 nm gold particles at the same concentration of gold nanoparticles (lower than $1.3*10^{10}$ particles/ml). This is because the smaller nanoparticles have larger surface area, which causes multiple reflections and scattering of the

excitation beam. By rotating the LCs' director, the results show that the absorption of the liquid crystal and the increased surface area critically dominate PL intensity.

Conflicts of Interest

The authors declare that they have no conflicts of interest.

Acknowledgments

This study was financially supported by Chung Shan Medical University Hospital (Contract no. CSH-2017-C-010).

References

[1] G. H. Brown, *Advances in Liquid Crystals*, Academic Press, London, UK, 444th edition, 1976.

[2] M. Čopič, J. E. Maclennan, and N. A. Clark, "Structure and dynamics of ferroelectric liquid crystal cells exhibiting thresholdless switching," *Physical Review E: Statistical, Nonlinear, and Soft Matter Physics*, vol. 65, no. 2, 2002.

[3] A. Kumar, J. Prakash, D. S. Mehta, A. M. Biradar, and W. Haase, "Enhanced photoluminescence in gold nanoparticles doped ferroelectric liquid crystals," *Applied Physics Letters*, vol. 95, no. 2, p. 023117, 2009.

[4] D. P. Singh, S. K. Gupta, T. Vimal, and R. Manohar, "Dielectric, electro-optical, and photoluminescence characteristics of ferroelectric liquid crystals on a graphene-coated indium tin oxide substrate," *Physical Review E: Statistical, Nonlinear, and Soft Matter Physics*, vol. 90, no. 2, Article ID 022501, 2014.

[5] L. Calucci, G. Ciofani, D. De Marchi et al., "Boron nitride nanotubes as T 2-weighted MRI contrast agents," *The Journal of Physical Chemistry Letters*, vol. 1, no. 17, pp. 2561–2565, 2010.

[6] F. V. Podgornov, A. M. Suvorova, A. V. Lapanik, and W. Haase, "Electrooptic and dielectric properties of ferroelectric liquid crystal/single walled carbon nanotubes dispersions confined in thin cells," *Chemical Physics Letters*, vol. 479, no. 4-6, pp. 206–210, 2009.

[7] Y.-S. Ha, H.-J. Kim, H.-G. Park, and D.-S. Seo, "Enhancement of electro-optic properties in liquid crystal devices via titanium nanoparticle doping," *Optics Express*, vol. 20, no. 6, pp. 6448–6455, 2012.

[8] Y. P. Piryatinskiï and O. V. Yaroshchuk, "Photoluminescence of pentyl-cyanobiphenyl in liquid-crystal and solid-crystal states," *Optics and Spectroscopy*, vol. 89, no. 6, pp. 860–866, 2000.

[9] J. W. Y. Lam, Y. Dong, J. Luo, K. K. L. Cheuk, Z. Xie, and B. Z. Tang, "Synthesis and photoluminescence of liquid crystalline poly(1-alkynes)," *Thin Solid Films*, vol. 417, no. 1-2, pp. 143–146, 2002.

[10] Y. P. Piryatinskiï, O. V. Yaroshchuk, L. A. Dolgov, T. V. Bidna, and D. Enke, "Photoluminescence of liquid-crystal azo derivatives in nanopores," *Optics and Spectroscopy*, vol. 97, no. 4, pp. 537–542, 2004.

[11] S. Kaur, S. P. Singh, A. M. Biradar, A. Choudhary, and K. Sreenivas, "Enhanced electro-optical properties in gold nanoparticles doped ferroelectric liquid crystals," *Applied Physics Letters*, vol. 91, no. 2, p. 023120, 2007.

[12] T. Härtung, P. Reichenbach, and L. M. Eng, "Near-field coupling of a single fluorescent molecule and a spherical gold nanoparticle," *Optics Express*, vol. 15, no. 20, pp. 12806–12817, 2007.

[13] J. Zhang and J. R. Lakowicz, "Enhanced luminescence of phenyl-phenanthridine dye on aggregated small silver nanoparticles," *The Journal of Physical Chemistry B*, vol. 109, no. 18, pp. 8701–8706, 2005.

[14] K. Palewska, A. Miniewicz, S. Bartkiewicz, J. Legendziewicz, and W. Strek, "Influence of electric field on photoluminescence of lanthanide-doped nematic liquid crystal," *Journal of Luminescence*, vol. 124, no. 2, pp. 265–272, 2007.

[15] S.-Y. Huang, C.-C. Peng, L.-W. Tu, and C.-T. Kuo, "Enhancement of luminescence of nematic liquid crystals doped with silver nanoparticles," *Molecular Crystals and Liquid Crystals*, vol. 507, pp. 301–306, 2009.

[16] K. Tanabe, Y. Suzui, M. Hasegawa, and T. Kato, "Full-color tunable photoluminescent ionic liquid crystals based on tripodal pyridinium, pyrimidinium, and quinolinium salts," *Journal of the American Chemical Society*, vol. 134, no. 12, pp. 5652–5661, 2012.

[17] H. Lu, L. Qiu, G. Zhang et al., "Electrically switchable photoluminescence of fluorescent-molecule-dispersed liquid crystals prepared via photoisomerization-induced phase separation," *Journal of Materials Chemistry C*, vol. 2, no. 8, pp. 1386–1389, 2014.

22

Effect of Background Magnetic Field on Type-II Superconductor under Oscillating Magnetic Field Simulated using Ginzburg-Landau Model

Hasnain Mehdi Jafri,[1] Congpeng Zhao,[1,2] Houbing Huang ⓘ,[1,3] and Xingqiao Ma ⓘ[1]

[1]Department of Physics, University of Science and Technology Beijing, Beijing 100083, China
[2]ASIC, China Center for Information Industry Development, Beijing 100083, China
[3]Advanced Research Institute of Multidisciplinary Science, Beijing Institute of Technology, Beijing 100081, China

Correspondence should be addressed to Xingqiao Ma; xqma@sas.ustb.edu.cn

Academic Editor: Joseph S. Poon

Cubic superconducting sample was simulated using time-dependent Ginzburg-Landau model under oscillating magnetic field with and without additional background static magnetic field. Vortex dynamics including entrance and exit from the sample was simulated. Magnetization and carrier concentration densities of the sample were studied as a function of external magnetic field variations. Anomalies in carrier concentration density were observed at certain values of the magnetic field which were correlated with the entrance and exit processes of vortices. Area swept by superconductor magnetization with magnetic field was observed to have a hysteresis-like behavior where area representing energy dissipated per cycle. This energy accumulation was suggested to cause instability in superconductor over the number of cycles and may result in thermal quenching. Temporal distribution of energy components showed consistency with the pattern observed for carrier concentration and magnetization under oscillating magnetic field. Rapid phase changes with magnetic oscillations resulted in oscillations in energy components, and irregular peaks and ripples in superconducting energy represent the situation of exit and entry of vortices. While the rise in interaction energy with cycles is referred to vortex relaxation time in a cycle, this energy is expected to accumulate and take other forms (e.g., heat) and is predicted to cause thermal quenching. In the presence of background static magnetic field, this energy dissipation was calculated to increase significantly while superconductor is subjected to oscillating magnetic field.

1. Introduction

In general, the behavior of superconductor in applied electric and magnetic field is attributed to the vortex dynamics in it, which generally results in energy dissipation (e.g., heat) and resulting thermal quenching of superconductivity. According to Bardeen-Cooper-Schrieffer (BCS) theory, spin singlet state (opposite spin electron pair), bound by phonon interaction, constitutes the ground state of condensate [1]. The proposed macroscopic quantum theory by Ginzburg and Landau (named Ginzburg-Landau (GL) theory) used Higgs mechanism of spontaneous symmetry breaking and quartic potential [2–5] to generate a local mass term of vector potential. Development of homogeneous ferromagnetic spins destroys the superconductivity if it exceeds critical magnetic field of the superconductor. This model successfully describes the Meissner-Ochsenfeld effect [4, 6–8]. The prediction of penetrating strong magnetic fields in type-II superconductors by Abrikosov [9] gave further credibility to GL model which was later verified experimentally [10–12]. Over the past 70 years, not only macroscopic properties of superconductors (such as categorizing superconductors in type-I and type-II, and description of vortex state in type-II superconductors [9]) but also mesoscopic superconducting samples were successfully described using this theory. GL model is probably the most accurate phenomenological model to describe macroscopic properties of superconductors [13, 14]. Gorkov [15] proved that close to critical temperatures microscopic

BCS theory also reduces to GL theory. Various numerical methods, including finite difference [16–18], finite element [19–21], and spectral method [22], have been developed for the solution of this model. In this report, we used the semi-implicit finite difference method with staggered grid scheme.

Generally, high-temperature superconductor (HTS) based magnetic Levitation (maglev) vehicle systems consist of HTS on a permanent magnet (PM) guideway [23–25]. In general, uniform magnetic field is considered for PM guideways but practically this is not the case; there are situations of magnetic field variations (e.g., cracks, magnetic contacts, and structural and magnetic defects) resulting in a variable magnetic field. The frequency of this variation depends on the speed of the HTS on PM guideway (i.e., speed of the vehicle). The magnetic field in such a case is not exactly oscillating (which was reported earlier [18]), but it is (practically) a constant magnetic field having small jitters (or oscillations) in its amplitude appearing due to imperfections in PM guideway. In this work, we studied such a magnetic field and compared the results with results reported earlier for perfectly oscillating magnetic fields [18]. To investigate this type of energy loss, we studied the behavior of type-II superconductor exposed to an oscillating magnetic field simulated using time-dependent Ginzburg-Landau (TDGL) equations [15, 26, 27] near critical temperature with and without an additional static background magnetic field. The dynamics of Abrikosov vortices in oscillating magnetic field is studied. A hysteresis-like behavior of sample magnetization was observed under oscillating magnetic fields, while background static magnetic field plays a vital role in thermal quenching of superconductor. The behavior of different energy components is also discussed in detail.

2. Ginzburg-Landau Theory

Primary variables in GL model are order parameter, ψ, and magnetic vector potential, A, occupying a superconductor sample in a three-dimensional region Ω having boundary Γ. In nondimensional state, these primary variables are linked to physical quantities, given as follows:

Density of superconducting charge carriers, n_s, $|\psi|^2$

Induced magnetic field, $\nabla \times A$

Current density, $J = \nabla \times \nabla \times A$.

Ginzburg-Landau free energy is taken from Gorkov and Eliashberg [28] given as follows:

$$G(\psi, A) = \frac{1}{2m}\left|(\hbar\nabla - \imath eA)\psi\right|^2 + \alpha|\psi|^2 \mp \frac{1}{2}\beta|\psi|^4 \tag{1}$$
$$+ \frac{1}{2\mu_o}\left|\nabla \times A - \mu_o B\right|^2,$$

where e is charge and m is effective mass of superconducting charge carriers, B is external magnetic field, $\hbar$ and $\imath$ are reduced Plank's constant (i.e., h/2π), and $\sqrt{-1}$, while α and β are phenomenological parameters depending on environmental factors (e.g., temperature).

We introduce London penetration depth ($\lambda = \sqrt{mc^2\beta/16\pi e^2|\alpha|}$) and coherence length ($\xi = h/2\pi\sqrt{2m|\alpha|}$) as length scale parameters. We introduce a length scale l and nondimensionalized physical quantities, special and time coordinates x ($x = x, y, z$) and t, order parameter ψ, magnetic vector potential A, and magnetic field B according to the following transformations:

$$x \longrightarrow lx,$$

$$\psi \longrightarrow \sqrt{\frac{|\alpha|}{\beta}}\,\psi,$$

$$A \longrightarrow \sqrt{\frac{8\pi\alpha^2 l^2}{\beta}}\,A, \tag{2}$$

$$B \longrightarrow \sqrt{\frac{8\pi|\alpha|^2}{\beta}}\,B,$$

$$\overline{\Phi} \longrightarrow \frac{|\alpha|}{2\gamma e}\,\overline{\Phi},$$

where γ is relaxation parameter. For simplicity, we choose length scale in terms of coherence length (i.e. $l = \xi$). Ginzburg-Landau parameter κ is defined as $\kappa = \lambda/\xi$. κ defines the type of superconductor, $\kappa < 1/\sqrt{2}$ for type-I and $\kappa > 1/\sqrt{2}$ for type-II superconductors. Minimizer of Ginzburg-Landau energy functional (1) satisfies Euler-Lagrange equations, generally known as Ginzburg-Landau (GL) equations:

$$\frac{\partial\psi}{\partial t} + \imath\overline{\Phi}\psi = \psi - |\psi|^2\psi - \left(\imath\nabla + \frac{1}{\kappa}A\right)^2\psi \tag{3}$$

$$-\sigma\left(\frac{\partial A}{\partial t} + \kappa\nabla\overline{\Phi}\right) = \nabla \times \nabla \times A + \frac{1}{\kappa^2}|\psi|^2 A \tag{4}$$
$$+ \frac{\imath}{2\kappa}\left(\psi^*\nabla\psi - \psi\nabla\psi^*\right).$$

σ is electrical conductivity defined as $\sigma = \sigma_n(2\pi\hbar/\gamma c^2)(l^2/\xi^2)$, while σ_n is normal conductance. External magnetic field is assumed to be uniformly distributed over the sample surface. GL equations were solved using natural boundary conditions, i.e., supercurrent across the boundaries is zero and magnetic field at the boundaries is given as follows:

$$\left(\imath\nabla + \frac{1}{\kappa}A\right)\psi.n = 0 \quad \text{on boundary } \Gamma \tag{5}$$

$$(\nabla \times A) \times n = B \times n \quad \text{on boundary } \Gamma, \tag{6}$$

where n is unit normal vector on the boundary Γ of the region Ω.

TDGL model is gauge invariant under transformation:

$$\psi' = \psi e^{\imath\kappa\chi}$$

$$A' = A + \nabla\chi \tag{7}$$

$$\overline{\Phi}' = \overline{\Phi} - \frac{\partial\chi}{\partial t},$$

where gauge χ is a function of space and time. For the present study, we choose zero electric potential gauge ($\overline{\Phi} = 0$) known as Coulomb Gauge. So (3) and (4) reduce to

$$\frac{\partial \psi}{\partial t} = \psi - |\psi|^2 \psi - \left(\iota \nabla + \frac{1}{\kappa} A \right)^2 \psi \qquad (8)$$

$$-\sigma \frac{\partial A}{\partial t} = \nabla \times \nabla \times A + \frac{1}{\kappa^2} |\psi|^2 A$$

$$+ \frac{\iota}{2\kappa} \left(\psi^* \nabla \psi - \psi \nabla \psi^* \right). \qquad (9)$$

Energy density $H_{tot} = H_{tot}(x,y,z,t)$ in superconductor is distributed into three parts: superconducting energy density $H_{sup} = H_{sup}(x,y,z,t)$, interaction energy density $H_{int} = H_{int}(x,y,z,t)$, and magnetic energy density $H_{mag} = H_{mag}(x,y,z,t)$ given as[18, 29]

$$H_{sup} = \frac{1}{\kappa^2} |\nabla \psi|^2 - |\psi|^2 + \frac{1}{2} |\psi|^4 \qquad (10)$$

$$H_{int} = \frac{\iota}{\kappa} A \left((\nabla \psi) \psi^* - \psi (\nabla \psi^*) \right) + |A|^2 |\psi|^2 \qquad (11)$$

$$H_{mag} = (B - \nabla \times A)^2, \qquad (12)$$

where length scales are taken in terms of London penetration depth (i.e. $l = \lambda$). Total energy (H_{tot}) density is the sum of these three energy densities, $H_{tot} = H_{sup} + H_{int} + H_{mag}$; total energy is the integral of total energy density over the region Ω, $G_{tot} = \int H_{tot}(x, y, z, t) d\Omega$.

Gauge invariant discretization [17, 30–33] is the most popular and widely used approach to solve TDGL equations, which is first-order accurate in time and second-order accurate in space, and other finite element [19, 20], finite difference [34, 35], and spectral method [22] have also been developed. Link variable schemes [16, 17, 36] are the most commonly used schemes to solve coupled GL equations. In the present work, in order to investigate carrier concentration, magnetization, and vortex dynamics in superconductor, we used coupled nonlinear TDGL equations (8) and (9) solved by finite difference scheme using link variables, described by Winiecki and Adams [16], implemented in three-dimensional code we have developed [18]. Equations (8) and (9) along with boundary condition equations (5) and (6) form the basis of the present work. The present work was performed close to critical temperature allowing the assumption of thermal suppression of surface barrier [37–39]. The frequency of oscillating magnetic field was set to $0.00278/t$ in nondimensionalized coordinates for the present work.

3. Results and Discussion

Three-dimensional, cubic superconductor domain of size $20\xi \times 20\xi \times 20\xi$, periodic along z-axis (along the direction of applied magnetic field), was discretized with grid size 0.5 ξ in each direction, and magnetic field was applied in z-axis direction with Ginzburg-Landau parameter $\kappa=4$. Initially, a constant magnetic field was applied, and on reaching

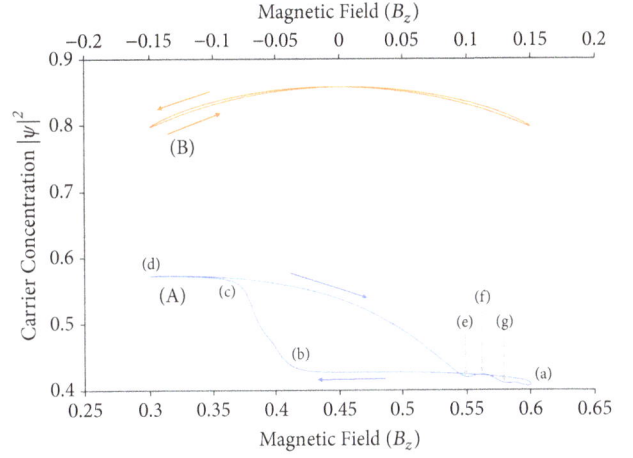

FIGURE 1: Variation of carrier concentration with magnetic field for cubic sample of side 20ξ each, with parameters $\kappa = 4$ and $\sigma = 1$ under magnetic fields (A) $B_z = 0.45\kappa + 0.15\kappa \cos\theta$ and (B) $B_z = 0.15\kappa \cos\theta$.

equilibrium state, it was allowed to oscillate as cosine wave. Cubic superconductor under two different types of magnetic field schemes, (A) $B_z = 0.45\kappa + 0.15\kappa \cos\theta$ and (B) $B_z = 0.15\kappa \cos\theta$ (where $\theta = \omega t = 2\pi f t$), was simulated to investigate the effect of static background magnetic field on the superconductor in the presence of an oscillating magnetic field. The system was initialized with the completely superconducting sample ($|\psi| = 1$) and no magnetic vector potential ($A = 0$). The system was relaxed to minimize Gibbs free energy. During this energy minimization process, vortices entered the sample and relaxed their positions to minimum energy. Magnetic field oscillations were introduced as the sample reached minimum energy state, resulting in a new series of entrance and leaving of vortices, corresponding to different positions of magnetic field wave. Figure 1 shows a variation of average carrier concentration of the sample with magnetic field. The pattern of carrier concentration is symmetric for positive and negative half cycles but nonsymmetric for increasing and decreasing parts of magnetic field oscillation, shown in Figure 1 (for magnetic field of type (B)), indicating a lag in vortex dynamics behind the oscillating magnetic field. For the situation of magnetic field of type (A), the system is initially subjected to $B_z = 0.6$, and after sufficient time, on reaching equilibrium state, a small oscillation with amplitude 0.15 in magnetic field was introduced.

During decaying half of magnetic field cycle (i.e., $|B_z| = 0.6 \longrightarrow 0.3$), carrier concentration does not increase as rapidly as decrease in magnetic field, till a certain value of magnetic field, where it is difficult for the sample to keep the vortices bound any further within the sample and a set of loosely bound vortices leave the sample, causing a sudden increase in carrier concentration shown by steps between positions (b) and (c) in Figure 1, carrier concentration peaks at minimum magnetic field, i.e., $B_z = 0.3$. During increasing half of magnetic field cycle (i.e., $|B_z| = 0.3 \longrightarrow 0.6$), penetration of magnetic field in large quantity starts from the sample edges till vortices are formed and they repel magnetic

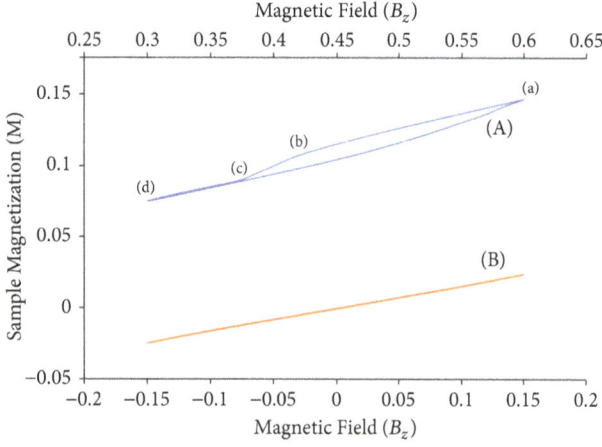

FIGURE 2: Magnetization ($M = \nabla \times A$) vs. magnetic field (B_z) plot for cubic sample of side 20ξ each, with parameters $\kappa=4$ and $\sigma=1$ under magnetic field oscillations of the type (A) $B_z=0.45\kappa + 0.15\kappa \cos \theta$ and (B) $B_z = 0.15\kappa \cos \theta$.

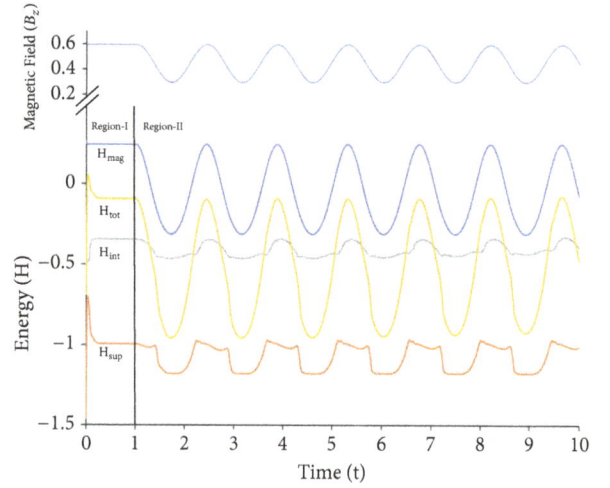

FIGURE 3: Temporal distribution of energy densities of cubic superconducting sample in static and oscillating magnetic fields, for cubic sample of side 20ξ each, with parameters $\kappa = 4$ and $\sigma = 1$, and $B_z = 0.45\kappa + 0.15\kappa \cos \theta$.

field penetrations from the edges. Relaxation of vortices causes an increase in carrier concentration indicated by sections (e) and (f) in Figure 1, and formation and relaxation of the first additional set of vortices are indicated by this small peak; the same process repeats itself for every subsequent set of vortices entering the sample indicated by (g) in Figure 1.

Energy dissipation (E_d) in magnetic oscillations is the area covered by hysteresis loop (M-H curve) of sample magnetization with magnetic field, which takes the following form:

$$E_d = \Omega \oint B_z . dM, \tag{13}$$

where M is sample magnetization ($M = \nabla \times A$), Ω is volume bounded by the superconductor, and B_z is the applied external magnetic field. This type of behavior of magnetic field trapped inside the superconductor lagging behind the magnetic field oscillations indicates the energy dissipation during the cycle. The area of loop, attributed to the energy dissipated during a cycle, was observed to increase with frequency of magnetic oscillations, due to limited velocities of vortices. Figure 2 shows variation of average sample magnetization ($\nabla \times A$) with external magnetic field oscillations at normalized frequency $f=0.00278/t$. It was observed that during decreasing and increasing half cycles of magnetic field oscillations, sample magnetization behaved in a similar way (comparing with plots behavior) to that of carrier concentration in Figure 1. During decreasing part of magnetic field sample magnetization had a rapid drop between (b) and (c) in Figure 2 at the same magnetic field intensity as observed between (b) and (c) in Figure 1. This behavior of a sudden drop in magnetization, between (b) and (c) in Figure 2, and sudden rise in carrier concentration, between (b) and (c) in Figure 1, represent the same process and are attributed to the process of a set of vortices leaving the sample. This trapped energy can be anticipated to transform in other forms of energy, mostly heat, which accumulates with a number of cycles, resulting in thermal quenching of the superconductor.

It can also be inferred that, at faster vehicle speeds, resulting in higher frequencies, this energy accumulation may result in a rapid quenching of superconductor. This energy loss during a cycle was observed to be ~10 times greater in the presence of additional static background magnetic field (type A) compared with that in its absence (type B), which can be seen by the area swept by both curves in Figure 2.

Temporal distribution of three energy components and total energy for the sample under the magnetic field of type (A) is shown in Figure 3. Initially, (at $t=0$) sample was in a nonequilibrium state with no magnetic field penetration and carrier concentration at maximum. Nonequilibrium state of magnetic field distribution across the boundary is the reason of large initial value of H_{mag}, but as the field penetration starts, H_{mag} rapidly drops to equilibrium values, because the number of vortices will remain the same before introduction of field oscillations and there would only be rearrangement of vortices in the sample, and H_{mag} do not change during the rearrangement of vortices. Initially, due to uniform penetration of vortices from the boundaries, H_{sup} increases rapidly because average carrier concentration decreases and the second term in (10) dominates. As the vortices are formed and start relaxing, H_{sup} starts decreasing forming an initial hump-like structure, indicating nonvortex to vortex transition. H_{int} is responsible for the interaction of magnetic and superconducting energies. The second term in (11) is responsible for the formation of vortices while the first term controls vortex dynamics inside the superconductor.

Figure 3 (region II) represents all the three component energy densities and total energy density with the magnetic field of type (A). As system reaches equilibrium, oscillations in magnetic field are introduced as cosine wave, in the range $0.6\kappa \geq B_z \geq 0.3\kappa$. In region II as B_z decreases, the balance between external magnetic field and vortex magnetization is disturbed, resulting in an increase in intervortex spacing and exit from the sample, and all the energies start oscillating with

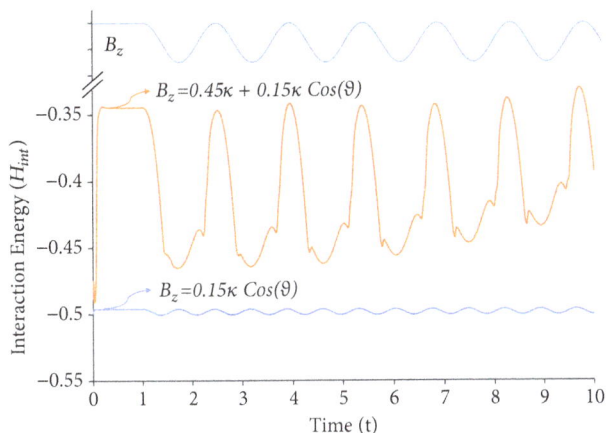

FIGURE 4: Temporal distribution of interaction energy densities of cubic superconducting sample in the magnetic field of types (A) $B_z = 0.45\kappa + 0.15\kappa \cos\theta$ and (B) $B_z = 0.15\kappa \cos\theta$, for cubic sample of side 20ξ each, with parameters $\kappa = 4$ and $\sigma = 1$.

the oscillations in B_z, and the process of repeated entrance and exit of vortices starts. Peak shape analysis of H_{sup} and H_{int} shows small kinks in energies, indicating the moments when outermost (or least bound) vortices leave the sample; after that, a sudden dip was observed in H_{sup} and H_{int} because the remaining vortices which are now free from force exerted by the vortices just left and relaxed themselves; this kink is also visible in total energy pattern, similar to the one observed at positions (b) and (c) in Figures 1 and 2. Interestingly, energy patterns of H_{sup} and H_{int} are not symmetric for increasing and decreasing half cycles of B_z, i.e., entrance and exit of vortices. Peak shape analysis shows a quick entrance and slow exit of vortices during increasing and decreasing half cycles of B_z, resulting in hysteresis-type behavior and energy accumulation which have already been discussed. An irregularity in H_{int} is observed after a number of cycles, due to energy accumulation. This effect was only observed in H_{int} due to its dependence on vortex dynamics, having limited velocities, resulting in H_{int} reaching equilibrium value with a delay compared to H_{mag} and H_{sup}. Repeated entrance and leaving of vortices over a cycle do not give sufficient time to the sample to completely attain equilibrium state, resulting in accumulation of energy in the sample in the form of interaction energy. This accumulated energy represented the area enclosed by magnetic field and sample magnetization shown in Figure 2, which may appear in the form of heat resulting in thermal quenching of superconductor in practical applications. This effect is significantly high (~10 times) for the samples under magnetic field of type (A) compared to the samples under magnetic field of type (B). Figure 4 shows the comparison of H_{int} of the systems with and without steady magnetic field (of types (A) and (B)) under the same oscillating component of magnetic field. It was observed (by the comparison of interaction energies in Figure 4) that, in the presence of steady background magnetic field, the energy dissipation (or the abnormalities in H_{int}) is ~10 times higher than that observed for the case without additional external magnetic field. Therefore, a static background magnetic field

plays a significant role in addition to oscillating magnetic field in thermal dissipation and subsequently quenching of superconductor moving on PM guideway in a Maglev vehicle system; the rate of such energy dissipation/accumulation is directly dependent on the speed of the vehicle.

4. Conclusions

Time-dependent Ginzburg-Landau model was used to investigate type-II superconductor under the oscillating magnetic field, with additional static background magnetic field. Finite difference scheme was used for cubic superconducting sample. Carrier concentration, magnetization, and energy in the presence and absence of static background magnetic field were studied. Nonuniform behavior in carrier concentration with the oscillating magnetic field was observed and was related to vortex dynamics on point-to-point basis. Lagging of average sample magnetization behind magnetic field was observed during magnetic field oscillations. Area of hysteresis loop was measured and was attributed to dissipated energy in the sample, accumulated over time. The same effect caused anomalies in interaction energy density of the sample under magnetic field oscillations. Ripples in superconducting energy density were observed due to entrance and exit of vortices during magnetic field oscillations, which is the result of the same effect of small peaks (and sudden rise) observed in average carrier concentration with increasing (and decreasing) external magnetic field magnitude. Comparison between energy patterns in the presence and absence of background steady magnetic field indicated that more energy is dissipated in the presence of background magnetic field than that in its absence, suggesting that, in practical devices (e.g., Maglev vehicles) when magnetic field is oscillating (magnetic guideways with defects), its necessary to shield the superconductor from background magnetic fields to avoid energy dissipation and thermal quenching of the superconductor.

Conflicts of Interest

The authors declare that there are no conflicts of interest regarding the publication of this paper.

Acknowledgments

This work was supported by National Science Foundation of China, Grant nos. 11174030 and 11504020. Author Hasnain Mehdi Jafri acknowledges the Higher Education Commission of Pakistan for the Ph.D. scholarship.

References

[1] J. Bardeen, L. N. Cooper, and J. R. Schrieffer, "Microscopic theory of superconductivity," *Physical Review A: Atomic, Molecular and Optical Physics*, vol. 106, article 162, 1957.

[2] P. W. Anderson, "Plasmons, gauge invariance, and mass," *Physical Review A: Atomic, Molecular and Optical Physics*, vol. 130, no. 1, pp. 439–442, 1963.

[3] F. Englert and R. Brout, "Broken symmetry and the mass of gauge vector mesons," *Physical Review Letters*, vol. 13, no. 9, article 321, 1964.

[4] G. S. Guralnik, C. R. Hagen, and T. W. B. Kibble, "Global conservation laws and massless particles," *Physical Review Letters*, vol. 13, no. 20, pp. 585–587, 1964.

[5] P. W. Higgs, "Broken symmetries and the masses of gauge bosons," *Physical Review Letters*, vol. 13, pp. 508-509, 1964.

[6] W. Meissner and R. Ochsenfeld, "Ein neuer Effekt bei Eintritt der Supraleitfähigkeit," *Naturwissenschaften*, vol. 21, no. 44, pp. 787-788, 1933.

[7] M. C. N. Fiolhais, H. Essén, C. Providencia, and A. B. Nordmark, "Magnetic field and current are zero inside ideal conductors," *Progress in Electromagnetics Research B*, no. 27, pp. 187–212, 2011.

[8] H. Essén and M. C. N. Fiolhais, "Meissner effect, diamagnetism, and classical physics-a review," *American Journal of Physics*, vol. 80, no. 2, pp. 164–169, 2012.

[9] A. A. Abrikosov, "Magnetic properties of superconductors of the second group," *Journal of Experimental and Theoretical Physics*, vol. 5, no. 6, 1957.

[10] P. L. Gammel, D. J. Bishop, G. J. Dolan et al., "Observation of hexagonally correlated flux quanta in YBa2Cu3O7," *Physical Review Letters*, vol. 59, no. 22, pp. 2592–2595, 1987.

[11] H. F. Hess, R. B. Robinson, R. C. Dynes, J. M. Valles Jr., and J. V. Waszczak, "Scanning-Tunneling-Microscope Observation of the Abrikosov Flux Lattice and the Density of States near and inside a Fluxoid," *Physical Review Letters*, vol. 62, no. 2, pp. 214–216, 1989.

[12] U. Essmann and H. Träuble, "The direct observation of individual flux lines in type II superconductors," *Physics Letters A*, vol. 24, no. 10, pp. 526-527, 1967.

[13] M. Karmakar, "Electrostatic potential in high-temperature superconducting cuprates: Extended ginzburg-landau theory," *Advances in Condensed Matter Physics*, vol. 2011, 2011.

[14] T. Kidanemariam and G. Kahsay, "Theoretical Study of Upper Critical Magnetic Field (HC2) in Multiband Iron Based Superconductors," *Advances in Condensed Matter Physics*, vol. 2016, Article ID 5470429, 10 pages, 2016.

[15] L. P. Gor'kov, "Microscopic derivation of the Ginzburg-Landau equations in the theory of superconductivity," *Journal of Experimental and Theoretical Physics*, vol. 9, no. 6, pp. 1364–1367, 1959.

[16] T. Winiecki and C. S. Adams, "A fast semi-implicit finite-difference method for the TDGL equations," *Journal of Computational Physics*, vol. 179, no. 1, pp. 127–139, 2002.

[17] W. D. Gropp, H. G. Kaper, G. K. Leaf, D. M. Levine, M. Palumbo, and V. M. Vinokur, "Numerical simulation of vortex dynamics in type-II superconductors," *Journal of Computational Physics*, vol. 123, no. 2, pp. 254–266, 1996.

[18] H. M. Jafri, X. Ma, C. Zhao et al., "Numerical simulation of vortex dynamics in type-II superconductors in oscillating magnetic field using time-dependent Ginzburg–Landau equations," *Journal of Physics: Condensed Matter*, vol. 29, no. 50, p. 505701, 2017.

[19] Q. Du, "Finite element methods for the time-dependent Ginzburg-Landau model of superconductivity," *Computers & Mathematics with Applications*, vol. 27, no. 12, pp. 119–133, 1994.

[20] Q. Du, M. D. Gunzburger, and J. S. Peterson, "Solving the Ginzburg-Landau equations by finite-element methods," *Physical Review B: Condensed Matter and Materials Physics*, vol. 46, no. 14, pp. 9027–9034, 1992.

[21] B. Li and Z. Zhang, "A new approach for numerical simulation of the time-dependent Ginzburg-Landau equations," *Journal of Computational Physics*, vol. 303, pp. 238–250, 2015.

[22] L. Q. Chen and J. Shen, "Applications of semi-implicit Fourier-spectral method to phase field equations," *Computer Physics Communications*, vol. 108, no. 2-3, pp. 147–158, 1998.

[23] M. Liu, S. Wang, J. Wang, and G. Ma, "Effect of AC magnetic field on the levitation force of YBCO bulk above NdFeB guideway," *Journal of Low Temperature Physics*, vol. 155, no. 3-4, pp. 169–176, 2009.

[24] J. Wang, S. Wang, Y. Zeng et al., "The first man-loading high temperature superconducting Maglev test vehicle in the world," *Physica C: Superconductivity and its Applications*, vol. 378-381, pp. 809–814, 2002.

[25] L. Zhang, J. Wang, S. Wang, and Q. He, "Influence of AC external magnetic field perturbation on the guidance force of HTS bulk over a NdFeB guideway," *Physica C: Superconductivity and its Applications*, vol. 459, no. 1-2, pp. 43–46, 2007.

[26] J. Barba-Ortega and E. Sardella, "Superconducting properties of a mesoscopic parallelepiped with anisotropic surface conditions," *Physics Letters A*, vol. 379, no. 47-48, pp. 3130–3135, 2015.

[27] J. Barba-Ortega, E. Sardella, and J. A. Aguiar, "Superconducting properties of a parallelepiped mesoscopic superconductor: A comparative study between the 2D and 3D Ginzburg-Landau models," *Physics Letters A*, vol. 379, no. 7, pp. 732–737, 2015.

[28] L. P. GoR'Kov and G. M. Eliashberg, "Generalization of the Ginzburg-Landau equations for non-stationary problems in the case of alloys with paramagnetic impurities," *Soviet Journal of Experimental and Theoretical Physics*, vol. 27, p. 328, 1968.

[29] T. S. Alstrøm, M. P. Sørensen, N. F. Pedersen, and S. r. Madsen, "Magnetic flux lines in complex geometry type-II superconductors studied by the time dependent Ginzburg-Landau equation," *Acta Applicandae Mathematicae*, vol. 115, no. 1, pp. 63–74, 2011.

[30] H. Frahm, S. Ullah, and A. T. Dorsey, "Flux dynamics and the growth of the superconducting phase," *Physical Review Letters*, vol. 66, no. 23, pp. 3067–3070, 1991.

[31] R. Kato, Y. Enomoto, and S. Maekawa, "Computer simulations of dynamics of flux lines in type-II superconductors," *Physical Review B: Condensed Matter and Materials Physics*, vol. 44, no. 13, pp. 6916–6920, 1991.

[32] M. MacHida and H. Kaburaki, "Direct simulation of the time-dependent Ginzburg-Landau equation for type-II superconducting thin film: Vortex dynamics and V-I characteristics," *Physical Review Letters*, vol. 71, no. 19, pp. 3206–3209, 1993.

[33] G. W. Crabtree, D. O. Gunter, H. G. Kaper, A. E. Koshelev, G. K. Leaf, and V. M. Vinokur, "Numerical simulations of driven vortex systems," *Physical Review B: Condensed Matter and Materials Physics*, vol. 61, no. 2, pp. 1446–1455, 2000.

[34] E. Coskun and M. K. Kwong, "Simulating vortex motion in superconducting films with the time-dependent Ginzburg-Landau equations," *Nonlinearity*, vol. 10, no. 3, pp. 579–593, 1997.

[35] J. F. Blackburn, A. Campbell, and E. K. H. Salje, "The force-free case in three-dimensional superconductors: A computational study," *Philosophical Magazine*, vol. 80, no. 8, pp. 1455–1471, 2000.

[36] J. Barba-Ortega, E. Sardella, and J. A. Aguiar, "Superconducting boundary conditions for mesoscopic circular samples," *Superconductor Science and Technology*, vol. 24, no. 1, 2011.

[37] R. Córdoba, T. I. Baturina, J. Sesé et al., "Magnetic field-induced dissipation-free state in superconducting nanostructures," *Nature Communications*, vol. 4, no. 1, p. 1437, 2013.

[38] C. P. Bean and J. D. Livingston, "Surface barrier in type-II superconductors," *Physical Review Letters*, vol. 12, no. 1, pp. 14–16, 1964.

[39] W. V. Pogosov, "Thermal suppression of surface barrier in ultrasmall superconducting structures," *Physical Review B: Condensed Matter and Materials Physics*, vol. 81, no. 18, 2010.

23

Electronic Structure and Magnetic Coupling of Pure and Mg-Doped KCuF$_3$

Fausto Cargnoni,[1] **Simone Cenedese,**[1] **Paolo Ghigna** ![ID],[2]
Mario Italo Trioni,[1] **and Marco Scavini** ![ID][1,3]

[1]*Consiglio Nazionale delle Ricerche, Istituto di Scienze e Tecnologie Molecolari, Via Golgi 19, 20133 Milano, Italy*
[2]*Dipartimento di Chimica, Università di Pavia and Unità INSTM di Pavia, Viale Taramelli 13, 27100 Pavia, Italy*
[3]*Dipartimento di Chimica, Università degli Studi di Milano, Via Golgi 19, 20133 Milano, Italy*

Correspondence should be addressed to Marco Scavini; marco.scavini@unimi.it

Academic Editor: Da-Ren Hang

We investigated the electronic and magnetic properties of KCuF$_3$ and KCu$_{0.875}$Mg$_{0.125}$F$_3$ crystals by means of Density Functional periodic computations at the B3LYP level of theory. We considered four possible magnetic ordering of the unpaired electrons on copper ions. Both materials are correctly predicted as being 1D antiferromagnetic insulators, and the superexchange parameters in the crystallographic *ab* planes and along the *c* direction measure +10 and -600 K, respectively. Residual spin polarization is found also on fluorine atoms, in agreement with literature results. We found a complete orbital ordering at Cu sites: in the copper reference frame d$_{xy}$, d$_{yz}$, d$_{xz}$, and d$_{z^2}$ orbitals contain about 2 electrons each, while the d$_{x^2-y^2}$ orbital is only partially filled. The perturbation induced by doping of KCuF$_3$ with Mg is very strong and localized on the first shell of F neighbours. Mg has a very small influence on the ordering of the 3d orbitals of copper and on the Cu-Cu magnetic superexchange parameters but reduces significantly the absolute energy differences between the antiferromagnetic ground state and the ferromagnetic phase, in agreement with the experiment. The absence of long range effects makes Mg a suitable dopant for the investigation of strongly correlated electronic systems by means of orbital dilution.

1. Introduction

Strong correlation of electrons in solids leads to a variety of phenomena that are still very far to be really understood. They range from superconductivity in cuprates, to colossal magnetoresistance in manganites, to heavy fermions formation in rare earth intermetallics. It is widely recognized that one of the reasons for the lack of understanding of these systems is that several degrees of freedom are intimately interconnected and that the identification of the leading interaction is a challenging task, a typical example being the interplay between orbital ordering (OO) and cooperative Jahn-Teller distortion (cJTd). As pointed out by Kugel and Khomskii in their seminal work [1], in presence of strong electron correlation orbitals are subject to exchange interaction and a preexisting OO tends to amplify any lattice instability. In this scenario, OO is the leading interaction that causes the setting up of cJTd. The pseudocubic perovskite KCuF$_3$ has always been considered as a model system for testing the Kugel-Khomskii model. However, the argument is still a matter of debate. LDA + DMFT calculations [2], for example, showed that the superexchange mechanisms in KCuF$_3$ are not strong enough to stabilize the cJTd up to the high temperature regime where it is experimentally still found to survive. While orbital degrees of freedom are not directly accessible in this system, the resonant X-ray Scattering (RXS) signal is dominated by the distortions in the F$^-$ ion positions [3, 4], and melting of cJTd has been observed in perovskite samples of composition KCu$_{1-x}$Mg$_x$F$_3$, with T increasing linearly at decreasing x [5, 6]. When extrapolated to x=0, a value of 0.166 eV is found, which can be considered as a reasonable estimate of the cJTd energy in the pure KCuF$_3$ compound. This experimental result is in nice agreement with the LDA + DMFT calculations [2] and supports the conclusion that the electron phonon coupling is the most important interaction in stabilizing the peculiar structure of $KCuF_3$. The subtle interplay between spin, orbital, and vibrational degrees of freedom of this material is still the subject of experimental and theoretical investigations [7–18], adopting a variety of techniques ranging from inelastic Raman and X-rays scattering and absorption, [11, 13], neutron scattering [10, 14], DFT based electronic structure computations [8], and lattice dynamics [7, 15] and, more recently, calculations within the variational Green's function formalism [17, 18].

With the aim of better understanding the relevance of these experimental findings and to gather them in the wider context of electronic structure of strongly correlated electron systems, here we present an *ab initio* study of Mg-doped $KCuF_3$. To this purpose, also the electronic and magnetic properties of pure $KCuF_3$ are investigated and discussed in comparison with the doped system and with previous calculations. Even for the pure compound, we present an accurate characterization of all the four possible magnetic structures, which has never been previously discussed in detail. However, the main target of our investigation is to characterize the structural, electronic, and magnetic properties of Mg-doped $KCuF_3$ and to determine how Mg doping affects the properties of this material.

The paper is organized as follows: in the next section we present the details of our computations, along with a brief excursion on the atomistic arrangement and the energetic in pure and Mg-doped $KCuF_3$. Then we discuss the electronic structure of these materials, using either the typical instruments of solid state physics and the Quantum Theory of Atoms in Molecules (QTAIM), which is based on a direct space analysis of the electron density distribution and provides a description closest to the chemists' point of view. Third, we present the magnetic coupling parameters obtained for the pure and the Mg-doped samples. The last section contains a general discussion about the effect of Mg doping in $KCuF_3$ together with conclusions and perspectives.

2. Computational Details and Structure Optimization

The electronic structure and magnetic coupling of pure and Mg-doped $KCuF_3$ have been determined by means of first principles periodic computations, as implemented in the CRYSTAL [19] code. For most computations we adopted the unrestricted B3LYP Hamiltonian, which proved to be successful in describing low-dimensional magnetic copper insulators [20]. Several other exchange-correlation functionals have been selected to perform different tests. The electronic wavefunctions have been expanded in terms of all electrons Gaussian basis sets designed for solid state computations. We assigned to potassium atoms a 21s13p3d basis set contracted to 5s4p1d [21], to copper a 21s13p5d set contracted to 6s5p2d [22], to fluorine a 12s5p set contracted to 4s3p [23], and to magnesium a 15s7p1d set contracted to 4s3p1d [24]. To obtain the wavefunctions of four different magnetic phases of $KCuF_3$, in each computation we locked properly the atomic spin of copper atoms during the first 30 cycles of the iterative procedure within the CRYSTAL code,

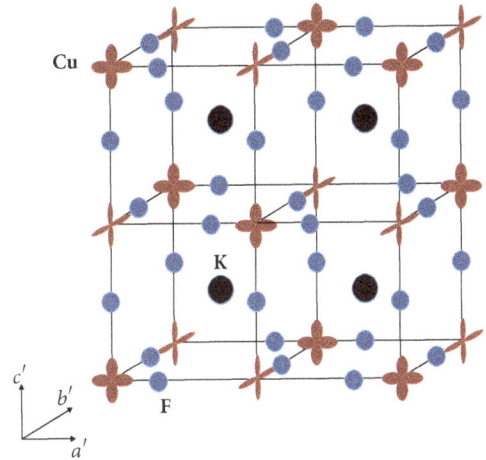

FIGURE 1: Stick-and-ball representation of $KCuF_3$ in the P*mmm* space group (see text for details). For the sake of clarity, only half a cell is shown in the figure ($0 \leq y \leq 0.5$). The shifts of the position of F ions in the ab plane, in respect to the middle of the Cu-Cu internuclear axis as in cubic perovskite, are enlarged to highlight the deformation of CuF_6 octahedra. The partially filled d_{x2-y2} orbitals responsible for orbital ordering are schematically represented in red.

TABLE 1: Crystallographic parameters of pure KCuF3 in the AF1 magnetic phase (I4/*mcm* space group).

Reference	a, b (Å)	c (Å)	x_F
this work	5.92	7.96	0.230
Experiment [25]	5.8569(6)	7.8487(8)	0.2276(1)

and then we let the wavefunction free to evolve towards the converged solution. The four magnetic phases considered in this study are all the possible combinations of ferromagnetic (FM) and antiferromagnetic (AF) order among neighbouring Cu atoms pairs in the ab plane and along the c axis (see Figure 1). In particular, in the AF1 phase there is AF order along the c axis and FM order in the ab plane. The AF2 phase presents FM order along the c axis and AF order in the ab plane. In the AF3 and in the F phases the order is entirely AF and FM, respectively.

As a first step, we optimized the structure of the AF1 phase of pure $KCuF_3$ within the I4/*mcm* space group. This corresponds to the optimization of two cell parameters ($a=b$, c), and of the relative position of fluorine atoms (x_F) on the internuclear axis of Cu-Cu neighbours in ab planes. The resulting data are reported in Table 1, and they are in fair agreement with the experimental results, relative errors in the estimate of a and c being about 1-1.5% [25]. The optimized position of F atoms (x_F=0.230) confirms that in the AF1 phase copper is surrounded by a distorted octahedron of fluorine atoms. The optimized Cu-F distances in ab planes are 1.93 and 2.26 Å and 1.99 Å along the c axis.

To compute the wavefunction of the four magnetic phases of pure and Mg-doped $KCuF_3$, we defined a supercell in the P*mmm* space group. The new cell vectors are defined as $a'=a+b$, $b'=a-b$, and $c'=c$, and the arrangement of the atoms within this choice is reported in Figure 1. Our calculations confirm that the AF1 phase is the ground state for pure $KCuF_3$. The AF3 phase, where the FM interactions in ab planes are turned into AF ones, presents a very small increase in energy, with E_{AF3}-E_{AF1}=12 meV/cell. A more marked energy increase is found when the AF interactions along the c axis are turned into FM ones: E_{AF2}-E_{AF1}=229 meV/cell, and E_F-E_{AF1}=234 meV/cell.

Doping of $KCuF_3$ with Mg atoms is described as a regular substitution of a single copper atom in each supercell, which corresponds to the overall composition of $KCu_{0.875}Mg_{0.125}F_3$. To account for the geometrical distortions introduced by Mg in the crystal network, we optimized the structure of the AF1 phase of Mg-doped $KCuF_3$. We optimized the cell parameters a' $(=b')$ and c', as well as the position of the 6 fluorine atoms closer to Mg. Though in principle a much larger set of coordinates should be considered to perform a complete optimization of the structure, we are confident that the perturbation induced in the electronic properties by Mg doping is properly accounted for by considering just the distortion of the first shell of neighbours surrounding the doping site. This assumption is strengthened by the results obtained as discussed in the following, as the perturbation induced by Mg doping is indeed very short-ranged. The geometrical parameters obtained for the AF1 phase have been adopted also for the AF2, AF3, and F magnetic arrangements. Test computations for the AF1 phase, conducted on a larger supercell, proved that no noticeable displacements occur for other atoms than the first fluorine neighbours of Mg.

The doping of $KCuF_3$ with a 12.5% percentage of Mg has no relevant effect on the cell parameters, at variance with experimental findings [26]. The dopant element reduces the distortion of the surrounding fluorine octahedron, and the three unique Mg-F distances measures are 1.94 and 2.20 Å in ab planes and 1.98 Å along c. It should be noted that in our calculations each Mg ion is surrounded by six Cu ions, while in real samples clusters of dopant ions can be formed, and they may affect the cell parameters in a more effective manner. However, the investigation of complex defect structures is beyond the scope of the present paper.

The changes of the Mg-F distances with respect to the corresponding Cu-F ones in the undoped material clearly reflect also on the environment of the neighbouring Cu atoms, and this has the effect of reducing also the distortions of the fluorine octahedra surrounding copper. Noticeably, Mg doping does not alter the ordering in energy of the four magnetic phases considered, but the energy differences become significantly smaller: E_{AF3}-E_{AF1}=4 meV/cell, E_{AF2}-E_{AF1}= 168 meV/cell, and E_F-E_{AF1}=184 meV/cell. The steep decrease of the energy difference between the AF1 and the F phases is in agreement with experimental determinations [26].

3. Electronic Structure

A very useful approach to get insight into the chemistry of a given material is represented by the Quantum Theory of Atoms in Molecules. This theory is based on the real space analysis of the electron density distribution, as a way to uniquely define the atomic and bonding properties of a given chemical system. The basic notions of this theory, several applications of the QTAIM to solid state systems,

and the formalism adopted by the authors can be found in [27–29]. The QTAIM computations have been carried out using the TOPOND code [30], interfaced with CRYSTAL [19]. Figures 2 and 3 report the electron density, the negative of the Laplacian of the electron density, and the spin density in selected crystallographic planes, while we collected in Table 2 the properties of Cu-F bonds in pure $KCuF_3$, along with the Cu-F bonding properties in $KCu_{0.875}Mg_{0.125}F_3$ for the three unique copper atoms closer to Mg (see also Figure 4). We reported just the results obtained for the AF1 phase, because no relevant change in Cu-F bonds occurs when the magnetic ordering of spins changes. Data in Table 2 are typical of interactions between closed shells, confirming that Cu-F bonds are indeed dominated by the large charge transfer from copper to fluorine (assuming the neutral atoms as reference). The value of the electron density at the *bond critical point (bcp)* is always small. The Laplacian at the *bcp* is large and positive, and hence it is dominated by its positive eigenvalue λ_3 corresponding to the curvature along the Cu-F internuclear axis. The *bcp* of Cu-F falls into a region of charge depletion; i.e., the electron density is not accumulated on the Cu-F axis but is removed from it, which is typical of interactions between closed shells. The differences among the Cu-F bonds can be entirely attributed to the different Cu-F internuclear distances: as Cu-F lengthens, ρ_b, $\nabla^2\rho_b$, and λ_{3b} decrease in absolute value. We observed no change in the nature of Cu-F bonds due to the presence of Mg into the lattice, and also in this case the variations of the bonding properties are strictly related to the changes of the internuclear Cu-F distance.

The atomic properties of pure $KCuF_3$ according to the QTAIM are collected in Table 3. We first note that the charge of copper atoms deviates significantly from the formal value of +2e, while the charge of potassium is quite close to +1e. Accordingly, the charge of fluorine atoms is far from the formal value of -1e, and its value is -0.76e for F atoms along the c axis and -0.78e for F atoms in ab planes. This indicates that the material is not fully ionic.

The local distribution of electronic density within the atomic basins can be rationalized analyzing its traceless quadrupole moment tensor, Q. The eigenvectors of Q associated with negative (positive) eigenvalues correspond to the directions of charge accumulation (depletion) with respect to an ideal spherical distribution. In pure and Mg-doped $KCuF_3$ potassium ions are nearly spherical, and the eigenvalues of Q are very small. Conversely, the charge distribution in the atomic basins of copper atoms presents a large anisotropy. The electron density is preferentially accumulated along the long Cu-F bonds at 2.26 Å, with the corresponding eigenvalue of Q that measures about -1, and it is removed from the directions of the shorter Cu-F bonds at 1.93 and 1.99 Å, with the eigenvalues of Q that measure +0.36 (Cu-F(c) at 1.99 Å) and +0.69 (Cu-F(ab) at 1.93 Å). As already observed in the literature [29–31], there is a nice agreement between orbital view and quadrupole moment tensor analysis in systems containing transition metals compounds. Consistently, in our material both approaches predict that the electron density of copper is removed from the axes of the short Cu-F bonds and accumulated along the longest bonds.

TABLE 2: Cu-F bonding properties in the AF1 phase of pure and Mg-doped $KCuF_3$.

System	bond	distance[a]	$\rho_b \cdot 10^{2\,b,c}$	$\nabla^2\rho_b \cdot 10^{2\,b,d}$	$\lambda_{3\,b} \cdot 10^{2\,b,e}$
$KCuF_3$	Cu-F (ab)	1.93	7.8	51	71
	Cu-F (ab)	2.26	3.7	15	23
	Cu-F (c)	1.99	6.5	41	58
$KCu_{0.875}Mg_{0.125}F_3$	Cu(a)-F (ab)	1.99	6.7	41	58
	Cu(a)-F (ab)	2.26	3.7	15	23
	Cu(a)-F (c)	1.99	6.5	40	57
	Cu(b)-F (ab)	1.93	7.7	51	71
	Cu(b)-F (ab)	2.25	3.8	16	24
	Cu(b)-F (c)	1.99	6.5	41	58
	Cu(c)-F (ab)	1.93	7.8	51	71
	Cu(c)-F (ab)	2.26	3.7	15	23
	Cu(c)-F (c)	2.00	6.4	40	56

[a] Values in Å; [b] the subscript b indicates that the properties are evaluated at the *bond critical point*; [c] ρ is the electron density; [d] $\nabla^2\rho$ is the Laplacian of the electron density; [e] λ_3 is the positive eigenvalue of the Hessian of the electron density.

TABLE 3: Atomic properties in pure $KCuF_3$ computed with the QTAIM approach.

System	Atom	q^a	V^b	Q^c
AF1	K	0.93	124	-0.02
				0.01
				0.01
	Cu	1.38	61	-1.05
				0.36
				0.69
	F(c)	-0.76	93	-0.49
				-0.49
				0.97
	F(ab)	-0.78	96	-0.44
				-0.41
				0.85
AF2	Cu	1.39	61	-1.00
				0.28
				0.72
AF3	Cu	1.39	61	-1.06
				0.36
				0.69
F	Cu	1.39	61	-1.02
				0.33
				0.69

[a] Atomic charge (electrons); [b] atomic volume (bohr³); [c] eigenvalues of the traceless quadrupole moment tensor (a.u.); positive (negative) values indicate that the electron density is preferentially removed (accumulated) along the direction of the corresponding eigenvector. The first eigenvalue for Cu atoms corresponds to the direction of the longest Cu-F bonds.

Quite interestingly, also the charge distribution of F anions is significantly anisotropic. The positive eigenvalues of Q (+0.97 for F(c), +0.85 for F(ab)) indicate that fluorine atoms remove electron density from the Cu-F bonding direction and accumulate charge away from copper neighbours. As can be seen in Table 3, copper atoms behave similarly in all the magnetic phases considered, and the same is found also for the other atoms of the crystals. The very small variations of the atomic properties observed among the four magnetic phases considered do not follow any recognizable trend.

The insertion of Mg atoms in the structure induces a strong local perturbation. In fact, the QTAIM charge of Mg is about +1.77e, significantly larger as compared to copper, and hence about 0.4e is delivered in the structure. Furthermore, the atomic volume of Mg is 33bohr³, much smaller than the value of 61bohr³ exhibited by Cu. Finally, the distribution of the electron density in the atomic basin of Mg atoms is almost spherical, with eigenvalues of Q comparable to those of K atoms.

The strong perturbation due to Mg doping is nearly entirely counterbalanced by the six fluorine atoms closer to Mg. The electron population of fluorine atoms increase by 0.10, 0.07, and 0.04e for F neighbours at 1.93, 1.98, and 2.20 Å, respectively. The atomic volumes increase accordingly: +6bohr³ for F atoms at 1.93 and +5bohr³ and +4bohr³ for F atoms at 1.98 and 2.20 Å. It is worth noting that also the distribution of electron density within the atomic basins of F atoms close to Mg becomes less anisotropic: the positive eigenvalue of Q for F neighbours in *ab* planes decreases from about +0.85 in the pure sample to about +0.55 in the doped compound and from +0.97 to +0.71 for F neighbours along the *c* axis. No relevant change in the atomic properties of other atoms is observed.

A complementary view to the real space description of the electron density distribution is given by the band structure and the analysis of the Density of States (DOS) obtained by the band structure. First of all, at the B3LYP level of theory our computations correctly predict that the system is an insulator whatever the magnetic phase considered, and Mg doping does not alter significantly the band gap. The system remains nonconducting even imposing an orbitally ordered solution to the undistorted cubic structure. Figure 1 of the SI section confirms that the valence states of pure and Mg-doped distorted $KCuF_3$ comes from a partial hybridization

$\rho(r)$

$-\nabla^2\rho(r)$

spin density$(\alpha\text{-}\beta)$

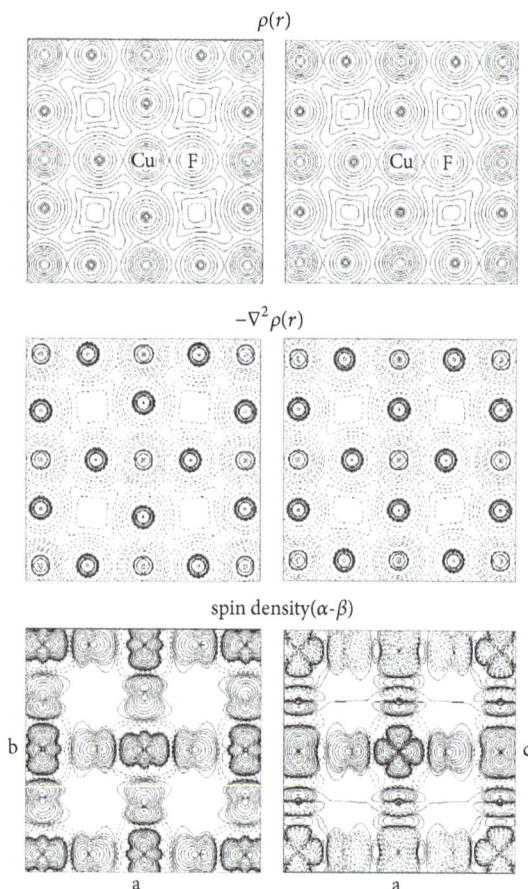

FIGURE 2: Contour plots of the electron density, the Laplacian of the electron density changed in sign and spin density in pure $KCuF_3$. Left panels refer to the ab and the right panels to the ac plane. Full and dashed lines indicate positive and negative values of the plotted function, respectively. Plots refer to the AF1 phase.

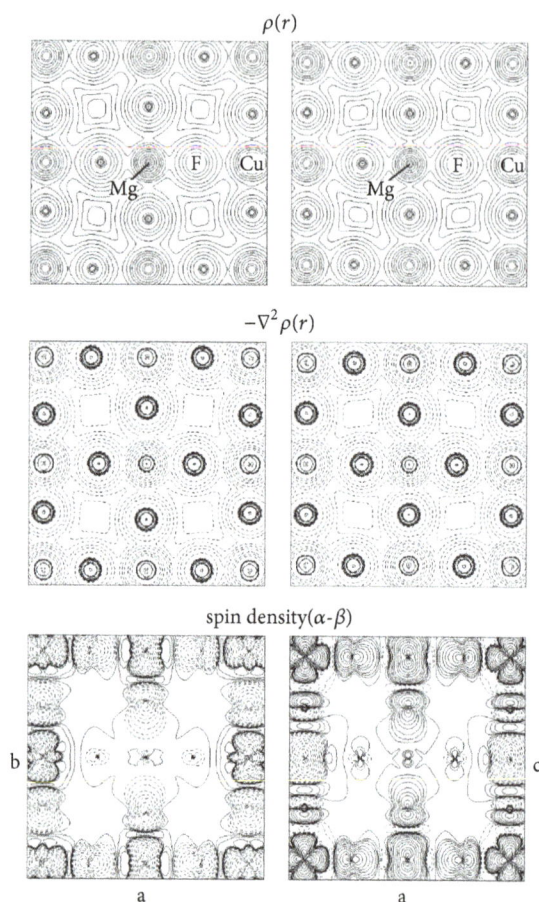

$\rho(r)$

$-\nabla^2\rho(r)$

spin density$(\alpha\text{-}\beta)$

FIGURE 3: Contour plots of the electron density, the Laplacian of the electron density changed in sign and spin density in Mg-doped $KCuF_3$. Left panels refer to the ab and the right panels to the ac plane. Full and dashed lines indicate positive and negative values of the plotted function, respectively. Plots refer to the AF1 phase.

between copper and fluorine orbitals, consistently with the picture of the mixed covalent-ionic character of Cu-F bonds, as evidenced above by QTAIM analysis. The contribution of fluorine orbitals to valence states becomes relevant at 2.25 eV below the Fermi level and increases moving to states at even lower energies. The Mg and K atoms act just as electron donors, and their orbitals do not concur to the formation of electronic states in the valence region.

In Figure 5 we reported the DOS projected onto the d orbitals of copper in the pure crystal with AF1 magnetic ordering. To analyze the population of individual orbitals within the copper d shell, we applied a rotation of the reference frame such that the z axis is directed along the longest Cu-F bonds at 2.26 Å, while x and y are aligned to the shorter Cu-F bonds. For the sake of clarity, the d_{xy}, d_{xz}, and d_{yz} are grouped together in Figure 5 and labelled "t_{2g}" (black curve) while $d_{x^2-y^2}$ (blue curve) and d_{z^2} (red curve) orbitals are drawn separately. Figure SI2 of the SI reports the DOS of individuals "t_{2g}" orbitals. It is easily seen that the d orbitals of copper are completely filled but for the $d_{x^2-y^2}$ one, which is empty in just one spin component neglecting a small contribution at about 5 eV below the Fermi level. This

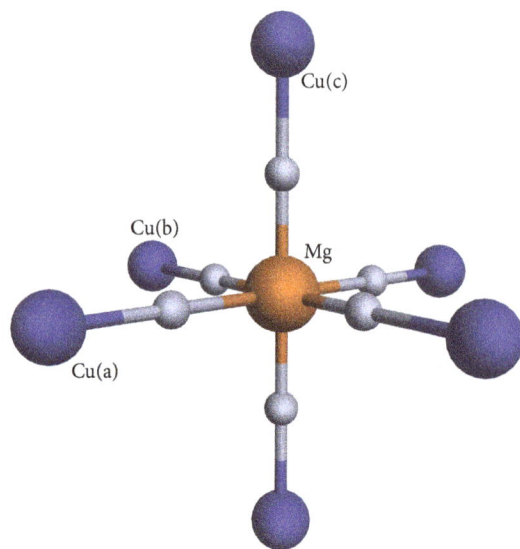

FIGURE 4: Stick-and-ball representation of the atoms surrounding Mg in the doped compound. Mg is 2.20 and 1.94 Å from fluorine atoms shared with Cu(a) and Cu(b), respectively.

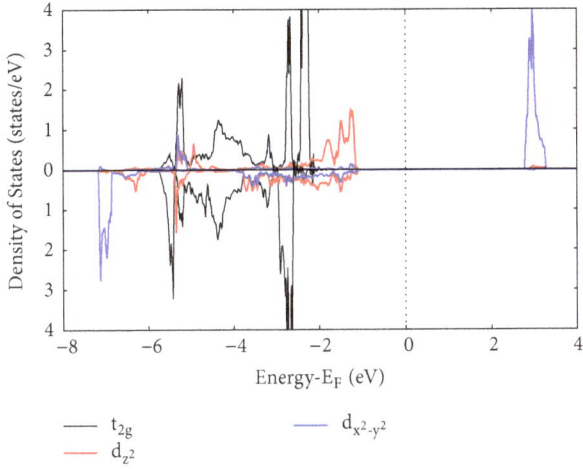

FIGURE 5: Projected density of states on copper atoms of pure $KCuF_3$ determined at the B3LYP level of theory for the AF1 phase. DOS in arbitrary units. Upper and bottom panels refer to different spin components. For the sake of clarity, d_{xy}, d_{xz}, and d_{yz} are grouped together and called "t_{2g}" (black curve) while $d_{x^2-y^2}$ (blue curve) and d_{z^2} (red curve) ones are drawn separately.

is consistent with the QTAIM analysis, in particular with the analysis of Q, indicating that copper distributes its electron density preferentially along the axis of the longest Cu-F bonds, while the axes of the Cu-F intermediate and short bonds are somewhat depleted. Furthermore, this confirms that at this level of theory the system is completely orbitally ordered, consistently with theoretical [2, 20, 32–34] and experimental [35, 36] results for this system. We further note that the DOS coming from t_{2g} orbitals (d_{xy}, d_{xz}, and d_{yz}) is mainly concentrated in sharp peaks, while the d_{z^2} one is closer to the Fermi level and distributed onto a wider range of energies. This picture does not change when different magnetic orderings are considered, and also the insertion of magnesium as dopant is not able to alter the filling and the ordering of copper orbitals. The only noticeable difference between the pure and doped materials is that the sharp peaks in the DOS of copper atoms become slightly broader (see Figure SI1).

As for the electron population, formally in $KCuF_3$ copper should be doubly positively charged and hence exhibit d^9 population. The Mulliken population analysis on the computed wavefunctions assigns to the d shell of copper a population of about 9.3 electrons in the pure and the Mg-doped crystals, which is consistent with the QTAIM outcome. The orbital population of copper does not depend on the spin ordering within the lattice, as we registered changes of about 0.01e among the different magnetic phases. The insertion of Mg into the crystal lattice and the relaxation of the position of fluorine atoms have no relevant effects onto the population of the d orbitals of copper, suggesting that (i) the perturbation due to Mg doping is extremely localized, (ii) slight modifications of the Jahn-Teller distortion around Cu have little effect on the orbital ordering, and therefore (iii) the Jahn-Teller distortion and the orbital ordering have different energy scales.

The orbital occupation of copper atoms implies that spin polarization around Cu is concentrated along the intermediate and short Cu-F bonds, as can be appreciated in Figure 2 for the AF1 phase. In the doped material no relevant spin polarization is found at the Mg sites, while neighbouring copper atoms remain essentially unchanged (Figure 3, AF1 phase). It is worth noting that a small amount of spin polarization is found also on fluorine atoms, and it depends strictly on the spin of neighbouring Cu atoms. In fact, copper neighbours at 1.93 Å in ab planes induce a spin population of about 0.08e on fluorine; the two Cu neighbours at 1.99 Å along c induce a polarization of 0.04e each, and the effect due to Cu atoms at 2.26 Å is negligible. As a result, in all magnetic phases fluorine atoms in ab planes have a net alfa-beta population of 0.08e, whose sign is the same of the closest copper atom. Conversely, F atoms along c have a zero net alfa-beta population in AF1 and AF3, and twice 0.04e = 0.08e in F and AF2. These simple rules apply also in the Mg-doped crystals, with the remark that no spin polarization comes from Mg atoms, as can be seen in Figure 3. Interestingly, these results are in agreement with LDA+U calculations [4], according to which Cu(3d)-F(2p) hybridization induces a finite spin density on the fluorine ions.

4. Magnetic Coupling

The magnetic coupling in pure and Mg-doped $KCuF_3$ has been investigated according to the Ising model [37, 38], which is derived from the phenomenological Heisenberg spin Hamiltonian for the isotropic interaction between localized magnetic moments. Its application to solid state systems is based on quite rough approximations, but it is widely used because it gives reliable results and has the advantage of relating the magnetic coupling parameters J to quantities directly available from first principles computations. We refer the interested reader to [20, 39, 40] for a thorough discussion on this subject and applications.

We assumed that in $KCuF_3$ the magnetic Hamiltonian contains just two relevant terms, describing the superexchange interaction between first neighbouring Cu pairs in ab planes and along the c axis. The total energy difference between two magnetic phases can then be expressed using the simple formula:

$$E_{PHASE2} - E_{PHASE1} = 2 \cdot J_{ab} \cdot S^2 \cdot z_{ab} \cdot n + 2 \cdot J_c \cdot S^2 \cdot z_c \cdot n \quad (1)$$

where E is the total energy per cell, J_i are the magnetic coupling parameters, S is the magnetic moment of Cu atoms (taken as the average between the values in the two phases), z_i is the number of first neighbours of a single Cu atom changing spin from one phase to the other, and n is the number of Cu atoms per cell. Considering the magnetic phases referred to as F, AF1, AF2, and AF3, there are four magnetic transitions that allow directly computing J_{ab} and J_c:

$$E_{AF1} - E_F = 2 \cdot J_{ab} \cdot S^2 \cdot 0 \cdot n + 2 \cdot J_c \cdot S^2 \cdot 2 \cdot n$$

$$= 4 \cdot n \cdot J_c \cdot S^2$$

TABLE 4: Magnetic coupling parameters between Cu pairs in ab planes (J_{ab}) and along the c axis (J_c).

Material	Transition	Jab (K)	Jc (K)
$KCuF_3$	F - AF1		-631
$KCuF_3$	AF2 - AF3		-620
$KCuF_3$	AF1 - AF3	17	
$KCuF_3$	F - AF2	7	
$KCu_{0.875}Mg_{0.125}F_3$	F - AF1		-655
$KCu_{0.875}Mg_{0.125}F_3$	AF2 - AF3		-611
$KCu_{0.875}Mg_{0.125}F_3$	AF1 - AF3	8	
$KCu_{0.875}Mg_{0.125}F_3$	F - AF2	28	

$$E_{AF3} - E_{AF2} = 2 \cdot J_{ab} \cdot S^2 \cdot 0 \cdot n + 2 \cdot J_c \cdot S^2 \cdot 2 \cdot n$$

$$= 4 \cdot n \cdot J_c \cdot S^2$$

$$E_{AF3} - E_{AF1} = 2 \cdot J_{ab} \cdot S^2 \cdot 4 \cdot n + 2 \cdot J_c \cdot S^2 \cdot 0 \cdot n$$

$$= 8 \cdot n \cdot J_{ab} \cdot S^2$$

$$E_{AF2} - E_F = 2 \cdot J_{ab} \cdot S^2 \cdot 4 \cdot n + 2 \cdot J_c \cdot S^2 \cdot 0 \cdot n$$

$$= 8 \cdot n \cdot J_{ab} \cdot S^2$$

$$(2)$$

In fact, in the phase transitions F to AF1 and AF2 to AF3 the magnetic coupling between Cu-Cu neighbours in *ab* planes remains unchanged, while the two neighbours of each copper atom directed along the *c* axis change spin. The opposite happens for the AF1 to AF3 and the F to AF2 transitions, where the four first neighbours of each Cu atom in *ab* planes change spin, while the ordering along the *c* axis is preserved.

When $KCuF_3$ is doped with Mg, the copper atoms become inequivalent, and hence the magnetic coupling cannot be described just in terms of the two magnetic parameters J_{ab} and J_c. However, as discussed in the previous section, the perturbation induced by Mg on the electronic properties of Cu atoms is extremely localized, and doping does not alter the properties of copper atoms nor the Cu-Cu internuclear distances. We then applied also in $KCu_{0.875}Mg_{0.125}F_3$ the formulae used for the pure compound. We averaged the magnetic moment S of all Cu atoms within the cell and considered just that the insertion of Mg into the lattice reduces the number of Cu-Cu superexchange interactions as compared to the pure compound. The computed magnetic coupling parameters J_{ab} and J_c in $KCu_{0.875}Mg_{0.125}F_3$ are thus effective values averaged over all the Cu-Cu interactions in the *ab* planes and along the *c* axis, respectively. The magnetic coupling parameters computed for pure and Mg-doped $KCuF_3$ at the B3LYP level of theory for the optimized structure are reported in Table 4.

It is well known that the evaluation of the magnetic coupling depends critically on the functionals adopted in the computation [41], and our computations confirm that a change in the percentage of Hartree-Fock exchange included

in the Hamiltonian induces dramatic changes in the evaluation of J terms. Even worse, literature studies prove that no systematic rule allows *a priori* defining the best functional for a given system. However, as will be discussed in the following, the B3LYP approach proves adequate for the materials under investigation in the present study.

First of all, B3LYP magnetic coupling parameters computed for the pure optimized structure compare well with the experimental data [35, 36, 42–46], and to vary the Hartree-Fock exchange contribution associated with the BECKE/LYP functionals leads to largely inadequate results, which even fail to reproduce the correct order of magnitude of the J terms. Data obtained on $KCuF_3$ with these same exchange and correlation functionals but different Hartree-Fock exchange contributions are collected in Tables 1SI and 2SI of the SI section.

We also checked that, at the B3LYP level of theory, the ideal undistorted structure is not an energy minimum, and whatever small perturbation in the cubic lattice induces relaxation to the distorted minimum. An orbitally ordered solution has been obtained also for the undistorted lattice, and its energy is slightly smaller than the nonorbitally ordered one. Test computations conducted with completely different functionals, namely, the PBE and the VBH ones, foresee that orbital ordering is not complete even in the distorted structure (the population of the d_{z^2} orbital decreases) and disappears when an undistorted geometry around copper atoms is applied. At the VBH level of theory the system is conducting not just in the cubic phase but also upon distortion of the lattice.

According to B3LYP results, the antiferromagnetic ordering of Cu spins along the *c* axis is largely favoured, while the ferromagnetic ordering is preferred in *ab* planes. Considering the F to AF1 transition or the AF2 to AF3 one gives very similar results for J_c, and the same happens for J_{ab} determined with the AF1 to AF3 or the AF2 to F phase transition. The computed value for J_c measures about -600 K, while the order of magnitude of J_{ab} is +10 K. These data compare very favourably with first principles computations on this same material conducted by Moreira et al. [20], and they are also consistent with the experiments [35, 36, 42–46] conducted on pure $KCuF_3$ samples.

Magnetic coupling parameters obtained for the Mg-doped material are very close to the values of the pure compound. The insertion of a single Mg atom on all 8 Cu sites clearly reduces the number of Cu-Cu superexchange interactions, thus decreasing the energy differences between AF1 and F magnetic phases, in agreement with the experiment [26], but their average strength remains nearly unchanged. These findings suggest that a 12.5% Mg doping of $KCuF_3$ does not alter the fundamental mechanism and the magnitude of the interactions building up the magnetic ordering of copper spins within the structure. The magnetic coupling between copper atoms (and hence the energy difference between different magnetic phases) is therefore strictly related to their local environment, i.e., to the number of superexchange interactions between neighbouring Cu atoms, regardless of the presence of Mg as doping element. This result is consistent with previous literature data obtained with first principles cluster simulations of the pure compound [47, 48].

5. General Discussion and Conclusions

In this paper we investigated the electronic and magnetic properties of pure $KCuF_3$ and of the Mg-doped $KCu_{0.875}Mg_{0.125}F_3$ compound, by means of first principles periodic Density Functional computations. Tests conducted with various functionals indicated that the B3LYP scheme is adequate for these systems.

Optimized structural parameters for the pure compound agree well with experimental data. Conversely, our approach does not reproduce the observed crystal cell variations upon substitution of Cu with Mg atoms. A reason for such discrepancy might be the clustering of Mg atoms in real samples, while in our study they are regularly distributed within the lattice. The structural effect of Mg insertion is just a reduction of the anisotropy of Mg-F distances upon relaxation.

Both in pure and doped compounds, we considered the four possible combinations of ferromagnetic (FM) and antiferromagnetic (AF) ordering between Next Nearest Neighbours (NNN) Cu-Cu pairs: the F, AF1, AF2, and AF3 phases. Our calculations correctly predict an insulating AF1 orbitally ordered ground state. Switching the FM interaction in the *ab* plane into AF has a very limited energy cost, while a marked energy increase is found when the AF interactions along the *c* axis are turned into FM ones. These energy differences reflect on the estimate of the magnetic coupling parameters J_{ab} and J_c computed according to the Ising model (about +10 K and -600 K, respectively) and nicely agree with available experimental results. Doping with Mg has a very limited effect on the estimate of J_{ab} and J_c. Overall, the properties of all atoms, with the exception of the six first neighbouring fluorine, seem to be unaffected by Mg insertion, suggesting that doping with Mg induces a strong but extremely localized perturbation. In particular, the bond properties, the absolute charges, and the electronic distribution in the 3d orbitals of Cu ions NNN of Mg are identical to the pure $KCuF_3$ compound. It is worth noticing that a complete orbital ordering of the 3d electrons is retained in the doped sample even for Cu(a) ion (see Figure 4), despite the significant elongation of the shortest Cu(a)-F bond from 1.93 to 1.99 Å. In other words the OO seems to be quite unaffected by structural distortions. This is in accord with the disentanglement of OO and cJTD foreseen by the model of Kugel and Khomskii [1]. Since Mg doping does not alter significantly the estimate of J_{ab} and J_c, the energy ordering of the four magnetic phases is maintained, and the lowering of the energy difference between the AF1 and F phases can be explained just considering the reduced number of Cu-F-Cu superexchange interactions in the unit cell.

In agreement with previous literature, either in the pure or in the doped compound, the 3d shell of Cu atoms is only partially filled, and residual spin polarization is found also on fluorine atoms. Copper atoms preferentially accumulate electron density along the two long Cu-F bonds, while they exhibit a depletion along the four shortest Cu-F bonds. The Cu-F interactions are dominated by the charge transfer from the metal to the halogen atom but retain a partially shared character, and hence $KCuF_3$ should not be considered a purely ionic material. This is confirmed by the relatively small charge of the Cu ion (+1.38e), which is quite far from the formal value of +2e. Mg delivers a larger amount of electron density towards the neighbouring atoms as compared to Cu, but this is almost entirely counterbalanced by neighbouring F atoms, which explains the moderate changes in the magnetic coupling parameters.

All the above findings converge in suggesting the extreme localization of the perturbation due to Mg doping in the $KCuF_3$ system. The absence of long range effects, which should affect the physical properties of the hosting compounds in a more complex way, makes Mg a suitable dopant for the investigation of strongly correlated electronic systems by means of orbital dilution, and this is particularly true in Cu based compounds, due to the very similar ionic radii of Cu^{2+} and Mg^{2+} in octahedral coordination (0.73 and 0.72 Å, respectively).

Conflicts of Interest

The authors declare that there are no conflicts of interest regarding the publication of this paper.

Acknowledgments

The authors gratefully acknowledge Dr. Davide Ceresoli for useful discussions

References

[1] K. I. Kugel and D. I. Khomskii, "The jahn-teller effect and magnetism: Transition metal compounds," *Soviet Physics—Uspekhi*, vol. 25, no. 4, pp. 621–641, 1982.

[2] E. Pavarini, E. Koch, and A. I. Lichtenstein, " Mechanism for Orbital Ordering in ," *Physical Review Letters*, vol. 101, no. 26, 2008.

[3] L. Paolasini, R. Caciuffo, A. Sollier, P. Ghigna, and M. Altarelli, "Coupling between spin and orbital degrees of freedom in KCuF3," *Physical Review Letters*, vol. 88, no. 10, pp. 1064031–1064034, 2002.

[4] N. Binggeli and M. Altarelli, "Orbital ordering, Jahn-Teller distortion, and resonant x-ray scattering in KCuF3," *Physical Review B: Condensed Matter and Materials Physics*, vol. 70, no. 8, pp. 1–85117, 2004.

[5] M. Scavini, M. Brunelli, C. Ferrero, C. Mazzoli, and P. Ghigna, "Experimental estimation of the cooperative Jahn-Teller energy in orbitally ordered KCu 0.8Mg 0.2F 3 perovskite," *The European Physical Journal B*, vol. 65, no. 2, pp. 187–190, 2008.

[6] P. Ghigna, M. Scavini, C. Mazzoli, M. Brunelli, C. Laurenti, and C. Ferrero, "Experimental disentangling of orbital and lattice energy scales by inducing cooperative Jahn-Teller melting in

KCu$_{1-x}$ Mg$_x$ F$_3$ solid solutions," *Physical Review B: Condensed Matter and Materials Physics*, vol. 81, no. 7, 2010.

[7] I. Leonov, D. Korotin, N. Binggeli, V. I. Anisimov, and D. Vollhardt, "Computation of correlation-induced atomic displacements and structural transformations in paramagnetic KCuF3 and LaMnO3," *Physical Review B: Condensed Matter and Materials Physics*, vol. 81, no. 7, 2010.

[8] C.-Z. Wang, D.-Y. Liu, H.-B. Tang, J. Liu, and L.-J. Zou, "Spin flop transitions under strong magnetic fields in orbital ordered KCuF3," *Physica B: Condensed Matter*, vol. 405, no. 5, pp. 1423–1427, 2010.

[9] E. Pavarini, "Lattice distortions in KCuF3: A paradigm shift?" *Annalen der Physik (Leipzig)*, vol. 523, no. 10, pp. 865–866, 2011.

[10] J.-S. Zhou, J. A. Alonso, J. T. Han, M. T. Fernández-Díaz, J.-G. Cheng, and J. B. Goodenough, "Jahn-Teller distortion in perovskite KCuF3 under high pressure," *Journal of Fluorine Chemistry*, vol. 132, no. 12, pp. 1117–1121, 2011.

[11] J. C. T. Lee, S. Yuan, S. Lal et al., "Two-stage orbital order and dynamical spin frustration in KCuF3," *Nature Physics*, vol. 8, no. 1, pp. 63–66, 2012.

[12] V. Gnezdilov, J. Deisenhofer, P. Lemmens et al., "Phononic and magnetic excitations in the quasi-one-dimensional Heisenberg antiferromagnet KCuF 3," *Low Temperature Physics*, vol. 38, no. 5, pp. 419–427, 2012.

[13] S. Yuan, M. Kim, J. T. Seeley et al., "Inelastic light scattering measurements of a pressure-induced quantum liquid in KCuF3," *Physical Review Letters*, vol. 109, no. 21, 2012.

[14] L. G. Marshall, J. Zhou, J. Zhang et al., " Unusual structural evolution in KCuF ," *Physical Review B: Condensed Matter and Materials Physics*, vol. 87, no. 1, 2013.

[15] D. Legut and U. D. Wdowik, "Vibrational properties and the stability of the KCuF3 phases," *Journal of Physics: Condensed Matter*, vol. 25, no. 11, 2013.

[16] V. V. Iglamov and M. V. Eremin, "Orbital order fluctuations in KCuF3," *Optics and Spectroscopy (English translation of Optika i Spektroskopiya)*, vol. 116, no. 6, pp. 828–831, 2014.

[17] K. Bieniasz, M. Berciu, M. Daghofer, and A. M. Oleś, "Green's function variational approach to orbital polarons in KCuF3," *Physical Review B: Condensed Matter and Materials Physics*, vol. 94, no. 8, 2016.

[18] K. Bieniasz, M. Berciu, and A. M. Oleś, "Orbiton-magnon interplay in the spin-orbital polarons of KCuF3 and LaMnO3," *Physical Review B: Condensed Matter and Materials Physics*, vol. 95, no. 23, 2017.

[19] R. Dovesi, V. R. Saunders, C. Roetti et al., *CRYSTAL06 User's Manual*, University of Torino, Turin, Italy, 2006.

[20] I. D. P. R. Moreira and R. Dovesi, "Periodic approach to the electronic structure and magnetic coupling in KCuF3, K2CuF4, and Sr2CuO 2Cl2 low-dimensional magnetic systems," *International Journal of Quantum Chemistry*, vol. 99, no. 5, pp. 805–823, 2004.

[21] R. Dovesi, C. Roetti, C. Freyria-Fava, M. Prencipe, and V. R. Saunders, "On the elastic properties of lithium, sodium and potassium oxide. An ab initio study," *Chemical Physics*, vol. 156, no. 1, pp. 11–19, 1991.

[22] K. Doll and N. M. Harrison, "Chlorine adsorption on the Cu(111) surface," *Chemical Physics Letters*, vol. 317, no. 3-5, pp. 282–289, 2000.

[23] R. Nada, C. R. A. Catlow, C. Pisani, and R. Orlando, "An ab-initio hartree-fock perturbed-cluster study of neutral defects in lif," *Modelling and Simulation in Materials Science and Engineering*, vol. 1, no. 2, pp. 165–187, 1993.

[24] L. Valenzano, Y. Noël, R. Orlando, C. M. Zicovich-Wilson, M. Ferrero, and R. Dovesi, "Ab initio vibrational spectra and dielectric properties of carbonates: Magnesite, calcite and dolomite," *Theoretical Chemistry Accounts*, vol. 117, no. 5-6, pp. 991–1000, 2007.

[25] R. H. Buttner, E. N. Maslen, and N. Spadaccini, "Structure, electron density and thermal motion of KCuF3," *Acta Crystallographica Section B: Structural Science*, vol. 46, no. 2, pp. 131–138, 1990.

[26] C. Oliva, M. Scavini, S. Cappelli, C. Bottalo, C. Mazzoli, and P. Ghigna, "Melting of orbital ordering in KMgxCu1- xF3 solid solution," *The Journal of Physical Chemistry B*, vol. 111, no. 21, pp. 5976–5983, 2007.

[27] C. Gatti, "Chemical bonding in crystals: new directions," *Zeitschrift für Kristallographie - Crystalline Materials*, vol. 220, no. 5/6, pp. 399–457, 2005.

[28] F. Cargnoni and M. Scavini, "Direct-space analysis of the electronic structure of the YBa2Cu3O6 and YBa2Cu3O7 crystals," *Canadian Journal of Chemistry*, vol. 80, no. 3, pp. 235–244, 2002.

[29] L. Bertini, F. Cargnoni, and C. Gatti, "Chemical insight into electron density and wave functions: Software developments and applications to crystals, molecular complexes and materials science," *Theoretical Chemistry Accounts*, vol. 117, no. 5-6, pp. 847–884, 2007.

[30] C. Gatti, *TOPOND-98 User's Manual* , CNR - CSRSRC, Italy, Milan, 1998.

[31] R. Bianchi, C. Gatti, V. Adovasio, and M. Nardelli, "Theoretical and Experimental (113 K) Electron-Density Study of Lithium Bis(tetramethylammonium) Hexanitrocobaltate(III)," *Acta Crystallographica Section B: Structural Science*, vol. 52, no. 3, pp. 471–478, 1996.

[32] A. I. Liechtenstein, V. I. Anisimov, and J. Zaanen, "Density-functional theory and strong interactions: orbital ordering in Mott-Hubbard insulators," *Physical Review B: Condensed Matter and Materials Physics*, vol. 52, no. 8, pp. R5467–R5470, 1995.

[33] M. D. Towler, R. Dovesi, and V. R. Saunders, "Magnetic interactions and the cooperative Jahn-Teller effect in KCuF3," *Physical Review B: Condensed Matter and Materials Physics*, vol. 52, no. 14, pp. 10150–10159, 1995.

[34] I. Leonov, N. Binggeli, D. Korotin, V. I. Anisimov, N. Stojić, and D. Vollhardt, "Structural relaxation due to electronic correlations in the paramagnetic insulator KCuF3," *Physical Review Letters*, vol. 101, no. 9, 2008.

[35] D. A. Tennant, R. A. Cowley, S. E. Nagler, and A. M. Tsvelik, "Measurement of the spin-excitation continuum in one-dimensional KCuF3 using neutron scattering," *Physical Review B: Condensed Matter and Materials Physics*, vol. 52, no. 18, pp. 13368–13380, 1995.

[36] D. A. Tennant, S. E. Nagler, D. Welz, G. Shirane, and K. Yamada, "Effects of coupling between chains on the magnetic excitation spectrum of KCuF3," *Physical Review B: Condensed Matter and Materials Physics*, vol. 52, no. 18, pp. 13381–13389, 1995.

[37] L. J. De Jongh and A. R. Miedema, "Experiments on simple magnetic model systems," *Advances in Physics*, vol. 23, no. 1, pp. 1–260, 1974.

[38] L. J. De Jongh and R. Block, "On the exchange interactions in some 3d-metal ionic compounds. I. The 180∘ superexchange in the 3d-metal fluorides XMF3 and X2MF4 (X=K, Rb, Tl; M=Mn, Co, Ni)," *Physica B+C*, vol. 79, no. 6, pp. 568–593, 1975.

[39] Y.-S. Su, T. A. Kaplan, S. D. Mahanti, and Y.-S. Harrison, "Crystal Hartree-Fock calculations for La2NiO4 and La2CuO4,"

Physical Review B: Condensed Matter and Materials Physics, vol. 59, no. 16, pp. 10521–10529, 1999.

[40] I. de P. R. Moreira and F. Illas, " Ab initio theoretical comparative study of magnetic coupling in ," *Physical Review B: Condensed Matter and Materials Physics*, vol. 55, no. 7, pp. 4129–4137, 1997.

[41] N. Wannarit, C. Pakawatchai, I. Mutikainen et al., "Hetero triply-bridged dinuclear copper(ii) compounds with ferromagnetic coupling: A challenge for current density functionals," *Physical Chemistry Chemical Physics*, vol. 15, no. 6, pp. 1966–1975, 2013.

[42] S. K. Satija, J. D. Axe, G. Shirane, H. Yoshizawa, and K. Hirakawa, "Neutron scattering study of spin waves in one-dimensional antiferromagnet KCuF3," *Physical Review B: Condensed Matter and Materials Physics*, vol. 21, no. 5, pp. 2001–2007, 1980.

[43] S. Kadota, I. Yamada, S. Yoneyama, and K. Hirakawa, "Formation of one-dimensional antiferromagnet in KCuF3 with the perovskite structure," *Journal of the Physical Society of Japan*, vol. 23, no. 4, pp. 751–756, 1967.

[44] K. Iio, H. Hyodo, K. Nagata, and I. Yamada, "Study of Magnetic Energy in One-Dimensional S=1/2 Heisenberg Antiferromagnet KCuF3 by Optical Birefringence," *Journal of the Physical Society of Japan*, vol. 44, no. 4, pp. 1393-1394, 1978.

[45] M. T. Hutchings, E. J. Samuelsen, G. Shirane, and K. Hirakawa, "Neutron-diffraction determination of the antiferromagnetic structure of KCuF3," *Physical Review A: Atomic, Molecular and Optical Physics*, vol. 188, no. 2, pp. 919–923, 1969.

[46] S. E. Nagler, D. A. Tennant, R. A. Cowley, T. G. Perring, and S. K. Satija, "Spin dynamics in the quantum antiferromagnetic chain compound KCuF3," *Physical Review B: Condensed Matter and Materials Physics*, vol. 44, no. 22, pp. 12361–12368, 1991.

[47] I. D. P. R. Moreira, F. Illas, and D. Maynau, "Local character of magnetic coupling in ionic solids," *Physical Review B: Condensed Matter and Materials Physics*, vol. 59, no. 10, pp. R6593–R6596, 1999.

[48] I. D. P. R. Moreira and F. Illas, "Ab initio study of magnetic interactions in KCuF3 and K2CuF4 low-dimensional systems," *Physical Review B: Condensed Matter and Materials Physics*, vol. 60, no. 8, pp. 5179–5185, 1999.

24

Research on a Micro-Nano Si/SiGe/Si Double Heterojunction Electro-Optic Modulation Structure

Song Feng ⓘ,[1,2] **Lian-bi Li ⓘ,**[1] **and Bin Xue**[1]

[1]School of Science, Xi'an Polytechnic University, Xi'an 710048, China
[2]State Key Laboratory of Functional Materials for Informatics, Shanghai Institute of Microsystem and Information Technology, Chinese Academy of Sciences, 865 Changning Road, Shanghai 200050, China

Correspondence should be addressed to Lian-bi Li; xpu_lilianbi@163.com

Academic Editor: Zaiquan Xu

The electro-optic modulator is a very important device in silicon photonics, which is responsible for the conversion of optical signals and electrical signals. For the electro-optic modulator, the carrier density of waveguide region is one of the key parameters. The traditional method of increasing carrier density is to increase the external modulation voltage, but this way will increase the modulation loss and also is not conducive to photonics integration. This paper presents a micro-nano Si/SiGe/Si double heterojunction electro-optic modulation structure. Based on the band theory of single heterojunction, the barrier heights are quantitatively calculated, and the carrier concentrations of heterojunction barrier are analyzed. The band and carrier injection characteristics of the double heterostructure structure are simulated, respectively, and the correctness of the theoretical analysis is demonstrated. The micro-nano Si/SiGe/Si double heterojunction electro-optic modulation is designed and tested, and comparison of testing results between the micro-nano Si/SiGe/Si double heterojunction micro-ring electro-optic modulation and the micro-nano Silicon-On-Insulator (SOI) micro-ring electro-optic modulation, Free Spectrum Range, 3 dB Bandwidth, Q value, extinction ratio, and other parameters of the micro-nano Si/SiGe/Si double heterojunction micro-ring electro-optic modulation are better than others, and the modulation voltage and the modulation loss are lower.

1. Introduction

With the rapid development of optoelectronic technology, optoelectronic devices have entered the nano era [1, 2], and many nanotechnology has been used in silicon photonics [3–5]. The reduced size of the device can improve the utilization rate of the chip area, but also the higher performance of the device is required, and the silicon epitaxial material is selected to make optoelectronic devices with higher performance. For example, the epitaxial InP material on silicon substrate is used to fabricate laser [6], the epitaxial Ge material is used to make detector [7], the epitaxial SiGe material is used to fabricate modulator [8], and so forth. In optoelectronic integration, the electro-optic modulator is a very important device, which is responsible for the conversion of optical signals and electrical signals [9, 10]. For the electro-optic modulator, the carrier density of waveguide region is one of the key parameters. When the injected carrier density of

waveguide region is greater, the change of refractive index is greater, and also the modulator is more easily modulated [11, 12]. The traditional method of increasing carrier density is to increase the external modulation voltage, but this way will increase the modulation loss and also is not conducive to photonics integration [8, 13]. So SiGe technology is one way to solve this problem in silicon photonics. Photonics Electronics Technology Research Association (PETRA) and University of Tokyo have made a high-speed and highly efficient Si optical modulator with strained SiGe layer and demonstrated highly efficient modulations of 0.67 and 0.81 V·cm for VπL at dc reverse bias voltages of −0.5 and −2 V$_{dc}$, respectively, and also demonstrated a high-speed operation of 25 Gbps for the Si-MOD at a wavelength of around 1.3 μm [14]. University of Toronto has made SiGe BiCMOS linear modulator drivers, and the measured differential gain and bandwidth are over 20 dB and 70 GHz, respectively, and P$_{1\,dB}$ is −2.5 dBm [15]. Ghent University demonstrated single-wavelength, serial

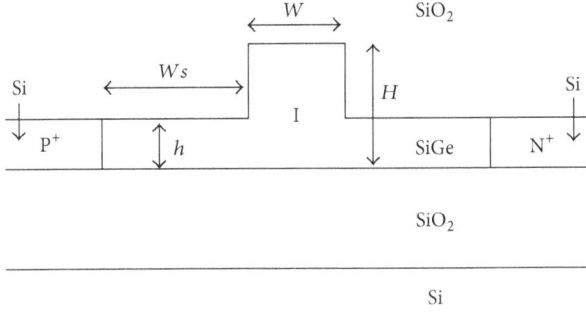

FIGURE 1: Si/SiGe/Si-OI structure.

and real-time 100 GB/s NRZ-OOK transmission over 500 m SSMF with a GeSi EAM implemented on a silicon photonics platform. The device was driven with 2 Vpp without 50 Ω termination, allowing a low-complexity solution for 400 GbE short-reach optical interconnects [16].

This paper presents a micro-nano Si/SiGe/Si double heterojunction electro-optic modulation structure, which can greatly improve the carrier injection concentration and decrease modulation voltage and loss. It is an ideal electro-optic modulator, which is conducive to the photonics integrated circuit miniaturization and integration and also provides a new way for the research of micro-nano optoelectronic devices based on silicon.

2. Structure and Energy Band

Based on the SOI PIN electro-optic modulation structure, a micro-nano Si/SiGe/Si double heterojunction electro-optic modulation structure is presented, which is abbreviated as Silicon/Silicon Germanium/Silicon-On-Insulator (Si/SiGe/Si-OI) structure, as shown in Figure 1. The bottom of the structure is On-Insulator (OI) material, the P+ and N+ regions in the top layer are silicon materials, and the I region is SiGe material in Figure 1. P+ region is doped by B element, and the doping concentration is 1×10^{19} cm^{-3}. N+ region is doped by P element, and the doping concentration is also 1×10^{19} cm^{-3}. I region is N type intrinsic SiGe, and the doping concentration is 1×10^{15} cm^{-3}. In Figure 1, structure parameters are $H = 220$ nm, $h = 50$ nm, $W = 400 \sim 600$ nm, and the width between active region and waveguide (Ws) is a variable value.

In order to facilitate comparison and analysis, two kinds of PIN modulation structures with Si and SiGe materials are established. One is the PIN modulation structure with Si based on the SOI PIN structure, which is abbreviated as SOI structure, as shown in Figure 2(a). The other is the PIN modulation structure with SiGe based on the Silicon Germanium-On-Insulator (SiGe-OI) PIN structure, which is abbreviated as SiGe-OI structure, as shown in Figure 2(b). The doping concentration and structure parameters of two structures are the same as Si/SiGe/Si double heterojunction electro-optic modulation structure.

Si/SiGe/Si-OI structure is composed of P type Si with the high doping concentration, N type SiGe with the low doping

concentration, and N type Si with the high doping concentration. Capital P and N represent Wide Band Gap Semiconductor, and lowercase n represents Narrow Band Gap Semiconductor. So the PIN structure can be equivalent as a Pn heterojunction and a nN heterojunction structure, and the band diagram is shown in Figure 3 under the equilibrium state.

When the PIN structure is working, the main carriers of n region are composed of electron from the P region and hole from the N region, and the barrier height is a key parameter who can influence injection efficiency of electron and hole. After heterojunction band of Si/SiGe/Si-OI structure is analyzed, the hole barrier height of Pn heterojunction of the Si/SiGe/Si-OI structure is represented by the following formula [17]:

$$-qV_D = -k_0 T \ln \frac{n_{i(\text{Si})} n_{i(\text{SiGe})}}{N_D N_A} - \frac{0.73x}{2}, \tag{1}$$

where qV_D is the barrier height, k_0 is the Boltzmann constant, T is the absolute temperature, $n_{i(\text{Si})}$ is the intrinsic carrier concentration of silicon, $n_{i(\text{SiGe})}$ is the intrinsic carrier concentration of Silicon Germanium, N_D is the doping concentration of n region, N_A is the doping concentration of P region, x is the Ge content. Formula (1) indicates that the hole barrier height of Pn heterojunction is related to the doping concentration of P region and n region, the band gap of materials, the temperature, and the Ge content. When the temperature is constant, the lower the doping concentration of P region and n region, the lower the hole barrier height. The narrower the band gap, the higher the intrinsic carrier concentration, and the lower the hole barrier height. The bigger the Ge content, the lower the hole barrier height.

Assuming that $N_A = 10^{19}$ cm^{-3}, $N_D = 10^{15}$ cm^{-3}, and the Ge content is 0.2, so the hole barrier height of Pn heterojunction can be calculated equal to $-qV_D = 0.7$ eV at room temperature. For SOI structure and SiGe-OI structure, the Pn junction is homojunction, and the hole barrier height can be analyzed according to the following formula [17]:

$$-qV_{D1,2} = E_{\text{FP}} - E_{\text{Fn}} = -kT \ln \frac{n_i^2}{N_D N_A}, \tag{2}$$

where E_{FP} is the Fermi level of the P region and E_{Fn} is the Fermi level of the n region. The hole barrier height of Pn homojunctions of the SOI structure can be calculated equal to $-qV_{D1} = 0.82$ eV, and the hole barrier height of Pn homojunctions of the SiGe-OI structure can be calculated equal to $-qV_{D2} = 0.733$ eV. In the same conditions, the hole barrier height of the Si/SiGe/Si-OI structure is lowest, and the hole barrier height of the SOI structure is highest. This result can be verified by Silvaco CAD software, as shown in Figure 4 [17].

The electron barrier height of nN heterojunction of the Si/SiGe/Si-OI structure can be represented by the following formula [17]:

$$qV_D' = kT \ln \frac{N_D' n_{i(\text{SiGe})}}{N_D n_{i(\text{Si})}} - \frac{0.73x}{2}, \tag{3}$$

where N_D' is the carrier concentration of N region and N_D is the carrier concentration of n region. Formula (3) indicates that

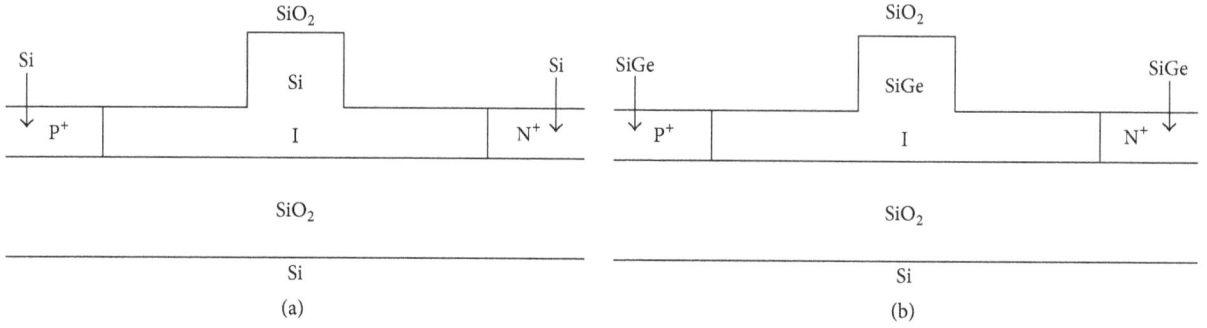

FIGURE 2: Two type modulation structures: (a) SOI structure and (b) SiGe-OI structure.

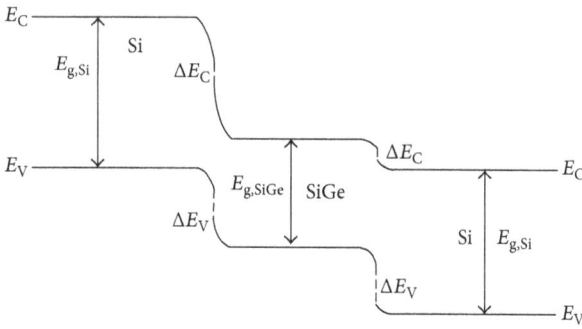

FIGURE 3: The band diagram of Si/SiGe/Si-OI structure under the equilibrium state.

that the electron barrier height of the nN heterojunction of Si/SiGe/Si-OI structure is also related to the doping concentration of N region and n region, the band gap of materials, the temperature, and the Ge content. When the temperature is constant, the lower the doping concentration of N region, the lower the hole barrier height. The higher the doping concentration of n region, the lower the hole barrier height. The wider band gap of n region, the lower the intrinsic carrier concentration, and the lower the hole barrier height. The narrower the band gap of N region, the higher the intrinsic carrier concentration, and the lower the hole barrier height. The smaller the Ge content, the lower the hole barrier height.

Assuming that $N'_D = 10^{19}$ cm^{-3}, $N_D = 10^{15}$ cm^{-3}, and the Ge content is 0.2, so the electron barrier height of nN heterojunction can be calculated equal to $qV'_D = 0.2$ eV at room temperature. For SOI structure and SiGe-OI structure, the nN junction is homojunction, and the electron barrier height can be analyzed according to the following formula [17]:

$$qV'_{D1,2} = E_{FN} - E_{Fn} = kT \ln \frac{N'_D}{N_D}. \tag{4}$$

The electron barrier height of nN homojunctions of the SOI structure can be calculated equal to $qV'_{D1} = 0.24$ eV, and the electron barrier height of nN homojunctions of the SiGe-OI structure can be calculated equal to $qV'_{D2} = 0.24$ eV. In the same conditions, the electron barrier height of the Si/SiGe/Si-OI structure is lowest, and the electron barrier height of the

SOI structure is highest. This result can be verified by Silvaco CAD software, as shown in Figure 4.

When the forward bias voltage is set between P region and N region, the original balance between carrier diffusion motion and drift motion is broken, and the PIN device is under nonequilibrium state. Because the doping concentration of the I region is very small, the resistance is very large. The doping concentration of the P region and the N region is very large, and the resistance is very small, so the external positive bias voltage is basically born in the I region. The forward bias voltage produces an electric field opposite to the Built-in Electric Field in the I region, which weakens the intensity of the electric field in the I region and reduces the space charge, so the barrier height decreases. Because the hole barrier height of Pn heterojunction and the electron barrier height of nN heterojunction of Si/SiGe/Si-OI structure all are lower than the barrier height of SiGe-OI structure and SOI structure, so the band of Si/SiGe/Si-OI structure is flattened first under the forward bias voltage. In other words, the Si/SiGe/Si-OI structure has the higher carrier injection at the same forward bias voltage. At 1 V forward voltage, the band simulations of three type modulation structures are shown in Figure 5 [17].

3. Result Analysis

In order to verify that the Si/SiGe/Si-OI structure has better characteristics, the carrier concentration of the SOI structure, the SiGe-OI structure, and the Si/SiGe/Si-OI structure are shown in Figure 6. Figure 6(a) shows the relationship between the electron concentration and modulation voltage for three structures, and Figure 6(b) shows the relationship between the hole concentration and modulation voltage. It can be seen from Figure 6 that, with the increase of modulation voltage, the carrier concentration of three kinds of modulation structure all are gradually increased. When the modulation voltage exceeds 0.6 V, the carrier concentration of the Si/SiGe/Si-OI structure is significantly greater than the carrier concentration of the SOI structure; when the modulation voltage is greater than 0.75 V, the carrier concentration of the Si/SiGe/Si-OI structure is significantly greater than the carrier concentration of SiGe-OI structure; when the modulation voltage is up to 0.9 V, the carrier concentration of the Si/SiGe/Si-OI structure is 6×10^{18} cm^{-3}, but the SOI structure

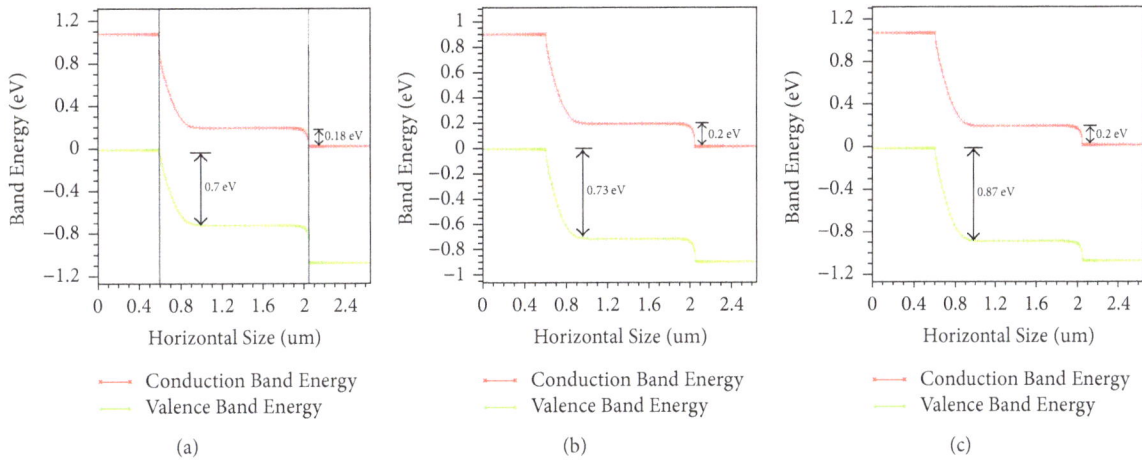

FIGURE 4: The band simulation of three type modulation structures: (a) Si/SiGe/Si-OI structure, (b) SiGe-OI structure, and (c) SOI structure.

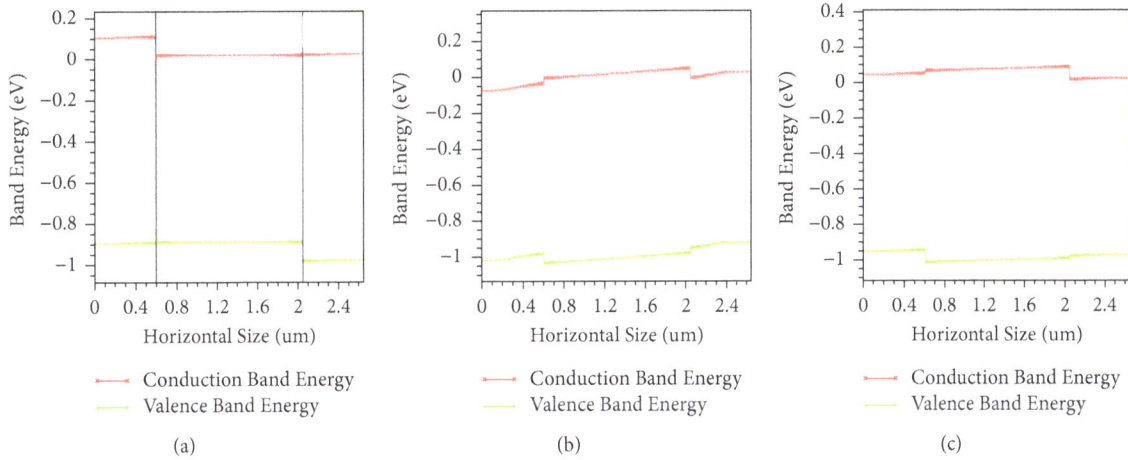

FIGURE 5: At 1 V forward voltage, the band simulation of three type modulation structures: (a) Si/SiGe/Si-OI structure, (b) SiGe-OI structure, and (c) SOI structure.

and the SiGe-OI structure reach the same carrier concentration at 2 V modulation voltage; when the modulation voltage reaches 2 V, the carrier concentration of the Si/SiGe/Si-OI structure could increase to $1.5 \times 10^{19} \, \text{cm}^{-3}$. Assuming the SOI structure and the SiGe-OI structure all work at 2 V modulation voltage, so the Si/SiGe/Si-OI structure can work only at 0.9 V modulation voltage. From that we can see the modulation voltage of electro-optic modulator effectively is reduced, and the injection efficiency of electro-optic modulator effectively is improved for the Si/SiGe/Si-OI structure.

In order to verify the correctness of the theoretical analysis and simulation results, we have made the micro-nano Si/SiGe/Si double heterojunction electro-optic modulation, as shown in Figure 7. For the modulation of the optical part, the micro-ring modulation structure is selected. Figure 7(a) shows schematic diagram of the micro-nano Si/SiGe/Si double heterojunction micro-ring electro-optic modulation, Figure 7(b) shows Laser Scanning Confocal Microscopy (LSCM) picture of device by LEXT-OLS4000, Figure 7(c) shows Scanning Electron Microscope (SEM) picture of

micro-ring structure by JSM-6700F, and Figure 7(d) shows SEM of device. The waveguide width is 450 nm, the ridge waveguide height is 220 nm, the planar waveguide height is 50 nm, the gap between waveguide and micro-ring is 200 nm, the micro-ring radius is 10 μm, the doping concentration of P+ region is $1 \times 10^{19} \, \text{cm}^{-3}$, the doping concentration of N+ region is $1 \times 10^{19} \, \text{cm}^{-3}$, and the doping concentration of I region is $1 \times 10^{15} \, \text{cm}^{-3}$. At the same processing conditions and device parameters, the micro-nano SOI micro-ring electro-optic modulation is also made for testing comparison.

The transmission spectrum of the micro-nano Si/SiGe/Si double heterojunction micro-ring electro-optic modulation has been tested, as shown in Figure 8. From that we can see the Free Spectrum Range (FSR) is 9.44 nm, and the 3 dB Bandwidth of the resonant peak is 0.12 nm, near the 1549 nm. The corresponding Q value is 12900, and the extinction ratio of the resonant peak is 11.2 dB.

At the same testing conditions, the transmission spectrum of the micro-nano SOI micro-ring electro-optic modulation has also been tested, as shown in Figure 9. From that

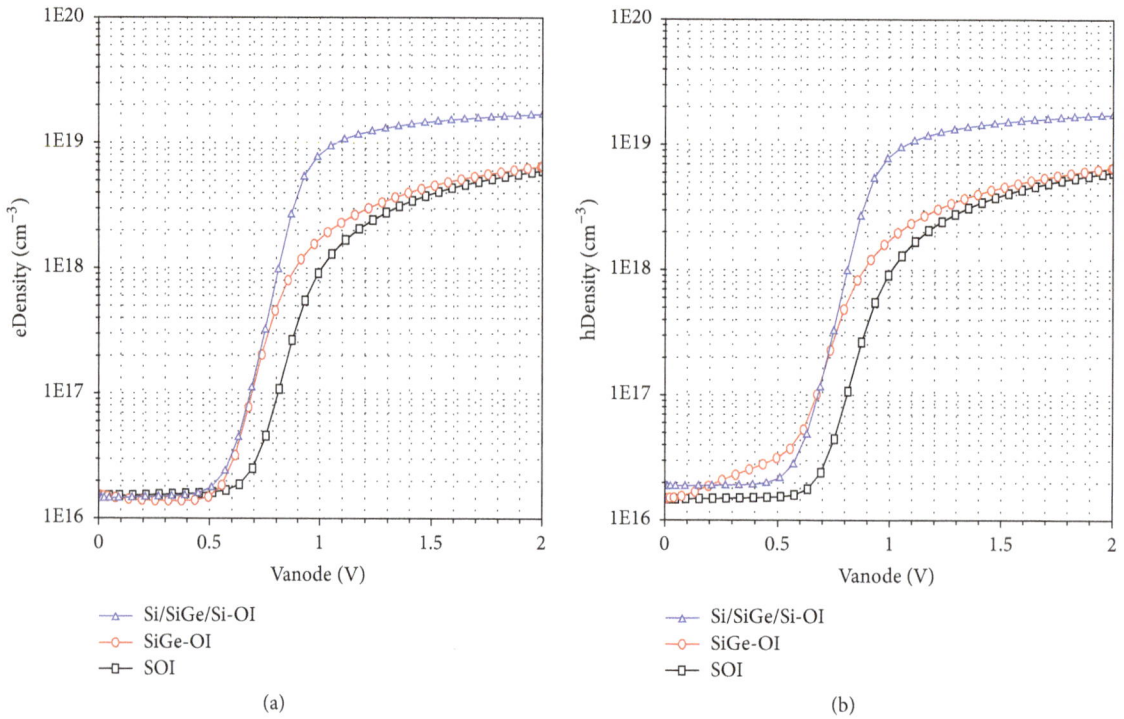

FIGURE 6: The relationship between the carriers concentration and modulation voltage: (a) electron concentration and (b) hole concentration for three structures.

we can see the Free Spectrum Range (FSR) is 9.41 nm, and the 3 dB Bandwidth of the resonant peak is 0.13 nm, near the 1548 nm. The corresponding Q value is 11900, and the extinction ratio of the resonant peak is 7.0 dB.

From the comparison of testing results between the micro-nano Si/SiGe/Si double heterojunction micro-ring electro-optic modulation and the micro-nano SOI micro-ring electro-optic modulation, we can see that Free Spectrum Range, 3 dB Bandwidth, Q value, extinction ratio, and other parameters of the micro-nano Si/SiGe/Si double heterojunction micro-ring electro-optic modulation are better than that of the micro-nano SOI micro-ring electro-optic modulation, and Q value and the extinction ratio are especially greater than that of the micro-nano SOI micro-ring electro-optic modulation.

The test result of the micro-nano Si/SiGe/Si double heterojunction micro-ring electro-optic modulation is shown in Figure 10. The forward bias voltages are 0, 1, 1.1, 1.2 V, respectively. From that we can see the resonant peak has obviously been blue-shifted after the forward bias voltage is set. When the forward bias voltage is 1 V, the value of blue-shift is 0.71 nm.

When the forward bias voltages are 0, 1.5, 1.8, 2 V, respectively, the transmission spectrum of the micro-nano SOI micro-ring electro-optic modulation has been tested, as shown in Figure 11. From that we can see the resonant peak has also obviously been blue-shifted after the forward bias voltage is set, but the modulation effect is worse than Si/SiGe/Si modulation. Only when the forward bias voltage is up to 2 V, the value of blue-shift is 0.71 nm.

From the comparison of testing results between the micro-nano Si/SiGe/Si double heterojunction micro-ring electro-optic modulation and the micro-nano SOI micro-ring electro-optic modulation, we can see that the modulation efficiency of the micro-nano Si/SiGe/Si double heterojunction micro-ring electro-optic modulation is better than that of the micro-nano SOI micro-ring electro-optic modulation at the same processing conditions and testing conditions. In other words, the modulation voltage of the micro-nano Si/SiGe/Si double heterojunction micro-ring electro-optic modulation is lower than that of the micro-nano SOI micro-ring electro-optic modulation, and the modulation power of the micro-nano Si/SiGe/Si double heterojunction micro-ring electro-optic modulation is lower too. So the testing results are consistent with previous theoretical analysis and simulation results.

4. Conclusions

This paper presents a micro-nano Si/SiGe/Si double heterojunction electro-optic modulation structure. Based on the band theory of single heterojunction, the barrier height of the new structure is quantitatively analyzed, and the formulas of barrier height are gotten using the numerical calculus and derivation. The reason of carrier injection enhancement in the new structure is analyzed. On the simulation platform, the new structure, SiGe-OI structure, and SOI structure are simulated. From the comparison of simulation results, when the modulation voltage is up to 0.9 V, the carrier concentration of the new structure is 6×10^{18} cm^{-3}, but the SOI structure and

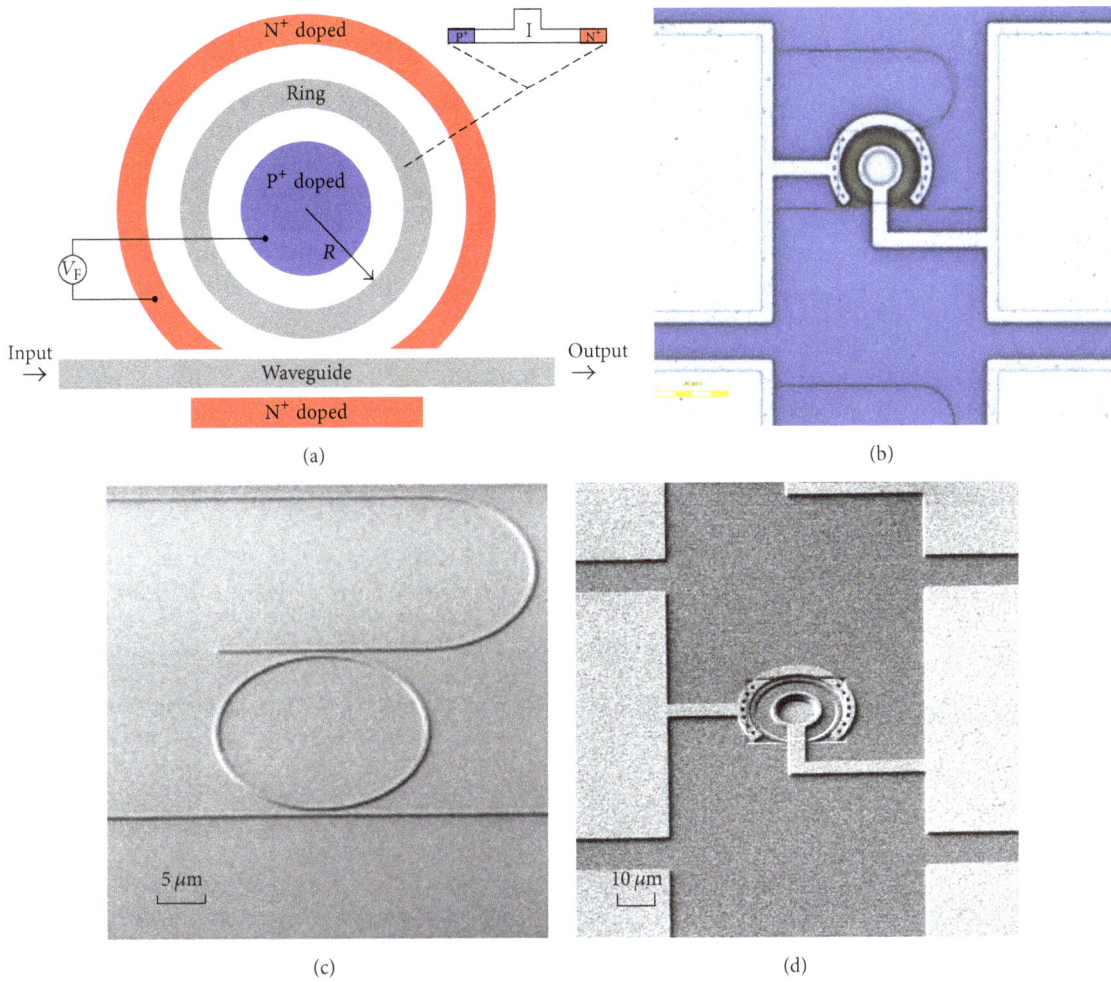

FIGURE 7: (a) Schematic diagram of device, (b) LSCM of device, (c) SEM of micro-ring structure, and (d) SEM of device.

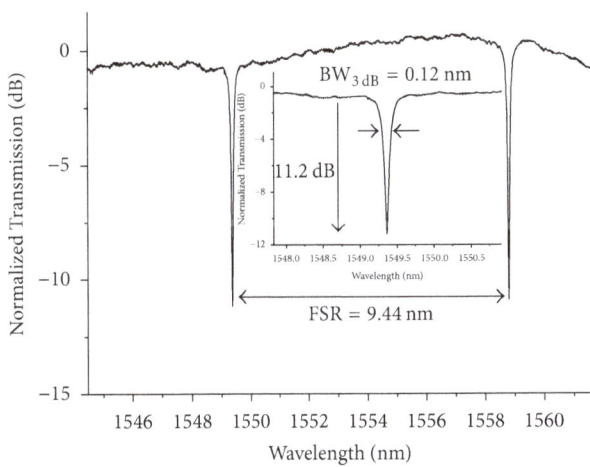

FIGURE 8: The transmission spectrum of the micro-nano Si/SiGe/Si double heterojunction micro-ring electro-optic modulation.

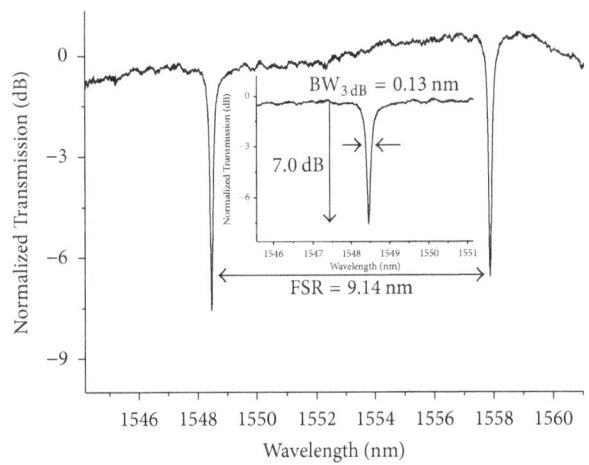

FIGURE 9: The transmission spectrum of the micro-nano SOI micro-ring electro-optic modulation.

the SiGe-OI structure reach the same carrier concentration at 2 V modulation voltage. Finally, the micro-nano Si/SiGe/Si double heterojunction electro-optic modulation is made and

tested. Comparison of testing results between the micro-nano Si/SiGe/Si double heterojunction micro-ring electro-optic

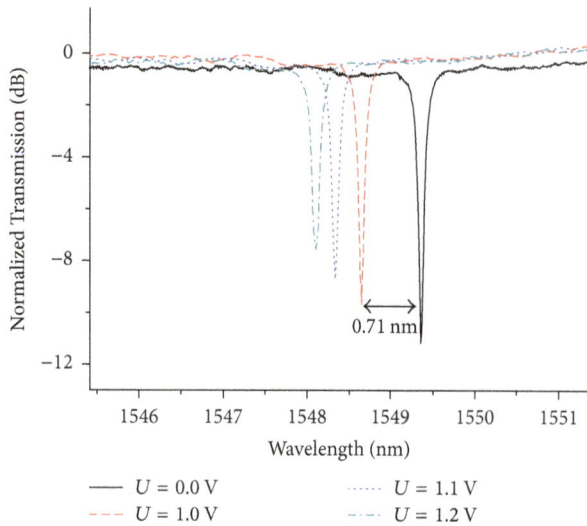

FIGURE 10: The static modulation of the micro-nano Si/SiGe/Si double heterojunction micro-ring electro-optic modulation.

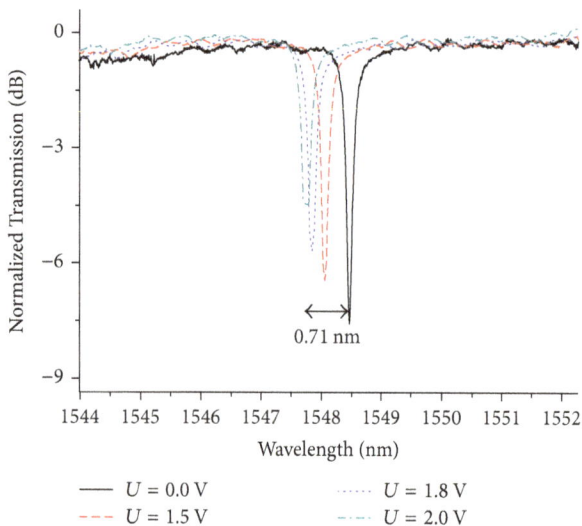

FIGURE 11: The static modulation of the micro-nano SOI micro-ring electro-optic modulation.

modulation and the micro-nano SOI micro-ring electro-optic modulation, Free Spectrum Range, 3 dB Bandwidth, Q value, extinction ratio, and other parameters of the micro-nano Si/SiGe/Si double heterojunction micro-ring electro-optic modulation are better than others, and the modulation voltage and the modulation power are lower. So, the micro-nano Si/SiGe/Si double heterojunction electro-optic modulation structure is an ideal device structure to replace the traditional SOI and SiGe-OI electro-optic modulators structure.

Conflicts of Interest

The authors declare that they have no conflicts of interest.

Acknowledgments

This work was financially supported by National Natural Science Foundation of China (Grant no. 61204080), Natural Science Foundation of Shaanxi Province of China (Grant no. 2017JM6075), Foundation of Shaanxi Provincial Education Department (Grant no. 17JK0335), State Key Laboratory of Functional Materials for Informatics (Grant no. SKL201804), and Shaanxi Province Ordinary University Key Disciplines Construction Projects of Special Funds (Grant no. (2008)169).

References

[1] K. Singh and D. Kaur, "Vertically arrays of Si-LaNiO3/BiFeO3/ Au core-shell nano-capacitors for prominent coupled electro-optic effect," *Journal of Applied Physics*, vol. 121, no. 11, Article ID 114103, 2017.

[2] K. Nozaki, S. Matsuo, T. Fujii et al., "Photonic-crystal nano-photodetector with ultrasmall capacitance for on-chip light-to-voltage conversion without an amplifier," *Optica*, vol. 3, no. 5, pp. 483–492, 2016.

[3] Q. Fang, J. Song, X. Luo et al., "Low Loss Fiber-to-Waveguide Converter with a 3-D Functional Taper for Silicon Photonics," *IEEE Photonics Technology Letters*, vol. 28, no. 22, pp. 2533–2536, 2016.

[4] Z. Zhang, J. Hu, H. Chen et al., "Low-crosstalk silicon photonics arrayed waveguide grating," *Chinese Optics Letters*, vol. 15, no. 4, Article ID 041301, 2017.

[5] Q. Fang, J. Hu, Z. Zhang et al., "A ring-mirrors-integrated silicon photonics arrayed waveguide grating," in *Proceedings of the 2017 Conference on Lasers and Electro-Optics Pacific Rim (CLEO-PR)*, pp. 1-2, Singapore, July 2017.

[6] J. Bowers, M. Davenport, and S. Liu, "A 20 GHz colliding pulse mode-locked heterogeneous InP-silicon laser," *Cleo: Science Innovations*, vol. SW4C.5, 2017.

[7] W. Hua and S.-X. Liu, "Nonlinear resonance phenomenon of one-dimensional Bose - Einstein condensate under periodic modulation," *Chinese Physics B*, vol. 23, no. 2, Article ID 020309, 2014.

[8] S. Akiyama, M. Imai, T. Baba et al., "Compact PIN-diode-based silicon modulator using side-wall-grating waveguide," *IEEE Journal of Selected Topics in Quantum Electronics*, vol. 19, no. 6, pp. 74–84, 2013.

[9] H. Zwickel, S. Wolf, C. Kieninger et al., "Silicon-organic hybrid (SOH) modulators for intensity-modulation / direct-detection links with line rates of up to 120 Gbit/s," *Optics Express*, vol. 25, no. 20, pp. 23784–23800, 2017.

[10] S. Heidmann, G. Ulliac, N. Courjal, and G. Martin, "Characterization and control of the electro-optic phase dispersion in lithium niobate modulators for wide spectral band interferometry applications in the mid-infrared," *Applied Optics*, vol. 56, no. 14, pp. 4153–4157, 2017.

[11] N. Hoppe, C. Rothe, A. Celik et al., "Single waveguide silicon-organic hybrid modulator," *Advances in Radio Science*, vol. 15, pp. 141–147, 2017.

[12] M. Lauermann, S. Wolf, P. C. Schindler et al., "40 GBd 16QAM signaling at 160 Gb/s in a silicon-organic hybrid modulator," *Journal of Lightwave Technology*, vol. 33, no. 6, Article ID 7015552, pp. 1210–1216, 2015.

[13] M. Streshinsky, R. Ding, A. Novack et al., "50 Gb/s silicon traveling wave Mach-Zehnder modulator near 1300 nm," in *Proceedings of the Optical Fiber Communications Conference and Exhibition*, vol. 14560428:1-3, 2014.

[14] J. Fujikata, M. Noguchi, Y. Kim et al., "High-speed and highly efficient Si optical modulator with strained SiGe layer," *Applied Physics Express*, vol. 11, no. 3, p. 032201, 2018.

[15] R. J. A. Baker, J. Hoffman, P. Schvan, and S. P. Voinigescu, "SiGe BiCMOS linear modulator drivers with 4.8-Vpp differential output swing for 120-GBaud applications," in *Proceedings of the 2017 IEEE Radio Frequency Integrated Circuits Symposium, RFIC 2017*, pp. 260–263, USA, June 2017.

[16] J. Verbist, M. Verplaetse, S. A. Srinivasan et al., "Real-Time 100 Gb/s NRZ-OOK transmission with a silicon photonics GeSi electro-Absorption modulator," in *Proceedings of the 6th IEEE Photonics Society Optical Interconnects Conference, OI 2017*, pp. 29-30, USA, June 2017.

[17] S. Feng, B. Xue, LB. Li et al., "Analysis of Si/SiGe/Si double heterojunction band about a noval structure of PIN electronic modulation," *Acta Physica Sinica*, vol. 65, no. 5, article 054201, Article ID 054201, 2016.

25

Amorphization Effect for Kondo Semiconductor CeRu$_2$Al$_{10}$

Yusuke Amakai,[1,2] Yasuhiro Shiojiri,[1] Kei Ishihara,[1] Hiroto Hitotsukabuto,[1] Shigeyuki Murayama,[1] Naoki Momono,[1,2] and Hideaki Takano[1]

[1]*Graduate School of Engineering, Muroran Institute of Technology, 27-1 Mizumoto-cho, Muroran, Hokkaido 050-8585, Japan*
[2]*Research Center for Environmentally Friendly Materials Engineering, Muroran Institute of Technology, 27-1 Mizumoto-cho, Muroran, Hokkaido 050-8585, Japan*

Correspondence should be addressed to Yusuke Amakai; a-rain@mmm.muroran-it.ac.jp

Academic Editor: Joseph S. Poon

We measured the magnetic susceptibility χ, electrical resistivity ρ, and specific heat C_p of a sputtered amorphous (a-)CeRu$_2$Al$_{10}$ alloy. χ value for a-CeRu$_2$Al$_{10}$ alloy follows a Curie-Weiss paramagnetic behavior in the high-temperature region, and magnetic transition was not observed down to 2 K. The effective paramagnetic moment p_{eff} is 1.19 μ_B/Ce-atom. The resistivity shows a typical disordered alloy behavior, that is, small temperature dependence for the whole temperature range. We observed an enhancement of ρ and C_p/T in the low-temperature region of $T < 10$ K. The enhancement in ρ is suppressed by applying a magnetic field. It is suggested that this behavior is caused by the Kondo effect.

1. Introduction

The ternary rare-earth compound, CeRu$_2$Al$_{10}$, exhibits an unusual antiferromagnetic phase transition at $T_0 \approx 27$ K [1, 2]. The resistivity for CeRu$_2$Al$_{10}$ exhibits a semiconducting behavior in the paramagnetic phase. However, this behavior is suppressed by substituting La in Ce-site of CeRu$_2$Al$_{10}$. In addition, T_0 decreases rapidly with an increase in La concentration and disappears when about half of Ce is substituted by La. The resistivity for Ce dilute region of La-substituted CeRu$_2$Al$_{10}$ exhibits a metallic behavior at high temperatures, and the resistivity exhibits a minimum in the low-temperature region. Tanida et al. proposed that the long-range magnetic order was suppressed randomly in Ce dilute region for La substitution of CeRu$_2$Al$_{10}$. The authors pointed out that the resistivity minimum can be explained by a typical impurity Kondo effect, where Ce exists as a magnetic impurity [3, 4]. Moreover, the semiconducting band gap of CeRu$_2$Al$_{10}$ is broken by La substitution.

However, several studies on the Kondo effect of structural-disordered Ce-alloys such as bulk metallic glasses and amorphous alloys, where Ce-atom is arranged randomly, have been recently performed. For example, in Ce-Al bulk metallic glasses, the tunable competition between the Kondo effect and the Ruderman-Kittel-Kasuya-Yoshida (RKKY) interaction with the variation in Ce-concentration and the magnetic field [5] is suggested. We studied the low-temperature properties of several binary amorphous Ce-alloys. Amorphous (a-)Ce-Mn and a-Ce-Ru alloys in Ce high-concentration region exhibit a large electronic specific heat coefficient γ (>200 mJ/molK^2) and T^2 law with a large coefficient A ($>0.02\ \mu\Omega$cm/K^2) in the low-temperature resistivity [6–10]. From these results, we show that an itinerant heavy-fermion state occurs at low temperatures as a Fermi-liquid ground state in the structure-disordered system after the formation of a dense Kondo state.

In this study, to investigate the influence of the structural-disordered effect on the electrical resistivity and magnetic properties for CeRu$_2$Al$_{10}$, we prepared a-CeRu$_2$Al$_{10}$ alloy and measured its magnetic susceptibility, resistivity, and specific heat. In addition, we prepared a-LaRu$_2$Al$_{10}$ alloy that does not have 4f-electrons with a rare-earth element for comparison.

2. Experimental

Bulk ingots of CeRu$_2$Al$_{10}$ and LaRu$_2$Al$_{10}$ were prepared by using the arc-melting method with a stoichiometric composition of Ce 99.9%, La 99.9% (Nippon Yttrium Co., Ltd.), Ru 99.95% (Rare Metallic Co., Ltd.), and Al 99.99% (Mitsuwa Chemicals Co., Ltd.), in Ar atmosphere. The amorphous alloy was prepared using a dc high-rate sputtering method with arc-melt ingots on a water-cooled Cu substrate (30 mmϕ). The sample thickness was ~200 μm. The structure of the obtained samples was confirmed using X-ray diffraction measurements. Measurements were performed with an as-sputtered film. The chemical compositions of the present amorphous alloys were determined using scanning electron microscope energy dispersive X-ray spectroscopy (SEM-EDS) to be $Ce_{10}Ru_{15}Al_{75}$ and $La_8Ru_{17}Al_{75}$ (suffixes represent at%). Thus, we will use the notations a-$CeRu_2Al_{10}$ and a-$LaRu_2Al_{10}$ for the samples, hereafter. The magnetic susceptibility was measured using a commercial SQUID magnetometer (Quantum Design MPMS) from 2 to 300 K. The electrical resistivity was measured by using a typical four-terminal method (Quantum Design PPMS) from 2 to 300 K. The resistivity measurements of a-$CeRu_2Al_{10}$ alloy were performed in a magnetic field (0, 20 kOe, 40 kOe, and 60 kOe) and the temperature range was 2–60 K. The specific heat was measured by PPMS from 2 to 300 K.

3. Results and Discussion

Figure 1 shows the X-ray diffraction patterns for a-$CeRu_2Al_{10}$ and a-$LaRu_2Al_{10}$ alloys. The diffraction patterns for both alloys exhibit two broad peaks at the center at about 25° and 42°, and definite Bragg peaks are not observed. Therefore, the samples are identified as amorphous materials.

Figure 2 shows the temperature dependence of the magnetic susceptibility χ (left axis) and the inverse susceptibility $1/(\chi-\chi_0)$ (right axis) for a-$CeRu_2Al_{10}$ and a-$LaRu_2Al_{10}$ alloys at $H = 10$ kOe. We calculated χ using the composition ratio of $Ce_{10}Ru_{15}Al_{75}$ obtained by SEM-EDS as 1 mol. χ value for a-$CeRu_2Al_{10}$ alloy increases monotonically with decreasing temperature, and a magnetic transition is not observed in the measurement temperature region. However, χ value for a-$LaRu_2Al_{10}$ alloy is almost independent of the temperature and is very small value (<10^{-6} emu/mol). $1/(\chi-\chi_0)$ value for a-$CeRu_2Al_{10}$ alloy exhibits a linear behavior in the high-temperature region of $T > 30$ K following the Curie-Weiss law,

$$\frac{1}{(\chi - \chi_0)} = \frac{1}{C}(T - \theta), \quad (1)$$

where χ_0 is a constant for the independence of temperature, C is the Curie constant, and θ is the Weiss temperature. χ_0 value for a-$CeRu_2Al_{10}$ alloy obtained from (1) was ~1.0×10^{-6} emu/mol. The value of θ is −20 K. C is obtained as

$$C = \frac{N_{Ce}\mu_{eff}^2}{3k_B}, \quad (2)$$

where N_{Ce} is the number of Ce atoms, μ_{eff} is the paramagnetic effective magnetic moment, and k_B is the Boltzmann constant. Here, μ_{eff} is calculated from Ce-concentration (10 at%) of composition ratio $Ce_{10}Ru_{15}Al_{75}$ obtained from SEM-EDS.

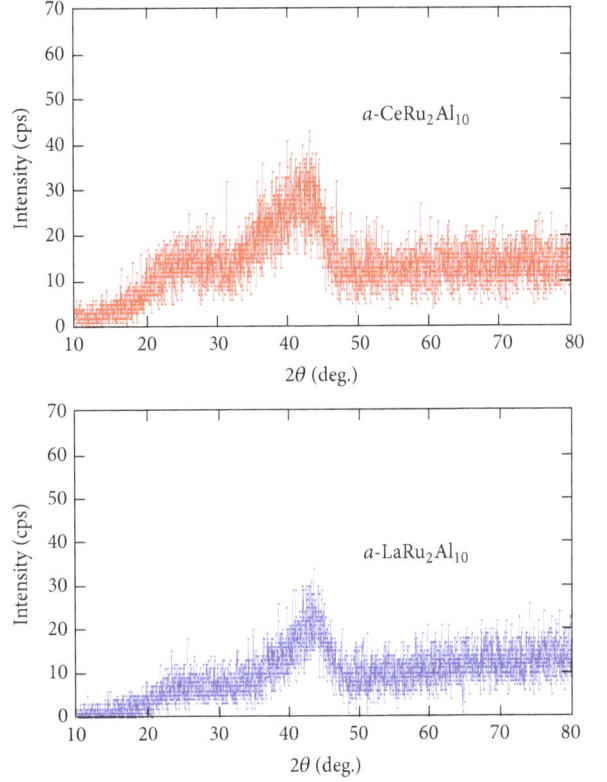

FIGURE 1: X-ray diffraction pattern of a-$CeRu_2Al_{10}$ and a-$LaRu_2Al_{10}$ alloys.

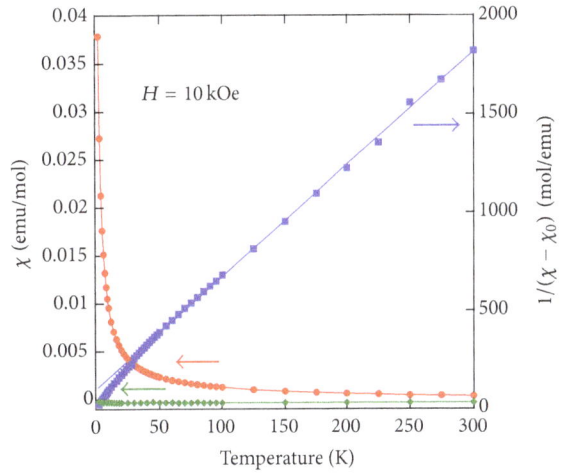

FIGURE 2: Temperature dependence of the magnetic susceptibility χ and inverse susceptibility $1/(\chi-\chi_0)$ for a-$CeRu_2Al_{10}$ alloy. The green points and line are the temperature dependence of the magnetic susceptibility χ for a-$LaRu_2Al_{10}$ alloy.

The estimated μ_{eff} obtained using (2) is 1.19 μ_B/Ce-atom. Since the effective magnetic moment expected for trivalent Ce ($J = 5/2$) is 2.54 μ_B, the obtained value of 1.19 μ_B/Ce is about half this value. Therefore, it is expected that about half of Ce of a-$CeRu_2Al_{10}$ alloy exists as nonmagnetic tetravalent Ce and the remaining half exists as magnetic trivalent Ce in the alloy.

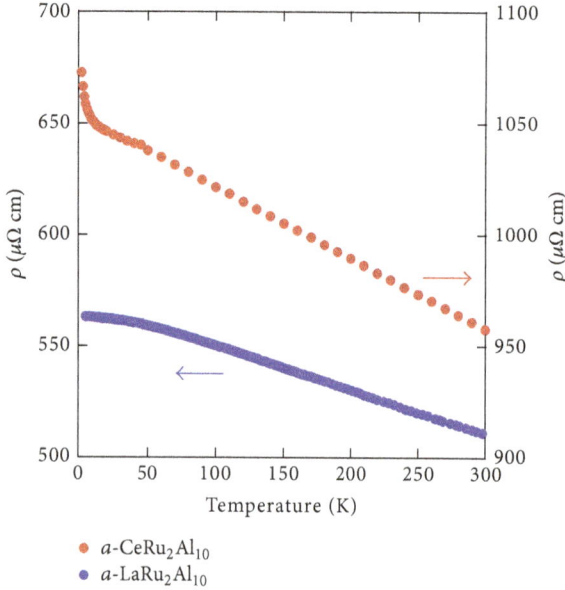

FIGURE 3: Temperature dependence of the resistivity ρ for a-CeRu$_2$Al$_{10}$ and a-LaRu$_2$Al$_{10}$ alloys.

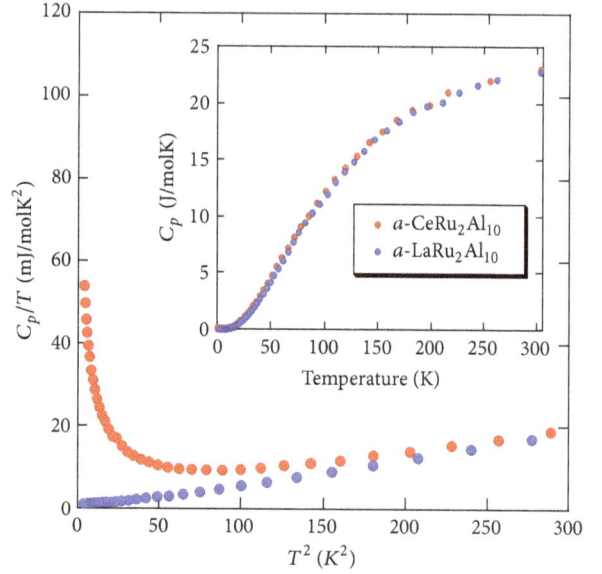

FIGURE 4: C_p/T versus T^2 plot for a-CeRu$_2$Al$_{10}$ and a-LaRu$_2$Al$_{10}$ alloys. The inset shows the temperature dependence of the specific heat C_p for a-CeRu$_2$Al$_{10}$ and a-LaRu$_2$Al$_{10}$ alloys.

Figure 3 shows the temperature dependence of the resistivity ρ for a-CeRu$_2$Al$_{10}$ and a-LaRu$_2$Al$_{10}$ alloys. The value of ρ at 300 K for a-CeRu$_2$Al$_{10}$ alloy is about 2 times greater than that of a-LaRu$_2$Al$_{10}$ alloy. ρ value for both alloys exhibits small temperature dependence less than 10% in the whole temperature region, and it increases with decreasing temperature. Such temperature dependence is one of the characteristics of disordered alloys. In contrast to ρ value of a-LaRu$_2$Al$_{10}$ alloy, which increases monotonically with decreasing temperature, ρ value of a-CeRu$_2$Al$_{10}$ alloy exhibits an increase in the low-temperature region. Generally, in the case of disordered alloys, the resistivity is larger than that for the crystalline counterparts owing to the random arrangement of atoms. However, the absolute value of the low-temperature ρ in a-CeRu$_2$Al$_{10}$ alloy is on about the same order as that for the polycrystalline CeRu$_2$Al$_{10}$. In addition, although ρ value of a-CeRu$_2$Al$_{10}$ alloy increases with decreasing temperature, a small temperature dependence similar to that for the typical disordered alloy was observed. Therefore, it is considered that the semiconductor band for crystalline CeRu$_2$Al$_{10}$ exhibited a metallic behavior as a result of amorphization.

Figure 4 shows the low-temperature specific heat C_p over T versus T^2 plots for a-CeRu$_2$Al$_{10}$ and a-LaRu$_2$Al$_{10}$ alloys. The inset shows the temperature dependence of C_p for the alloys. C_p values of both alloys are almost in agreement in the whole temperature region, as shown in the inset. The low-temperature specific heat of a usual metal can be expressed by

$$C_p = \gamma T + \beta T^3, \qquad (3)$$

where γT is the electronic specific heat term and βT^3 is the phonon specific heat term. This relation can also be applied to amorphous alloys [11]. As shown in Figure 4, C_p/T of a-LaRu$_2$Al$_{10}$ alloy follows a linear relation as a function

of T^2 below 50 K^2. However, C_p/T of a-CeRu$_2$Al$_{10}$ alloy increases rapidly with decreasing T^2 below 100 K^2. The value of C_p/T at the lowest temperature is ~54 mJ/molK^2. The temperature region where the enhancement in C_p/T occurs agrees with the temperature region of the upturn for ρ. Such enhancement in C_p/T and ρ at low-temperature does not occur for a-LaRu$_2$Al$_{10}$. Therefore, it is considered to be due to the magnetic contribution of the 4f-electron of Ce. In order to clarify the magnetic contribution of the 4f-electron of Ce for a-CeRu$_2$Al$_{10}$, we measured the magnetoresistance of a-CeRu$_2$Al$_{10}$.

Figure 5 shows the magnetic field dependence of the transverse magnetoresistance $\Delta\rho/\rho$ at 2 K, 10 K, and 60 K for a-CeRu$_2$Al$_{10}$ alloy. We calculated $\Delta\rho/\rho$ as follows:

$$\frac{\Delta\rho}{\rho} = \frac{\rho(H) - \rho(0)}{\rho(0)}, \qquad (4)$$

where $\rho(H)$ is the resistivity in the magnetic field and $\rho(0)$ is the resistivity at zero field. $\Delta\rho/\rho$ increases negative with increasing magnetic field strength. $\Delta\rho/\rho$ at 2 K exhibits a large negative enhancement compared with that at other temperatures. Figure 6 shows the temperature dependence of $\Delta\rho/\rho$ at 20 kOe, 40 kOe, and 60 kOe for a-CeRu$_2$Al$_{10}$ alloy. The $\Delta\rho/\rho$ in the high-temperature region ($T > 10$ K) is almost independent of the temperature for all the magnetic fields. However, $\Delta\rho/\rho$ in the low-temperature region increases rapidly with decreasing temperature at $T < 10$ K for all the fields. Thus, the enhancement of ρ at low temperatures is suppressed by the applied magnetic field.

The present a-CeRu$_2$Al$_{10}$ alloy seems reasonable that the magnetic order of the crystalline counterpart have disappeared because they lose a long-range crystallographic order by amorphization. However, the local structure of

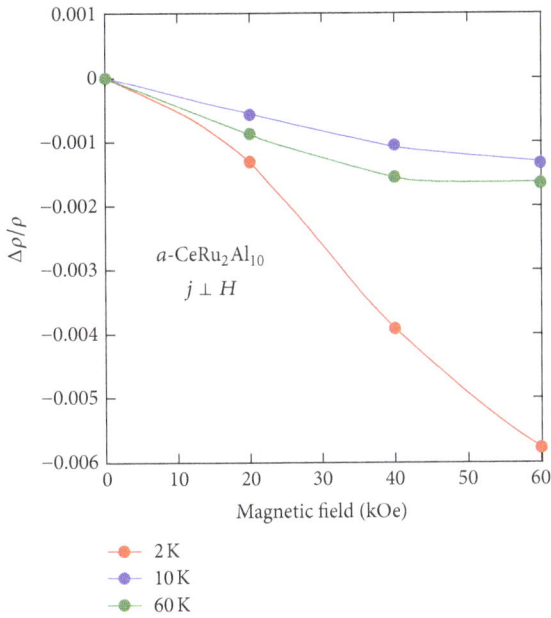

FIGURE 5: Field dependence of the magnetoresistance $\Delta\rho/\rho$ at various temperatures for a-CeRu$_2$Al$_{10}$ alloy.

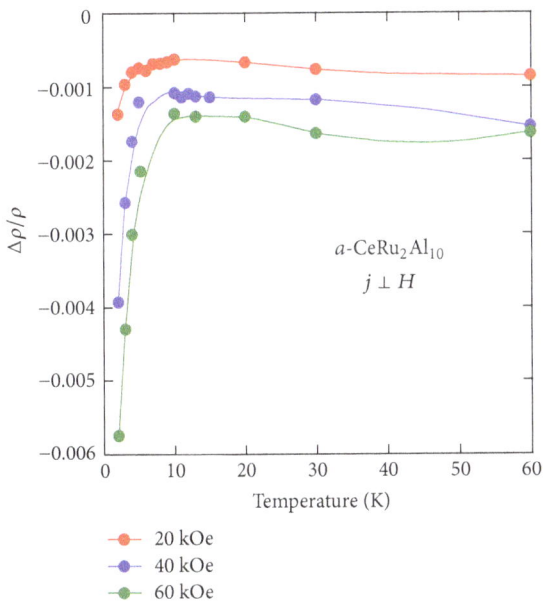

FIGURE 6: Temperature dependence of the magnetoresistance $\Delta\rho/\rho$ under various magnetic fields for a-CeRu$_2$Al$_{10}$ alloy.

the amorphous alloy is generally considered to be similar to that of the crystalline counterpart. Even in a-CeRu$_2$Al$_{10}$ alloy, a local structure similar to that of the crystalline CeRu$_2$Al$_{10}$ is expected to be realized. In addition, based on μ_{eff}, we have shown that half of Ce in a-CeRu$_2$Al$_{10}$ alloy is in the trivalent state. Ce^{3+} ions in a-CeRu$_2$Al$_{10}$ alloy are distributed randomly in the amorphous sample. In this case, it is considered that the random distribution of Ce^{3+} in the disordered structure realizes the impurity Kondo effect at low temperatures. Thus, the enhancement of

ρ and C_p/T is observed in the low-temperature region for a-CeRu$_2$Al$_{10}$ alloy. Since the Kondo scattering is suppressed by the applying a magnetic field, $\Delta\rho/\rho$ increases negatively. Therefore, the negative increase of $\Delta\rho/\rho$ for a-CeRu$_2$Al$_{10}$ alloy at low temperatures and high magnetic fields indicates that the scattering center of the resistivity is due to Kondo scattering.

4. Conclusion

We prepared a-CeRu$_2$Al$_{10}$ and a-LaRu$_2$Al$_{10}$ alloys and measured their susceptibility, resistivity, and specific heat. χ value for a-CeRu$_2$Al$_{10}$ alloy follows the Curie-Weiss law in the high-temperature region of $T > 30$ K with no magnetic transition. The effective paramagnetic moment and the Weiss temperature are 1.19 μ_B/Ce-atom and –20 K, respectively. ρ value for a-CeRu$_2$Al$_{10}$ and a-LaRu$_2$Al$_{10}$ alloys shows a small temperature dependence. However, an enhancement of ρ was observed in the low-temperature region for a-CeRu$_2$Al$_{10}$ alloy. C_p/T value of a-CeRu$_2$Al$_{10}$ alloy increases rapidly with decreasing T^2 below 100 K^2. The magnetoresistance $\Delta\rho/\rho$ for a-CeRu$_2$Al$_{10}$ alloy increases rapidly with decreasing temperature for $T < 10$ K. Therefore, it is considered that a-CeRu$_2$Al$_{10}$ alloy is formed the impurity Kondo state in the low-temperature region.

Conflicts of Interest

The authors declare that there are no conflicts of interest regarding the publication of this paper.

Acknowledgments

The authors would like to thank Dr. S. Yamanaka of the Muroran Institute of Technology for allowing the use of the SEM-EDS. The authors wish to thank Dr. Y. Kawamura of the Muroran Institute of Technology for the useful discussion on crystalline CeRu$_2$Al$_{10}$.

References

[1] A. M. Strydom, "Thermal and electronic transport in CeRu$_2$Al$_{10}$: evidence for a metal-insulator transition," *Physica B: Condensed Matter*, vol. 404, no. 19, pp. 2981–2984, 2009.

[2] T. Nishioka, Y. Kawamura, T. Takesaka et al., "Novel phase transition and the pressure effect in YbFe$_2$Al$_{10}$-type CeT$_2$Al$_{10}$ (T = Fe, Ru, Os)," *Journal of the Physical Society of Japan*, vol. 78, no. 12, Article ID 123705, 2009.

[3] H. Tanida, D. Tanaka, M. Sera et al., "Possible long-range order with singlet ground state in CeRu$_2$Al$_{10}$," *Journal of the Physical Society of Japan*, vol. 79, no. 4, Article ID 043708, 2010.

[4] H. Tanida, D. Tanaka, M. Sera, T. Nishioka, and M. Matsumura, "Long range order in CeRu2Al10," *Journal of the Physical Society of Japan*, vol. 80, Article ID SA023, 2011.

[5] Q. S. Zeng, C. R. Rotundu, W. L. Mao et al., "Low temperature transport properties of Ce-Al metallic glasses," *Journal of Applied Physics*, vol. 109, no. 11, Article ID 113716, 2011.

[6] Y. Amakai, S. Murayama, Y. Obi, H. Takano, and K. Takanashi, "Evidence of a heavy fermion state in the disordered Ce-alloy

system without translation symmetry," *Physical Review B - Condensed Matter and Materials Physics*, vol. 79, no. 24, Article ID 245126, 2009.

[7] Y. Amakai, S. Murayama, Y. Obi, H. Takano, N. Momono, and K. Takanashi, "Thermal expansion of structure-disordered heavy-fermion Ce alloys," *Journal of the Physical Society of Japan*, vol. 80, Article ID SA057, 2011.

[8] Y. Obi, S. Murayama, Y. Amakai, Y. Okada, and K. Asano, "Heavy fermion like behavior and superconductivity in amorphous Ce$_x$Ru$_{100-x}$ alloy," *Physica B: Condensed Matter*, vol. 378-380, pp. 857-858, 2006.

[9] Y. Amakai, S. Murayama, Y. Obi, H. Takano, N. Momono, and K. Takanashi, "Magnetic properties of structure-disordered heavy fermion Ce-Ru alloys," *Journal of Physics: Conference Series*, vol. 150, no. 4, Article ID 042004, 2009.

[10] Y. Amakai, D. Yoshii, S. Murayama et al., "La substitution effect to the heavy-fermion state in structure-disordered Ce-Ru alloys," *Journal of Physics: Conference Series*, vol. 391, no. 1, Article ID 012002, 2012.

[11] W. Yang, H. Liu, X. Yang, and L. Dou, "Low temperature specific heat of amorphous alloys," *Journal of Low Temperature Physics*, vol. 160, 148 pages, 2010.

26

First-Principles Calculations on Atomic and Electronic Properties of Ge/4H-SiC Heterojunction

Bei Xu,[1] **Changjun Zhu,**[1] **Xiaomin He,**[2] **Yuan Zang,**[2] **Shenghuang Lin,**[3] **Lianbi Li**◉**,**[1] **Song Feng**◉**,**[1] **and Qianqian Lei**[1]

[1]*School of Science, Xi'an Polytechnic University, Xi'an 710048, China*
[2]*Department of Electronic Engineering, Xi'an University of Technology, Xi'an 710048, China*
[3]*Department of Applied Physics, The Hong Kong Polytechnic University, Hung Hom, Hong Kong*

Correspondence should be addressed to Lianbi Li; xpu_lilianbi@163.com

Academic Editor: Mohindar S. Seehra

First-principles calculation is employed to investigate atomic and electronic properties of Ge/SiC heterojunction with different Ge orientations. Based on the density functional theory, the work of adhesion, relaxation energy, density of states, and total charge density are calculated. It is shown that Ge(110)/4H-SiC(0001) heterointerface possesses higher adhesion energy than that of Ge(111)/4H-SiC(0001) interface, and hence Ge/4H-SiC(0001) heterojunction with Ge[110] crystalline orientation exhibits more stable characteristics. The relaxation energy of Ge(110)/4H-SiC(0001) heterojunction interface is lower than that of Ge(111)/4H-SiC(0001) interface, indicating that Ge(110)/4H-SiC(0001) interface is easier to form at relative low temperature. The interfacial bonding is analysed using partial density of states and total charge density distribution, and the results show that the bonding is contributed by the Ge-Si bonding.

1. Introduction

SiC semiconductor has become one of the most excellent materials for ultraviolet-sensitive devices owing to its wide bandgap [1, 2]. However, it is not sensitive to the infrared and visible light region. Ge/SiC heterojunction was employed to solve the problem, in which the Ge layer of micronanostructure was used as an absorption layer for nearinfrared (NIR) light [3]. By using the Ge/SiC heterojunction, SiC-based NIR light-operated device could be realized. The Ge/4H-SiC heterostructures are prepared by using low pressure chemical vapor deposition (LPCVD) on 4H-SiC(0001) substrates. Details of the growth process could be found in [4–6]. However, the lattice mismatch between Ge(111) primitive cell ($a_{Ge(111)}$ = 4.000 Å) and 4H-SiC(0001) primitive cell ($a_{4H-SiC(0001)}$ = 3.078 Å) is as large as 23.0%, which can cause distortion or even dislocation near the interface, leading to a poor crystalline quality of the Ge epilayer. Hence, it is necessary and imperative to investigate the atomic and electronic properties of the Ge/SiC heterojunction.

First-principles calculation based on density functional theory (DFT) has been widely used as an important microscopic study method in recent years. The first-principles calculation can be implemented to predict material properties and, consequently, a lot of valuable results have been achieved. Li et al. [7] used the first-principles method to investigate the interface adhesion energy, interface energy, interface fracture toughness, and electronic structure of the β-SiC(111)/α-Ti(0001) heterojunction. Six kinds of C-terminated β-SiC(111)/α-Ti(0001) models were established to study the effect of stack position and inclination angle on interface bonding and fracture toughness. Lin et al. [8] investigated the atomic structures and electronic properties of interfaces between aluminum and four kinds of ceramics with different orientations. They discovered that aluminum metal carbide interface is more stable than aluminum metal nitrides interface and, moreover, the (111) interfaces were found to possess the largest adhesion energy. He et al. [9, 10] studied the Si(111)/6H-SiC(0001) heterojunction by using the first-principles. It is found that

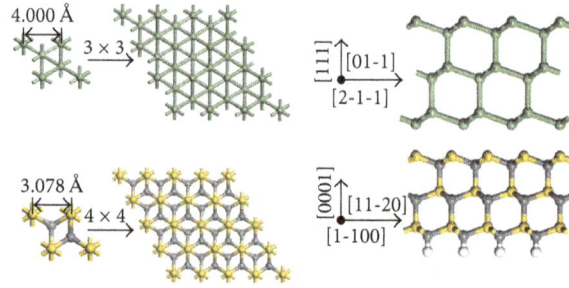

FIGURE 1: A schematic of Ge(111)/4H-SiC(0001) heterointerface model.

the Si-terminated Si(111)/6H-SiC(0001) heterojunction has higher adhesion energy and lower relaxation degree than C-terminated Si(111)/6H-SiC(0001) heterojunction. Xu et al. [11] have studied interfacial properties and electronic structure of Al(111)/4H-SiC(0001) interface.

In this paper, we present first-principles calculations of adhesion energy, relaxation energy, density of states, and total charge density of Ge(111)/4H-SiC(0001) interface and Ge(110)/4H-SiC(0001) interface, while analysing the electronic structure, geometry property, and the corresponding physical picture. Furthermore, the first-principles methods are used to investigate the structure of Ge/SiC heterointerface, which can provide a theoretical basis for the growth of Ge/SiC heterojunctions in experiment.

2. Methods

All the calculations in this work were implemented by using the Cambridge Serial Total Energy Package (CASTEP) Code [12, 13], which are based on the density functional theory (DFT) [14]. Generalized gradient approximation (GGA) of Perdew–Burke–Ernzerhof (PBE) scheme was employed to describe the exchange-correlation functional [15]. By comparing the lattice constants of GGA(PBE) and local density approximation (LDA) [16] with Caperlay-Alder Perdew-Zunger (CA-PZ) approximation algorithms, it is shown that the deviation of GGA(PBE) is smaller than that of LDA(CA-PZ). Therefore, the GGA-PBE function is implemented in the following Ge/4H-SiC(0001) heterojunction calculation. In order to make the system stable and the calculation speed optimal, plane wave cut-off energy was selected as 550 eV for a bulk, a surface, and an interface. The sampling of irreducible edge of Brillouin zone was performed with a regular Monkhorst-Pack grid with $7 \times 7 \times 7$ k points for the bulk and $5 \times 5 \times 1$ k points for the surface and interface, respectively. The SCF convergence threshold was 2.0×10^{-6} eV/atom, and the convergence tolerance for energy was selected as 2.0×10^{-5} eV/atom. The force tolerance, stress, and displacement tolerance were set as 0.05 eV/Å, 0.1 GPa, and 0.002 Å, respectively. To avoid interaction between surface atoms, a vacuum layer of 13 Å was selected for each surface and interface system.

3. Results and Discussions

3.1. Ge/4H-SiC Heterojunction Model. Figure 1 displays the interface structure of the Ge(111)/4H-SiC(0001) heterojunction based on the TEM characterizations [3]. The primitive cells of Ge(111) surface and 4H-SiC(0001) surface possess lattice constants of $[01-1]_{Ge} = 4.000$ Å, $[11-20]_{SiC} = 3.078$ Å. The lattice matching is $3:4$ of Ge to SiC with a residual mismatch of 2.60% in the two parallel orientations using the smallest supercell mismatch. In order to saturate suspension bonding, H atoms are employed to passivate the surface. Figure 2 shows the Ge(110)/4H-SiC(0001) heterojunction. The primitive cells of Ge(110) surface and 4H-SiC(0001) surface with constants lattice of $[001]_{Ge} = 5.658$ Å, $[1-10]_{Ge} = 4.000$ Å, $[10-10]_{SiC} = 5.331$ Å, and $[-12-10]_{SiC} = 3.078$ Å are cleaved due to the Ge[110] growth orientation on 4H-SiC(0001). The lattice matching is revealed as $1:1$ Ge to SiC with a residual mismatch of −5.78% and $3:4$ Ge to SiC with a lattice mismatch of 2.60% in the two parallel orientations. The interlayer distances of Ge(111)/4H-SiC(0001) interface and Ge(110)/4H-SiC(0001) interface are optimized by energy calculation before evaluating the interfacial properties of heterostructures. The functional relationship between energy and interlayer spacing is shown in Figure 3. Both of the Ge(111)/4H-SiC(0001) and Ge(110)/4H-SiC(0001) heterostructures have the same optimized interlayer distances of 2.30 Å. Similar conclusions are given in [17].

Because of the large lattice mismatch strain, the lattice mismatch between 4H-SiC and Ge is totally accommodated by misfit dislocations (MD) rather than by uniform elastic strains [1, 2]. The lattice mismatch of the Ge/4H-SiC interfaces can be calculated, as shown in Table 1. The Ge(111)/4H-SiC(0001) interface has the same $3:4$ Ge-to-SiC matching mode with a residual mismatch of 2.60% along both the Ge[01-1] and Ge[2-1-1] orientations. In contrast, the situation of the Ge(110)/4H-SiC(0001) interface is different, along Ge[1-10] orientation, the Ge-to-SiC matching mode is still $3:4$; along the vertical orientation of Ge[001], the Ge-to-SiC mode changes to $1:1$ and the residual mismatch changes to −5.78% correspondingly. The MD densities of the Ge(111)/4H-SiC(0001) interface and Ge(110)/4H-SiC(0001) interface are as low as 5.334×10^{14} cm^{-2} and 1.523×10^{14} cm^{-2},

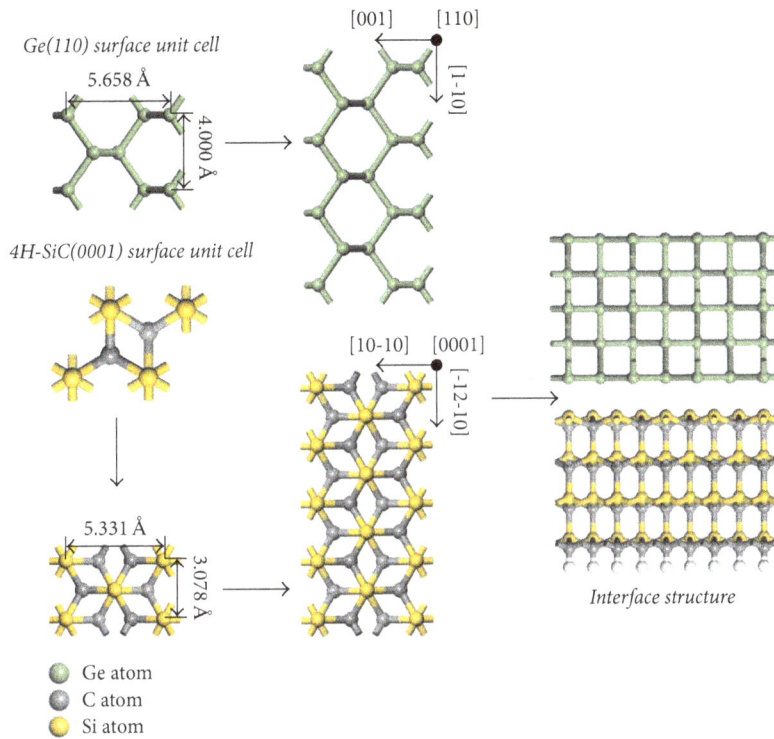

FIGURE 2: A schematic of Ge(110)/4H-SiC(0001) heterointerface model.

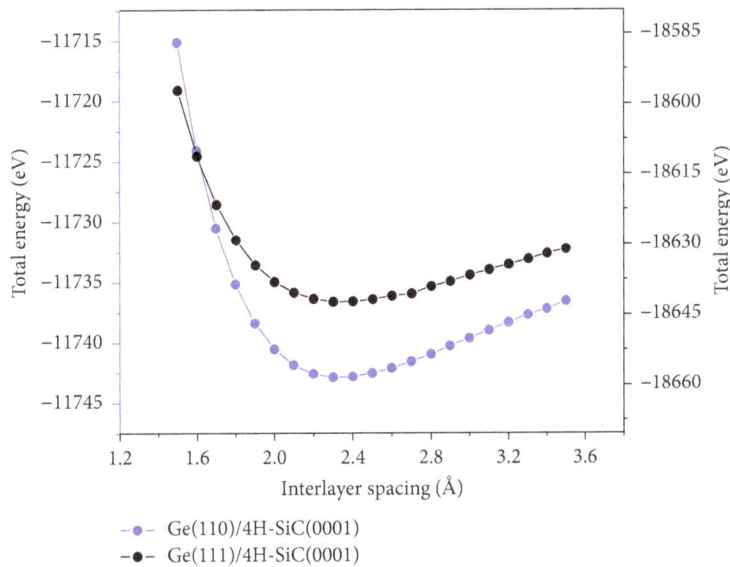

FIGURE 3: The interlayer spacing of Ge(110)/4H-SiC(0001) and Ge(111)/4H-SiC(0001) heterointerfaces.

respectively, as shown in Table 1. In addition, the Ge(110)/4H-SiC(0001) interface has fewer defects than the Ge(111)/4H-SiC(0001) interface.

3.2. Heterointerfaces Properties

3.2.1. Adhesion Energy and Relaxation Energy. To gain an insight into the binding strength of the interface, we calculated the work of adhesion (W_{ad}), which is defined as the reversible work to separate an interface into two free surfaces given by the difference in total energy between the interface and its initial isolated slabs according to the following formula [18–20]:

$$W_{ad} = \frac{(E_{Ge} + E_{SiC} - E_{Ge/4H\text{-}SiC})}{(NA)}, \quad (1)$$

where E_{Ge} and E_{SiC} are the total energy of Ge slab and SiC slab, where one slab remained and the other is replaced by

TABLE 1: The lattice mismatch of the Ge/4H-SiC heterostructures calculated with the domain matching model.

Growth orientation	Ge-to-SiC matching mode		Residual mismatch		MD density
Ge(111)/4H-SiC(0001)	Ge[01-1] SiC[11-20]	Ge[2-1-1] SiC[1-100]	Ge[01-1] SiC[11-20]	Ge[2-1-1] SiC[1-100]	5.334×10^{14} cm^{-2}
	3 : 4	3 : 4	2.60%	2.60%	
Ge(110)/4H-SiC(0001)	Ge[001] SiC[10-10]	Ge[1-10] SiC[-12-10]	Ge[001] SiC[10-10]	Ge[1-10] SiC[-12-10]	1.523×10^{14} cm^{-2}
	1 : 1	3 : 4	−5.78%	2.60%	

TABLE 2: Unrelaxed and relaxed E_{Ge}, $E_{4H\text{-}SiC(0001)}$, $E_{Ge/4H\text{-}SiC}$, and W_{ad}.

	$E_{Ge(111)}$ (eV)	$E_{4H\text{-}SiC(0001)}$ (eV)	$E_{Ge(111)/4H\text{-}SiC(0001)}$ (eV)	W_{ad} (J/m^2)
Unrelaxed	−5777.006	−12844.091	−18642.163	0.104
Relaxed	−5777.551	−12845.745	−18644.926	0.106
	$E_{Ge(110)}$ (eV)	$E_{4H\text{-}SiC(0001)}$ (eV)	$E_{Ge(110)/4H\text{-}SiC(0001)}$ (eV)	W_{ad} (J/m^2)
Unrelaxed	−3207.981	−8523.457	−11742.871	0.193
Relaxed	−3208.026	−8524.842	−11746.039	0.222

TABLE 3: Relaxation energies of Ge(111)/4H-SiC(0001) and Ge(110)/4H-SiC(0001) interfaces.

Heterojunction	E'_{total} (eV)	E_{total} (eV)	$E_{relaxion}$ (eV/atom)
Ge(111)/4H-SiC(0001)	−18642.163	−18644.926	−0.017
Ge(110)/4H-SiC(0001)	−11742.871	−11746.039	−0.030

vacuum in the same supercell, respectively. $E_{Ge/4H\text{-}SiC}$ denotes the total energy of the interface system, N is the number of atoms at the interface in the model, and A is the interfacial area. Based on (1), the variable values are obtained and listed in Table 2.

In addition, the relaxation energy $E_{relaxion}$ can be determined by an expression as follows:

$$E_{relaxion} = \frac{\left(E_{total} - E'_{total}\right)}{N}, \quad (2)$$

where E'_{total} and E_{total} are the total energies of the unrelaxed and relaxed interface systems, respectively, and N is the number of atoms in the system. Based on (2), the variable values are obtained and listed in Table 3.

Table 2 shows that the bonding energy of the unrelaxed interface is smaller than that of the relaxed one, indicating that the relaxed interface is more stable. It is also shown that the adhesion energy of Ge(110)/4H-SiC(0001) interface is higher than that of the Ge(111)/4H-SiC(0001) interface, indicating that Ge(110)/4H-SiC(0001) heterointerface is more energetically stable than Ge(111)/4H-SiC(0001) heterointerface. As shown in Table 3, the relaxation energy of Ge(110)/4H-SiC(0001) interface is lower than that of Ge(111)/4H-SiC(0001) interface, suggesting that Ge(110) films are easier to deposit on 4H-SiC(0001) substrates at relative low temperatures, which is consistent with the conclusions in [6].

The influence of relaxation on the atom positions at the Ge/4H-SiC interfaces is investigated. Figures 4(a) and 4(b) represent the atom-stacking structures of the post-optimized Ge(110)/4H-SiC(0001) interface and Ge(111)/4H-SiC(0001) interface, respectively. It is shown that the position of atoms near the interface deviates from the original position to some extent, displaying certain displaces. To quantitatively compare the extent of relaxation between Ge(110)/4H-SiC(0001) and Ge(111)/4H-SiC(0001) interface, corresponding variations of XYZ coordinates and variations of distance were calculated, as shown in Figures 4(c)–4(j). The first and second layers of atoms at the interface severely deviate from the equilibrium position. Approaching to the bulk materials, the deviations decrease drastically, suggesting that, as the interface formed, merely one or two layers of atoms at the interface were significantly influenced. In the meantime, one can also observe that at the interface the variation of Ge atoms is larger than that of SiC atoms, indicating that the relaxation occurs mainly on the Ge side. It is shown that the variation of atoms in Figure 4(d) is larger than that in Figure 4(c), which is attributed to the fact that the lattice mismatch in the Y direction is greater than that in the X direction at the Ge(110)/4H-SiC(0001) interface. However, the variation of the atoms in Figure 4(g) is almost the same as that in Figure 4(h), since the lattice mismatch in the Y direction is commensurate to that in the X direction at the Ge(111)/4H-SiC(0001) interface.

3.2.2. Electronic Structure and Bonding. In order to understand the essence of bonds of Ge(111)/4H-SiC(0001) and Ge(110)/4H-SiC(0001) interfaces, the total charge density and charge density difference of Ge(111)/4H-SiC(0001) and Ge(110)/4H-SiC(0001) interfaces are calculated, as shown in Figures 5(a)–5(d), respectively. High charge accumulation

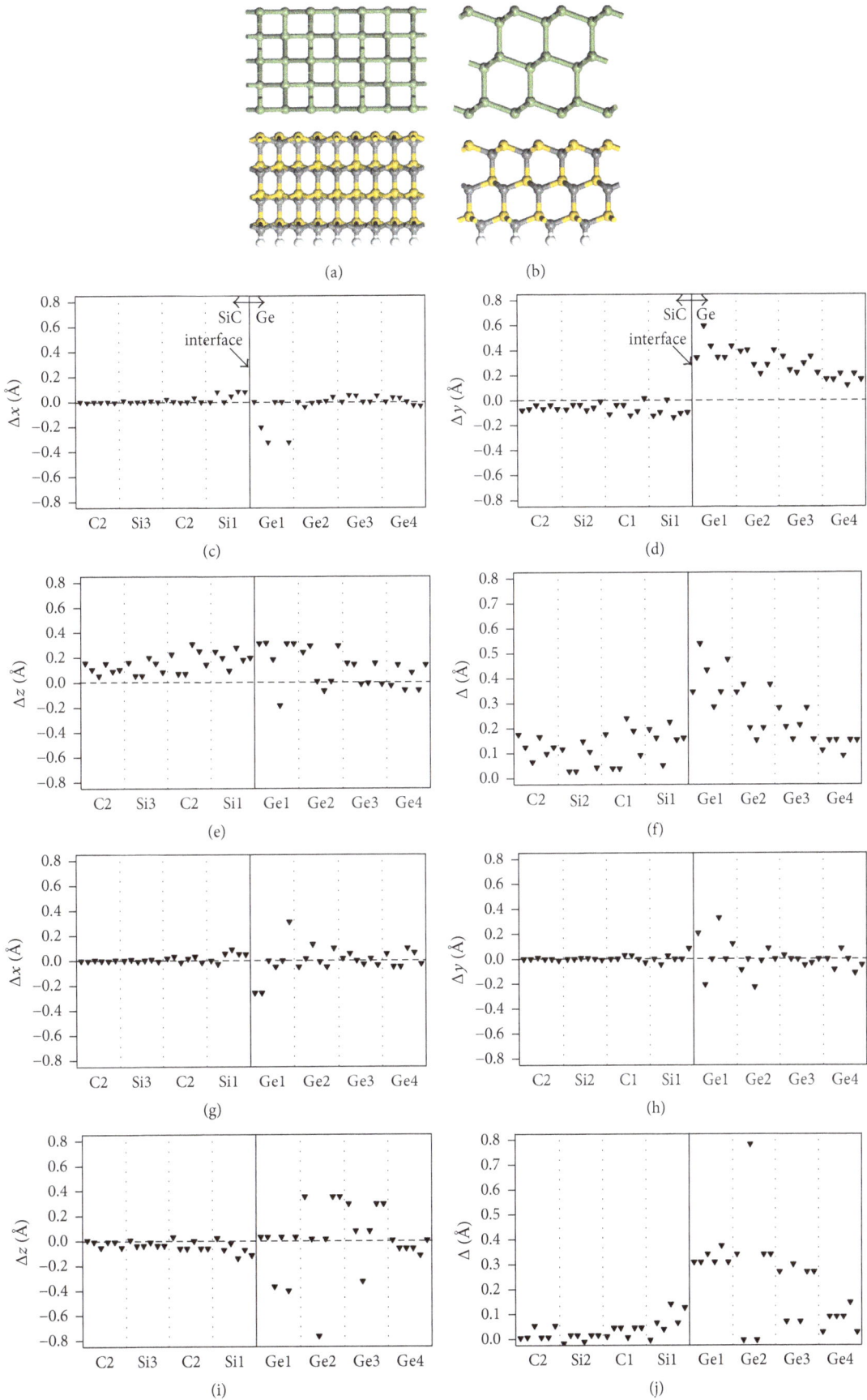

FIGURE 4: Ge(110)/4H-SiC(0001) heterointerface and variations of XYZ coordinates and variation of distance are shown in (a) and (c–f), respectively. Ge(111)/4H-SiC(0001) heterointerface and variations of XYZ coordinates and variation of distance are shown in (b) and (g–j), respectively.

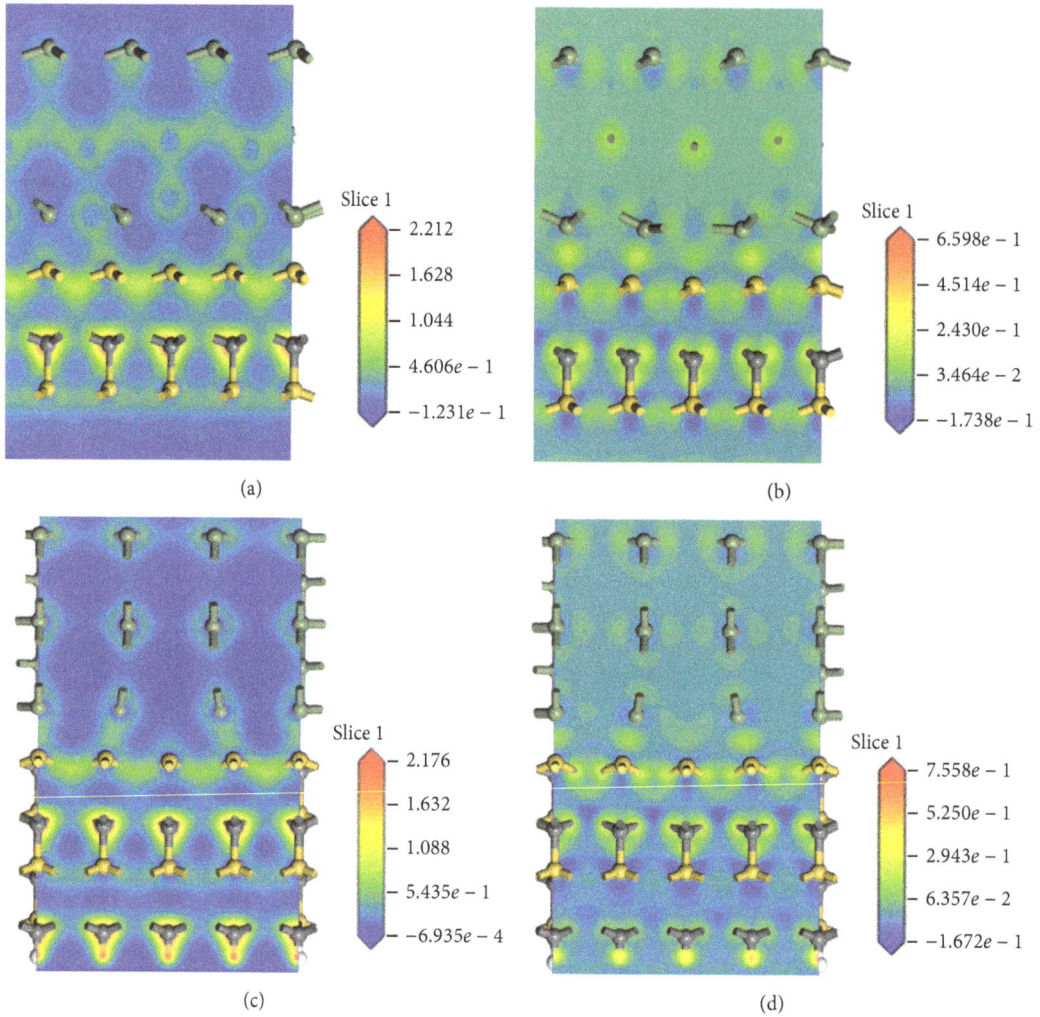

FIGURE 5: Total charge density for Ge(111)/4H-SiC(0001) (a) and Ge(110)/4H-SiC(0001) heterojunction (c). The charge density difference for Ge(111)/4H-SiC(0001) (b) and Ge(110)/4H-SiC(0001) heterojunction (d).

appears between interfacial Ge and Si atom at the Ge(111)/4H-SiC(0001) interface, indicating that Ge-Si bonding is formed at the interface. The charge density difference of Ge(111)/4H-SiC(0001) interface is displayed in Figure 5(b). The blue areas represent the depletion region of the charges, and the yellow areas show the accumulation region of the charge. It is shown that an extremely large amount of charges, which mainly come from the bulk materials near the interface, accumulated at the Ge(111)/4H-SiC(0001) interface, suggesting the formation of Ge-Si bonding at interface due to the transfer of bulk materials. For Ge(110)/4H-SiC(0001) interfaces, there are lots of charge accumulation between Ge atom of Ge crystal and Si atom of SiC crystal in interface, as shown in Figures 5(c) and 5(d). Therefore, the existence of Ge-Si bonding is proved.

An insight into the bonding properties of Ge(110)/4H-SiC(0001) interface and Ge(111)/4H-SiC(0001) interface are provided by calculating the partial density of states (PDOS) of the interface. The partial density of states (PDOS) of Ge(110)/4H-SiC(0001) interface and Ge(111)/4H-SiC(0001) interface are shown in Figures 6(a) and 6(b), respectively.

It reveals that relaxation merely occurs in one or two layers of atoms near the interface by analysing atoms position at the interface. For comparison, hereby, the partial density of states (PDOS) of Ge bulk and 4H-SiC bulk are displayed in Figures 6(a) and 6(b) as well for comparison, and, as a result, several distinct features can be observed. Firstly, compared to the case of bulk materials, the first and second layers exhibit delocalization and lower density, suggesting that electrons in the first and second layers are transferred into the interface and involved in the formation of bonding. Secondly, the distribution of density of states of the first and second layers shifts from the low energy region to the high energy region as compared to the case of bulk materials, which is largely caused by the rearrangement of the atoms and the formation of the interface. Finally, the distribution of density of states of the first and second layers on the Ge side severely deviates from that of the bulk, showing similarity to the distribution of density of states on the SiC side. Similar situation occurs on the distribution of density of states of the first and second layers on the SiC side, indicating that the distribution of

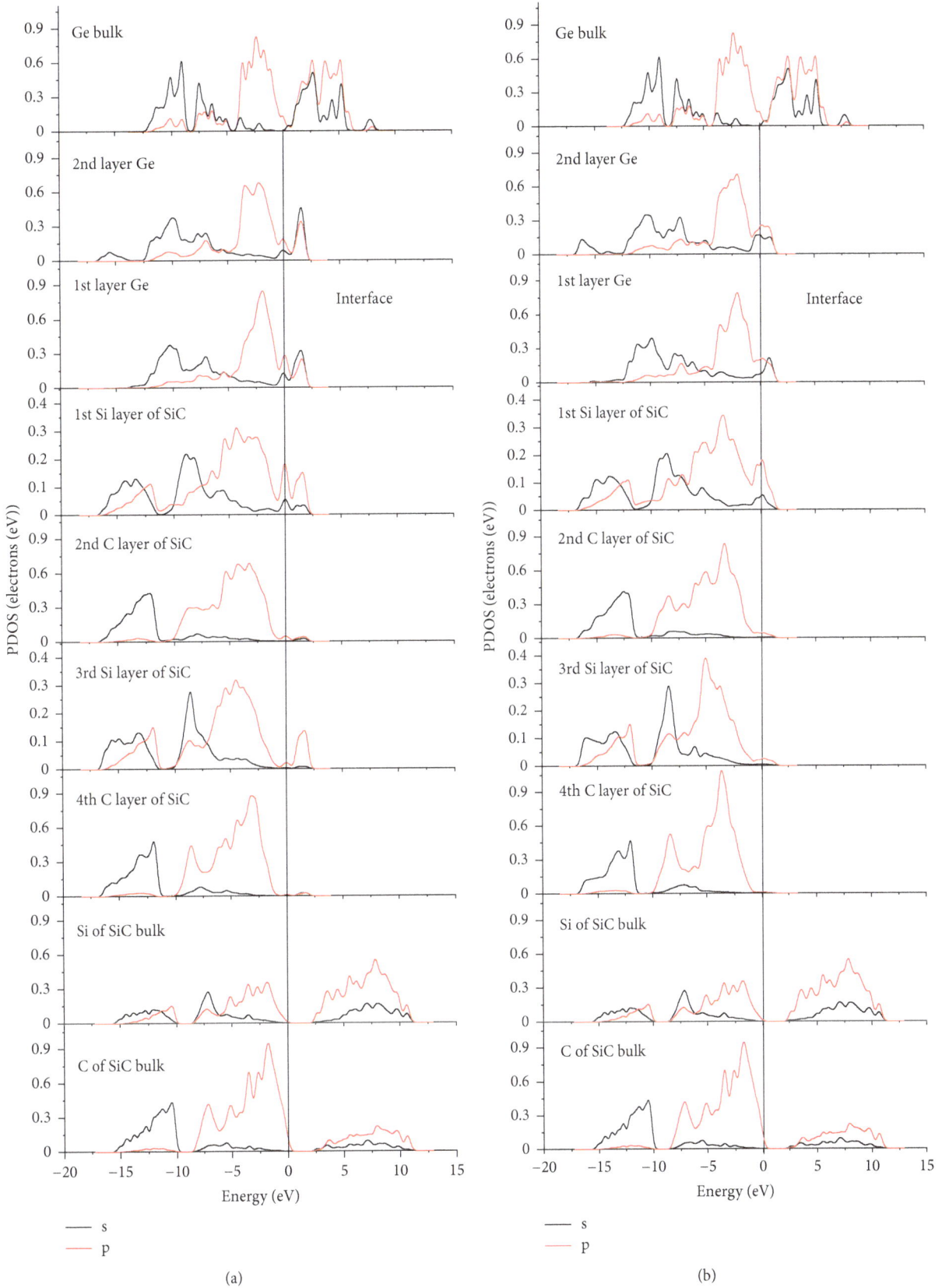

FIGURE 6: Partial density of state (PDOS) of Ge(110)/4H-SiC(0001) (a) and Ge(111)/4H-SiC(0001) heterojunctions (b).

density of states at the interface is influenced by the bulk materials on both sides. As shown in Figure 6(a), on the Ge side, the significant peaks appear in the ranges from -12.5 eV to -6 eV, -6 eV to -5 eV, -5 eV to 1 eV, and 1 eV to 2.5 eV. The densities of states from -12.5 eV to -6 eV and 1 eV to 2.5 eV originate mainly from the Ge-4s, -5 eV to 1 eV is mainly from the Ge-4p, and the Ge-4s and Ge-4p are mixed to the density of states from -6 eV to -5 eV, indicating the presence of Ge-Ge bonds. The density of states from -16 eV to -13 eV is mainly originated from the C-2s and Si-3s, -13 eV to -10 eV is mainly associated with the C-2s and Si-3p, -10 eV to -7.5 eV is mostly related to the C-2p and Si-3s, and -7.5 eV to 0 eV is largely originated from the C-2p and Si-3p. By comparing with the bulk material, the distribution of density of states at the heterointerface shifts toward low energy slightly. Furthermore, by comparing the Ge(110)/4H-SiC(0001) and Ge(111)/4H-SiC(0001) heterointerfaces, for the first Ge layer of Ge slab and the first Si layer of 4H-SiC slab, several distinct resonance peaks appear in the range of -4 eV to 0 eV as well. As shown in Figure 6(b), Ge(111)/4H-SiC(0001) heterointerface significant resonance peaks appear as well. The peaks mainly originate from the orbital hybridization of Si-3p and Ge-4P, indicating the formation of Ge-Si bond at the interface.

4. Conclusions

First-principles calculations are utilized to gain an insight into the interfacial properties of Ge/4H-SiC. The stability, electronic structure, and bonding properties of Ge(111)/4H-SiC(0001) and Ge(110)/4H-SiC(0001) are mainly studied. The works of adhesion of Ge(110)/4H-SiC(0001) and Ge(111)/4H-SiC(0001) interfaces are 0.222 J/m^2 and 0.106 J/m^2, respectively. The work of adhesion of Ge(110)/4H-SiC(0001) interface is higher than that of the Ge(111)/4H-SiC(0001) interface, leading to that Ge(110)/4H-SiC(0001) interface is more stable than Ge(111)/4H-SiC(0001) interface. Ge(110)/4H-SiC(0001) interface is easier to form at low temperatures due to its lower relaxation energy as compared with Ge(111)/4H-SiC(0001) interface. Calculations on the electronic structure and PDOS indicate that the Ge-Si bonds have been formed at the interface, which are mainly due to the orbital hybridization of Si-3p and Ge-4P.

Conflicts of Interest

The authors declare that they have no conflicts of interest.

Acknowledgments

This work was supported financially by the National Natural Science Foundation of China (Grants nos. 51402230 and 21503153), the Project Supported by Natural Science Basic Research Plan in Shaanxi Province of China (Grants nos. 2015JM6282 and 2017JM6075), Scientific Research Program Funded by Shaanxi Provincial Education Department (Grants nos. 14JK1302 and 17JK0335), and Doctoral Scientific Research Foundation of Xi'an University of Technology (Grant no. Y201605).

References

[1] L. B. Li, Z. M. Chen, Y. Zang, L. X. Song, Y. L. Han, and Q. Chu, "Epitaxial growth of Si/SiC heterostructures with different preferred orientations on 6H-SiC(0001) by LPCVD," *CrystEngComm*, vol. 18, no. 30, pp. 5681–5685, 2016.

[2] L. B. Li, Z. M. Chen, Y. Zang, and S. Feng, "Atomic-scale characterization of Si(110)/6H-SiC(0001) heterostructure by HRTEM," *Materials Letters*, vol. 163, pp. 47–50, 2016.

[3] Q. Chu, L. Li, C. Zhu, Y. Zang, S. Lin, and Y. Han, "Preparation of SiC/Ge/graphene heterostructure on 4H-SiC(0001)," *Materials Letters*, vol. 211, pp. 133–137, 2018.

[4] Y. L. Han, H. B. Pu, Y. Zang, and L. B. Li, "Epitaxial growth of Ge film on 6H-SiC(0001) by LPCVD," *Optoelectronics and Advanced Materials – Rapid Communications*, vol. 10, no. 9-10, pp. 737–739, 2016.

[5] K. Alassaad, V. Soulière, F. Cauwet et al., "Ge incorporation inside 4H-SiC during homoepitaxial growth by chemical vapor deposition," *Acta Materialia*, vol. 75, pp. 219–226, 2014.

[6] P. M. Gammon, A. Pérez-Tomás, M. R. Jennings et al., "Interface characteristics of n-n and p-n Ge/SiC heterojunction diodes formed by molecular beam epitaxy deposition," *Journal of Applied Physics*, vol. 107, no. 12, p. 124512, 2010.

[7] J. Li, Y. Q. Yang, L. L. Li et al., "Interfacial properties and electronic structure of β-SiC(111)/α-Ti(0001): A first principle study," *Journal of Applied Physics*, vol. 113, Article ID 023516, 2013.

[8] Z. Lin, X. Peng, T. Fu et al., "Atomic structures and electronic properties of interfaces between aluminum and carbides/nitrides: A first-principles study," *Physica E: Low-dimensional Systems and Nanostructures*, vol. 89, pp. 15–20, 2017.

[9] X. M. He, Z. M. Chen, H. B. Pu et al., "First-principles calculations on Si (220) located 6H—SiC (10$\bar{1}$0) surface with different stacking sites," *Chinese Physics B*, vol. 23, no. 10, Article ID 106802, 2014.

[10] X.-M. He, Z.-M. Chen, L. Huang, and L.-B. Li, "First-principles calculations on atomic and electronic properties of Si(111)/6H-SiC(0001) heterojunction," *Modern Physics Letters B*, vol. 29, no. 29, Article ID 1550182, 2015.

[11] X. Xu, H. Wang, M. Zha, C. Wang, Z. Yang, and Q. Jiang, "Effects of Ti, Si, Mg and Cu additions on interfacial properties and electronic structure of Al(111)/4H-SiC(0001) interface: A first-principles study," *Applied Surface Science*, vol. 437, pp. 103–109, 2018.

[12] H. Xiong, Z. Liu, H. Zhang, Z. Du, and C. Chen, "First principles calculation of interfacial stability, energy and electronic properties of SiC/ZrB2 interface," *Journal of Physics and Chemistry of Solids*, vol. 107, pp. 162–169, 2017.

[13] J. Yang, J. Huang, D. Fan, and S. Chen, "First-principles investigation on the electronic property and bonding configuration of NbC (111)/NbN (111) interface," *Journal of Alloys and Compounds*, vol. 689, pp. 874–884, 2016.

[14] J. Martinez, S. B. Sinnott, and S. R. Phillpot, "Adhesion and diffusion at TiN/TiO2 interfaces: A first principles study," *Computational Materials Science*, vol. 130, pp. 249–256, 2017.

[15] W. H. Lee and X. H. Yao, "First principle investigation of phase transition and thermodynamic properties of SiC," *Computational Materials Science*, vol. 106, pp. 76–82, 2015.

[16] M. Piasecki, M. G. Brik, and I. V. Kityk, "Tl4CdI6 – Wide band gap semiconductor: First principles modelling of the structural, electronic, optical and elastic properties," *Materials Chemistry and Physics*, vol. 163, pp. 562–568, 2015.

[17] N. Chandran, M. Sall, J. Arvanitidis et al., "On the Formation of Graphene by Ge Intercalation of a 4H-SiC Surface," *Materials Science Forum*, vol. 821-823, pp. 961–964, 2015.

[18] J. Li, Y. Yang, G. Feng, X. Luo, Q. Sun, and N. Jin, "First-principles study of stability and properties on β-SiC/TiC(111) interface," *Journal of Applied Physics*, vol. 114, no. 16, Article ID 163522, 2013.

[19] D. Yin, X. Peng, Y. Qin, and Z. Wang, "Electronic property and bonding configuration at the TiN(111)/VN(111) interface," *Journal of Applied Physics*, vol. 108, no. 3, Article ID 033714, 2010.

[20] D. Y. Dang, L. Y. Shi, J. L. Fan, and H. R. Gong, "First-principles study of W-TiC interface cohesion," *Surface and Coatings Technology*, vol. 276, pp. 602–605, 2015.

27

Investigation of the Atomic Structure of Ge-Sb-Se Chalcogenide Glasses

M. Fabian ⓘ,[1,2] **N. Dulgheru,**[3] **K. Antonova,**[4] **A. Szekeres,**[4] **and M. Gartner**[3]

[1]*Centre for Energy Research, Hungarian Academy of Sciences, H-1525 Budapest P.O.B. 49, Hungary*
[2]*Wigner Research Centre for Physics, Hungarian Academy of Sciences, H-1525 Budapest P.O.B. 49, Hungary*
[3]*Institute of Physical Chemistry "Ilie Murgulescu", Romanian Academy, 202 Splaiul Independentei, 060021 Bucharest, Romania*
[4]*Institute of Solid State Physics, Bulgarian Academy of Sciences, Tzarigradsko Chaussee 72, 1784 Sofia, Bulgaria*

Correspondence should be addressed to M. Fabian; fabian.margit@energia.mta.hu

Academic Editor: Jörg Fink

Glasses with composition of $Ge_xSb_{40-x}Se_{60}$ (x= 40, 35, 32, 27, 20, 15 at. %) have been synthesized. Neutron and X-ray diffraction techniques were used to study the atomic glassy structure, and Reverse Monte Carlo (RMC) simulations were applied to model the 3-dimensional atomic configurations and thorough mapping of the atomic parameters, such as first and second neighbour distances, coordination numbers, and bond-angle distributions. The results are explained with formation of $GeSe_4$ and $SbSe_3$ structural units, which correlate with the Ge/Sb ratio. For all the studied compositions, the Ge-Se, Sb-Se, Ge-Ge, and Se-Se bonds are significant. RMC simulations reveal the presence of Ge-Sb and Sb-Sb bonds, being dependent on Ge/Sb ratio. All atomic compositions satisfy formal valence requirements, i.e., Ge is fourfold coordinated, Sb is threefold coordinated, and Se is twofold coordinated. By increasing the Sb content, both the Se-Ge-Se bonds angle of $107\pm3°$ and Se-Sb-Se bonds angle of $118\pm3°$ decrease, respectively, indicating distortion of the structural units. Far infrared Fourier Transform spectroscopic measurements conducted in the range of 50-450 cm^{-1} at oblique (75°) incidence radiation have revealed clear dependences of the IR band's shift and intensity on the glassy composition, showing features around x=27 at.% supporting the topological phase transition to a stable rigid network consisting mainly of $SbSe_3$ pyramidal and $GeSe_4$ tetrahedral clusters. These results are in agreement with the Reverse Monte Carlo models, which define the Ge and Sb environment.

1. Introduction

In recent years a great deal of interest has been devoted to the optical studies of chalcogenide glasses because of their wide area of applications as electrical and optical components as well as optical fibers in the infrared optical region, which are the basis of many applications [1–4]. Ternary Ge-Sb-Se glasses are promising materials for infrared optical fibers application because of their high transparency in the key region and good thermal, mechanical, and chemical properties. The physicochemical, elastic-plastic, and optical properties of the ternary Ge-Sb-Se chalcogenide system have been intensively studied [5–13], and the influence of Ge/Sb ratio on the optical, electronic, and microstructural properties of Ge-Sb-Se system has been considered. A common and useful parameter for the explanation of the compositional

dependence on the physical properties is the coordination number Z of covalent bonds per atom, characterizing the structural atomic units [5, 13, 14]. For the compositions under investigation, a value of Z=2.65-2.67 corresponding to x=25-27 at.% has been obtained as critical for a transition from 2D cross-linked chains arrangements to formation of 3D network of tetrahedral $GeSe_4$ and pyramidal $SbSe_3$ when the x value increases [5, 13].

Still one of the main questions remains how the structure of Ge environment changes, when the Ge atoms are replaced by Sb atoms. A comprehensive answer can be obtained by combining different methods and techniques, such as neutron diffraction (ND), high-energy X-ray diffraction (XRD), and far infrared (IR) Fourier Transform spectroscopy, which offers comprehensive information about the atomic structure. In addition, theoretical modeling of such experimental results using Reverse Monte Carlo (RMC) method allows building three-dimensional structure models and, thus, obtaining a more detailed description of the atomic-scale glassy structure. In connection with this, we have successfully performed experiments combining the ND and XRD techniques and RMC simulation method for the study of ternary As-Se-Te [15] and quaternary Ge-Sb-S(Se)-Te glasses [16–18]. Using the experience gained, we have started to investigate new Ge-Sb-Se system. The compositional dependence of the structure and physical and optical properties of ternary Ge-Sb-Se system is of interest, especially considering the effects of average coordination number, Z, of covalent bonds per atom.

In this paper, we focused on the atomic-scale structure characterization of nonstoichiometric $Ge_xSb_{40-x}Se_{60}$ compositions, where x varied between 15 and 40 at. %. For examining the alteration of the glassy structure with changes in composition, we applied an approach in which ND and high-energy XRD techniques with Reverse Monte Carlo (RMC) modeling were combined. The procedure of RMC modeling is considered in details in Section 3.1. By this way, characteristic parameters, such as structure factors, partial atomic pair correlation functions, coordination numbers, and bond-angle distribution functions, are established. Recently, the chemical short order in $Ge_{20}Sb_xSe_{80-x}$ (x=5, 15, 20) compositions has been studied by the nuclear techniques and extended X-ray absorption fine structure measurements [19] as one of the compositions, namely, $Ge_{20}Sb_{20}Se_{40}$, is among the compositions we study herein. Our results obtained by the combination of different methods would widen the knowledge about these materials and would contribute to a better understanding of the structural changes by composition.

In addition, we performed Fourier Transform infrared (FTIR) reflectance measurements in the terahertz spectral range in order to get more information about the microstructure and basic chemical bonding of the studied compositions. So far, few far IR spectral measurements have been carried out for studying the long-wavelength multiphonon edge of these materials. Raman and FTIR spectra have been reported in [5] and [8, 10, 20, 21], respectively, as in the latter PE pellets are used resulting in a not so good spectrum even when they are converted in absorption units. To avoid this, we carried out the FTIR measurements on thin (~ 1 μm) films deposited on quartz substrates. Using incidence radiation under a large angle of 75° the obtained spectra are more informative about the composition microstructure with variation of x value. The results of this paper correspond to the conclusions presented by other authors and are in agreement with our earlier publications [5] on bulk samples served here as parent materials for thin films preparation.

2. Experimental Section

2.1. Sample Preparation. The binary $Ge_{40}Se_{60}$ and ternary $Ge_xSb_{40-x}Se_{60}$ glasses with x=15, 20, 27, 32, and 35 at. % were prepared from elements of 99.999 % purity. Appropriate quantities from Ge, Se, and Sb were sealed into quartz ampoules after evacuation down to a pressure of 10^{-3} Pa. The bulk glasses were synthesized by the conventional melt-quenching method in a rotary furnace at 950°C. After

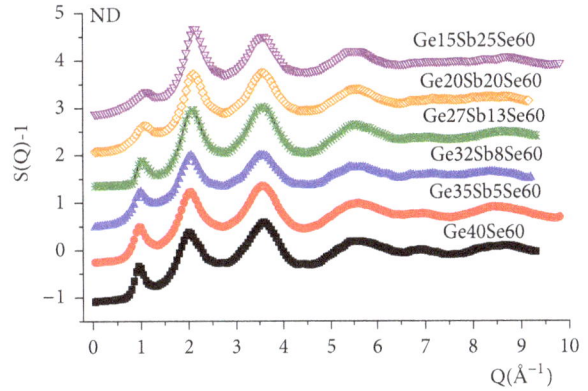

FIGURE 1: Neutron diffraction structure factor, $S(Q)$, of the chalcogenide samples with composition of $Ge_xSb_{40-x}Se_{60}$ (x=40, 35, 32, 27, 20, 15 at.%) is displayed: experimental data (colour marks) and RMC simulation (solid lines). For better clarity, the curves are shifted vertically.

homogenization for 24 h, the melts were quenched in air.

Part of the bulk samples was powdered. At the ND and XRD studies, powdered material was used as specimen with ~ 3 g/each composition. The powdered glasses served also as parent material at evaporation of thin films for Fourier Transform infrared (FTIR) spectroscopic studies.

The thin films were deposited onto quartz substrates by vacuum thermal evaporation of the powders from the corresponding parent $Ge_xSb_{40-x}Se_{60}$ materials at a residual pressure of 10^{-4} Pa in the chamber. The thickness of the films (~1 μm) was controlled in situ by MIKI FFV quartz sensor device.

2.2. Neutron and X-Ray Diffraction Experiments. The neutron diffraction measurements were performed at room temperature in the 2-axis 'PSD' monochromatic neutron diffractometer (λ_0=1.068 Å; Q=0.45-9.8 Å^{-1}) [22] at the 10 MW Budapest Research Reactor, using thermal neutrons. The powder specimens (~3 g/each) were filled in cylindrical vanadium sample holder of 8 mm diameter. The structure factors, $S(Q)$, were evaluated from the raw experimental data. The overall run of the ND experimental curves is similar for the investigated samples, but characteristic differences especially in the low Q-range can be observed. Figure 1 presents the $S(Q)$ from neutron diffraction experiments, where the results of the Reverse Monte Carlo model calculation are also drawn, but this will be discussed later, in Section 3.

The X-ray diffraction experiments were carried out at the beam line BW5 at HASYLAB, DESY [23]. The powdered samples were filled into quartz capillary tubes of 2 mm in diameter (wall thickness of ~0.02 mm) and measured at room temperature. The energy of the radiation was 109.5 keV (λ_0=0.113 Å). The high-energy synchrotron X-ray radiation makes it possible to reach diffraction data up to high-Q values. In this study the XRD structure factors were obtained up to 18 Å^{-1}, as for higher Q-values the experimental data became rather noisy. Because the atomic parameters of the

TABLE 1: Neutron and X-ray diffraction weighting factors, w_{ij}(%) at Q=1.05 Å^{-1}, of the partial atomic pairs in $Ge_{15}Sb_{25}Se_{60}$ sample.

$Ge_{15}Sb_{25}Se_{60}$					
Ge-Ge	Ge-Sb	Ge-Se	Sb-Sb	Sb-Se	Se-Se
ND weighting factor (%)					
2.75	6.24	21.43	3.54	24.30	41.74
XRD weighting factor (%) at Q=1.05 Å^{-1}					
1.59	8.49	13.56	11.32	36.16	28.88

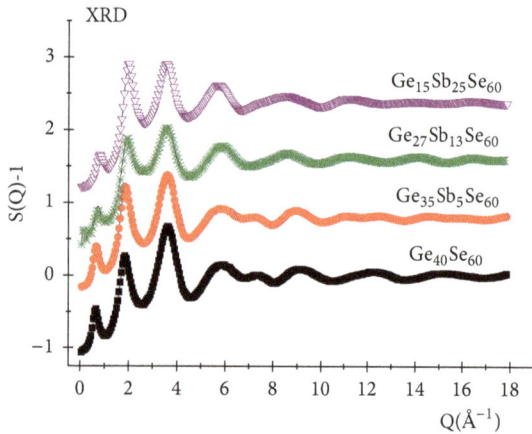

FIGURE 2: X-ray diffraction structure factor, $S(Q)$, of the ternary $Ge_xSb_{40-x}Se_{60}$ with composition of x=35, 27, 15 at.% and the binary $Ge_{40}Se_{60}$ glasses is displayed: experimental data (colour marks) and RMC simulation (solid lines). For better clarity, the curves are shifted vertically.

components are similar, the XRD structure factors were similar to ND ones. We consider here the XRD data for four samples which shows well the concentration dependence within the series. The XRD experimental $S(Q)$ data together with the RMC simulation results (details of the RMC modeling will be discussed in Section 3) are presented in Figure 2.

The shape and character of the ND (Figure 1) and XRD (Figure 2) structure factors testify to a fully amorphous glassy structure in nature. However, there are some fine differences between the ND and XRD spectra. The differences in the overall run of these spectra lie in the different values of the weighting factors, w_{ij}, of the partial structure factors, $S_{ij}(Q)$, defined as

$$S(Q) = \sum_{i,j}^{k} w_{ij} S_{ij}(Q), \tag{1}$$

$$w_{ij}^{ND} = \frac{c_i c_j b_i b_j}{\left[\sum_{i,j}^{k} c_i b_j\right]^2}, \tag{2}$$

$$w_{ij}^{XRD}(Q) = \frac{c_i c_j f_i(Q) f_j(Q)}{\left[\sum_{i,j}^{k} c_i f_i(Q)\right]^2}, \tag{3}$$

where c_i, c_j are the molar fractions of the components, b_i, b_j are the coherent neutron and $f_i(Q)$, $f_j(Q)$ are the X-ray scattering amplitudes, and k is the number of elements

in the sample. For the binary $Ge_{40}Se_{60}$ sample k is 2, thus $k(k+1)/2$=3, and for the ternary Ge-Sb-Se samples k is 3, thus $k(k+1)/2$=6, and, therefore, 6 different atom pairs are present in the studied series. The neutron scattering amplitude of an element is constant in the entire Q-range [24], while the X-ray scattering amplitude is Q-dependent [25] and for each atom it differs in a somewhat different way. To illustrate the differences, the comparison of the corresponding weighting factors, w_{ij}, for the two radiations is shown in Table 1, where w_{ij}^{ND} and $w_{ij}^{XRD}(Q)$ at Q=1.05 Å^{-1} are given for the $Ge_{15}Sb_{25}Se_{60}$ sample.

It can be seen that the Ge-Se and Se-Se atom pairs have a significant contribution in the neutron experiment, while the Sb-Sb and Sb-Se atom pairs have a dominant weight in the X-ray experiment. On the other hand, the differences appeared for the Ge-Ge and Ge-Sb atom pairs are not more significant than those for the other ones. Taking into consideration all these characteristics, we have concluded that the two radiations give complementary information, and both types of measurements are needed to obtain a real structure for the investigated samples.

2.3. Fourier Transform Infrared Spectrophotometry. In order to detect the basic chemical bonding in the $Ge_xSb_{40-x}Se_{60}$ compositions Fourier Transform infrared measurements were performed in the terahertz spectral range of 450-50 cm^{-1} of the polarized light. FTIR spectra were measured in close to reflectance-absorbance geometry with an incident radiation angle of 75°, permitting maximal optical path through the sample. A Bruker Vertex 70 instrument and a reflectance accessory Bruker A513/Q were used. The resolution was 2 cm^{-1} and the number of scans was 100. A gold standard mirror was used for the spectral normalization.

3. Results and Discussion

3.1. Reverse Monte Carlo Simulation. The experimental diffraction $S(Q)$ data has been simulated by the RMC method [26], which is a widely used effective tool to model disordered structures. The RMC minimizes the squared difference between the experimental $S(Q)$ and the calculated one from a 3D atomic configuration. From the RMC configuration the partial pair correlation functions, $(g_{ij}(r))$, the coordination numbers, and the bond-angle distributions were calculated using the software package RMC^{++} [27].

In order to investigate the possible effect of the starting configuration on the results, the following model was built

TABLE 2: Cut-off distances (Å) for atom pairs used in the final RMC run.

Atom pairs	Cut-off distances (Å)					
	$Ge_{40}Se_{60}$	$Ge_{35}Sb_5Se_{60}$	$Ge_{32}Sb_8Se_{60}$	$Ge_{27}Sb_{13}Se_{60}$	$Ge_{20}Sb_{20}Se_{60}$	$Ge_{15}Sb_{25}Se_{60}$
Ge-Se	2.15	2.05	2.15	2.15	2.10	2.125
Sb-Se	-	2.10	2.125	2.15	2.175	2.17
Ge-Sb	-	2.25	2.25	2.20	2.25	2.25
Ge-Ge	2.30	2.20	2.20	2.20	2.20	2.25
Se-Se	2.00	2.20	2.20	2.15	2.15	2.175
Sb-Sb	-	2.35	2.35	2.345	2.35	2.30

up. The initial configuration was generated by random distribution of 10 000 atoms in a cubic simulation box. The experimentally measured density was 0.032, 0.030, 0.031, 0.038, 0.039, and 0.038 atoms/Å^3 [28] corresponding to box edges of 31.90 Å, 34.66 Å, 34.29 Å, 32.05 Å, 31.76 Å, and 32.05 Å for the $Ge_{40}Se_{60}$, $Ge_{35}Sb_5Se_{60}$, $Ge_{32}Sb_8Se_{60}$, $Ge_{27}Sb_{13}Se_{60}$, $Ge_{20}Sb_{20}Se_{60}$, and $Ge_{15}Sb_{25}Se_{60}$ samples, respectively.

The Ge, Sb, and Se elements possess comparable neutron and X-ray scattering amplitudes, which makes it rather difficult to identify the atomic positions. In order to avoid unreasonable short-range orders for all atom pairs, in the RMC simulation procedure constraints were applied to the minimum interatomic distances between atom pairs (cut-off distances). The starting cut-off distances were taken from the previous studies for the similar contents [15, 17] and from the literature based on the results of similar or very close glassy compositions [29, 30]. Several RMC runs have been performed by modifying slightly the cut-off distances for each atom pairs. The final cut-off distances, used further in the RMC modeling, are summarized in Table 2. From our previous study and from literature data, we expect that the main glassy network is built up by the GeSe$_{4/2}$ tetrahedra and the SbSe$_{3/2}$ pyramid units. Therefore, it is suggested that the atomic structure of the ternary glasses is built up by chain connection of GeSe$_{4/2}$ tetrahedra and SbSe$_{3/2}$ pyramid through a 2-coordinated Se atoms. Thus, the 4-coordinated Ge unit connects through a Se atom to a 3-coordinated Sb unit and, therefore, the first neighbour distance of Ge-Sb is less realistic. In the literature, there are studies [29, 31–33], which support this suggestion. Considering the values of the weighting factors in Table 1, one can see that Ge-Sb bond's weighting factors ($w_{ND,Ge-Sb}$=6.24% and $w_{XRD,Ge-Sb}$=8.49%) are lower than those for Ge-Se ($w_{ND,Ge-Se}$=21.43% and $w_{XRD,Ge-Se}$=13.56%) and Sb-Se ($w_{ND,Sb-Se}$=24.30% and $w_{XRD,Sb-Se}$=36.16%), which bonds have well known first neighbour distances. From these follows that Ge-Sb bonds most probably have the second neighbour distance. Nevertheless, in the RMC modeling, we did not exclude any first neighbour distances and Figure 3 presents all partial pair correlation functions obtained from the RMC calculations.

The experimental structure factors ($S(Q)$) are compared with those accumulated from the data of the RMC models of the corresponding glass compositions. For each sample, close to 20 RMC configurations were obtained in the intervals of more than 1 500 000 accepted moves of atoms inside of the

simulation box. Good agreement between the experimental and calculated diffraction data was achieved. The comparison between the experimental and calculated structure factors are illustrated in Figures 1 and 2. The measured neutron diffraction data in Figure 1 shows significant dependence on the composition, especially at low-Q values. For all compositions, the first and the most intensive peaks were situated at 1.1 and 2.1 Å^{-1}, while the next less intensive ones are at 3.55 Å^{-1} and at 5.5 Å^{-1}. Figure 2 displays the XRD $S(Q)$ spectra, where the first intensive peak is observed at 0.6 Å^{-1}, a second intensive peak at 1.8 Å^{-1}, and two other broaden and much less intensive peaks at 3.5 Å^{-1} and at 5.7 Å^{-1}.

In order to get comprehensive structural information about the studied series, we have investigated all homopolar and heteropolar atomic distances. The partial atomic pair correlation functions, $g_{ij}(r)$, were revealed from the RMC simulation with a good reproducibility and acceptable statistics. The obtained partial pair distribution functions are presented in Figure 3. The first peaks are well defined and, thus, we could perform an unambiguous evaluation of distances.

It was found that Ge-Se, Sb-Se, Ge-Ge, and Se-Se bonds are significant in the studied samples. The peaks related to the first interatomic distances appear at 2.35±0.01 Å, 2.54±0.01 Å, 2.47±0.03 Å, and 2.32±0.02 Å, respectively, as these peaks appear at the same position within the error for all the studied compositions. Because of the allowed cut-off distances (Table 2), first interatomic distances between Ge-Sb and Sb-Sb atoms were found at 2.55±0.05 Å and 2.57±0.03 Å, but the corresponding first peak's intensity was very low (Figures 3(c) and 3(e)). This suggests that second neighbour distance between Ge and Sb is more realistic. Second characteristic peaks were obtained for Ge-Sb, Ge-Ge, Sb-Sb, and Se-Se atom pairs, which are displayed at 3.80±0.05 Å, 3.85±0.05 Å, 3.80±0.05 Å, and 3.80±0.05 Å, respectively.

The partial atomic pair correlation functions are summarized in Table 3. These values, with exceptions of Ge-Sb and Ge-Ge, are close to the data given in [19], but the differences could be due to the annealing of samples in that work, which altered their glassy structure.

The great advance of the RMC method is that the coordination number, CN_{ij}, can be obtained from the final particle configurations. From the partial pair distribution functions we calculated the number of nearest neighbours for Ge, Sb, and Se atoms using the corresponding bond cut-off distances.

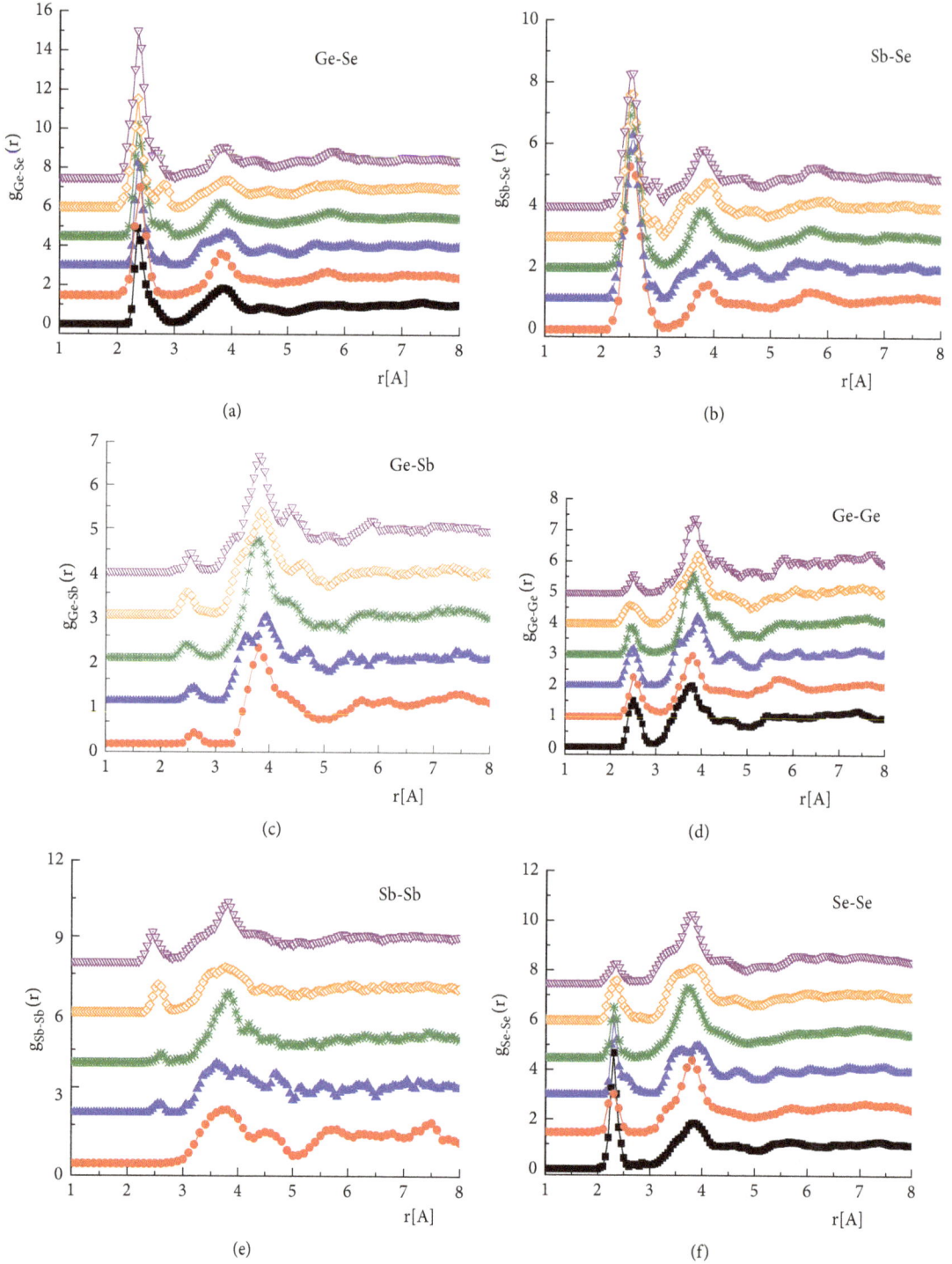

FIGURE 3: Partial atom pair correlation functions, $g_{ij}(r)$ for the $Ge_{40}Se_{60}$ (black square), $Ge_{35}Sb_5Se_{60}$ (red circle), $Ge_{32}Sb_8Se_{60}$ (blue triangle), $Ge_{27}Sb_{13}Se_{60}$ (green crosses), $Ge_{20}Sb_{20}Se_{60}$ (orange square), and $Ge_{15}Sb_{25}Se_{60}$ (purple reverse triangle) compositions, obtained by RMC modeling of the Ge-Se (a), Sb-Se (b), Ge-Sb (c), Ge-Ge (d), Sb-Sb (e), and Se-Se (f) atom pairs. For better clarity, the curves are shifted vertically.

It is necessary to specify the range of r within which atoms are counted as neighbours and which determine the coordination shells. We introduced the quantities r_1 and r_2, where r_1 and r_2 are the positions of minimum values on the lower and upper side of the corresponding peaks, respectively. From the literature, we know some details related to the coordination numbers of the present atoms: concerning the Ge atoms, they are 4-coordinated, the Sb atoms are 3-coordinated, and Se atoms expected to be close to 2-coordinated sites. To obtain more information about the atomic structure, we carried out

TABLE 3: Interatomic distances, r_{ij} (Å), in chalcogenide samples obtained from the RMC simulation.

Atom pairs	Interatomic distances, r_{ij} (Å)					
	$Ge_{40}Se_{60}$	$Ge_{35}Sb_5Se_{60}$	$Ge_{32}Sb_8Se_{60}$	$Ge_{27}Sb_{13}Se_{60}$	$Ge_{20}Sb_{20}Se_{60}$	$Ge_{15}Sb_{25}Se_{60}$
Ge-Se	2.34/3.85±0.02	2.35/3.82±0.02	2.36/3.90±0.02	2.35/3.80±0.02	2.35/3.85±0.02	2.35/3.82±0.02
Sb-Se	-	2.54/3.88±0.02	2.55/3.90±0.05	2.55/3.80±0.02	2.53/3.90±0.02	2.55/3.80±0.03
Ge-Sb	-	2.60/3.80±0.05	2.60/3.55/3.80±0.05	2.50/3.80±0.05	2.50/3.85±0.05	2.52/3.80±0.05
Ge-Ge	2.50/3.80±0.02	2.50/3.80±0.02	2.47/3.90±0.03	2.45/3.80±0.02	2.45/3.90±0.03	2.50/3.80±0.02
Se-Se	2.30/3.85±0.02	2.30/3.80±0.05	2.30/3.60/3.90±0.05	2.30/3.75±0.05	2.30/3.83±0.05	2.33/3.80±0.05
Sb-Sb	-	3.80/4.35±0.05	2.57/3.85±0.05	2.60/3.80±0.05	2.60/3.75±0.05	2.55/3.80±0.05

TABLE 4: Average coordination numbers, CN_{ij}, calculated from the RMC simulation. In brackets are indicated the intervals of r within which the atoms are counted as neighbours and, thus, the actual coordination numbers are calculated. The error is ~5% for Ge-Se, Sb-Se, Se-Ge and ~10% for Ge-Sb, Sb-Ge, Se-Sb, Ge-Ge, Sb-Sb, Se-Se. The total coordination distributions are signed by Ge-X for Ge, Sb-X for Sb, and Se-X for Se atom.

Atom pairs	Coordination Number, CN_{ij}					
	$Ge_{40}Se_{60}$	$Ge_{35}Sb_5Se_{60}$	$Ge_{32}Sb_8Se_{60}$	$Ge_{27}Sb_{13}Se_{60}$	$Ge_{20}Sb_{20}Se_{60}$	$Ge_{15}Sb_{25}Se_{60}$
Ge-Se	2.97 (r_1:2.15-r_2:2.90)	3.10 (r_1:2.2- r_2:2.85)	3.22 (r_1:2.15- r_2:2.8)	3.45 (r_1:2.15- r_2:2.95)	3.63 (r_1:2.15- r_2:2.95)	3.84 (r_1:2.1- r_2:2.9)
Ge-Sb	-	0.05 (r_1:2.4- r_2:2.85)	0.02 (r_1:2.4- r_2:2.8)	0.03 (r_1:2.2- r_2:2.75)	0.05 (r_1:2.3-2.75)	0.07 (r_1:2.3- r_2:2.7)
Sb-Se	-	2.97 (r_1:2.2-r_2:3.0)	2.80 (r_1:2.15- r_2:3.1)	2.78 (r_1:2.15- r_2:3.0)	2.64 (r_1:2.2- r_2:2.95)	2.59 (r_1:2.15- r_2:3.0)
Sb-Ge	-	0.03 (r_1:2.4- r_2:2.85)	0.05 (r_1:2.4- r_2:2.8)	0.04 (r_1:2.2- r_2:2.75)	0.09 (r_1:2.3- r_2:2.75)	0.11 (r_1:2.3- r_2:2.7)
Se-Ge	1.98 (r_1:2.15- r_2:2.90)	1.81 (r_1:2.2- r_2:2.85)	1.72 (r_1:2.15- r_2:2.8)	1.55 (r_1:2.15- r_2:2.95)	1.21 (r_1:2.2- r_2:2.85)	0.96 (r_1:2.1- r_2:2.9)
Se-Sb	-	0.21 (r_1:2.2-3.0)	0.31 (r_1:2.15-3.1)	0.58 (r_1:2.15-3.0)	0.88 (r_1:2.15-2.95)	1.08 (r_1:2.15-3.0)
Ge-Ge	0.74 (r_1:2.25- r_2:2.90)	0.55 (r_1:2.2- r_2:2.95)	0.52 (r_1:2.15- r_2:2.8)	0.40 (r_1:2.2- r_2:2.8)	0.23 (r_1:2.2- r_2:2.8)	0.05 (r_1:2.25- r_2:2.75)
Sb-Sb	-	0.00	0.03 (r_1:2.35- r_2:2.75)	0.02 (r_1:2.45- r_2:2.75)	0.3 (r_1:2.35- r_2:2.85)	0.15 (r_1:2.2- r_2:2.75)
Se-Se	0.04 (r_1:2.1- r_2:2.6)	0.02 (r_1:2.1- r_2:2.65)	0.02 (r_1:2.15- r_2:2.6)	0.03 (r_1:2.15- r_2:2.6)	0.05 (r_1:2.1- r_2:2.65)	0.08 (r_1:2.1- r_2:2.6)
Ge-X	3.71	3.69	3.76	3.88	3.91	3.96
Sb-X	-	3.00	2.88	2.84	2.85	2.85
Se-X	2.02	2.02	2.05	2.16	2.14	2.12

constrained simulations, in which Ge and Sb atoms were forced to have four and three neighbours, respectively. The corresponding average coordination numbers, CN_{ij}, obtained from the RMC simulations, are tabulated in Table 4. The coordination numbers show a clear dependence with the Ge/Sb ratio. The results suggest the presence of two basic structural units, namely, the $GeSe_4$ tetrahedral units with fourfold coordinated Ge atom and $SbSe_3$ pyramids with threefold coordinated Sb atom.

In Figure 4 the average coordination number distributions of the possible chemical bonds, calculated from the RMC modeling, are presented. It can be seen that the average coordination number around Ge atoms is close to four atoms, as is expected by the formation of tetrahedral units in the network.

However, with increasing Sb content the Ge-Se coordination number, CN_{Ge-Se}, slightly increases from 3.10 to 3.84 atoms, which indicates a small degree of angle distortion affecting the tetrahedral surrounding. This may be caused by the formation of Se-Ge bonds, as the coordination of Se atoms increases from 2.02 to 2.16. The average CN_{Ge-X} coordination number increases from 3.69 to 3.96 with increasing Sb content. This suggests that the chalcogenide network consists of stable tetrahedral Ge units. The CN_{Sb-Se} coordination number continuously decreases from 2.97 to 2.50 with increasing Sb concentration. The Sb-X average coordination number slightly varied from 3.00 to 2.85. The Se-Ge coordination number continuously decreases from 1.98 to 0.96 with increasing Sb concentration, while the Se-Sb coordination number increases from 0.21 to 1.08;

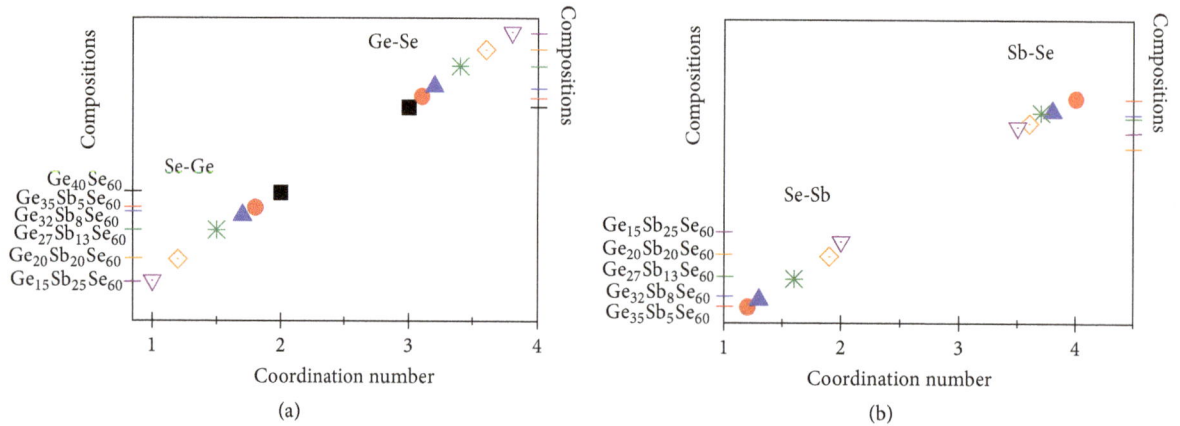

FIGURE 4: Coordination number distributions, CN_{ij}, for the Ge-Se and Se-Ge bonds (a) and Sb-Se and Se-Sb bonds (b) in the corresponding compositions of $Ge_{40}Se_{60}$ (black square), $Ge_{35}Sb_5Se_{60}$ (red circle), $Ge_{32}Sb_8Se_{60}$ (blue triangle), $Ge_{27}Sb_{13}Se_{60}$ (green crosses), $Ge_{20}Sb_{20}Se_{60}$ (orange square), and $Ge_{15}Sb_{25}Se_{60}$ (purple reverse triangle).

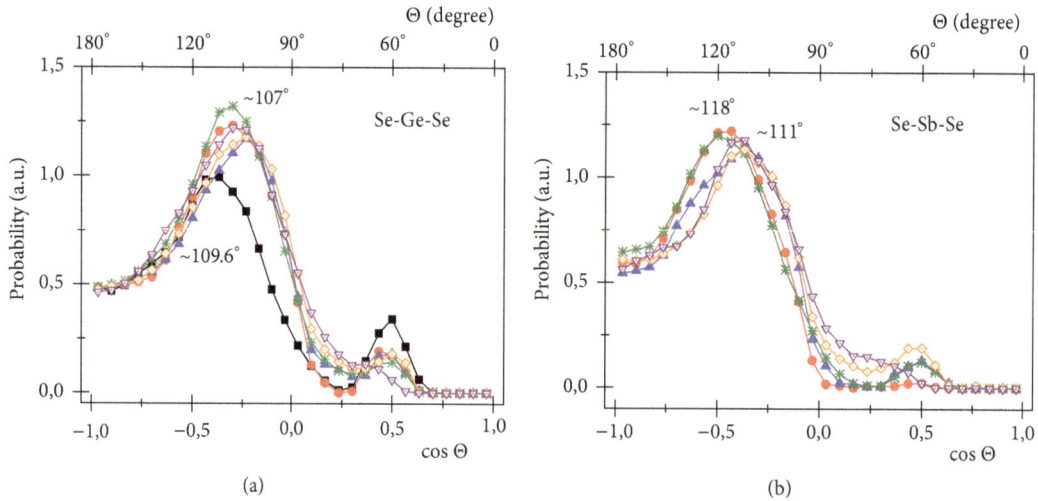

FIGURE 5: Bond-angle distributions of Se-Ge-Se (a) and Se-Sb-Se (b) configurations for the corresponding compositions of $Ge_{40}Se_{60}$ (black square), $Ge_{35}Sb_5Se_{60}$ (red circle), $Ge_{32}Sb_8Se_{60}$ (blue triangle), $Ge_{27}Sb_{13}Se_{60}$ (green crosses), $Ge_{20}Sb_{20}Se_{60}$ (orange square), and $Ge_{15}Sb_{25}Se_{60}$ (purple reverse triangle).

nevertheless, the concentration of Se atoms is kept constant.

Figure 5 shows the bond-angle distributions for the Ge and Sb atoms with Se-Ge-Se and Se-Sb-Se three-particle bond angles. We calculated the three-particle bond-angle distributions using the final atomic configuration of the RMC algorithm, plotted both as the function of $\cos(\Theta)$ (scale below) and Θ (upper scale), where Θ represents the actual bond angle.

For the Se-Ge-Se bonds, the peak positions are at $107\pm3°$ (for the ternary samples) [in agreement with [34]] and at $109.6\pm2°$ (for the binary sample) [35] which are very close to the tetrahedral angle of $109.47°$. The peak distribution is very similar for all the studied samples. The broad distribution is quite asymmetric, but with the increase of Sb concentration, a shift down can be observed suggesting that the tetrahedral environment has become distorted. The distribution of Se-Sb-Se bonding angles is broad and asymmetric, the average

angles being $118\pm3°$ and $111\pm5°$. The broad distribution suggests that 3-fold Sb atoms are present. The Se-Sb-Se bond angles distribution also shows similar characteristics for all samples, as with the increase of Sb concentration a shift down to the $111\pm5°$ can be observed implying considerable distortion in Sb_3 planar geometry. In most of the cases, a weak peak around $60\pm3°$ appears, strongly dependent on glassy composition, which is due to the occasional presence of threefold rings [35].

The present RMC simulation of the ND and XRD data points out that the Sb atoms incorporated in $Ge_{40}Se_{60}$ glass bind covalently to Se and form $SbSe_3$ - trigonal units. On the other hand, Se atoms, which are also connected to Ge, form $GeSe_4$ tetrahedral structural units. These findings are in agreement with several spectroscopic studies [5, 8, 36].

The atomic parameters support the formation of Ge-tetrahedral and Sb-trigonal units in $Ge_{40}Se_{60}$ glass structure. When the Sb content gradually increases, strong Sb units are

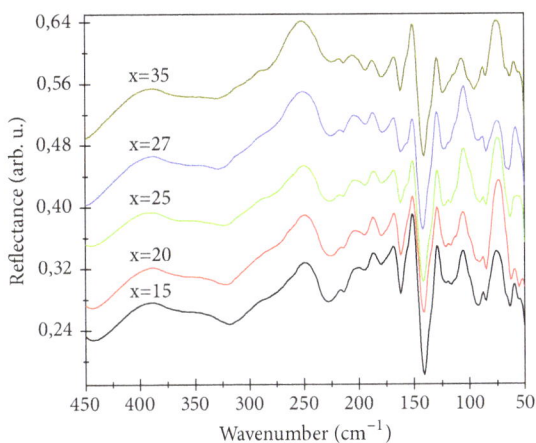

FIGURE 6: Far FTIR reflectance spectra of $Ge_xSb_{40-x}Se_{60}$ films at different x values.

formed with light influence of the tetrahedral Ge network, which causes only negligible changing of the partial pair distribution functions, coordination numbers, and bond-angle distributions. The Ge-Sb bonds show a broad distribution at 3.8 Å, which suggests the existence of connection between Ge-tetrahedral and Sb-pyramidal units through Se atoms. Nevertheless, the Ge-Ge bonds indicate connections between the Ge-tetrahedral, even if in the literature only theoretical bonding is reported. Based on these findings we suppose that a few numbers of Ge-clusters exist, which are connected to the Sb-pyramids. From the values of bond lengths and coordination numbers, we can conclude that the network structure is constructed through connection of tetrahedral $GeSe_4$ and trigonal $SbSe_3$ structural units. The average coordination numbers vary with glassy composition, as it follows from Figure 4. In all the studied samples, the concentration of the Se atoms is constant. With increasing of Sb concentration, coordination of Ge atoms becomes close to 4, while the coordination of Sb atoms becomes close to 3, both coordination sites being close to theoretical ones. These results are in agreement with the literature and support formation of tetrahedral Ge and pyramidal Sb units [19, 29, 37]. For the Se atoms, the coordination distribution changes in opposite direction, and with increasing of Sb concentration the Se-Ge coordination numbers decrease and the Se-Sb coordination numbers increase. This indicates that Se atom prefers to bind to Sb atom rather than Ge atom.

From the above considerations we can conclude that the changes in the distribution of the chemical bonds from the $Ge_xSb_{40-x}Se_{60}$ compositions with average coordination numbers Z < 2.67 to the compositions with Z > 2.67 are responsible for the structural rearrangement of the glassy network [5, 6, 13]

3.2. FTIR Spectra Analysis.

The FTIR reflectance spectra in dependence on the index x (Ge content) are presented in Figure 6. They show considerably rich set of bands because unlike the crystal structure, the lack of symmetry in amorphous material changes the selection rules and, thus, some

forbidden IR modes become active. The spectra present well pronounced reflectance-absorbance bands of collective low-frequency vibrations. Although the spectral change with the Ge content is very small, there are bands which can be assigned to the vibrations affected by the x variation.

The band at 250 cm^{-1} is characteristic for Ge_xSe_{100-x} binary materials [12, 38]. With increasing Ge and with introducing Sb in our $Ge_xSb_{40-x}Se_{60}$ samples the band becomes larger and more intensive. As we have already established that in $Ge_xSb_{40-x}Se_{60}$ glasses structural transformation occurs at x= 25-27 at.% [6], we assume that $GeSe_2$ and $GeSe_4$ units coexist [5, 10, 13] together with Se-Se bond vibration (shoulder at 288 cm^{-1}) [8]. With increasing the Ge content and, respectively, decreasing the content of heavier Sb atoms, these bands move to shorter wavelengths, as seen in Figure 7, where a peculiarity appears at x=27 at.%, corresponding to an average coordination number of Z=2.67.

The broad and structured band at 160-210 cm^{-1} is also sensitive to the x variation. Its components at 168, 187, and 203 cm^{-1} are reduced in intensity by increasing the Ge content. In [5, 39] the authors have discussed Raman spectra in this spectral region assigning the measured frequencies to the dynamics of the tetrahedral $GeSe_4$ vibrations with addition of Sb atoms and to the appearance of pyramidal $SbSe_3$ vibrations. Moreover, the heteropolar Sb-Se and Ge-Se bonds vibrational modes have close position at 190 cm^{-1} and 200 cm^{-1}, respectively [5, 38]. The distinct peak around 168 cm^{-1} can be connected with formation of Ge-Ge bonds by cross-linked Ge atoms in the $Ge_2Se_{6/2}$ chains (Se_3Ge-$GeSe_3$ units) [5, 40, 41]. We are convinced that these modes are responsible for forming the complex band as IR active due to the lack of symmetry in our amorphous samples, as it was obtained by FTIR transmission measurements in [10, 21].

The band picked at 150 cm^{-1} with a shoulder at 157 cm^{-1} does not shift with x but its intensity varies (see Figure 8). It could be related to the Sb-Se (at small x) and to the Ge-Se (at higher x) nonsymmetric vibrations, respectively, [8, 21]. At the highest value of x=35 at.%, the band is slightly enlarged thus showing the coexistence of these vibrations.

The main bands in the terahertz range, 106 cm^{-1} and 74 cm^{-1}, belong to the vibrations of $GeSe_4$ tetrahedral and respectively to $SbSe_3$ pyramidal units [8, 21]. Because of these complex structures, they are surrounded with many weak peaks and shoulders. In Figure 8, it can be seen that the band intensities as function of x evolve in opposite directions, following the variation of the Ge and Sb content, respectively.

Bands in the FTIR spectra, located in the 250-320 cm^{-1} region (Figure 6) have very low intensity and a tendency to vanish by increasing the Ge at.% and decreasing the Sb at.%, respectively. They may be attributed to several vibrations of Se-Se, Ge-Ge, and Ge-Sb bonds in modified $GeSe_4$ - structural units [21, 38, 41].

In the 450-350 cm^{-1} spectral range, a very broad band, centred at 420 cm^{-1} and with a weak shoulder around 340 cm^{-1}, is observed. It is independent on the glassy composition and is related to vibration of oxygen impurity atoms in Ge-O bonding [42, 43]. The presence of oxygen-related impurity bonds in these Ge-Sb-Se films has been detected by infrared

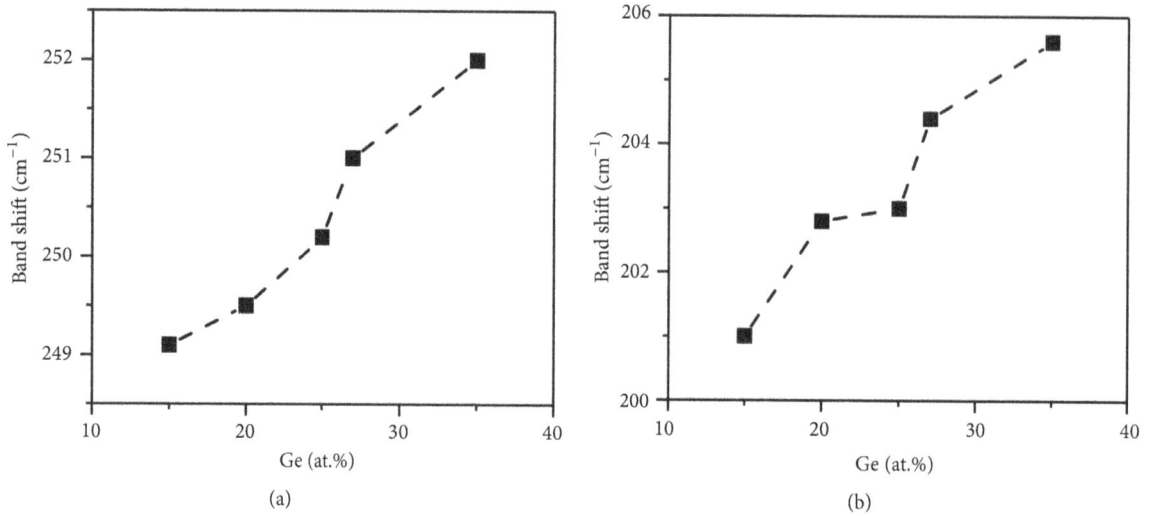

FIGURE 7: Shift of the bands at 250 cm^{-1} (a) and 200 cm^{-1} (b) with variation of x value.

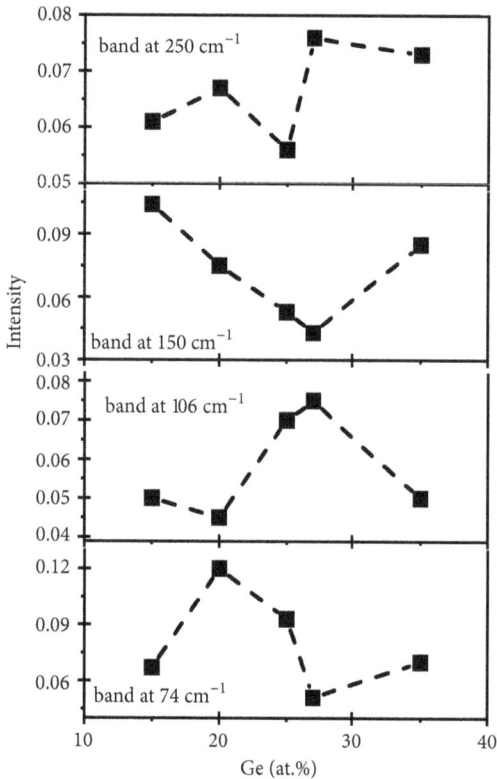

FIGURE 8: Dependence of bands intensity on the x values.

ellipsometric measurements performed in the 400- 4000 cm^{-1} spectral range [44].

From the study of the basic chemical bonds vibrational modes in chalcogenide $Ge_xSb_{40-x}Se_{60}$ films considering the FTIR reflection spectra at oblique (75°) incidence radiation, we obtained specific dependences of the band's shift and intensity on the film composition. Figures 7 and 8 show

features around x=27 at.%, thus confirming the earlier published model of the topological phase transition to a stable rigid network consisting mainly of $SbSe_3$ pyramidal and $GeSe_4$ tetrahedral clusters. These results support the RMC modeling, which reveals that through connection of these $GeSe_4$ and $SbSe_3$ structural units, it is constructed a Ge-Sb-Se network structure.

4. Conclusions

Binary $Ge_{40}Se_{60}$ and ternary $Ge_xSb_{40-x}Se_{60}$ glasses with x=35, 32, 27, 20, and 15 at.% compositions have been synthesized. Their structural properties have been investigated by neutron and X-ray diffraction measurements and RMC simulation procedures, in which all the possible chemical bonds in the glasses are taken into consideration. All specimens are fully amorphous in nature as confirmed by the ND and XRD patterns. It is revealed that addition of Sb atom to the binary Ge-Se glass does not change significantly the basic Ge-tetrahedral structural units but leads to their distortion. With increasing of Sb concentration, well-defined Sb-pyramidal units are formed in the ternary glass structure. The $Ge_xSb_{40-x}Se_{60}$ glassy network builds up from $GeSe_{4/2}$ tetrahedra and the $SbSe_{3/2}$ pyramid units, connected through 2-coordinated Se atoms. Therefore, we may conclude that the "structural unit"-based description of the RMC results is one possible way of interpreting the measured diffraction data.

The FTIR reflection spectra of chalcogenide $Ge_xSb_{40-x}Se_{60}$ films studied in the terahertz spectral range at oblique (75°) incidence radiation showed characteristic bands related to Sb-Se and Ge-Se chemical bonds vibration in $SbSe_3$ pyramidal and $GeSe_4$ tetrahedral units, respectively. The peculiarities in the compositional dependences of the band positions and intensities confirm the existence of a topological phase transition in the $Ge_xSb_{40-x}Se_{60}$ glasses around a critical Ge content of x=27 at.%., corresponding to an average coordination number Z=2.67.

Conflicts of Interest

The authors declare that they have no conflicts of interest.

Acknowledgments

The authors would like to thank Dr. Martin von Zimmermann for assistance in using BW5 beamline, under Hasylab Project No. I-20100116EC. The authors also acknowledge thanks to V. Pamukchieva for synthesizing the studied materials. The support from the Hungarian, Bulgarian, and Romanian Academies under the 2014-2017 Collaboration Agreements is highly appreciated. K. Antonova and A. Szekeres thank the European Regional Development Fund, Ministry of Economy of Bulgaria, Operational Programme "Development of the Competitiveness of the Bulgarian Economy" 2007–2013, Contract No. BG161PO003-1.2.04-0027-C0001, for purchasing the BRUKER spectrophotometer Vertex 70. N. Dulgheru and M. Gartner gratefully acknowledge the support of the research infrastructure under POS-CCEO 2.2.1 INFRANANOCHEM - project, No. 19/01.03.2009.

References

[1] J. A. Savage, *Infrared Optical Materials and Their Antireflection Coatings*, Bristol, UK, 1st edition, 1985.

[2] A. M. Andriesh, "Properties of chalcogenide glasses for optical waveguides," *Journal of Non-Crystalline Solids*, vol. 77–78, pp. 1219–1228, 1985.

[3] T. Kanamori, Y. Terunuma, S. Takahashi, and T. Miyashita, "Chalcogenide Glass Fibers for Mid-Infrared Transmission," *Journal of Lightwave Technology*, vol. 2, no. 5, pp. 607–613, 1984.

[4] G. G. Devyatykh and E. M. Dianov, "Research in KRS-5 and chalcogenide glass fibers," in *Proceedings of the SPIE 0484, Infrared Optical Materials and Fibers III*, vol. 484, pp. 105–107, 1984.

[5] Z. G. Ivanova, V. Pamukchiva, and M. Vlcek, "On the structural phase transformations in $Ge_xSb_{40-x}Se_{60}$ glasses," *Journal of Non-Crystalline Solids*, vol. 293-295, pp. 580–585, 2001.

[6] V. Pamukchieva, A. Szekeres, E. Savova, and E. Vlaikova, "Compositional dependence of the optical parameters of $Ge_xSb_{40-x}Se_{60}$ glasses," *Journal of Non-Crystalline Solids*, vol. 242, pp. 110–114, 1998.

[7] R. Svoboda, D. Brandová, and J. Málek, "Crystallization behavior of $GeSb_2Se_4$ chalcogenide glass," *Journal of Non-Crystalline Solids*, vol. 388, pp. 46–54, 2014.

[8] N. Sharma, S. Sharda, V. Sharma, and P. Sharma, "Far-infrared investigation of ternary Ge–Se–Sb and quaternary Ge–Se–Sb–Te chalcogenide glasses," *Journal of Non-Crystalline Solids*, vol. 375, pp. 114–118, 2013.

[9] A. K. Varshneya and D. Mauro, "Microhardness, indentation toughness, elasticity, plasticity, and brittleness of Ge–Sb–Se chalcogenide glasses," *Journal of Non-Crystalline Solids*, vol. 353, pp. 1291–1297, 2007.

[10] D. R. Goyal and A. S. Maan, "Far-infrared absorption in amorphous $Sb_{15}Ge_xSe_{85-x}$ glasses," *Journal of Non-Crystalline Solids*, vol. 183, pp. 182–185, 1995.

[11] S. Y . Shin, J. M. Choi, J. Seo et al., "The effect of doping Sb on the electronic structure and the device characteristics of Ovonic Threshold Switches based on Ge-Se," *Scientific Reports*, vol. 4, 2014.

[12] P. Sharma, V. S. Rangra, P. Sharma, and S. C. Katyal, "Far-infrared study of amorphous $Ge_{0.17}Se_{0.83-x}Sb_x$ chalcogenide glasses," *Journal of Alloys and Compounds*, vol. 480, pp. 934–937, 2009.

[13] S. Mahadevan and A. Giridhar, "Floppy to rigid transition and chemical ordering in Ge—Sb(As)—Se glasses," *Journal of Non-Crystalline Solids*, vol. 143, pp. 52–58, 1992.

[14] K. Tanaka, "Structural phase transitions in chalcogenide glasses," *Physical Review B: Condensed Matter and Materials Physics*, vol. 39, no. 2, pp. 1270–1279, 1989.

[15] M. Fábián, E. Sváb, V. Pamukchieva, A. Szekeres, S. Vogel, and U. Ruett, "Study of As_2Se_3 and As_2Se_2Te glass structure by neutron- and X-ray diffraction methods," *Journal of Physics: Conference Series*, vol. 253, no. 1, Article ID 012053, 2010.

[16] V. Pamukchieva, A. Szekeres, K. Todorova, E. Svab, and M. Fabian, "Compositional dependence of the optical properties of new quaternary chalcogenide glasses of Ge-Sb-(S,Te) system," *Optical Materials*, vol. 32, pp. 45–48, 2009.

[17] V. Pamukchieva, A. Szekeres, K. Todorova et al., "Evaluation of basic physical parameters of quaternary Ge–Sb–(S,Te) chalcogenide glasses," *Journal of Non-Crystalline Solids*, vol. 355, pp. 2485–2490, 2009.

[18] M. Fabian, E. Svab, V. Pamukchieva et al., "Reverse Monte Carlo modeling of the neutron and X-ray diffraction data for new chalcogenide Ge-Sb-S(Se)-Te glasses," *Journal of Physics and Chemistry of Solids*, vol. 74, pp. 1355–1362, 2013.

[19] I. Pethes, R. Chahal, V. Nazabal et al., "Chemical Short-Range Order in Selenide and Telluride Glasses," *The Journal of Physical Chemistry B*, vol. 120, pp. 9204–9214, 2016.

[20] P. Kumar, V. Modgil, and V. S. Rangra, "The Far-Infrared Study of Ge Modified Sn–Se–Pb Chalcogenide Glasses," *Journal of Non-Oxide Glasses*, vol. 6, pp. 27–35, 2014.

[21] S. Sharda, N. Sharma, P. Sharma, and V. Sharma, "Finger prints of chemical bonds in Sb–Se–Ge and Sb–Se–Ge–In glasses: A Far-IR study," *Journal of Non-Crystalline Solids*, vol. 362, pp. 136–139, 2013.

[22] E. Sváb, G. Mészáros, and F. Deák, "Neutron Powder Diffractometer at the Budapest Research Reactor," *Materials Science Forum*, vol. 228-231, pp. 247–252, 1996.

[23] H. Poulsen, J. Neuefeind, H. B. Neumann, J. R. Schneider, and M. D. Zeidler, "Amorphous silica studied by high energy X-ray diffraction," *Journal of Non-Crystalline Solids*, vol. 188, pp. 63–74, 1995.

[24] A.C. Hannon, "ISIS Disordered Materials Database 2006," https://www.isis.stfc.ac.uk/Pages/Disordered-Materials.aspx, (accessed 15 July 2018).

[25] D. Waasmaier and A. Kirfel, "New analytical scattering-factor functions for free atoms and ions," *Acta Crystallographica*, vol. 51, pp. 416–431, 1995.

[26] R. L. McGreevy and L. Pusztai, "Reverse Monte Carlo Simulation: A New Technique for the Determination of Disordered Structures," *Molecular Simulation*, vol. 1, pp. 359–367, 1988.

[27] O. Gereben, P. Jovari, L. Temleitner, and L. Pusztai, "A new version of the RMC++ Reverse Monte Carlo programme, aimed at investigating the structure of covalent glasses," *Journal of Optoelectronics and Advanced Materials*, pp. 3021–3027, 2007.

[28] V. Pamukchieva, "Private communication," 2015.

[29] M. A. Paesler, D. A. Baker, G. Lucovsky, A. E. Edwards, and P. C. Taylor, "EXAFS study of local order in the amorphous chalcogenide semiconductor $Ge_2Sb_2Te_5$," *Journal of Physics and Chemistry of Solids*, vol. 68, pp. 873–877, 2007.

[30] I. Petri, P. S. Salmon, and H. E. Fischer, "Defects in a Disordered World: The Structure of Glassy $GeSe_2$," *Physical Review Letters*, vol. 84, pp. 2413–2416, 2000.

[31] F. Kakinuma, T. Fukunaga, and K. Suzuki, "Structural study of $Ge_xSb_{40-x}S_{60}$ (x = 10, 20 and 30) glasses," *Journal of Non-Crystalline Solids*, vol. 353, pp. 3045–3048, 2007.

[32] L. Changgui, L. Zhuobin, Y. Lei et al., "Network Structure in GeS_2–Sb_2S_3 Chalcogenide Glasses: Raman Spectroscopy and Phase Transformation Study," *The Journal of Physical Chemistry C*, vol. 116, pp. 5862–5867, 2012.

[33] M. Fraenkl, B. Frumarova, V. Podzemna et al., "How silver influences the structure and physical properties of chalcogenide glass $(GeS_2)_{50}(Sb_2S_3)_{50}$," *Journal of Non-Crystalline Solids*, vol. 499, pp. 412–419, 2018.

[34] P. Jóvári, I. Kaban, J. Steiner, B. Beuneu, A. Schöps, and M. A. Webb, "Local order in amorphous $Ge_2Sb_2Te_5$ and $GeSb_2Te_4$," *Physical Review B: Condensed Matter and Materials Physics*, vol. 77, Article ID 035202, 2008.

[35] M. T. M. Shatnawi, "Reverse Monte Carlo Modeling of the Rigidity Percolation Threshold in Ge_xSe_{1-x} Glassy Networks," *New Journal of Glass and Ceramics*, vol. 5, no. 3, pp. 31–43, 2015.

[36] I. Quiroga, C. Corredor, F. Bellido, J. Vazquez, P. Villares, and R. J. Garay, "Infrared studies of a $Ge_{0.20}Sb_{0.05}Se_{0.75}$ glassy semiconductor," *Journal of Non-Crystalline Solids*, vol. 196, pp. 183–186, 1996.

[37] W.-H. Wei, R.-P. Wang, X. Shen, L. Fang, and B. Luther-Davies, "Correlation between structural and physical properties in Ge-Sb-Se glasses," *The Journal of Physical Chemistry C*, vol. 117, pp. 16571–16576, 2013.

[38] T. Ohsaka, "Infrared spectra of glassy Se containing small amounts of S, Te, As, or Ge," *Journal of Non-Crystalline Solids*, vol. 17, pp. 121–128, 1975.

[39] F. Verger, V. Nazabal, F. Colas et al., "RF sputtered amorphous chalcogenide thin films for surface enhanced infrared absorption spectroscopy," *Optical Materials Express*, vol. 3, pp. 2112–2131, 2013.

[40] E. Baudet, C. Cardinaud, A. Girard et al., "Structural analysis of RF sputtered Ge-Sb-Se thin films by Raman and X-ray photoelectron spectroscopies," *Journal of Non-Crystalline Solids*, vol. 444, pp. 64–72, 2016.

[41] Y. Chen, T. Xu, X. Shen et al., "Optical and structure properties of amorphous Ge–Sb–Se films for ultrafast all-optical signal processing," *Journal of Alloys and Compounds*, vol. 580, pp. 578–583, 2013.

[42] A. M. Hofmeister, J. Horigan, and J. M. Campbell, "Infrared spectra of GeO_2 with the rutile structure and prediction of inactive modes for isostructural compounds," *American Mineralogist*, vol. 75, pp. 1238–1248, 1990.

[43] G. S. Henderson, D. R. Neuville, B. Cochain, and L. Cormier, "The structure of GeO_2–SiO_2 glasses and melts: A Raman spectroscopy study," *Journal of Non-Crystalline Solids*, vol. 355, no. 8, 2009.

[44] N. Dulgheru, M. Gartner, M. Anastasescu et al., "Influence of compositional variation on the optical and morphological properties of Ge—Sb—Se films for optoelectronics application," *Infrared Physics & Technology*, vol. 93, pp. 260–270, 2018.

Permissions

All chapters in this book were first published in ACMP, by Hindawi Publishing Corporation; hereby published with permission under the Creative Commons Attribution License or equivalent. Every chapter published in this book has been scrutinized by our experts. Their significance has been extensively debated. The topics covered herein carry significant findings which will fuel the growth of the discipline. They may even be implemented as practical applications or may be referred to as a beginning point for another development.

The contributors of this book come from diverse backgrounds, making this book a truly international effort. This book will bring forth new frontiers with its revolutionizing research information and detailed analysis of the nascent developments around the world.

We would like to thank all the contributing authors for lending their expertise to make the book truly unique. They have played a crucial role in the development of this book. Without their invaluable contributions this book wouldn't have been possible. They have made vital efforts to compile up to date information on the varied aspects of this subject to make this book a valuable addition to the collection of many professionals and students.

This book was conceptualized with the vision of imparting up-to-date information and advanced data in this field. To ensure the same, a matchless editorial board was set up. Every individual on the board went through rigorous rounds of assessment to prove their worth. After which they invested a large part of their time researching and compiling the most relevant data for our readers.

The editorial board has been involved in producing this book since its inception. They have spent rigorous hours researching and exploring the diverse topics which have resulted in the successful publishing of this book. They have passed on their knowledge of decades through this book. To expedite this challenging task, the publisher supported the team at every step. A small team of assistant editors was also appointed to further simplify the editing procedure and attain best results for the readers.

Apart from the editorial board, the designing team has also invested a significant amount of their time in understanding the subject and creating the most relevant covers. They scrutinized every image to scout for the most suitable representation of the subject and create an appropriate cover for the book.

The publishing team has been an ardent support to the editorial, designing and production team. Their endless efforts to recruit the best for this project, has resulted in the accomplishment of this book. They are a veteran in the field of academics and their pool of knowledge is as vast as their experience in printing. Their expertise and guidance has proved useful at every step. Their uncompromising quality standards have made this book an exceptional effort. Their encouragement from time to time has been an inspiration for everyone.

The publisher and the editorial board hope that this book will prove to be a valuable piece of knowledge for researchers, students, practitioners and scholars across the globe.

List of Contributors

V. D. Lakhno
Keldysh Institute of Applied Mathematics, RAS, Miusskaya Sq. 4, Moscow 125047, Russia

Chuen-Lin Tien
Department of Electrical Engineering, Feng Chia University, Taichung 40724, Taiwan
Ph.D. Program of Electrical and Communications Engineering, Feng Chia University, Taichung 40724, Taiwan

Rong-Ji Lin
Ph.D. Program of Electrical and Communications Engineering, Feng Chia University, Taichung 40724, Taiwan

Shang-Min Yeh
Department of Optometry, Central Taiwan University of Science and Technology, Taichung 40601, Taiwan

Ibrahim Avgin
Department of Electrical and Electronic Engineering, Ege University, Bornova, 3500 Izmir, Turkey

David Huber
Physics Department, University of Wisconsin-Madison, Madison, WI 53706, USA

Javed Ahmad, Syed Hamad Bukhari, M. Tufiq Jamil and Mehr Khalid Rehmani
Department of Physics, Bahauddin Zakariya University, Multan 60800, Pakistan

Hammad Ahmad
Nanoscience and Technology Department, National Center for Physics, Quaid-i-Azam University Campus, Islamabad 45320, Pakistan

Tahir Sultan
Department of Civil Engineering, Bahauddin Zakariya University, Multan 60800, Pakistan

N. N. Syrbu, A. V. Tiron and N. P. Bejan
Department of Telecommunication, Technical University of Moldova, Chis, inău, Moldova

V. V. Zalamai
Laboratory of Materials for Photovoltaics and Photonics, Institute of Applied Physics, Academy of Sciences of Moldova, Chişinău, Moldova

Victor Atanasov
Department of Condensed Matter Physics, Sofia University, 5 Boul. J. Bourchier, 1164 Sofia, Bulgaria

Jun Guo, Xiaoyu Dai and Yuanjiang Xiang
SZU-NUS Collaborative Innovation Center for Optoelectronic Science and Technology, Key Laboratory of Optoelectronic Devices and Systems of Ministry of Education and Guangdong Province, College of Optoelectronic Engineering, Shenzhen University, Shenzhen 518060, China

Leyong Jiang
College of Physics and Information Science, Hunan Normal University, Changsha 410081, China

Zhiwei Zheng
SZU-NUS Collaborative Innovation Center for Optoelectronic Science and Technology, Key Laboratory of Optoelectronic Devices and Systems of Ministry of Education and Guangdong Province, College of Optoelectronic Engineering, Shenzhen University, Shenzhen 518060, China
College of Physics and Information Science, Hunan Normal University, Changsha 410081, China

E. V. Voronina, A. G. Ivanova and A. V. Pyataev
Institute of Physics, Kazan Federal University, Kazan 420008, Russia

A. K. Arzhnikov
Physical-Technical Institute, Ural Branch of Russian Academy of Sciences, Izhevsk 426000, Russia

A. I. Chumakov
European Synchrotron Radiation Facility, 38043 Grenoble Cedex 9, France

N. I. Chistyakova
Faculty of Physics, M. V. Lomonosov Moscow State University, Moscow 119991, Russia

A. V. Korolev
Institute of Metal Physics, Ural Branch of Russian Academy of Sciences, Yekaterinburg 620108, Russia

Armenak A. Osipov and Leyla M. Osipova
Institute of Mineralogy of UB RAS, 456317, Miass, Russia

Egor Yu. Kaniukov, Dzmitry V. Yakimchuk, Victoria D. Bundyukova, Alena E. Shumskaya and Sergey E. Demyanov
Cryogenic Research Division, Scientific-Practical Materials Research Center NAS of Belarus, Minsk 220072, Belarus

Abdulkarim A. Amirov
Center for Functionalized Magnetic Materials (FunMagMa) & Institute of Physics Mathematics and Informational Technologies, Immanuel Kant Baltic Federal University, Kaliningrad 236016, Russia
Amirkhanov Institute of Physics Daghestan Scientific Center, Russian Academy of Sciences, Makhachkala 367003, Russia

V. Ya. Degoda and M. Alizadeh
Taras Shevchenko National University of Kyiv, 64 Volodymyrs'ka Street, 01601 Kyiv, Ukraine

N. O. Kovalenko
Institute for Single Crystals NAS of Ukraine, 61001, Nauki Ave, Kharkiv, Ukraine

N. Yu. Pavlova
National Pedagogical Dragomanov University, 9 Pyrogova Street, 01601 Kyiv, Ukraine

G. Gulyamov
Namangan Engineering Pedagogical Institute, 160103 Namangan, Uzbekistan

U. I. Erkaboev and A. G. Gulyamov
Physico-Technical Institute, NGO "Physics-Sun", Academy of Sciences of Uzbekistan, 100084 Tashkent, Uzbekistan

Sujata Tarafdar
Condensed Matter Physics Research Center (CMPRC), Physics Department, Jadavpur University, Kolkata 700 032, India

Yuri Yu. Tarasevich
Laboratory of Mathematical Modeling, Astrakhan State University, Astrakhan 414056, Russia

Moutushi Dutta Choudhury
Centre for Advanced Studies in Condensed Matter and Solid State Physics, Department of Physics, Savitribai Phule Pune University, Pune 411 007, India

Tapati Dutta
Condensed Matter Physics Research Center (CMPRC), Physics Department, Jadavpur University, Kolkata 700 032, India
Physics Department, St. Xavier's College, Kolkata 700 016, India

Duyang Zang
Functional Soft Matter and Materials Group (FS2M), Key Laboratory of Material Physics and Chemistry under Extraordinary Conditions, School of Science, Northwestern Polytechnical University, Xi'an, Shaanxi 710129, China

Eva Klemenčič
Faculty of Natural Sciences and Mathematics, University of Maribor, Maribor, Slovenia

Mitja Slavinec
Faculty of Natural Sciences and Mathematics, University of Maribor, Maribor, Slovenia
Academic Scientific Union of Pomurje (PAZU), Murska Sobota, Slovenia

Limin Zhang, Ning Li, Hui Xing and Rong Zhang
Shaanxi Key Laboratory of Condensed Matter Structures and Properties and Key Laboratory of Space Applied Physics and Chemistry, Ministry of Education, School of Science, Northwestern Polytechnical University, 127West Youyi Road, Xi'an, Shaanxi 710072, China

Kaikai Song
School of Mechanical, Electrical & Information Engineering, Shandong University at Weihai, 180Wenhua Xilu, Weihai, Shandong 264209, China

Salem Neily and Sami Dhouibi
Laboratoire de Physique de la Matiére Condensée et Nanosciences LR 11 ES 40, Faculté des Sciences, Université de Monastir, rue de l'Environnement, 5019 Monastir, Tunisia

Roland Bonnet
Universit´e Grenoble Alpes, CNRS, SIMaP, 38000 Grenoble, France

I. N. Yakovkin
Institute of Physics of National Academy of Sciences of Ukraine, Prospect Nauki 46, Kiev 03028, Ukraine

Yuriy Garbovskiy
UCCS BioFrontiers Center and Department of Physics, University of Colorado Colorado Springs, 1420 Austin Bluffs Parkway, Colorado Springs, CO 80919, USA

Chuen-Lin Tien
Department of Electrical Engineering, Feng Chia University, Taichung 40724, Taiwan
Ph.D. Program of Electrical and Communications Engineering, Feng Chia University, Taichung 40724, Taiwan

Rong-Ji Lin and Shu-Hui Su
Ph.D. Program of Electrical and Communications Engineering, Feng Chia University, Taichung 40724, Taiwan

Chi-Ting Horng
Department of Ophthalmology, Fooyin University Hospital, Pingtung, Taiwan
Department of Pharmacy, Tajen University, Pingtung, Taiwan

Hsin-Yu Yao
Department of Ophthalmology, Kaohsiung Armed Forces General Hospital, Kaohsiung 802, Taiwan

Shang-Min Yeh
Department of Optometry, Central Taiwan University of Science and Technology, Taichung 40601, Taiwan

Chi-Huang Chang
Department of Ophthalmology, Chung Shan Medical University Hospital, Taichung 402, Taiwan
Department of Medicine, Chung Shan Medical University, Taichung 402, Taiwan

Rong-Ji Lin
Ph.D. Program of Electrical and Communications Engineering, Feng Chia University, Taichung 40724, Taiwan
Department of Optometry, Da-Yeh University, Changhua 515, Taiwan

Chuen-Lin Tien
Ph.D. Program of Electrical and Communications Engineering, Feng Chia University, Taichung 40724, Taiwan

Department of Electrical Engineering, Feng Chia University, Taichung 40724, Taiwan

Shang-Min Yeh
Department of Optometry, Central Taiwan University of Science and Technology, Taichung 40601, Taiwan

Hasnain Mehdi Jafri and XingqiaoMa
Department of Physics, University of Science and Technology Beijing, Beijing 100083, China

Congpeng Zhao
Department of Physics, University of Science and Technology Beijing, Beijing 100083, China
ASIC, China Center for Information Industry Development, Beijing 100083, China

Houbing Huang
Department of Physics, University of Science and Technology Beijing, Beijing 100083, China
Advanced Research Institute of Multidisciplinary Science, Beijing Institute of Technology, Beijing 100081, China

Fausto Cargnoni, Simone Cenedese and Mario Italo Trioni
Consiglio Nazionale delle Ricerche, Istituto di Scienze e Tecnologie Molecolari, Via Golgi 19, 20133 Milano, Italy

Paolo Ghigna
Dipartimento di Chimica, Università di Pavia and Unità INSTM di Pavia, Viale Taramelli 13, 27100 Pavia, Italy

Marco Scavini
Consiglio Nazionale delle Ricerche, Istituto di Scienze e Tecnologie Molecolari, Via Golgi 19, 20133 Milano, Italy
Dipartimento di Chimica, Università degli Studi di Milano, Via Golgi 19, 20133 Milano, Italy

Lian-bi Li and Bin Xue
School of Science, Xi'an Polytechnic University, Xi'an 710048, China

Song Feng
State Key Laboratory of Functional Materials for Informatics, Shanghai Institute of Microsystem and Information Technology, Chinese Academy of Sciences, 865 Changning Road, Shanghai 200050, China

Yasuhiro Shiojiri, Kei Ishihara, Hiroto Hitotsukabuto, Shigeyuki Murayama and Hideaki Takano
Graduate School of Engineering, Muroran Institute of Technology, 27-1 Mizumoto-cho, Muroran, Hokkaido 050-8585, Japan

Yusuke Amakai and Naoki Momono
Graduate School of Engineering, Muroran Institute of Technology, 27-1 Mizumoto-cho, Muroran, Hokkaido 050-8585, Japan
Research Center for Environmentally Friendly Materials Engineering, Muroran Institute of Technology, 27-1 Mizumoto-cho, Muroran, Hokkaido 050-8585, Japan

Bei Xu, Changjun Zhu, Lianbi Li, Song Feng and Qianqian Lei
School of Science, Xi'an Polytechnic University, Xi'an 710048, China

Xiaomin He and Yuan Zang
Department of Electronic Engineering, Xi'an University of Technology, Xi'an 710048, China

Shenghuang Lin
Department of Applied Physics,The Hong Kong Polytechnic University, Hung Hom, Hong Kong

M. Fabian
Centre for Energy Research, Hungarian Academy of Sciences, H-1525 Budapest, Hungary
Wigner Research Centre for Physics, Hungarian Academy of Sciences, H-1525 Budapest Hungary

N. Dulgheru and M. Gartner
Institute of Physical Chemistry "IlieMurgulescu", Romanian Academy, 202 Splaiul Independentei, 060021 Bucharest, Romania

K. Antonova and A. Szekeres
Institute of Solid State Physics, Bulgarian Academy of Sciences, Tzarigradsko Chaussee 72, 1784 Sofia, Bulgaria

Index

www.ingramcontent.com/pod-product-compliance
Lightning Source LLC
Chambersburg PA
CBHW082043190326
41458CB00010B/3445